Curves and Surfaces
for Computer Graphics

To the memory of Pierre Etienne Bézier (1910–1999).

Only for you, children of doctrine and learning, have we
written this work. Examine this book, ponder the meaning
we have dispersed in various places and gathered again;
what we have concealed in one place we have disclosed
in another, that it may be understood by your wisdom.

–Heinrich Agrippa von Nettesheim, *De Occulta Philosophia,* (1531)

David Salomon

Curves and Surfaces
for Computer Graphics

With 207 Figures, 12 in Full Color

 Springer

David Salomon (Emeritus)
Department of Computer Science
California State University, Northridge
Northridge, CA 91330-8281
U.S.A.
dsalomon@csun.edu

ISBN-10: 0-387-24196-5   e-ISBN: 0-387-28452-4   Printed on acid-free paper.
ISBN-13: 978-0-387-24196-8

Printed in the United States of America.   (HAM)

9 8 7 6 5 4 3 2 1   SPIN 11360285

springeronline.com

Preface

Images are all around us. We see them in color and in high resolution. In fact, the natural images we see with our eyes seem perfectly smooth, with no jagged edges and no graininess. Computer graphics, on the other hand, deals with images that consist of small dots, pixels. When we first hear of this feature of computer graphics, we tend to dismiss the entire field as trivial. It seems intuitively obvious that an image that consists of dots would always look artificial, rough, and inferior to what we see with our eyes. Yet state-of-the-art computer-generated images are often difficult or impossible to distinguish from their real counterparts, even though they are discrete, made of pixels, and not continuous.

A similar dichotomy exists in painting. Many painters try to mimic nature and paint smooth and continuous pictures. Others choose to be pointillists. They paint by placing many small dots on their canvas. The most important pointillist was the 19th century French impressionist Georges Seurat.

> Georges Seurat (1859–1891) was a leader in the late 19th century neo-impressionism movement, a school of painting that uses tiny brushstrokes of contrasting colors to achieve a delicate play of light and create subtle changes in form (Figure C.1). Seurat used this technique, which became known as pointillism or divisionism, to create huge paintings that are made entirely of small dots of pure color. The dots are too small to be distinguished when looking at the work in its entirety, but they make his paintings shimmer with brilliance. His most well-known works are *Une Baignade* (1883–84) and *Un dimanche après-midi à l'Île de la Grande Jatte* (1884–86). The art critic Arsène Alexandre had this to say about the latter painting: "Everything was so new in this immense painting—the conception was bold and the technique one that nobody had ever seen or heard before. This was the famous pointillism."

Even though it generates discrete images made of dots, the field of computer graphics has been extremely successful. It has started from nothing in the 1960s and has since attracted many workers and researchers. They developed general techniques and specialized algorithms to generate and manipulate images and thereby turned computer graphics into the useful, practical discipline it is today.

The chief aim of computer graphics is to display and print realistic-looking images. This task is achieved by computing the outer surface of the object or objects to be displayed, and rendering it by simulating the way it is seen in real life. Most real objects are visible because they reflect light, so the main task of rendering is to simulate light reflection. (Relatively few objects are visible because of light that they generate. A completely transparent object is visible by the light it refracts. Most objects, however, do not generate light and are not transparent. They are seen because they reflect some of the light that falls on them.)

Rendering is therefore an important part of computer graphics, but this book is concerned with the computations of surfaces. In order to render a real object, such as a teapot or a car, its surface has first to be calculated and stored in the computer as a mathematical expression. This expression is a model of the real object, which is why the process of generating the model is known as *geometric modeling*. The rendering algorithm then scans the surface point by point, computes the normal vector to the surface at every point, and uses the normal to compute the amount and color of light reflected from the point.

The book also deals with curves, because an understanding of curves is a key to understanding surfaces. Most mathematical methods for curves can be extended to surfaces, which is why this book covers various approaches to curve design and shows that many curve methods can be generalized to surfaces.

The most important term in the field of curve and surface design is *interpolation*. It comes from the Latin *inter* (between) and *polare* (to polish) and it means to compute new values that lie between (or that are an average of) certain given values. A typical algorithm for curves starts with a set of points and employs interpolation to compute a smooth curve that passes through the points. Such points are termed data points and they define an interpolating curve. Some methods start with both points and vectors and compute a curve that passes through the points and at certain points also moves in the directions of the given vectors.

Another important term in this field is *approximation*. Certain curve and surface methods start with points and perhaps also vectors and compute a curve or a surface that passes close to the points but not necessarily through them. Such points are known as control points and the curve or the surface defined by them is referred to as an approximating curve or surface. Most chapters of this book describe interpolation or approximation methods.

Chapter 1 presents the basic theory of curves and surfaces. It discusses the all-important parametric representation and covers basic concepts such as curvature, tangent vectors, normal vectors, curve and surface continuity, and Cartesian products.

Chapter 2 introduces the simplest curves and surfaces. Straight lines, flat planes, triangles, and bilinear and lofted surfaces are presented and illustrated with examples.

Chapter 3 discusses polynomial interpolation. Given a set of points, the problem is to compute a polynomial that passes through them. This problem is then extended to a surface patch that passes through a given two-dimensional set of points. The chapter starts with the important parametric cubic (PC) curves. It continues with the general method of Lagrange interpolation and its relative, the Newton interpolation method. Simple polynomial surfaces are presented, followed by Coons surfaces, a family of simple surface patches based on polynomials.

The mathematically-elegant Hermite interpolation technique is the topic of Chapter 4. The chapter discusses cubic and higher-order Hermite curve segments, special and degenerate hermite segments, Hermite interpolation curves, the Ferguson surface patch, the Coons surface patch, the bicubic surface patch, and Gordon surfaces. A few less-important topics are also touched upon.

The important concept of splines is covered in Chapter 5. Spline methods for curves and surfaces are more practical than polynomial methods and several spline methods are based on Hermite interpolation. The main topics in this chapter are cubic splines (several varieties are discussed), cardinal splines, Kochanek–Bartels splines, spline surface patches, and cardinal spline patches.

Chapter 6 is devoted to Bézier methods for curves and surfaces. The Bernstein form of the Bézier curve is introduced, followed by techniques for fast computation of the curve and by a list of the properties of the curve. This leads to a discussion of how to smoothly connect Bezier segments. The de Casteljau construction of the Bézier curve is described next. It is followed by the technique of blossoming and by methods for subdividing the curve, for degree elevation and for controlling its tension. Sometimes one wants to interpolate a set of points by a Bézier curve and this problem is also discussed. Rational Bézier curves have important advantages and are assigned a separate section.

The chapter continues with material on Bézier surfaces. The topics discussed are rectangular Bézier surfaces and their smooth joining, triangular Bézier surfaces and their smooth joining, and the Gregory surface patch and its tangent vectors.

The last of the "interpolation/approximation" chapters is Chapter 7, on the all-important B-spline technique. B-spline curve topics are the quadratic uniform B-spline curve, the cubic uniform B-spline curve, multiple control points, cubic B-splines with tension, higher-degree uniform B-splines, interpolating B-splines, open uniform B-spline, nonuniform B-splines, matrix form of the nonuniform B-spline curve, subdividing the B-spline curve, and NURBS. The B-spline surface topics are uniform B-spline surfaces, an interpolating bicubic patch, and a quadratic-cubic B-spline patch.

Subdivision methods for curves and surfaces are discussed in Chapter 8. These methods are also based on interpolation, but are different from the traditional interpolation methods discussed in the preceding chapters. The following important techniques are described in this chapter: The de Casteljau refinement process, Chaikin's algorithm, the quadratic uniform B-spline curve, the cubic uniform B-spline curve, bi-quadratic B-spline patches, bicubic B-spline patches, Doo–Sabin subdivision methods, Catmull–Clark surfaces, and Loop subdivision surfaces.

Chapter 9 presents the various types of sweep surfaces. This is a completely different approach to surface design and representation. A sweep surface is generated by constructing a curve and moving it along another curve, while optionally also rotating and scaling it, to create a surface patch. A special case of sweep surfaces is surfaces of revolution. They are created when a curve is rotated about an axis.

Appendix A is a short discussion of conic sections, a family of simple curves that have special applications.

Appendix B discusses simple methods for the approximate representation of circles by polynomials.

Appendix C is a small collection of color images, most of which appear elsewhere in the book in grayscale.

Finally, Appendix D discusses several useful and interesting commands and techniques employed in the various *Mathematica* code listings sprinkled throughout the book.

History of Curves and Surfaces

Section 2.4 discusses lofted surfaces but does not explain the reason for this unusual name. Historically, shipbuilders were among the first to mechanize their operation by developing mathematical models of surfaces. Ships tend to be big and the only dry place in a shipyard large enough to store full-size drawings of ship parts was in the sail lofts above the shipyard's dry dock. Certain parts of a ship's surface are flat in one direction and curved in the other direction, so such surfaces became known as lofted.

In the 1960s, both car and aircraft manufacturers became interested in applying computers to automate the design of their vehicles. Traditionally, artists and designers had to make clay models of each part of the surface of a car or airplane and these models were later used by the production people to produce stamp molds. With the help of the computer it became possible to design a section of surface on the computer in an interactive process, then have the computer drive a milling machine to actually make the part.

The box on page 175 mentions the work of Pierre Bézier at Renault and Paul de Casteljau at Citroën, the contributions of Steven Coons to Ford Motors and William Gordon and Richard F. Riesenfeld to General Motors, and the efforts of James Ferguson in constructing airplane surfaces.

As a result of these developments in the 1960s and 1970s, the area of computer graphics that deals with curves and surfaces has become known, in 1974, as computer assisted geometric design (CAGD). Several sophisticated CAGD software systems have been developed in the 1980s for general use in manufacturing and in other fields such as chemistry (to model molecules), geoscience (for specialized maps), and architecture (for three-dimensional models of buildings).

Hardware developments in the 1980s made it possible to use CAGD techniques in the 1990s to produce computer-generated special effects for movies (an example is *Jurassic Park*), followed by full-length movies, such as *Toy story*, *Finding Nemo*, and *Shrek*, that were entirely generated by computer.

A detailed survey of the history of this field can be found in [Farin 04]. Several first-person historical accounts by pioneers in this field are collected in [Rogers 01].

Resources for Curves and Surfaces

As is natural to expect, the World Wide Web has many resources for CAGD. In addition to the many texts available in this field, the journals *CAD* and *CAGD* carry state-of-the-art papers and articles. See [CAD 04] and [CAGD 04]. Following is a list of some of the most important resources for computer graphics, not just CAGD, current as of mid-2005.

▪   `http://www.siggraph.org/` is the official home page of SIGGRAPH, the special interest group for graphics, one of many SIGs that are part of the ACM.

▪   The Web page `http://www.siggraph.org/conferences/fundamentals` has useful course notes from SIGGRAPH conferences.

- The Web page `http://www.faqs.org/faqs/graphics/faq/` by John Grieggs has answers to frequently-asked questions on graphics, as well as pointers to other resources. It hasn't been updated since 1995.

- See `http://www.cis.ohio-state.edu/~parent/book/outline.html` for the latest version of Richard Parent's book on computer animation.

- `http://mambo.ucsc.edu/psl/cg.html` is a jumping point to many sites that deal with computer graphics.

- A similar site is `http://www.cs.rit.edu/~ncs/graphics.html` that also has many links to CG sites.

- `http://ls7-www.cs.uni-dortmund.de/cgotn/` is a very extensive site of computer-graphics-related pointers.

- *IEEE Computer Graphics and Applications* is a technical journal carrying research papers and news. See `http://computer.org/cga`.

- *Animation Magazine* is a monthly publication covering the entire animation field, computer and otherwise. Located at `http://www.bcdonline.com/animag/`.

- *Computer Graphics World* is a monthly publication concentrating on news, see `http://cgw.pennnet.com/home.cfm`.

- An Internet search for `CAD` or `CAGD` returns many sites.

Software Resources

Those who want to experiment with curves and surfaces can either write their own software (most likely in OpenGL) or learn to use one of several powerful software packages available either commercially or as freeware. Here are the ones found by the author in mid 2005.

- Mathematica, from [Wolfram Research 05], is the granddaddy of all mathematical software. It has facilities for numerical computations, symbolic manipulations, and graphics. It also has all the features of a very high-level programming language.

- Matlab (matrix lab), from [Mathworks 05] is a similar powerful package that many find easier to use.

- Blender is powerful software that computes and displays many types of curves and surfaces. It has powerful tools for animation and game design and is available for several platforms from [Blender 05].

- DesignMentor is a free software package that computes and displays curves, surfaces, and Voronoi regions and triangulations. It is available from [DesignMentor 05].

- Wings3D, from [Wings3D 05], is free software that constructs subdivision surfaces.

- GIMP is a free image manipulation program for tasks such as photo retouching, image composition, and image authoring. It is available from [GIMP 05] for many operating systems, in many languages, but it does not compute curves and surfaces.

A Word on Notation

It is common practice to represent nonscalar quantities such as points, vectors, and matrices with boldface. Below are examples of the notation used here:

x, y, z, t, u, v — Italics are used for scalar quantities such as coordinates and parameters.

$\mathbf{P}, \mathbf{Q}_i, \mathbf{v}, \mathbf{M}$ — Boldface is used for points, vectors, and matrices.

$\vec{\mathbf{CP}}$ — An alternative notation for vectors, used when the two endpoints of the vector are known.

$\mathbf{P}(t), \mathbf{P}(u, v)$ — Boldface with arguments is used for nonscalar functions such as curves and surfaces.

$\begin{pmatrix} a_{11} & a_{12} \\ a_{21} & a_{22} \end{pmatrix}$ — Parentheses (sometimes square brackets) are used for matrices.

$\begin{vmatrix} a_{11} & a_{12} \\ a_{21} & a_{22} \end{vmatrix}$ — Vertical bars are used for determinants.

$|\mathbf{v}|$ — The absolute value (length) of vector $\mathbf{v}$.

$\mathbf{A}^T$ — The transpose of matrix $\mathbf{A}$.

$x^*, \mathbf{P}^*$ — The transformed values of scalars and points.

$f^u(u), \mathbf{P}^t(t), \mathbf{P}^{tt}(t)$ — The (first or second) derivatives of scalar and vector functions. For third and higher derivatives, a prime is usually used.

$\dfrac{df(u)}{du}, \dfrac{d\mathbf{P}(t)}{dt}$ — Alternative notation for derivatives.

$\dfrac{df^2(u)}{du^2}, \dfrac{d\mathbf{P}^2(t)}{dt^2}$ — Alternative notation for higher-order derivatives.

$\dfrac{\partial f(u, v)}{\partial u}, \dfrac{\partial \mathbf{P}(u, v)}{\partial v}$ — Partial derivatives.

$f(x)|_{x_0}$ or $f(x_0)$ — Value of function $f(x)$ at point x_0.

$\displaystyle\sum_{i=1}^{n} x_i$ — The sum $x_1 + x_2 + \cdots + x_n$.

$\displaystyle\prod_{i=1}^{n} x_i$ — The product $x_1 x_2 \ldots x_n$.

⬦ **Exercise 1:** What is the meaning of $(\mathbf{P}_1, \mathbf{P}_2, \mathbf{P}_3, \mathbf{P}_4)$?

Readership of the Book

The book aims at mathematically mature readers (i.e., those who can deal comfortably with mathematical abstractions), who are familiar with computers and computer graphics, and are looking for a mathematically-easy presentation of geometric modeling. The material presented here requires no previous knowledge of curves, splines, or surfaces. The key ideas are introduced slowly, are examined, when possible, from several points of view, and are illustrated by figures, examples, and (solved) exercises. The discussion must involve some mathematics, but it is nonrigorous and therefore easy to grasp. The mathematical background required includes polynomials, matrices, vector operations, and elementary calculus. The following features enhance the usefulness of the book:

■ The powerful *Mathematica* software system is used throughout the book to implement the various concepts discussed. When a figure is computed in *Mathematica*, the code is listed with the figure. These codes are meant to be readable rather than efficient and fast, and are therefore easy to read and to modify even by inexperienced *Mathematica* users.

■ The book has many examples. Experience shows that examples are important for a thorough understanding of the kind of material discussed in this book. The conscientious reader should follow each example carefully and try to work out variations of the examples. Many examples also include *Mathematica* code.

■ Many exercises are sprinkled throughout the text. These are also important and should be worked out. The answers are also provided, but should be consulted only to verify the reader's own answer, or as a last resort.

■ The book aims to be practical, not theoretical. After reading and understanding a topic, the reader should be able to design and implement the concepts discussed there. The few mathematical proofs found in the book are simple, and there is no attempt to present an overall theory encompassing all curves and surfaces. The following advice by Proust is adhered to:

> A book in which there are theories is like an article from which the price mark has not been removed.
> —Marcel Proust, *Time Regained* (1921).

Currently, the book's Web site is part of the author's Web site, which is located at `http://www.ecs.csun.edu/~dsalomon/`. Domain name `DavidSalomon.name` has been reserved and will always point to any future location of the Web site. The author's email address is `dsalomon@csun.edu`, but any email sent to email address ⟨*anyname*⟩`@DavidSalomon.name` will reach the author.

I would like to thank Garry Helzer for his *Mathematica* implementation of the triangular Bézier surfaces. Figures 6.31 and C.2 were computed with this code. The *Mathematica* notebook for this code is available in the book's Web site.

This book is dedicated to the memory of Pierre Bézier, but the field of computer aided geometric design (CAGD) is the creation of many dedicated researchers, programmers, users, and authors. Let us remember their contributions.

Over the projected image he worked with a pointillist technique, using
infinitesimal gradations of color, covering the whole spectrum
dot by dot, so that he always began from a blindingly bright
nucleus and ended at absolute black, or vice versa, depending on
the mystical or cosmological concept he wanted to express.

—Umberto Eco, *Foucault's Pendulum* (1988)

Lakeside, California   David Salomon

Contents

Contents

Contents

My kind publishers announced, some time ago, a table of contents, which included chapters on jay and fish-hawk, panther, and musquash, and a certain savage old bull moose that once took up his abode too near my camp for comfort. My only excuse for their non-appearance is that my little book was full before their turn came. They will find their place, I trust, in another volume presently.

—William J. Long, *Secret of the Woods* (1901)

1
Basic Theory

1.1 Points and Vectors

Real life methods for constructing curves and surfaces often start with points and vectors, which is why we start with a short discussion of the properties of these mathematical entities. The material in this section applies to both two-dimensional and three-dimensional points and vectors, while the examples are given in two-dimensions.

Points and vectors are different mathematical entities. A point has no dimensions; it represents a location in space. A vector, on the other hand, has no well-defined location and its only attributes are direction and magnitude. People tend to confuse points and vectors because it is natural to associate a point $\mathbf{P}$ with the vector $\mathbf{v}$ that points from the origin to $\mathbf{P}$ (Figure 1.1a). This association is useful, but the reader should bear in mind that $\mathbf{P}$ and $\mathbf{v}$ are different.

Both points and vectors are represented by pairs or triplets of real numbers, but these numbers have different meanings. A point with coordinates $(3, 4)$ is located 3 units to the right of the y axis and 4 units above the x axis. A vector with components $(3, 4)$, however, points in direction $4/3$ (it moves 3 units in the x direction for every 4 units in the y direction, so its slope is $4/3$) and its magnitude is $\sqrt{3^2 + 4^2} = 5$. It can be located anywhere.

In mathematics, entities are always associated with operations. An entity that cannot be operated on is generally not useful. Thus, we discuss operations on points and vectors. The first operation is to multiply a point $\mathbf{P}$ by a real number α. The product $\alpha\mathbf{P}$ is a point on the line connecting $\mathbf{P}$ to the origin (Figure 1.1b). Note that this line is infinite and $\alpha\mathbf{P}$ can be located anywhere on it, depending on the value of α.

The next operation is subtracting points. Let $\mathbf{P}_0 = (x_0, y_0)$ and $\mathbf{P}_1 = (x_1, y_1)$ be two points. The difference $\mathbf{P}_1 - \mathbf{P}_0 = (x_1 - x_0, y_1 - y_0) = (\Delta x, \Delta y)$ is well defined. It is the vector (the direction and distance) from $\mathbf{P}_0$ to $\mathbf{P}_1$ (Figure 1.1b).

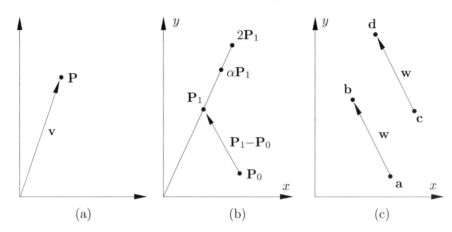

Figure 1.1: Operations on Points.

Figure 1.1c shows two pairs of points **a b** and **c d**. Points **a** and **c** are different and so are **b** and **d**. The vectors **b** − **a** and **d** − **c**, however, are identical

Example: The two points $\mathbf{P}_0 = (5, 4)$ and $\mathbf{P}_1 = (2, 6)$ are subtracted to produce the pair $\mathbf{P}_1 - \mathbf{P}_0 = (-3, 2)$. The new pair is a vector, because it represents a direction and a distance. To get from $\mathbf{P}_0$ to $\mathbf{P}_1$, we need to move -3 units in the x direction and 2 units in the y direction. Similarly, $\mathbf{P}_0 - \mathbf{P}_1$ is the direction from $\mathbf{P}_1$ to $\mathbf{P}_0$. The distance between the points is $\sqrt{(-3)^2 + 2^2}$. These properties do not depend on the particular coordinate axes used. If we translate the origin—or, equivalently, translate the points—m units in the x direction and n units in the y direction, the points will have new coordinates, but the difference will not change. The same property (the difference of points being independent of the coordinate axes) holds after rotation, scaling, shearing, and reflection: the so-called *affine transformations* (or mappings). This is why the operation of subtracting two points is affinely invariant. (Note that the product $\alpha \mathbf{P}$ is also affinely invariant.)

The sum of a point and a vector is well defined and is a point. Figure 1.2a shows the two sums $\mathbf{P}_1^* = \mathbf{P}_1 + \mathbf{v}$ and $\mathbf{P}_2^* = \mathbf{P}_2 + \mathbf{v}$. It is easy to see that the relative positions of $\mathbf{P}_1^*$ and $\mathbf{P}_2^*$ are the same as those of $\mathbf{P}_1$ and $\mathbf{P}_2$. Another way to look at the sum $\mathbf{P} + \mathbf{v}$ is to observe that it moves us away from $\mathbf{P}$, which is a point, in a certain direction and by a certain distance, thereby bringing us to another point. Yet another way of showing the same thing is to rewrite the relation $\mathbf{a} - \mathbf{b} = \mathbf{v}$ as $\mathbf{a} = \mathbf{b} + \mathbf{v}$, which shows that the sum of point $\mathbf{b}$ and vector $\mathbf{v}$ is a point $\mathbf{a}$.

Given any two points $\mathbf{P}_0$ and $\mathbf{P}_2$, the expression $\mathbf{P}_0 + \alpha(\mathbf{P}_2 - \mathbf{P}_0)$ is the sum of a point and a vector, so it is a point that we can denote by $\mathbf{P}_1$. The vector $\mathbf{P}_2 - \mathbf{P}_0$ points from $\mathbf{P}_0$ to $\mathbf{P}_2$, so adding it to $\mathbf{P}_0$ produces a point on the line connecting $\mathbf{P}_0$ to $\mathbf{P}_2$. Thus, we conclude that the three points $\mathbf{P}_0$, $\mathbf{P}_1$, and $\mathbf{P}_2$ are collinear. Note that the expression $\mathbf{P}_1 = \mathbf{P}_0 + \alpha(\mathbf{P}_2 - \mathbf{P}_0)$ can be written $\mathbf{P}_1 = (1 - \alpha)\mathbf{P}_0 + \alpha\mathbf{P}_2$, showing that $\mathbf{P}_1$ is a linear combination of $\mathbf{P}_0$ and $\mathbf{P}_2$. In general, any of three collinear points can be written as a linear combination of the other two. Such points are not independent.

⋄ **Exercise 1.1:** Given the three points $\mathbf{P}_0 = (1, 1)$, $\mathbf{P}_1 = (2, 2.5)$, and $\mathbf{P}_2 = (3, 4)$, are they collinear?

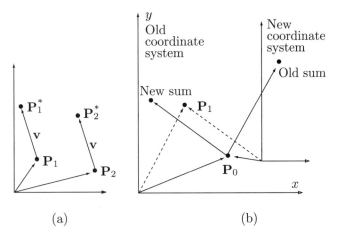

Figure 1.2: (a) Adding a Point and a Vector. (b) Adding Points.

⋄ **Exercise 1.2:** What can we say about four collinear points?

The next operation to consider is the sum of points. In general this operation is not well defined. We intuitively feel that adding two points should be done like adding vectors. The lines connecting the points with the origin should be added, to produce a sum vector. In fact, as Figure 1.2b shows, this operation depends on the coordinate axes. Moving the origin (or moving the points) will move the sum of the vectors a different distance or in a different direction, thereby changing the sum of the points. This is why the sum of points is, in general, undefined.

Example: Given the two points $(5, 3)$ and $(7, -2)$, we add them to produce $(12, 1)$. We now move the two points one unit to the left to become $(4, 3)$ and $(6, -2)$. Their new sum is $(10, 1)$, a point located two units to the left of the original sum.

There is, however, one important special case where the sum of points is well defined, the so-called *barycentric sum*. If we multiply each point by a weight and if the weights add up to 1, then the sum of the weighted points is affinely invariant, i.e., it is a valid point. Here is the (simple) proof: If $\sum_{i=0}^{n} w_i = 1$, then

$$
\begin{aligned}
\sum_{i=0}^{n} w_i \mathbf{P}_i &= \mathbf{P}_0 + \sum_{i=1}^{n} w_i \mathbf{P}_i - (1 - w_0)\mathbf{P}_0 \\
&= \mathbf{P}_0 + w_1 \mathbf{P}_1 + w_2 \mathbf{P}_2 + \cdots + w_n \mathbf{P}_n - (w_1 + \cdots + w_n)\mathbf{P}_0 \\
&= \mathbf{P}_0 + w_1(\mathbf{P}_1 - \mathbf{P}_0) + w_2(\mathbf{P}_2 - \mathbf{P}_0) + \cdots + w_n(\mathbf{P}_n - \mathbf{P}_0) \\
&= \mathbf{P}_0 + \sum_{i=1}^{n} w_i(\mathbf{P}_i - \mathbf{P}_0).
\end{aligned}
\tag{1.1}
$$

This is the sum of the point $\mathbf{P}_0$ and the vector $\sum_{i=1}^{n} w_i(\mathbf{P}_i - \mathbf{P}_0)$, and we already know that the sum of a point and a vector is a point.

Notice that the proof above does not assume that the weights are nonnegative and barycentric weights can in fact be negative. A little experiment may serve to convince the sceptics. Given two points (a, b) and (c, d) we construct the barycentric sum $(x, y) = -0.5(a, b) + 1.5(c, d)$. If we now translate both points by the vector (α, β), the sum is modified to

$$-0.5(a + \alpha, b + \beta) + 1.5(c + \alpha, d + \beta) = -0.5(a, b) + 1.5(c, d) + (\alpha, \beta) = (x, y) + (\alpha, \beta).$$

The barycentric sum (x, y) is translated by the same vector.

Mathematically-savvy readers may be familiar with the concept of normalization. Given a set of weights w_i that add up to $\alpha \neq 1$, they can be normalized by dividing each weight by the sum α. Thus, if we need a barycentric sum of certain quantities P_i and we are given nonbarycentric weights w_i, we can compute

$$\sum_{i=1}^{n} \frac{w_i}{\sum_{j=1}^{n} w_j} P_i = \sum_{i=1}^{n} \left(\frac{w_i}{\alpha} \right) P_i = \sum_{i=1}^{n} r_i P_i,$$

where the new, normalized weights r_i are barycentric.

Barycentric sums are common in curve and surface design. This book has numerous examples of curves and surfaces that are constructed as weighted sums of points, and they all must be barycentric. When a curve consists of a non-barycentric weighted sum of points, its shape depends on the particular coordinate system used. The shape changes when either the curve or the coordinate axes are moved or are affinely transformed. Such a curve is ill conditioned and cannot be used in practice.

The Isotropic Principle

Given a curve that's constructed as the sum

$$\mathbf{P}(t) = \sum w_i \mathbf{P}_i + \sum u_i \mathbf{v}_i,$$

where $\mathbf{P}_i$ are points and $\mathbf{v}_i$ are vectors, the curve is independent of the particular coordinate system used if and only if the weights w_i are barycentric. There is no similar requirement for the u_i weights. Notice that the points can be data points, control points, or any other points. The vectors can be tangents, second derivatives or any other vectors, but the statement above is always true. This statement is sometimes known as the *isotropic principle*.

A special case is the barycentric sum of two points $(1 - t)\mathbf{P}_0 + t\mathbf{P}_1$. This is a point on the line from $\mathbf{P}_0$ to $\mathbf{P}_1$. In fact, the entire straight segment from $\mathbf{P}_0$ to $\mathbf{P}_1$ is obtained when t is varied from 0 to 1 (Figure 1.3a). To see this, we write $\mathbf{P}(t) = (1 - t)\mathbf{P}_0 + t\mathbf{P}_1$. Clearly, $\mathbf{P}(0) = \mathbf{P}_0$ and $\mathbf{P}(1) = \mathbf{P}_1$. Also, since $\mathbf{P}(t) = t(\mathbf{P}_1 - \mathbf{P}_0) + \mathbf{P}_0$, $\mathbf{P}(t)$ is a linear function of t, which implies a straight line in t. The tangent vector is the derivative $\frac{d\mathbf{P}}{dt}$ and it is the constant $\mathbf{P}_1 - \mathbf{P}_0$, the direction from $\mathbf{P}_0$ to $\mathbf{P}_1$. Notice that this derivative is a vector, not a number. Selecting $t = 1/2$ yields $\mathbf{P}(0.5) = 0.5\mathbf{P}_1 + 0.5\mathbf{P}_0$, the midpoint between $\mathbf{P}_0$ and $\mathbf{P}_1$.

The concept of barycentric weights is so useful that the two numbers $1 - t$ and t are termed the *barycentric coordinates* of point $\mathbf{P}(t)$ with respect to $\mathbf{P}_0$ and $\mathbf{P}_1$.

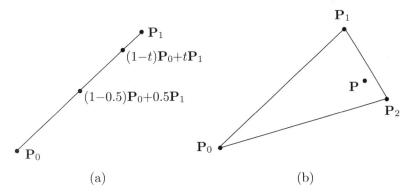

Figure 1.3: Line and Triangle.

The word *barycentric* seems to have first been used in [Dupuy 48]. It is derived from *barycenter*, meaning "center of gravity," because such weights are used to calculate the center of gravity of an object. Barycentric weights have many uses in geometry in general and in curve and surface design in particular.

Another useful example is the barycentric coordinates of a two-dimensional point with respect to the three corners of a triangle. Imagine a triangle with corners $\mathbf{P}_0$, $\mathbf{P}_1$, and $\mathbf{P}_2$ (Figure 1.3b). Any point $\mathbf{P}$ inside the triangle can be expressed as the weighted combination

$$\mathbf{P} = u\mathbf{P}_0 + v\mathbf{P}_1 + w\mathbf{P}_2, \quad \text{where} \quad u + v + w = 1. \tag{1.2}$$

The proof is that Equation (1.2) can be written explicitly as three equations in the three unknowns u, v, and w:

$$\begin{aligned}
P_x &= uP_{0x} + vP_{1x} + wP_{2x}, \\
P_y &= uP_{0y} + vP_{1y} + wP_{2y}, \\
1 &= u + v + w.
\end{aligned} \tag{1.3}$$

The solutions are unique provided that the three equations are independent. ◀

◇ **Exercise 1.3:** Show that Equation (1.3) consists of three independent equations if the three points $\mathbf{P}_0$, $\mathbf{P}_1$, and $\mathbf{P}_2$ are independent.

◇ **Exercise 1.4:** Show that the barycentric coordinates of point $\mathbf{P}_0$ with respect to $\mathbf{P}_0$, $\mathbf{P}_1$, and $\mathbf{P}_2$ are $(1, 0, 0)$. Also discuss the barycentric coordinates of points outside the triangle.

Example: Let $\mathbf{P}_0 = (1, 1)$, $\mathbf{P}_1 = (2, 3)$, $\mathbf{P}_2 = (5, 1)$, and $\mathbf{P} = (2, 2)$. Equation (1.3) becomes

$$(2, 2) = u(1, 1) + v(2, 3) + w(5, 1); \quad u + v + w = 1,$$

or

$$2 = u + 2v + 5w,$$
$$2 = u + 3v + w, \qquad \text{which yield} \qquad \begin{cases} u = 3/8, \\ v = 1/2, \\ w = 1/8. \end{cases}$$
$$1 = u + v + w,$$

⋄ **Exercise 1.5:** For a given triangle, calculate the (x, y, z) coordinates of the point with barycentric coordinates $(1/3, 1/3, 1/3)$. This point is called the *centroid* and is one of many centers that can be defined for a triangle. (Imagine cutting the triangle out of a piece of cardboard. If you try to support it at the centroid, it will balance.)

(This material is useful for the triangular Bézier surface patches described in Section 6.23.)

The barycentric combination is the most fundamental operation on points; so much so that it is used to define affine transformations. The definition is: a transformation of points in space is affine if it leaves barycentric combinations invariant. Hence, if $\mathbf{P} = \sum w_i \mathbf{P}_i$ and $\sum w_i = 1$, and if $\mathbf{T}$ is an affine transformation, then $\mathbf{TP} = \sum w_i \mathbf{TP}_i$. All common geometric transformations—such as scaling, shearing, rotation, and reflection—are affine.

Note: The difference of two points is a vector. We can consider such a difference a weighted sum where the weights add up to zero (they are $+1$ and -1). It turns out that a weighted sum of points where the weights add up to zero is a vector. To prove this, let

$$\mathbf{Q} = \sum_{i=1}^{n} w_i \mathbf{P}_i, \qquad \text{where} \qquad \sum w_i = 0,$$

and let $\mathbf{P}$ be a point. The sum $\mathbf{R} = \mathbf{Q} + \mathbf{P}$ is barycentric (since its coefficients add up to 1) and is therefore a point. The difference $\mathbf{R} - \mathbf{P} = \mathbf{Q}$ is a difference of points and is therefore a vector.

Note: Multiplying a point by a number produces a point, so if $\mathbf{P}$ is a point, then $-\mathbf{P}$ is also a point. It is located on the line connecting $\mathbf{P}$ with the origin, on the other side of the origin from $\mathbf{P}$. Once this is understood, we notice that the sum of points $\mathbf{P} + \mathbf{Q}$ can be written as the difference of points $\mathbf{P} - (-\mathbf{Q})$. This difference is, of course, the vector from point $-\mathbf{Q}$ to point $\mathbf{P}$ (Figure 1.4), so we conclude that the sum $\mathbf{P} + \mathbf{Q}$ of two points is well defined but is not very useful, since it tells us something about the relative positions of $\mathbf{P}$ and $-\mathbf{Q}$, not of $\mathbf{P}$ and $\mathbf{Q}$. Assuming that Figure 1.4 depicts the points $\mathbf{Q} = (-5, -1)$ and $\mathbf{P} = (4, 3)$, the sum $\mathbf{P} + \mathbf{Q}$ equals $(-5, -1) + (4, 3) = (-1, 2)$. This shows that in order to get from point $-\mathbf{Q}$ to point $\mathbf{P}$, we need to move one negative step in the x direction for every two steps in the y direction.

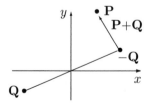

Figure 1.4: Adding Two Points.

◇ **Exercise 1.6:** Let **P** and **Q** be points and let **v** and **w** be vectors. What is the sum **P** − **Q** + **v** + **w**?

1.1.1 Operations on Vectors

The notation $|\mathbf{P}|$ indicates the magnitude (or absolute value) of vector **P**. Vector addition is defined by adding the individual elements of the vectors being added: $\mathbf{P} + \mathbf{Q} = (P_x, P_y, P_z) + (Q_x, Q_y, Q_z) = (P_x + Q_x, P_y + Q_y, P_z + Q_z)$. This operation is both commutative $\mathbf{P} + \mathbf{Q} = \mathbf{Q} + \mathbf{P}$ and associative $\mathbf{P} + (\mathbf{Q} + \mathbf{T}) = (\mathbf{P} + \mathbf{Q}) + \mathbf{T}$. Subtraction of vectors $(\mathbf{P} - \mathbf{Q})$ is done similarly and results in the vector from **Q** to **P**.

Vectors can be multiplied in three different ways as follows:

1. The product of a real number α by a vector **P** is denoted by $\alpha\mathbf{P}$ and produces the vector $(\alpha x, \alpha y, \alpha z)$. It changes the magnitude of **P** by a factor α, but does not change its direction.

2. The dot product of two vectors is denoted by $\mathbf{P} \bullet \mathbf{Q}$ and is defined as the scalar

$$(P_x, P_y, P_z)(Q_x, Q_y, Q_z)^T = \mathbf{PQ}^T = P_xQ_x + P_yQ_y + P_zQ_z.$$

This also equals $|\mathbf{P}||\mathbf{Q}|\cos\theta$, where θ is the angle between the vectors. The dot product of perpendicular vectors (also called *orthogonal vectors*) is therefore zero. The dot product is commutative, $\mathbf{P} \bullet \mathbf{Q} = \mathbf{Q} \bullet \mathbf{P}$.

The triple product $(\mathbf{P} \bullet \mathbf{Q})\mathbf{R}$ is sometimes useful. It can be represented as

$$
\begin{aligned}
(\mathbf{P} \bullet \mathbf{Q})\mathbf{R} &= (P_xQ_x + P_yQ_y + P_zQ_z)(R_x, R_y, R_z) \\
&= \big((P_xQ_x + P_yQ_y + P_zQ_z)R_x, (P_xQ_x + P_yQ_y + P_zQ_z)R_y, \\
&\qquad (P_xQ_x + P_yQ_y + P_zQ_z))R_z \\
&= (Q_x, Q_y, Q_z)\begin{pmatrix} P_xR_x & P_yR_x & P_zR_x \\ P_xR_y & P_yR_y & P_zR_y \\ P_xR_z & P_yR_z & P_zR_z \end{pmatrix} \\
&= \mathbf{Q}(\mathbf{PR}),
\end{aligned}
\tag{1.4}
$$

where the notation $(\mathbf{PR})$ stands for the 3×3 matrix of Equation (1.4).

3. The cross product of two vectors (also called the *vector product*) is denoted by $\mathbf{P} \times \mathbf{Q}$ and is defined as the vector

$$(P_2Q_3 - P_3Q_2, -P_1Q_3 + P_3Q_1, P_1Q_2 - P_2Q_1).\tag{1.5}$$

It is easy to show that $\mathbf{P} \times \mathbf{Q}$ is perpendicular to both **P** and **Q**.

◇ **Exercise 1.7:** Show it!

The following expressions show how $\mathbf{P} \times \mathbf{Q}$ can be expressed by means of a determinant:

$$\mathbf{P} \times \mathbf{Q} = \begin{vmatrix} \mathbf{i} & \mathbf{j} & \mathbf{k} \\ P_1 & P_2 & P_3 \\ Q_1 & Q_2 & Q_3 \end{vmatrix} = \mathbf{i}\begin{vmatrix} P_2 & P_3 \\ Q_2 & Q_3 \end{vmatrix} - \mathbf{j}\begin{vmatrix} P_1 & P_3 \\ Q_1 & Q_3 \end{vmatrix} + \mathbf{k}\begin{vmatrix} P_1 & P_2 \\ Q_1 & Q_2 \end{vmatrix}$$

$$= (P_2Q_3 - P_3Q_2, -P_1Q_3 + P_3Q_1, P_1Q_2 - P_2Q_1),$$

or, alternatively, by means of a matrix

$$= (Q_1, Q_2, Q_3) \begin{pmatrix} 0 & P_3 & -P_2 \\ -P_3 & 0 & P_1 \\ P_2 & -P_1 & 0 \end{pmatrix}. \tag{1.6}$$

◇ **Exercise 1.8:** The cross-product $\mathbf{P} \times \mathbf{Q}$ is perpendicular to both $\mathbf{P}$ and $\mathbf{Q}$. In what direction does it point?

The cross-product is not commutative and is not associative. It is, however, distributive with respect to addition or subtraction of vectors. Hence, $\mathbf{P} \times (\mathbf{Q} \pm \mathbf{T}) = \mathbf{P} \times \mathbf{Q} \pm \mathbf{P} \times \mathbf{T}$.

The magnitude of $\mathbf{P} \times \mathbf{Q}$ equals $|\mathbf{P}||\mathbf{Q}| \sin \theta$, where θ is the angle between the two vectors. The cross-product, therefore, has a simple geometric interpretation. Its magnitude equals the area of the parallelogram defined by the two vectors.

◇ **Exercise 1.9:** Given that $\mathbf{P} \times \mathbf{Q} = 0$, what does it tell us about the vectors involved?

◇ **Exercise 1.10:** Derive the vector line equation for the straight segment between two given points $\mathbf{P}_1$ and $\mathbf{P}_2$.

1.1.2 The Scalar Triple Product

The scalar triple product of three vectors, $\mathbf{P}$, $\mathbf{Q}$, and $\mathbf{R}$, is defined as

$$S = \mathbf{P} \bullet (\mathbf{Q} \times \mathbf{R}) = P_1(Q_2R_3 - Q_3R_2) + P_2(Q_3R_1 - Q_1R_3) + P_3(Q_1R_2 - Q_2R_1)$$

$$= \begin{vmatrix} P_1 & P_2 & P_3 \\ Q_1 & Q_2 & Q_3 \\ R_1 & R_2 & R_3 \end{vmatrix}. \tag{1.7}$$

Interchanging two rows in a determinant changes its sign, so interchanging rows twice leaves the determinant unchanged. This is why the triple product is not affected by a cyclic permutation of its three components. We can therefore write

$$S = \mathbf{P} \bullet (\mathbf{Q} \times \mathbf{R}) = \mathbf{Q} \bullet (\mathbf{R} \times \mathbf{P}) = \mathbf{R} \bullet (\mathbf{P} \times \mathbf{Q}).$$

The triple product has a simple geometric interpretation. It equals the volume of the parallelepiped defined by the three vectors. An important corollary is: if the three vectors are coplanar, then the parallelepiped defined by them has zero volume, implying that their scalar triple product is zero. This property is used in Section 2.2.1 to determine whether or not a given polygon is planar.

1.1.3 Projecting a Vector

A common and useful operation on vectors is projecting a vector $\mathbf{a}$ on another vector $\mathbf{b}$. The idea is to break vector $\mathbf{a}$ up into two perpendicular components $\mathbf{c}$ and $\mathbf{d}$, such that $\mathbf{c}$ is in the direction of $\mathbf{b}$.

Figure 1.5a shows that $\mathbf{a} = \mathbf{c} + \mathbf{d}$ and $|\mathbf{c}| = |\mathbf{a}| \cos \alpha$. On the other hand, $\mathbf{a} \bullet \mathbf{b} = |\mathbf{a}||\mathbf{b}| \cos \alpha$, yielding the magnitude of $\mathbf{c}$:

$$|\mathbf{c}| = |\mathbf{a}| \frac{(\mathbf{a} \bullet \mathbf{b})}{|\mathbf{a}||\mathbf{b}|} = \frac{(\mathbf{a} \bullet \mathbf{b})}{|\mathbf{b}|}. \tag{1.8}$$

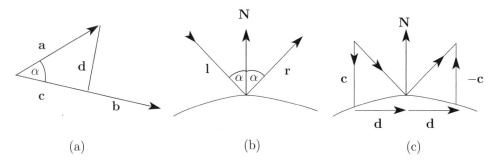

Figure 1.5: Projecting a Vector.

The direction of **c** is identical to the direction of **b**, so we can write vector **c** as

$$\mathbf{c} = |\mathbf{c}|\frac{\mathbf{b}}{|\mathbf{b}|} = \frac{(\mathbf{a} \bullet \mathbf{b})}{|\mathbf{b}|^2}\mathbf{b}. \tag{1.9}$$

Example: Given vectors $\mathbf{a} = (2,1)$ and $\mathbf{b} = (1,0)$, we compute the projection of a on b.

$$\mathbf{c} = \frac{(\mathbf{a} \bullet \mathbf{b})}{|\mathbf{b}|^2}\mathbf{b} = \frac{2\times 1 + 1\times 0}{1^2 + 0^2}(2,0) = (4,0), \qquad \mathbf{d} = \mathbf{a} - \mathbf{c} = (-2,1).$$

⋄ **Exercise 1.11:** The projection method works also for three-dimensional vectors. Given vectors $\mathbf{a} = (2,1,3)$ and $\mathbf{b} = (1,0,-1)$, calculate the projection of $\mathbf{a}$ on $\mathbf{b}$.

Summary: The following operations have been discussed in this section:

$$\text{point} - \text{point} = \text{vector}, \quad \text{scalar} \times \text{point} = \text{point}, \quad \text{vector} \pm \text{vector} = \text{vector},$$
$$\text{scalar} \times \text{vector} = \text{vector}, \quad \text{point} + \text{vector} = \text{point}, \quad \text{vector} \bullet \text{vector} = \text{scalar},$$
$$\text{vector} \times \text{vector} = \text{vector}.$$

The operation point + point is left undefined (since it is not useful). A barycentric sum of points is a point, and a weighted sum of points where the weights add up to zero is a vector.

From the dictionary

Vector: (1) A variable quantity that can be resolved into components. (2) A straight line segment whose length is magnitude and whose orientation in space is direction. (3) Any agent (person or animal or microorganism) that carries and transmits a disease.

1.2 Parametric Blending

Parametric blending is a family of techniques that make it possible to vary the value of some quantity in small steps, without any discontinuities. Blending can be thought of as averaging or interpolating. The following are examples:

1. Numbers. The average of the two numbers 15 and 18 is $(15 + 18)/2 = 16.5$. This can also be written as $0.5 \times 15 + 0.5 \times 18$, which can be interpreted as the *blend*, or the weighted sum, of the two numbers, where each is assigned a weight of 0.5. When the weights are different, such as $0.9 \times 15 + 0.1 \times 18$, the result is a blend of 90% 15 and 10% 18.

2. Points. If $\mathbf{P}_1$ and $\mathbf{P}_2$ are points, then the expression $\alpha \mathbf{P}_1 + \beta \mathbf{P}_2$ is a blend of the two points, in which α and β are the weights (or the coefficients). If α and β are nonnegative and $\alpha + \beta = 1$, then the blend is a point on the straight segment connecting $\mathbf{P}_1$ and $\mathbf{P}_2$.

3. Rotations. A rotation in three dimensions is described by means of the rotation angle (one number) and the axis of rotation (three numbers). These four numbers can be combined into a mathematical entity called quaternion and two quaternions can also be blended, resulting in a smooth sequence of rotations that proceeds in small, equal steps from an initial rotation to a final one. This type of blending is useful in computer animation.

4. Curve construction. Given a number of points, a curve can be created as a weighted sum of the points. It has the form $\sum w_i(t)\mathbf{P}_i$, where the weights $w_i(t)$ are barycentric. Such a curve is a *blend* of the points. For each value of t, the blend is different, but we have to make sure that the sum of the weights is always 1. It is possible to blend vectors, in addition to points, as part of the curve, and the weights of the vectors don't have to satisfy any particular requirement. Most of the curve methods described in this book generate a curve as a blend of points, vectors, or both.

A special case of curve construction is the linear blending of two points, which can be expressed as $(1 - t)\mathbf{P}_1 + t\,\mathbf{P}_2$ for $0 \leq t \leq 1$ [this is Equation (2.1)].

5. Surfaces. Using the same principle, points, vectors, and curves can be blended to form a surface patch.

6. Images. Various types of image processing, such as sharpening, blurring, and embossing, are performed by blending an image with a special mask image.

7. It is possible to blend points in nonlinear ways. An intuitive way to get, for example, quadratic blending is to square the two weights of the linear blend. However, the result, which is $\mathbf{P}(t) = (1 - t)^2 \mathbf{P}_1 + t^2 \mathbf{P}_2$, depends on the particular coordinate axes used, since the two coefficients $(1 - t)^2$ and t^2 are not barycentric. It turns out that the sum $(1 - t)^2 + 2t(1 - t) + t^2$ equals 1. As a result, we can use quadratic blending to blend three points, but not two.

Similarly, if we try a cubic blend by simply writing $\mathbf{P}(t) = (1 - t)^3 \mathbf{P}_1 + t^3 \mathbf{P}_2$, we end up with the same problem. Cubic blending can be achieved by adding four terms with weights t^3, $3t^2(1 - t)$, $3t(1 - t)^2$, and $(1 - t)^3$.

We therefore conclude that Bézier methods (Chapter 6) can be used for blending. The Bézier curve is a result of blending several points with the Bernstein polynomials, which add up to unity. Quadratic and cubic blending are special cases of the Bézier blending (or the Bézier interpolation).

1.3 Parametric Curves

As mentioned in the Preface, the main aim of computer graphics is to display an arbitrary surface so that it looks real. The first step toward this goal is an understanding of curves. Once we have an algorithm to calculate and display any curve, we may try to extend it to a surface.

In practice, curves (and surfaces) are specified by the user in terms of points and are constructed in an interactive process. The user starts by entering the coordinates of points, either by scanning a rough image of the desired shape and digitizing certain points on the image, or by drawing a rough shape on the screen and selecting certain points with a pointing device such as a mouse. After the curve has been drawn, the user may want to modify its shape by moving, adding, or deleting points. Such points can be employed in two different ways:

1. We may want the curve to pass through them. Such points are called *data points* and the curve is called an interpolating curve.

2. We may want the points to control the shape of the curve by exerting a "pull" on it. A point may pull part of the curve toward it, allowing the user to change the shape of the curve by moving the point. Generally, however, the curve does not pass through the point. Such points are called *control points* and the curve is called an approximating curve.

A mathematical function $y = f(x)$ can be plotted as a curve. Such a function is the *explicit* representation of the curve. The explicit representation is not general, since it cannot represent vertical lines and is also single-valued. For each value of x, only a single value of y is normally computed by the function.

The *implicit* representation of a curve has the form $F(x, y) = 0$. It can represent multivalued curves (more than one y value for an x value). A common example is the circle, whose implicit representation is $x^2 + y^2 - R^2 = 0$.

The explicit and implicit curve representations can be used only when the function is known. In practical applications—where complex curves such as the shape of a car or of a toaster are needed—the function is normally unknown, which is why a different approach is required.

The curve representation used in practice is called the *parametric representation*. A two-dimensional parametric curve has the form $\mathbf{P}(t) = \big(f(t), g(t)\big)$ or $\mathbf{P}(t) = \big(x(t), y(t)\big)$. The functions f and g become the (x, y) coordinates of any point on the curve, and the points are obtained when the parameter t is varied over a certain interval $[a, b]$, normally $[0, 1]$.

A simple example of a two-dimensional parametric curve is $\mathbf{P}(t) = (2t - 1, t^2)$. When t is varied from 0 to 1, the curve proceeds from the initial point $\mathbf{P}(0) = (-1, 0)$ to the final point $\mathbf{P}(1) = (1, 1)$. The x coordinate is linear in t and the y coordinate varies as t^2.

The first derivative $\frac{d\mathbf{P}(t)}{dt}$ is denoted by $\mathbf{P}^t(t)$, or by $\dot{\mathbf{P}}$, or by $(P_x^t(t), P_y^t(t))$. This derivative is the tangent vector to the curve at any point. The derivative is a vector and not a point because it is the limit of the difference $(\mathbf{P}(t + \Delta) - \mathbf{P}(t))/\Delta$, and the difference of points is a vector. As a vector, the tangent possesses a direction (the direction of the curve at the point) and a magnitude (which indicates the speed of the curve at the point). The tangent, however, is not the slope of the curve. The tangent is a pair of numbers, whereas the slope is a single number. The slope equals $\tan\theta$, where θ is the angle between the tangent vector and the x axis. The slope of a two-dimensional parametric curve is obtained by

$$\frac{dy}{dx} = \frac{\frac{dy}{dt}}{\frac{dx}{dt}} = \frac{P_y^t(t)}{P_x^t(t)}.$$

Example: The curve $\mathbf{P}(t) = (x(t), y(t)) = (1 + t^2/2, t^2)$. Its tangent vector is $\mathbf{P}^t(t) = (t, 2t)$ and the slope is $2t/t = 2$. The slope is constant, which indicates that the curve is a straight line. This is also easy to see from the tangent vector. The direction of this vector is always the same since it can be described by saying "for every t steps in the x direction, move $2t$ steps in the y direction."

Example: A circle. Because of its high symmetry, a circle can be represented in different ways. We list four different parametric representations of a circle of radius R centered on the origin.

1. $\mathbf{P}(t) = R(\cos t, \sin t)$, where $0 \le t \le 2\pi$. This is identical to the polar representation.

2. Substituting $t = \tan(u/2)$ yields $\mathbf{P}(t) = R[(1 - t^2)/(1 + t^2), 2t/(1 + t^2)]$. When $0 \le t \le 1$, this generates the first quadrant from $(R, 0)$ to $(0, R)$ (see also Figure 1.6a).

3. $\mathbf{P}(t) = R(t, \pm\sqrt{1 - t^2})$. When $0 \le t \le 1$ this generates the first quadrant from $(0, R)$ to $(R, 0)$ and, simultaneously, the third quadrant from $(0, -R)$ to $(-R, 0)$.

4. $\mathbf{P}(t) = (0.441, -0.441)t^3 + (-1.485, -0.162)t^2 + (0.044, 1.603)t + (1, 0)$. When $0 \le t \le 1$, this generates (approximately) the first quadrant from $(1, 0)$ to $(0, 1)$.

(See also circle example in Section 6.15, and Equation (Ans.31).)

⋄ **Exercise 1.12:** Explain how representation 4 is derived.

⋄ **Exercise 1.13:** Figure 1.6b shows a polygon inscribed in a circle. It is clear that adding sides to the polygon brings it closer to the circle. Calculate the difference $R - d$ as a function of n, the number of polygon sides.

The particle paradigm: Better insight into the behavior of parametric functions can be gained by thinking of the curve $\mathbf{P}(t) = (x(t), y(t))$ as a path traced out by a hypothetical particle. The parameter t can then be interpreted as time and the first two derivatives $\mathbf{P}^t(t)$ and $\mathbf{P}^{tt}(t)$ can be interpreted as the velocity and acceleration of the particle, respectively. It turns out that different parametric representations of the same curve may have different "speeds." The particle represented by $(\cos t, \sin t)$, for example, "moves" along the circle at speed $\mathbf{P}^t(t) = (-\sin t, \cos t)$, which is constant since $|\mathbf{P}^t(t)| = \sqrt{\sin^2 t + \cos^2 t} = 1$. The particle of circle representation 2, on the other hand, moves at the variable velocity

$$\mathbf{P}^t(t) = \left(\frac{-4t}{(1 + t^2)^2}, \frac{2(1 - t^2)}{(1 + t^2)^2}\right).$$

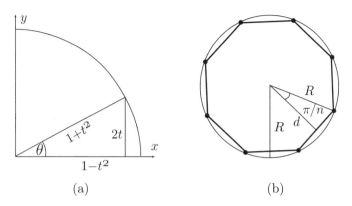

(a)   (b)

Figure 1.6: (a) A Parametric Representation.

(b) A Polygon Inscribed in a Circle.

◇ **Exercise 1.14:** Show that this velocity does vary with t.

◇ **Exercise 1.15:** What three-dimensional curve is described by the parametric function $(\cos t, \sin t, t)$? (Hint: see Section 2.4.1).

See also page 354 for the parametric representations of the sphere, the ellipsoid, and of the torus as a small circle rotating around a larger circle.

1.4 Properties of Parametric Curves

Generally, it is impossible to tell much about the behavior of a parametric curve $\mathbf{P}(t) = (x(t), y(t))$ by examining the two components $x(t)$ and $y(t)$ separately. Each of the two functions may have features that do not exist in the combination. The reverse is also true—the combined curve may have features not found in any of the two components.

Here is an example of two smooth curves whose combination is a parametric plane curve with a cusp (a sharp corner). The following two curves are polynomials in t:

$$x(t) = -18t^2 + 18t + 2, \qquad y(t) = -16t^3 + 24t^2 - 12t + 5, \quad \text{where} \quad 0 \le t \le 1.$$

They are smooth, since their derivatives $x'(t) = -36t + 18$ and $y'(t) = -48t^2 + 48t - 12$ are continuous in the range $0 \le t \le 1$. However, the combined curve

$$\mathbf{P}(t) = (0, -16)t^3 + (-18, 24)t^2 + (18, -12)t + (2, 5)$$

has a sharp corner (a cusp or a kink), because its tangent vector

$$\mathbf{P}^t(t) = 3(0, -16)t^2 + 2(-18, 24)t + (18, -12)$$

satisfies $\mathbf{P}^t(0.5) = (0, 0)$.

◇ **Exercise 1.16:** Find two curves $x(t)$ and $y(t)$, each with a cusp, such that the combined curve $\mathbf{P}(t) = (x(t), y(t))$ is smooth.

The parametric curves used in computer graphics are normally based on polynomials, since polynomials are simple functions that are easy to calculate and are flexible enough to create many different shapes. However, in principle, any functions can be used to create a parametric curve. Here is an example that uses the smooth sine and cosine curves to create the nonsmooth parametric curve shown on the right. It is defined by the simple expression

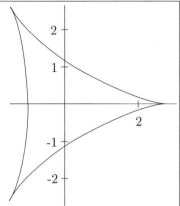

$$\mathbf{P}(t) = (2\cos(t) + \cos(2t), 2\sin(t) - \sin(2t)),$$

where $0 \le t \le 2\pi$. This curve has cusps at $t = 0$, $t = 0.261799$, and $t = 0.523599$. Another example of a parametric curve that's not a simple polynomial is the circular Bézier curve, Equation (4.141) of [Salomon 99].

⋄ **Exercise 1.17:** Find three curves $x(t)$, $y(t)$, and $z(t)$, each a cubic polynomial, such that the combined curve $\mathbf{P}(t) = (x(t), y(t), z(t))$ is not a cubic polynomial.

Note. A word about the notation used here. We have used the letter $\mathbf{P}$ to denote both points and curves. The same letter is later used to denote surfaces. In spite of using the same letter, the notation is unambiguous. It is always easy to tell what a particular $\mathbf{P}$ stands for by counting the number of free parameters. Something like $\mathbf{P}(u, w)$ denotes a surface since it depends on two variable parameters, whereas $\mathbf{P}(0, w)$ is a curve and $\mathbf{P}(u_0, 1)$ (for a fixed u_0) is a point.

One important feature of curves is *independence of the coordinate axes*. We don't want the curve to change shape when the coordinate axes (or the points defining the curve) are moved rigidly or rotated. Here is an example of how such a thing can happen. Consider the parametric curve

$$\mathbf{P}(t) = (1 - t)^3 \mathbf{P}_0 + t^3 \mathbf{P}_1 = \left((1 - t)^3 x_0 + t^3 x_1, (1 - t)^3 y_0 + t^3 y_1\right).$$

It is easy to see that $\mathbf{P}(0) = \mathbf{P}_0$ and $\mathbf{P}(1) = \mathbf{P}_1$ (the curve passes through the two points). What kind of a curve is $\mathbf{P}(t)$? The tangent vector of our curve is

$$\left(\frac{dx}{dt}, \frac{dy}{dt}\right) = \left(-3(1 - t)^2 x_0 + 3t^2 x_1, -3(1 - t)^2 y_0 + 3t^2 y_1\right).$$

To calculate the slope, we have to select actual points. We start with the two points $\mathbf{P}_0 = (0, 0)$ and $\mathbf{P}_1 = (5, 6)$. The slope of the curve is

$$\frac{dy}{dx} = \frac{dy}{dt} \bigg/ \frac{dx}{dt} = \frac{-3(1 - t)^2 0 + 3t^2 \times 6}{-3(1 - t)^2 0 + 3t^2 \times 5} = \frac{6}{5} = \text{constant},$$

so the curve is a straight line.

Next, we translate both points by the same amount $(0, -1)$, so the new points are $\mathbf{P}_0 = (0, -1)$ and $\mathbf{P}_1 = (5, 5)$. The new slope is

$$\frac{3(1-t)^2 + 15t^2}{15t^2} = \frac{1}{5}\left(\frac{1}{t} - 1\right) + 1.$$

It is no longer constant and therefore the curve is no longer a straight line (Figure 1.7). The curve has changed its shape just because its endpoints have been moved!

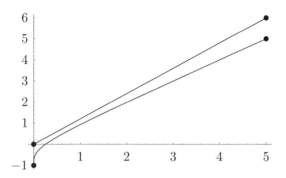

```
(* non-barycentric weights example *)
Clear[p0,p1,g1,g2,g3,g4];
p0={0,0}; p1={5,6};
g1=ParametricPlot[(1-t)^3 p0+t^3 p1,{t,0,1},PlotRange->All, Compiled->False,
DisplayFunction->Identity];
g3=Graphics[{AbsolutePointSize[4], {Point[p0],Point[p1]} }];
p0={0,-1}; p1={5,5};
g2=ParametricPlot[(1-t)^3 p0+t^3 p1,{t,0,1},PlotRange->All, Compiled->False,
PlotStyle->AbsoluteDashing[{2,2}], DisplayFunction->Identity];
g4=Graphics[{AbsolutePointSize[4], {Point[p0],Point[p1]} }];
Show[g2,g1,g3,g4, DisplayFunction->$DisplayFunction, DefaultFont->{"cmr10", 10}];
```

Figure 1.7: Effect of Nonbarycentric Weights.

It turns out that a curve of the form $\mathbf{P}(t) = \sum_{i=0}^{n} w_i(t)\mathbf{P}_i$, is independent of the particular coordinate axes used if $\sum_{i=0}^{n} w_i(t) = 1$. This is arguably the most important property of barycentric weights.

It is easy to extend the concept of parametric curves to three dimensions (space curves) with two minor differences (1) $\mathbf{P}(t)$ should be of the form $\big(x(t), y(t), z(t)\big)$ and (2) the slope of a three-dimensional curve is undefined. Such a curve has a tangent vector $d\mathbf{P}/dt$, but not a slope.

$\diamond$ **Exercise 1.18:** Show that the parametric curve

$$\mathbf{P}(t) = \mathbf{P} + 2\alpha(\mathbf{Q} - \mathbf{P})t + (1 - 2\alpha)(\mathbf{Q} - \mathbf{P})t^2, \quad 0 \le t \le 1 \qquad (1.10)$$

(where α is any real number) is a straight line, even though it is a polynomial of degree 2 in t. Note that the curve goes from point $\mathbf{P}$ to point $\mathbf{Q}$.

1.4.1 Uniform and Nonuniform Parametric Curves

So far, we have assumed that the parameter t of a parametric curve $\mathbf{P}(t) = (x(t), y(t))$ varies in the interval $[0, 1]$. It is also possible to vary t in other ranges, and such curves may be useful in special applications. This idea arises naturally when we try to fit a curve to a given set of data points. One question that should be answered in such a case is what value should the parameter t have at each point. It turns out that this is both a problem and an opportunity. A practical, interactive algorithm for curve design should make it possible to treat the values of t at the data points as parameters, and therefore to produce an entire family of curves, all of whose members pass through the given data points (but behave differently between points). This gives the designer an extra tool that can be used to construct the right curve.

The two approaches to this problem are (1) increment t by one for each point and (2) increment t by different values. The former approach yields a *uniform* parametric curve, while the latter results in a *nonuniform* parametric curve. Uniform parametric curves are normally easy to calculate and they produce good results when the points are roughly equally spaced. However, when the spacing of the points is very different, a uniform curve may look strange and unnatural, even though it passes through all the data points. This is when a nonuniform parametric curve should be used.

If the spacings of the points are far from uniform, it is common to increase the value of t at point $\mathbf{P}_i$ by the distance $|\mathbf{P}_i - \mathbf{P}_{i-1}|$. Notice that this distance is the chord length from point $\mathbf{P}_{i-1}$ to point $\mathbf{P}_i$. If this convention is used, then t starts at zero and is assigned the accumulated chord length at every data point. If the curve does not oscillate much between data points, the chord length is a good approximation to the arc length of the curve, with the result that t is assigned, in such a case, values that are close to the arc length. A curve $\mathbf{P}(s)$ where the parameter is the arc length s has a tangent vector $\mathbf{P}^s(s)$ of magnitude one (it's a unit vector). If we express such a curve as $\mathbf{P}(s) = (x(s), y(s))$, then $(x^s(s), y^s(s))$ is a unit vector, which implies that $|x^s(s)| \leq 1$ and $|y^s(s)| \leq 1$. This, in turn, means that the slopes of both curves $x(s)$ and $y(s)$ are bounded between -1 and $+1$, so the two curves are never too steep and are generally well behaved.

1.4.2 Curve Continuity

In practice, a complete curve is often made up of segments, so it is important to understand how individual segments can be connected. There are two types of curve continuities: *geometric* and *parametric*. If two consecutive segments meet at a point, the total curve is said to have G^0 geometric continuity. (It may look as in Figure 1.8a.) If, in addition, the directions of the tangent vectors of the two segments are the same at the point, the curve has G^1 geometric continuity at the point. The two segments connect smoothly (Figure 1.8b). In general, a curve has geometric continuity G^n at a join point if every pair of the first n derivatives of the two segments have the same direction at the point. If the same derivatives also have identical magnitudes at the point, then the curve is said to have C^n parametric continuity at the point.

We can refer to C^0, C^1, and C^2 as point, tangent, and curvature continuities, respectively. Figure 1.9 illustrates the geometric meanings of the three types. In part C^0 of the figure, the curve is continuous at the interior point, but its tangent is not. The curve changes its direction abruptly at the point; it has a kink. In part C^1, both

(a) (b) (c)

Figure 1.8: (a) G^0 Continuity (a Sharp Corner). (b) G^1 Continuity (a Smooth Connection). (c) G^2 Continuity (a Tight Curve).

the curve and its tangent are continuous at the interior point, but the curve changes its shape at the point from a straight line (zero curvature) to a curved line (nonzero curvature). Thus, the curvature is discontinuous at the point. In part C^2 the curve starts curving before it reaches the interior point, in order to preserve its curvature at the point. Generally, high continuity results in a smoother curve.

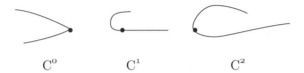

C^0 C^1 C^2

Figure 1.9: Three Curve Continuities.

A C^k continuity is more restrictive than G^k, so a curve that has C^k continuity at a join point also has G^k continuity at the point, but there is an exception. Imagine two segments connecting at a point, where both have tangent vectors of $(0,0,0)$ at the point. The vectors are identical, so the curve has C^1 continuity at the point. However, Exercise 5.3 (page 146) shows that the two segments may move in different directions at the point, in which case the curve will not have G^1 continuity.

Parameter Substitution

Instead of naming the parameter t, we can give it a different name. Moreover, we can use a function of t as the parameter. It can be shown that if $g(t)$ is a function that increases monotonically with t (i.e., if $t_2 > t_1$ implies $g(t_2) > g(t_1)$), then the curve $\mathbf{P}(g(t))$ will have the same shape as $\mathbf{P}(t)$ (although $g(t)$ will normally have to vary in a different range than t).

For two-dimensional curves, the substitution does not affect the slope of the curve since

$$\frac{\frac{dy(g)}{dg}}{\frac{dx(g)}{dg}} \Big/ \frac{\frac{dg(t)}{dt}}{\frac{dg(t)}{dt}} = \frac{\frac{dy(t)}{dt}}{\frac{dx(t)}{dt}} = \frac{dy(t)}{dx(t)}.$$

The reason for having two types of continuities has to do with parameter substitution (see box). Given a curve segment $\mathbf{P}(t)$ where $0 \le t \le 1$, we can substitute $T = t^2$.

The new segment $\mathbf{Q}(T) = \mathbf{Q}(t^2)$, where $0 \leq T \leq 1$, is identical in shape to $\mathbf{P}(t)$. The two identical curves must, of course, have the same tangents. However, their calculated tangent vectors have different magnitudes because

$$\frac{d\mathbf{Q}(t^2)}{dt} = 2t\frac{d\mathbf{Q}(t)}{dt} = 2t\frac{d\mathbf{P}(t)}{dt}.$$

This is why we separate the direction and the magnitude of the tangent vectors when considering curve continuities. If the directions of the tangent vectors are equal, they produce a smooth join and we call this case G^1 continuity (which is often all that is required in practice).

Example: Consider the two straight segments $\mathbf{P}(t) = (8t, 6t)$ and $\mathbf{Q}(t) = (4(t + 2), 3(t+2))$. The first goes from $(0,0)$ to $(8,6)$ and the second goes from $(8,6)$ to $(12,9)$. Their tangent vectors are $\mathbf{P}^t(t) = (8,6)$ and $\mathbf{Q}^t(t) = (4,3)$. The segments connect smoothly at $(8,6)$ (in fact, they look like one straight segment), but their tangent vectors are different at that point! Thus, the total curve has G^1 continuity at point $(8,6)$, but not C^1 continuity.

It is interesting to note, however, that the unit tangent vectors **are** equal at the joint. The magnitude of $\mathbf{P}^t(t)$ is $\sqrt{8^2 + 6^2} = 10$ and that of $\mathbf{Q}^t(t) = \sqrt{4^2 + 3^2} = 5$. The two unit tangent vectors are therefore equal $(8/10, 6/10) = (4/5, 3/5)$. Thus, the unit tangent vector provides a better measure of the direction of the curve than the tangent vector itself. Another natural vector that's associated with every point of a smooth curve is the curvature, a basic concept that's discussed in Section 1.6.

A curve whose tangent vector and curvature vector (Section 1.6.6) are everywhere continuous is said to have G^2 (second-order geometric) continuity.

> You can do anything you like with me except paint me, Hughie dear. I have to draw the line somewhere. But that's just what you *can't* do—draw a line, I mean. I like you in every way, as you well know, except as a painter. You would have been a good painter if you had never painted—did I invent that?
>
> —L. P. Hartley, *The Hireling*

1.5 PC Curves

Parametric curves used in computer graphics are based on polynomials. A polynomial of degree one has the form $\mathbf{P}_1(t) = \mathbf{A}t + \mathbf{B}$ and is, therefore, a straight line so it can only be used in limited cases. A parametric polynomial of degree 2 (quadratic) has the form $\mathbf{P}_2(t) = \mathbf{A}t^2 + \mathbf{B}t + \mathbf{C}$ and is always a parabola (see next paragraph and Appendix A). A polynomial of degree 3 (cubic) has the form $\mathbf{P}_3(t) = \mathbf{A}t^3 + \mathbf{B}t^2 + \mathbf{C}t + \mathbf{D}$ and is the simplest curve that can have complex shapes and can also be a space curve. (The complexity of this polynomial is limited, though. It can have at most one loop, and, if it does not have a loop, it can have at most two inflection points, see Section 1.6.8). Polynomials of higher degrees are sometimes needed, but they generally wiggle too much and are difficult to control. They also have more coefficients, so they require more input data to determine all the coefficients. As a result, a complete curve is often constructed from segments, each a parametric cubic polynomial (also called a PC). The complete curve is a piecewise polynomial curve, sometimes also called a *spline* (see definition on page 141).

Plane curves described by degree-2 polynomials are conic sections, but this is true only for the implicit representation. A plane curve described parametrically by a degree-2 polynomial can only be a parabola. Given such a curve $\mathbf{P}(t) = \mathbf{a}\,t^2 + \mathbf{b}\,t + \mathbf{c}$ we observe that it has a single value for any value of t and that it grows without limit when t becomes very large (positive or negative). Thus, when t approaches $\pm\infty$, $\mathbf{P}(t)$ also approaches ∞ or $-\infty$ (depending on the sign of $\mathbf{a}$) but there is only one branch that goes toward ∞ and one branch that goes toward $-\infty$. We therefore conclude that $\mathbf{P}(t)$ cannot be an ellipse because ellipses are finite, and it cannot be a hyperbola because these curves approach $\pm\infty$ in two directions. It must therefore be a parabola. A more rigorous proof, using parameter substitution, can be found in [Gallier 00], page 66.

Figure 1.10 shows seven data points and two curves that fit them. The dashed curve is a polynomial of degree 6; the solid curve is a spline. It is easy to see that the polynomial oscillates, whereas the spline curve is tight and is therefore more pleasing to the eye.

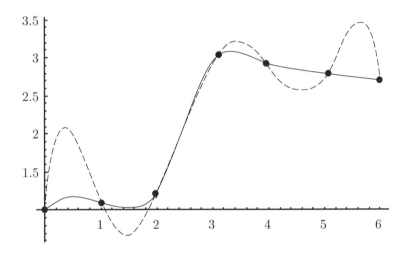

```
Clear[points];
points={{0,1},{1,1.1},{2,1.2},{3,3},{4,2.9},{5,2.8},{6,2.7}};
InterpolatingPolynomial[points,x];
Interpolation[points,InterpolationOrder->3];
Show[ListPlot[points,Prolog->AbsolutePointSize[5]],
 Plot[%%,{x,0,6},PlotStyle->Dashing[{0.05,0.05}]],
 Plot[%[x],{x,0,6}]]
```

Figure 1.10: Polynomial and Spline Fit.

⋄ **Exercise 1.19:** Show that a quadratic polynomial must be a plane curve.

⋄ **Exercise 1.20:** Why does a high-degree polynomial wiggle?

Question: The word "quad" comes from Latin for "four," so why is a degree-2 polynomial called quadratic? While we are at it, why is a degree-3 polynomial called cubic?

Answer: A square of side length n has four sides (it is quadratic), but its area is n^2 and this is associated with a degree-2 polynomial, which has terms up to x^2. Similarly, a cube of side length n has volume n^3, which is why the term "cubic" has become associated with a degree-3 polynomial.

A single PC segment is determined by means of points (data or control) or tangent vectors. Continuity considerations are also used sometimes to constrain the curve. Regardless of the input data, the segment always has the form $\mathbf{P}(t) = \mathbf{A}t^3 + \mathbf{B}t^2 + \mathbf{C}t + \mathbf{D}$. Thus, four unknown coefficients have to be calculated, which requires four equations. The equations must depend on four known quantities, points or vectors, that we denote by $\mathbf{G}_1$ through $\mathbf{G}_4$. The PC segment is expressed as the product

$$\mathbf{P}(t) = (t^3, t^2, t, 1) \begin{pmatrix} m_{11} & m_{12} & m_{13} & m_{14} \\ m_{21} & m_{22} & m_{23} & m_{24} \\ m_{31} & m_{32} & m_{33} & m_{34} \\ m_{41} & m_{42} & m_{43} & m_{44} \end{pmatrix} \begin{pmatrix} \mathbf{G}_1 \\ \mathbf{G}_2 \\ \mathbf{G}_3 \\ \mathbf{G}_4 \end{pmatrix} = \mathbf{T}(t) \cdot \mathbf{M} \cdot \mathbf{G},$$

where $\mathbf{M}$ is the basis matrix that depends on the method used and $\mathbf{G}$ is the geometry vector, consisting of the four given quantities. The segment can also be written as the weighted sum

$$\begin{aligned} \mathbf{P}(t) &= (t^3 m_{11} + t^2 m_{21} + t m_{31} + m_{41})\mathbf{G}_1 + (t^3 m_{12} + t^2 m_{22} + t m_{32} + m_{42})\mathbf{G}_2 \\ &\quad + (t^3 m_{13} + t^2 m_{23} + t m_{33} + m_{43})\mathbf{G}_3 + (t^3 m_{14} + t^2 m_{24} + t m_{34} + m_{44})\mathbf{G}_4 \\ &= B_1(t)\mathbf{G}_1 + B_2(t)\mathbf{G}_2 + B_3(t)\mathbf{G}_3 + B_4(t)\mathbf{G}_4 = \mathbf{B}(t) \cdot \mathbf{G} = \mathbf{T}(t) \cdot \mathbf{N} \cdot \mathbf{G}, \end{aligned}$$

where $\mathbf{B}(t)$ equals the product $\mathbf{T}(t) \cdot \mathbf{M}$ and the $B_i(t)$ are the weights. They are also called the *blending functions*, since they blend the four given quantities. If any of the quantities being blended are points, their weights should be barycentric. In the case where all four quantities are points, this requirement implies that the sum of the elements of matrix $\mathbf{M}$ should equal 1 (because the 16 elements of $\mathbf{M}$ are also the elements of the $B_i(t)$'s).

A PC segment can also be written in the form

$$\mathbf{P}(t) = \mathbf{A}t^3 + \mathbf{B}t^2 + \mathbf{C}t + \mathbf{D} = (t^3, t^2, t, 1) \begin{pmatrix} A_x & A_y & A_z \\ B_x & B_y & B_z \\ C_x & C_y & C_z \\ D_x & D_y & D_z \end{pmatrix} = \mathbf{T}(t) \cdot \mathbf{C},$$

where $\mathbf{A} = (A_x, A_y, A_z)$ and similarly for $\mathbf{B}$, $\mathbf{C}$, and $\mathbf{D}$. Its first derivative is

$$\frac{d\mathbf{P}(t)}{dt} = \frac{d\mathbf{T}(t)}{dt} \cdot \mathbf{C} = (3t^2, 2t, 1, 0)\mathbf{C}$$

and this is the tangent vector of the curve. This vector points in the direction of the tangent to the curve, but its magnitude is also important. It describes the *speed* of the curve.

In physics, if the function $x(t)$ describes the position of an object at time t, then $dx(t)/dt$ describes its velocity, and $d^2x(t)/dt^2$ gives its acceleration. This is also true for curves, but the speed in question is not the speed of drawing the curve on the screen! Rather, it is the distance covered on the curve when t is incremented in equal steps (see the particle paradigm of Section 1.3).

This concept is important in computer animation. Imagine a camera moving along the curve while t is incremented in equal steps. The speed of the camera at a point is given by the magnitude of the tangent vector at that point. If we want the camera to move at a constant speed, all tangent vectors must have the same magnitude. For this to happen, the tangent vector must be independent of t, a constant. This implies that the second derivative (the acceleration) is the zero vector, and the curve itself must be a linear function of t, a straight line. Any other curve has a tangent vector that depends on t, implying that the curve itself moves at variable speed.

1.5.1 Fast Computation of a PC

This section employs the method of *forward differences*, together with the Taylor series representation, to speed up the calculation of a point on a parametric curve $\mathbf{P}(t)$. Once this method is implemented, an entire curve can be drawn in a loop where t is incremented from 0 to 1 in small, equal steps of Δ. In iteration $i + 1$, a point $\mathbf{P}([i + 1]\Delta)$ is computed and is connected to the previous point $\mathbf{P}(i\Delta)$ by a short, straight segment. Section 6.3 applies this method to the Bézier curve.

The principle of forward differences is to find a quantity $\mathbf{dP}$ such that $\mathbf{P}(t + \Delta) = \mathbf{P}(t) + \mathbf{dP}$ for any value of t. If such a $\mathbf{dP}$ can be found, then it is enough to calculate $\mathbf{P}(0)$, and use forward differences to compute

$$\mathbf{P}(0 + \Delta) = \mathbf{P}(0) + \mathbf{dP},$$
$$\mathbf{P}(2\Delta) = \mathbf{P}(\Delta) + \mathbf{dP} = \mathbf{P}(0) + 2\mathbf{dP},$$
$$\vdots$$
$$\mathbf{P}([i + 1]\Delta) = \mathbf{P}(i\Delta) + \mathbf{dP} = \mathbf{P}(0) + (i + 1)\,\mathbf{dP}.$$

The point is that $\mathbf{dP}$ should not depend on t. If $\mathbf{dP}$ turns out to depend on t, then as we advance t from 0 to 1, we have to use different values of $\mathbf{dP}$, slowing down the calculations. The fastest way to calculate the curve is to precalculate $\mathbf{dP}$ before the loop starts and repeatedly add this precalculated value to $\mathbf{P}(0)$ inside the loop.

We calculate $\mathbf{dP}$ from the *Taylor series* representation of the curve. The Taylor series of a function $f(t)$ at a point $f(t + \Delta)$ is the infinite sum

$$f(t + \Delta) = f(t) + f'(t)\Delta + \frac{f''(t)\Delta^2}{2!} + \frac{f'''(t)\Delta^3}{3!} + \cdots.$$

In order to avoid dealing with an infinite sum, we limit our discussion to the popular PC curves. The mathematical treatment for any other type of curve (a different-degree polynomial or a nonpolynomial) is similar, although normally more complex. A general PC curve has the form $\mathbf{P}(t) = \mathbf{a}t^3 + \mathbf{b}t^2 + \mathbf{c}t + \mathbf{d}$, so only its first three derivatives are nonzero. These derivatives are

$$\mathbf{P}^t(t) = 3\mathbf{a}t^2 + 2\mathbf{b}t + \mathbf{c}, \quad \mathbf{P}^{tt}(t) = 6\mathbf{a}t + 2\mathbf{b}, \quad \mathbf{P}^{ttt}(t) = 6\mathbf{a},$$

so the Taylor series representation produces

$$\begin{aligned}
\mathbf{dP} &= \mathbf{P}(t+\Delta) - \mathbf{P}(t) \\
&= \mathbf{P}^t(t)\Delta + \frac{\mathbf{P}^{tt}(t)\Delta^2}{2} + \frac{\mathbf{P}^{ttt}(t)\Delta^3}{6} \\
&= 3\mathbf{a}\,t^2\Delta + 2\mathbf{b}\,t\Delta + \mathbf{c}\Delta + 3\mathbf{a}\,t\Delta^2 + \mathbf{b}\Delta^2 + \mathbf{a}\Delta^3.
\end{aligned}$$

This seems a failure since $\mathbf{dP}$ is a function of t (it should therefore be denoted by $\mathbf{dP}(t)$ instead of just $\mathbf{dP}$) and is also slow to calculate. However, the original PC curve $\mathbf{P}(t)$ is a degree-3 polynomial, whereas $\mathbf{dP}(t)$ is only a degree-2 polynomial. This suggests a way out of our difficulty. We can try to express $\mathbf{dP}(t)$ by means of the Taylor series, similar to what we did with the original curve $\mathbf{P}(t)$. This should result in a forward difference $\mathbf{ddP}(t)$ that's a polynomial of degree 1 in t. The quantity $\mathbf{ddP}(t)$ can, in turn, be represented by another Taylor series to produce a forward difference $\mathbf{dddP}$ that's a degree-0 polynomial in t, i.e., a constant. Once this is done, we hope to end up with an algorithm of the form

Compute $\mathbf{P}(0)$, dP, ddP, and dddP;
P = $\mathbf{P}(0)$;
<u>for</u> t:=0 <u>to</u> 1 <u>step</u> Δt <u>do</u>
PN:=P+dP; dP:=dP+ddP; ddP:=ddP+dddP;
line(P,PN);
P:=PN;
<u>endfor</u>;

The quantity $\mathbf{ddP}(t)$ is obtained by

$$\mathbf{dP}(t+\Delta) = \mathbf{dP}(t) + \mathbf{ddP}(t) = \mathbf{dP}(t) + \mathbf{dP}^t(t)\Delta + \frac{\mathbf{dP}(t)^{tt}\Delta^2}{2},$$

yielding

$$\begin{aligned}
\mathbf{ddP}(t) &= \mathbf{dP}^t(t)\Delta + \frac{\mathbf{dP}(t)^{tt}\Delta^2}{2} \\
&= (6\mathbf{a}\,t\Delta + 2\mathbf{b}\Delta + 3\mathbf{a}\Delta^2)\Delta + \frac{6\mathbf{a}\Delta\Delta^2}{2} \\
&= 6\mathbf{a}\,t\Delta^2 + 2\mathbf{b}\Delta^2 + 6\mathbf{a}\Delta^3.
\end{aligned}$$

Finally, $\mathbf{dddP}$ is similarly obtained by $\mathbf{ddP}(t+\Delta) = \mathbf{ddP}(t) + \mathbf{dddP} = \mathbf{ddP}(t) + \mathbf{ddP}^t(t)\Delta$, yielding $\mathbf{dddP} = \mathbf{ddP}^t(t)\Delta = 6\mathbf{a}\Delta^3$, a constant.

The four quantities involved in the calculation of the curve are therefore

$$\mathbf{P}(t) = \mathbf{a}t^3 + \mathbf{b}t^2 + \mathbf{c}t + \mathbf{d},$$
$$\mathbf{dP}(t) = 3\mathbf{a}\,t^2\Delta + 2\mathbf{b}\,t\Delta + \mathbf{c}\Delta + 3\mathbf{a}\,t\Delta^2 + \mathbf{b}\Delta^2 + \mathbf{a}\Delta^3,$$
$$\mathbf{ddP}(t) = 6\mathbf{a}\,t\Delta^2 + 2\mathbf{b}\Delta^2 + 6\mathbf{a}\Delta^3,$$
$$\mathbf{dddP} = 6\mathbf{a}\Delta^3.$$

They have to be calculated at $t = 0$ before the loop starts, then each iteration computes the first three quantities from those of the previous iteration ($\mathbf{dddP}$ doesn't depend on t). Here are the details

$$\mathbf{P}(0) = \mathbf{d}, \quad \mathbf{dP}(0) = \mathbf{a}\Delta^3 + \mathbf{b}\Delta^2 + \mathbf{c}\Delta, \quad \mathbf{ddP}(0) = 6\mathbf{a}\Delta^3 + 2\mathbf{b}\Delta^2, \quad \mathbf{dddP} = 6\mathbf{a}\Delta^3.$$
$$\mathbf{P}(\Delta) = \mathbf{a}\Delta^3 + \mathbf{b}\Delta^2 + \mathbf{c}\Delta + \mathbf{d} = \mathbf{P}(0) + \mathbf{dP}(0),$$
$$\mathbf{dP}(\Delta) = \mathbf{a}\Delta^3 + 2\mathbf{b}\Delta^2 + \mathbf{c}\Delta + 3\mathbf{a}\Delta^3 + \mathbf{b}\Delta^2 + \mathbf{a}\Delta^3 = \mathbf{dP}(0) + \mathbf{ddP}(0),$$
$$\mathbf{ddP}(\Delta) = 6\mathbf{a}\Delta^3 + 2\mathbf{b}\Delta^2 + 6\mathbf{a}\Delta^3 = \mathbf{ddP}(0) + \mathbf{dddP},$$

$$\cdots$$

$$\mathbf{P}([i+1]\Delta) = \mathbf{P}(i\Delta) + \mathbf{dP}(i\Delta),$$
$$\mathbf{dP}([i+1]\Delta) = \mathbf{dP}(i\Delta) + \mathbf{ddP}(i\Delta),$$
$$\mathbf{ddP}([i+1]\Delta) = \mathbf{ddP}(i\Delta) + \mathbf{dddP}.$$

Thus, each iteration computes a point $\mathbf{P}([i+1]\Delta)$ on the curve by performing six simple operations, three additions and three assignments. No multiplications are needed.

1.5.2 Subdividing a Parametric Curve

Parametric curves are defined by means of points (data or control) and sometimes also vectors. Editing such a curve is normally done by moving points around and by adding new points. Intuitively, it is clear that adding points allows for finer control of the shape of the curve. On the other hand, adding points results in a curve that's a high-degree polynomial, and such polynomials tend to oscillate. Also, more points implies more calculations to compute and display the curve.

It therefore seems that a reasonable method to obtain the right curve is to start with a few points, and if these are not enough to obtain the desired shape of the curve, to add a point (or a few points) at a time until the desired shape is achieved.

This section discusses a different approach whereby the correct curve is achieved by subdividing a parametric curve into two segments. Together, the two segments have the same shape as the original curve, but they are defined by more entities (points or vectors), thereby making it possible to fine-tune the curve. This approach is applied in Section 6.8 to the Bézier curve. Section 1.12 extends this approach to surface patches.

> The control of large numbers is possible, and like unto that of small numbers, if we subdivide them.
>
> —Sun Tze

We limit our discussion to cubic curves, but the method illustrated here applies to polynomial curves of any degree. Let

$$\mathbf{P}(t) = (t^3, t^2, t, 1)\mathbf{M}\begin{pmatrix}\mathbf{P}_0\\\mathbf{P}_1\\\mathbf{P}_2\\\mathbf{P}_3\end{pmatrix} \qquad (1.11)$$

be any cubic parametric curve defined by four nonscalar entities (points or vectors) where the parameter t varies from 0 to 1. We construct the two halves $\mathbf{P}_1(t)$ and $\mathbf{P}_2(t)$ of this curve by varying the parameter in the intervals $[0, 0.5]$ and $[0.5, 1]$ (Section 6.8 shows how the unequal ranges $[0, \alpha]$ and $[\alpha, 1]$ can be used instead).

Each of the two new curves should have the same shape as half of the original curve. Each half should therefore be written as an expression similar to Equation (1.11) but based on a new set of entities $\mathbf{Q}_i$ computed from the original set $\mathbf{P}_i$. To construct the first half $\mathbf{P}_1(t)$, we define a new parameter $u = 2t$. When t varies in the range $[0, 0.5]$, u varies from 0 to 1. The first half of the curve is obtained from Equation (1.11) by substituting $t = u/2$

$$\begin{aligned}\mathbf{P}_1(u) &= (u^3/8, u^2/4, u/2, 1)\mathbf{M}\begin{pmatrix}\mathbf{P}_0\\\mathbf{P}_1\\\mathbf{P}_2\\\mathbf{P}_3\end{pmatrix}\\[2mm]&= (u^3, u^2, u, 1)\begin{pmatrix}\tfrac{1}{8} & 0 & 0 & 0\\0 & \tfrac{1}{4} & 0 & 0\\0 & 0 & \tfrac{1}{2} & 0\\0 & 0 & 0 & 1\end{pmatrix}\mathbf{M}\begin{pmatrix}\mathbf{P}_0\\\mathbf{P}_1\\\mathbf{P}_2\\\mathbf{P}_3\end{pmatrix}\\[2mm]&= (u^3, u^2, u, 1)\mathbf{LM}\begin{pmatrix}\mathbf{P}_0\\\mathbf{P}_1\\\mathbf{P}_2\\\mathbf{P}_3\end{pmatrix}\\[2mm]&= (u^3, u^2, u, 1)\mathbf{M}\begin{pmatrix}\mathbf{Q}_0\\\mathbf{Q}_1\\\mathbf{Q}_2\\\mathbf{Q}_3\end{pmatrix}.\end{aligned} \qquad (1.12)$$

The last line of Equation (1.12) expresses $\mathbf{P}_1(u)$ in terms of new entities $\mathbf{Q}_i$. It shows that these entities can be calculated from the equation

$$\mathbf{M}\begin{pmatrix}\mathbf{Q}_0\\\mathbf{Q}_1\\\mathbf{Q}_2\\\mathbf{Q}_3\end{pmatrix} = \mathbf{LM}\begin{pmatrix}\mathbf{P}_0\\\mathbf{P}_1\\\mathbf{P}_2\\\mathbf{P}_3\end{pmatrix}, \text{ whose solution is } \begin{pmatrix}\mathbf{Q}_0\\\mathbf{Q}_1\\\mathbf{Q}_2\\\mathbf{Q}_3\end{pmatrix} = \mathbf{M}^{-1}\mathbf{LM}\begin{pmatrix}\mathbf{P}_0\\\mathbf{P}_1\\\mathbf{P}_2\\\mathbf{P}_3\end{pmatrix}. \qquad (1.13)$$

⋄ **Exercise 1.21:** Why does $\mathbf{P}_1(t)$ have the same shape as the first half of $\mathbf{P}(t)$?

The second half, $\mathbf{P}_2(t)$ is calculated similarly. We first define a new parameter $u = 2t - 1$. When t varies in the range $[0.5, 1]$, u varies from 0 to 1. The second half of the curve is obtained from Equation (1.11) by substituting $t = (u+1)/2$:

$$\mathbf{P}_2(u) = \left((u+1)^3/8, (u+1)^2/4, (u+1)/2, 1\right)\mathbf{M}\begin{pmatrix} \mathbf{P}_0 \\ \mathbf{P}_1 \\ \mathbf{P}_2 \\ \mathbf{P}_3 \end{pmatrix}$$

$$= (u^3, u^2, u, 1)\begin{pmatrix} \frac{1}{8} & 0 & 0 & 0 \\ \frac{3}{8} & \frac{1}{4} & 0 & 0 \\ \frac{3}{8} & \frac{2}{4} & \frac{1}{2} & 0 \\ \frac{1}{8} & \frac{1}{4} & \frac{1}{2} & 1 \end{pmatrix}\mathbf{M}\begin{pmatrix} \mathbf{P}_0 \\ \mathbf{P}_1 \\ \mathbf{P}_2 \\ \mathbf{P}_3 \end{pmatrix}$$

$$= (u^3, u^2, u, 1)\mathbf{RM}\begin{pmatrix} \mathbf{P}_0 \\ \mathbf{P}_1 \\ \mathbf{P}_2 \\ \mathbf{P}_3 \end{pmatrix}$$

$$= (u^3, u^2, u, 1)\mathbf{M}\begin{pmatrix} \mathbf{Q}_4 \\ \mathbf{Q}_5 \\ \mathbf{Q}_6 \\ \mathbf{Q}_7 \end{pmatrix}. \tag{1.14}$$

The new entities $\mathbf{Q}_i$ are calculated for this second half by

$$\begin{pmatrix} \mathbf{Q}_4 \\ \mathbf{Q}_5 \\ \mathbf{Q}_6 \\ \mathbf{Q}_7 \end{pmatrix} = \mathbf{M}^{-1}\mathbf{RM}\begin{pmatrix} \mathbf{P}_0 \\ \mathbf{P}_1 \\ \mathbf{P}_2 \\ \mathbf{P}_3 \end{pmatrix}. \tag{1.15}$$

Given matrix $\mathbf{M}$ and four entities $\mathbf{P}_i$, the eight new entities $\mathbf{Q}_i$ can be calculated from Equations (1.13) and (1.15). The generalization of this method to higher-degree curves is straightforward. As an example, we apply this method to the cubic Bézier curve, Equation (6.8). Matrix $\mathbf{M}$ and its inverse are

$$\mathbf{M} = \begin{pmatrix} -1 & 3 & -3 & 1 \\ 3 & -6 & 3 & 0 \\ -3 & 3 & 0 & 0 \\ 1 & 0 & 0 & 0 \end{pmatrix}, \qquad \mathbf{M}^{-1} = \begin{pmatrix} 0 & 0 & 0 & 1 \\ 0 & 0 & \frac{1}{3} & 1 \\ 0 & \frac{1}{3} & \frac{2}{3} & 1 \\ 1 & 1 & 1 & 1 \end{pmatrix}.$$

The matrix products of Equations (1.13) and (1.15) now become

$$\mathbf{M}^{-1}\mathbf{LM} = \begin{pmatrix} 1 & 0 & 0 & 0 \\ \frac{1}{2} & \frac{1}{2} & 0 & 0 \\ \frac{1}{4} & \frac{2}{4} & \frac{1}{4} & 0 \\ \frac{1}{8} & \frac{3}{8} & \frac{3}{8} & \frac{1}{8} \end{pmatrix}, \qquad \mathbf{M}^{-1}\mathbf{RM} = \begin{pmatrix} \frac{1}{8} & \frac{3}{8} & \frac{3}{8} & \frac{1}{8} \\ 0 & \frac{1}{4} & \frac{2}{4} & \frac{1}{4} \\ 0 & 0 & \frac{1}{2} & \frac{1}{2} \\ 0 & 0 & 0 & 1 \end{pmatrix}. \tag{1.16}$$

The eight new entities (which in this case are control points) are

$$\mathbf{Q}_0 = \mathbf{P}_0,$$

$$\mathbf{Q}_1 = \frac{1}{2}\mathbf{P}_0 + \frac{1}{2}\mathbf{P}_1 = \frac{1}{2}(\mathbf{P}_0 + \mathbf{P}_1),$$

$$\mathbf{Q}_2 = \frac{1}{4}\mathbf{P}_0 + \frac{2}{4}\mathbf{P}_1 + \frac{1}{4}\mathbf{P}_2 = \frac{1}{2}\left(\frac{1}{2}(\mathbf{P}_0 + \mathbf{P}_1) + \frac{1}{2}(\mathbf{P}_1 + \mathbf{P}_2)\right),$$

$$\mathbf{Q}_3 = \frac{1}{8}\mathbf{P}_0 + \frac{3}{8}\mathbf{P}_1 + \frac{3}{8}\mathbf{P}_2 + \frac{1}{8}\mathbf{P}_3$$
$$= \frac{1}{2}\left(\frac{1}{2}\left(\frac{1}{2}(\mathbf{P}_0 + \mathbf{P}_1) + \frac{1}{2}(\mathbf{P}_1 + \mathbf{P}_2)\right) + \frac{1}{2}\left(\frac{1}{2}(\mathbf{P}_1 + \mathbf{P}_2) + \frac{1}{2}(\mathbf{P}_2 + \mathbf{P}_3)\right)\right),$$

$$\mathbf{Q}_4 = \frac{1}{8}\mathbf{P}_0 + \frac{3}{8}\mathbf{P}_1 + \frac{3}{8}\mathbf{P}_2 + \frac{1}{8}\mathbf{P}_3$$
$$= \frac{1}{2}\left(\frac{1}{2}\left(\frac{1}{2}(\mathbf{P}_0 + \mathbf{P}_1) + \frac{1}{2}(\mathbf{P}_1 + \mathbf{P}_2)\right) + \frac{1}{2}\left(\frac{1}{2}(\mathbf{P}_1 + \mathbf{P}_2) + \frac{1}{2}(\mathbf{P}_2 + \mathbf{P}_3)\right)\right),$$

$$\mathbf{Q}_5 = \frac{1}{4}\mathbf{P}_1 + \frac{2}{4}\mathbf{P}_2 + \frac{1}{4}\mathbf{P}_3 = \frac{1}{2}\left(\frac{1}{2}(\mathbf{P}_1 + \mathbf{P}_2) + \frac{1}{2}(\mathbf{P}_2 + \mathbf{P}_3)\right),$$

$$\mathbf{Q}_6 = \frac{1}{2}\mathbf{P}_1 + \frac{1}{2}\mathbf{P}_2 = \frac{1}{2}(\mathbf{P}_1 + \mathbf{P}_2),$$

$$\mathbf{Q}_7 = \mathbf{P}_3.$$

Section 6.8 shows a different approach, using the mediation operator, to the problem of subdividing a curve. That approach is applied to the Bézier curve.

1.6 Curvature and Torsion

The first derivative $\mathbf{P}^t(t)$ of a parametric curve $\mathbf{P}(t)$ is the tangent vector of the curve. In this section, we denote the unit tangent vector at point $\mathbf{P}(i)$ by $\mathbf{T}(i)$. Thus,

$$\mathbf{T}(i) = \frac{\mathbf{P}^t(i)}{|\mathbf{P}^t(i)|}.$$

The tangent vector is an example of an *intrinsic property* of a curve. An intrinsic property of a geometric figure depends only on the figure and not on the particular choice of the coordinate axes. Any geometric figure may have intrinsic and extrinsic properties. A triangle has three angles and a quadrilateral has four edges, regardless of the choice of coordinates. The tangent vector of a curve, as well as its curvature, does not depend on the particular coordinate system used. In contrast, the slope of a curve depends on the particular coordinates chosen, which makes it an extrinsic property of the curve.

⋄ **Exercise 1.22:** Give a few more intrinsic and extrinsic properties of geometric figures.

This section discusses the important intrinsic properties of parametric curves. They include the principal vectors (the tangent, normal, and binormal vectors), the principal planes (the osculating, rectifying, and normal planes), and the concepts of curvature and torsion. These properties are all local and they vary from point to point on the curve.

They are therefore functions of the parameter t. Notice that these properties exist for all curves, but the discussion here is limited to parametric curves.

> Newton was seeking better methods—more general—for finding the slope of a curve at any particular point, as well [as] another quantity, related but once removed, the degree of curvature, rate of bending, "the crookedness in lines." He applied himself to the tangent, the straight line that grazes the curve at any point. The straight line that the curve would become at that point, if it could be seen through an infinitely powerful microscope.
>
> —James Gleick, *Isaac Newton* (2003)

1.6.1 Normal Plane

The normal plane to a curve $\mathbf{P}(t)$ at point $\mathbf{P}(i)$ is the plane that's perpendicular to the tangent $\mathbf{P}^t(i)$ and contains point $\mathbf{P}(i)$. If $\mathbf{Q}$ is an arbitrary point on the normal plane, then Figure 1.11 shows that $(\mathbf{Q} - \mathbf{P}(i)) \bullet \mathbf{P}^t(i) = 0$. This can be written $\mathbf{Q} \bullet \mathbf{P}^t(i) - \mathbf{P}(i) \bullet \mathbf{P}^t(i) = 0$ or

$$x \cdot x_i^t + y \cdot y_i^t + z \cdot z_i^t - (x_i \cdot x_i^t + y_i \cdot y_i^t + z_i \cdot z_i^t) = 0, \tag{1.17}$$

an expression that has the familiar form $Ax + By + Cz + D = 0$ (Section 2.2.2).

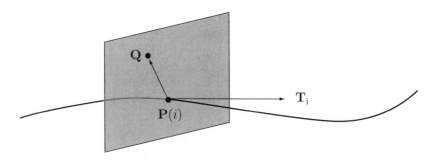

Figure 1.11: The Normal Plane.

1.6.2 Principal Normal Vector

Another important vector associated with a curve is the *principal normal vector* $\mathbf{N}(t)$. This unit vector is normal to the curve (and is therefore contained in the normal plane and is also perpendicular to the tangent vector), but it is called the principal normal since it points in a special direction, the direction in which the curve is turning. The principal normal vector points toward a point called the *center of curvature* of the curve. To express $\mathbf{N}(t)$ in terms of the curve and its derivatives, we select two nearby points, t and $t + \Delta t$, on the curve. The tangent vectors at the two points are $\mathbf{a} = \mathbf{P}^t(t)$ and $\mathbf{b} = \mathbf{P}^t(t + \Delta t)$, respectively. If we subtract them as in Figure 1.12a, we get $\mathbf{c} = \mathbf{b} - \mathbf{a}$. The difference vector $\mathbf{c}$ can be interpreted in two ways. On one hand, we can say that it is a small change in the tangent vector $\mathbf{P}^t(t)$, so we can denote it $\Delta \mathbf{P}^t(t)$. On the other hand, since the tangent vector can be interpreted as the velocity of the curve, any changes in it can be interpreted as acceleration, that is, the second derivative $\mathbf{P}^{tt}(t)$.

Thus, we can write $\mathbf{c} = \Delta\mathbf{P}^t(t) = \mathbf{P}^{tt}(t)$. The two vectors $\mathbf{a} = \mathbf{P}^t(t)$ and $\mathbf{b} = \mathbf{P}^t(t+\Delta t)$ define a plane and the principal normal vector lies at the intersection of this plane and the normal plane. Our task is therefore to compute a vector that is perpendicular to the tangent $\mathbf{a} = \mathbf{P}^t(t)$ and that is contained in the plane defined by $\mathbf{a}$ and $\mathbf{b}$.

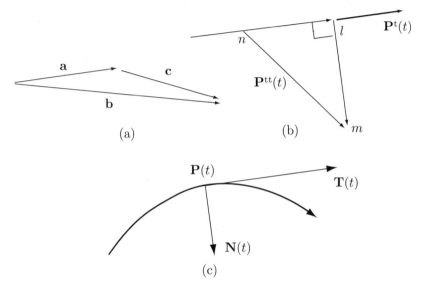

Figure 1.12: The Principal Normal Vector.

Figure 1.12b shows vector $\vec{nl}$, which is the projection of $\mathbf{P}^{tt}(t)$ (vector $\vec{nm}$) onto $\mathbf{P}^t(t)$. Equation (1.8) tells us that the length of $\vec{nl}$ is

$$\frac{\mathbf{P}^{tt}(t) \bullet \mathbf{P}^t(t)}{|\mathbf{P}^t(t)|}.$$

Since $\vec{nl}$ is in the direction of $\mathbf{P}^t(t)$, we can write the vector $\vec{nl}$ as

$$\vec{nl} = \frac{\mathbf{P}^{tt}(t) \bullet \mathbf{P}^t(t)}{|\mathbf{P}^t(t)|} \cdot \frac{\mathbf{P}^t(t)}{|\mathbf{P}^t(t)|} = \frac{\mathbf{P}^{tt}(t) \bullet \mathbf{P}^t(t)}{|\mathbf{P}^t(t)|^2}\mathbf{P}^t(t).$$

We denote the vector $\vec{lm}$ by $\mathbf{K}(t)$ and compute it from the relation $\vec{nl} + \vec{lm} = \vec{nm} = \mathbf{P}^{tt}(t)$:

$$\mathbf{K}(t) = \mathbf{P}^{tt}(t) - \vec{nl} = \mathbf{P}^{tt}(t) - \frac{\mathbf{P}^{tt}(t) \bullet \mathbf{P}^t(t)}{|\mathbf{P}^t(t)|^2}\mathbf{P}^t(t). \qquad (1.18)$$

The principal normal vector $\mathbf{N}(t)$ is a unit vector in the direction of $\mathbf{K}(t)$, so it is given by

$$\mathbf{N}(t) = \frac{\mathbf{K}(t)}{|\mathbf{K}(t)|}.$$

⋄ **Exercise 1.23:** What can we say about the nature of the principal normal vector of a straight line?

⋄ **Exercise 1.24:** Calculate the principal normal vector of the PC curve $\mathbf{P}(t) = (-1, 0)t^3 + (1, -1)t^2 + (1, 1)t$. Notice that this curve is Equation (4.10), so we know that it goes from $(0, 0)$ to $(1, 0)$ with start and end tangents $(1, 1)$, $(0, -1)$, respectively. Use this to check your results.

1.6.3 Binormal Vector

The third important vector associated with a curve is the *binormal vector* $\mathbf{B}(t)$. It is defined as the vector perpendicular to both the tangent and principal normal, so its definition is simply $\mathbf{B}(t) = \mathbf{T}(t) \times \mathbf{N}(t)$. Notice that it is a unit vector. Since the binormal is perpendicular to the tangent, it is contained in the normal plane. The three vectors $\mathbf{T}(t)$, $\mathbf{N}(t)$, and $\mathbf{B}(t)$ therefore constitute an orthogonal coordinate system that moves along the curve as t varies, except at cusps, where they are undefined.

1.6.4 The Osculating Plane

Imagine three points h, i, and j, located close to each other on a curve. If they are not collinear, they define a plane. Now, move h and j independently closer and closer to i. As these points move, the plane may change. The plane obtained at the limit is called the *osculating plane* at point i (Figure 1.13). It contains the tangent vector $\mathbf{T}(i)$ and the principal normal $\mathbf{N}(i)$. If $\mathbf{Q}$ is an arbitrary point on the osculating plane, then the plane equation is given by the determinant $|(\mathbf{Q} - \mathbf{P}(i)) \, \mathbf{P}^t(i) \, \mathbf{P}^{tt}(i)| = 0$, which can be written explicitly as

$$(x - x_i)(y_i^t z_i^{tt} - y_i^{tt} z_i^t) - (y - y_i)(x_i^t z_i^{tt} - x_i^{tt} z_i^t) + (z - z_i)(x_i^t y_i^{tt} - x_i^{tt} y_i^t) = 0.$$

Another way to obtain the plane equation is to use the fact that point $\mathbf{P}(i)$ and vectors $\mathbf{T}(i)$ and $\mathbf{N}(i)$ are contained in the osculating plane. Any general point $\mathbf{Q}$ in the osculating plane can, therefore, be expressed as $\mathbf{Q} = \mathbf{P}(i) + \alpha \mathbf{T}(i) + \beta \mathbf{N}(i)$, where α and β are real parameters. The osculating plane of a plane curve is, of course, the plane of the curve. The osculating plane of a straight line is undefined.

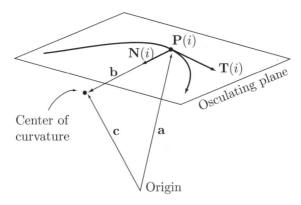

Figure 1.13: The Osculating Plane.

Incidentally, two curves joined at a point have C^2 continuity (Section 1.4.2) at the point if they have the same osculating planes and the same curvature vectors at the point.

◇ **Exercise 1.25:** (1) Calculate the Bézier curve for the four points $\mathbf{P}_0 = (0,0,0)$, $\mathbf{P}_1 = (1,0,0)$, $\mathbf{P}_2 = (2,1,0)$, and $\mathbf{P}_3 = (3,0,1)$. [Those unfamiliar with this curve should use Equation (6.8).] Notice that this is a space curve since the first three points are in the $z = 0$ plane, while the fourth one is outside that plane. (2) Calculate the (unnormalized) principal normal vector of the curve and find its values for $t = 0, 0.5$, and 1. (3) Calculate the osculating plane of the curve and find its equations for $t = 0, 0.5$, and 1 as above.

1.6.5 Rectifying Plane

The plane perpendicular to the principal normal vector of a curve is called the rectifying plane of the curve. If the curve is $\mathbf{P}(t)$, $\mathbf{N}(t)$ is its principal normal, and $\mathbf{Q}$ is an arbitrary point on the rectifying plane, then the equation of the rectifying plane at point $\mathbf{P}(i)$ is $[\mathbf{Q} - \mathbf{P}(i)] \bullet \mathbf{N}(i) = 0$. Another equation is obtained when we realize that both the tangent and binormal vectors are contained in the rectifying plane. A general point on this plane can therefore be expressed as $\mathbf{Q} = \mathbf{P}(i) + \alpha\mathbf{T}(i) + \beta\mathbf{B}(i)$.

Figure 1.14 shows the three unit vectors and three planes associated with a particular point $\mathbf{P}(i)$ on a curve. They constitute intrinsic properties of the curve and together they form the *moving trihedron* of the curve, which can be considered a local coordinate system for the curve. The three vectors constitute the local coordinate axes and the three planes divide the space around point $\mathbf{P}(i)$ into eight octants. The curve passes through the normal plane and is tangent to both the osculating and rectifying planes.

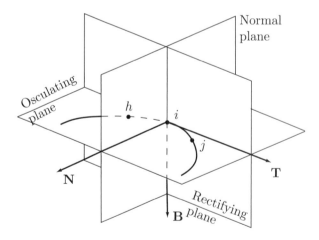

Figure 1.14: The Moving Trihedron.

1.6.6 Curvature

The curvature of a curve is a useful entity, so it deserves to be rigorously defined. Intuitively, the curvature should be a number that measures how much the curve deviates from a straight line at any point. It should be large in areas where the curve wiggles, oscillates, or makes a sudden direction change; it should be small in areas where the curve is close to a straight line. It is also useful to associate a direction with the curvature, i.e., to make it a vector.

Given a parametric curve $\mathbf{P}(t)$ and a point $\mathbf{P}(i)$ on it, we calculate the first two derivatives $\mathbf{P}^t(i)$ and $\mathbf{P}^{tt}(i)$ of the curve at the point. We then construct a circle that has these same first and second derivatives and move it so it grazes the point. This is called the *osculating circle* of the curve at the point. The curvature is now defined as the vector $\kappa(i)$ whose direction is from point $\mathbf{P}(i)$ to the center of this circle and whose magnitude is the reciprocal of the radius of the circle.

Using differential geometry, it can be shown that the vector

$$\frac{\mathbf{P}^t(t) \times \mathbf{P}^{tt}(t)}{|\mathbf{P}^t(t)|^3}$$

has the right magnitude. However, this vector is perpendicular to both $\mathbf{P}^t(t)$ and $\mathbf{P}^{tt}(t)$, so it is perpendicular to the osculating plane. To bring it into the plane, we need to cross-product it with $\mathbf{P}^t(t)/|\mathbf{P}^t(t)|$, so the result is

$$\kappa(t) = \frac{\mathbf{P}^t(t) \times \mathbf{P}^{tt}(t) \times \mathbf{P}^t(t)}{|\mathbf{P}^t(t)|^4}. \tag{1.19}$$

Figure 1.13 shows that the curvature (vector $\mathbf{b}$) is in the direction of the binormal $\mathbf{N}(t)$, so it can be expressed as $\kappa(t) = \rho(t)\mathbf{N}(t)$ where $\rho(t)$ is the *radius of curvature* at point $\mathbf{P}(t)$.

Given a curve $\mathbf{P}(t)$ with an arc length $s(t)$, we assume that $d\mathbf{P}/ds$ is a *unit* tangent vector:

$$\frac{d\mathbf{P}(t)}{ds} = \frac{d\mathbf{P}(t)}{dt}\frac{ds(t)}{dt} = \frac{\mathbf{P}^t(t)}{s^t(t)}. \tag{1.20}$$

Equation (1.20) shows the following:

1. $d\mathbf{P}(t)/ds$ and $\mathbf{P}^t(t)$ point in the same direction. Therefore, since $d\mathbf{P}(t)/ds$ is a unit vector, we get

$$\frac{d\mathbf{P}(t)}{ds} = \frac{\mathbf{P}^t(t)}{|\mathbf{P}^t(t)|}.$$

2. $s^t(t) = |\mathbf{P}^t(t)|$.

We now derive the expression for curvature from a different point of view. The curvature k is defined by $d^2\mathbf{P}(t)/ds^2 = k\mathbf{N}$, where $\mathbf{N}$ is the unit principal normal vector (Section 1.6.2). The problem is to express k in terms of the curve $\mathbf{P}(t)$ and its derivatives, not involving the (normally unknown) function $s(t)$. We start with

$$\frac{d^2\mathbf{P}(t)}{ds^2} = \frac{d}{ds}\left(\frac{\mathbf{P}^t(t)}{|\mathbf{P}^t(t)|}\right) = \frac{\frac{d}{dt}\left(\frac{\mathbf{P}^t(t)}{|\mathbf{P}^t(t)|}\right)}{s^t(t)}$$

$$= \frac{\frac{\mathbf{P}^{tt}(t)}{|\mathbf{P}^t(t)|} - \frac{\mathbf{P}^t(t)}{|\mathbf{P}^t(t)|^2}\cdot\frac{d|\mathbf{P}^t(t)|}{dt}}{|\mathbf{P}^t(t)|}. \tag{1.21}$$

The identity $\mathbf{A} \bullet \mathbf{A} = |\mathbf{A}|^2$ is true for any vector $\mathbf{A}(t)$ and it implies

$$\mathbf{A}(t) \bullet \mathbf{A}^t(t) = |\mathbf{A}(t)| \frac{d|\mathbf{A}(t)|}{dt}.$$

When we apply this to the vector $\mathbf{P}^t(t)$, we get

$$\frac{d^2\mathbf{P}(t)}{ds^2} = \frac{\mathbf{P}^{tt}(t)}{\mathbf{P}^t(t) \bullet \mathbf{P}^t(t)} - \frac{\mathbf{P}^t(t) \bullet \mathbf{P}^{tt}(t)}{\left(\mathbf{P}^t(t) \bullet \mathbf{P}^t(t)\right)^2}\mathbf{P}^t(t), \tag{1.22}$$

which can also be written

$$k\mathbf{N} = \frac{d^2\mathbf{P}(t)}{ds^2} = \frac{\mathbf{P}^t(t) \times \left(\mathbf{P}^{tt}(t) \times \mathbf{P}^t(t)\right)}{\left(\mathbf{P}^t(t) \bullet \mathbf{P}^t(t)\right)^2}. \tag{1.23}$$

1.6.7 Torsion

Torsion is a measure of how much a given curve deviates from a plane curve. The torsion $\tau(i)$ of a curve at a point $\mathbf{P}(i)$ is defined by means of the following two quantities:

1. Imagine a point h close to i. The curve has rectifying planes at points h and i (Figure 1.15). Denote the angle between them by θ.

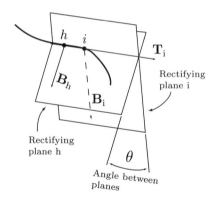

Figure 1.15: Torsion.

2. Denote by s the arc length from point h to point i.

The torsion of the curve at point i is defined as the limit of the ratio θ/s when h approaches i. Figure 1.15 shows how the rectifying plane rotates about the tangent as we move on the curve from h to i. The torsion can be expressed by means of the derivatives of the curve and by means of the curvature

$$\tau(t) = \frac{|\mathbf{P}^t(t)\,\mathbf{P}^{tt}(t)\,\mathbf{P}^{ttt}(t)|}{|\mathbf{P}^t(t) \times \mathbf{P}^t(t)|^2} = \frac{|\mathbf{P}^t(t)\,\mathbf{P}^{tt}(t)\,\mathbf{P}^{ttt}(t)|}{|\mathbf{P}^t(t)|^6}\rho(t)^2.$$

(The numerator is a determinant and the denominator is an absolute value. This expression is meaningful only when $\rho(t) < \infty$.) The torsion of a plane curve is zero.

It is interesting to note that a curve can be fully defined by specifying its curvature and torsion as functions of its arc length s. The functions $\kappa = f(s)$ and $\tau = g(s)$ uniquely define the shape of a curve (although not its location in space). An alternative is the single (implicit) function $F(\kappa, \tau, s) = 0$.

An alternative representation can be derived for a plane curve. Assume that $\mathbf{P}(t) = (x(t), y(t))$ is a curve in the xy plane. Figure 1.16 shows that its shape can be determined if its start point $\mathbf{P}(0)$ and its slope (or, equivalently, angle θ) are known as functions of the arc length s. Since θ is the angle between the tangent and the x axis, functions $x(s)$ and $y(s)$ must satisfy

$$\frac{dx}{ds} = \cos\theta, \qquad \frac{dy}{ds} = \sin\theta.$$

Differentiating produces

$$\frac{d^2x}{ds^2} = -\sin\theta\frac{d\theta}{ds} = -\frac{dy}{ds}\frac{d\theta}{ds}, \quad \frac{d^2y}{ds^2} = \cos\theta\frac{d\theta}{ds} = \frac{dx}{ds}\frac{d\theta}{ds}. \tag{1.24}$$

Figure 1.16 also shows that $d\theta/ds$ is the magnitude of the curvature κ, so the conclusion is that, given the curvature $\kappa(s)$ of a curve as a function of its arc length, the two functions $x(s)$ and $y(s)$ can be calculated, either analytically, or point by point numerically, from the differential equations (1.24).

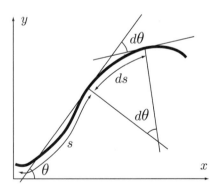

Figure 1.16: A Plane Curve.

◇ **Exercise 1.26:** Given $\kappa(s) = R$ (a constant), solve Equation (1.24) for $x(s)$ and $y(s)$. What kind of a curve is this?

1.6.8 Inflection Points

An inflection point is a point on a curve where the curvature is zero. On a straight line, every point is an inflection point. On a typical curve, an inflection point is created when the curve reverses its direction of turning (for example, from a clockwise direction to a counterclockwise direction). From the definition of curvature [Equation (1.19)] it follows that an inflection point satisfies

$$0 = |\mathbf{P}^t(t) \times \mathbf{P}^{tt}(t)| = \sqrt{(\mathbf{P}^t(t) \times \mathbf{P}^{tt}(t)) \bullet (\mathbf{P}^t(t) \times \mathbf{P}^{tt}(t))}.$$

Therefore,

$$(\mathbf{P}^t(t) \times \mathbf{P}^{tt}(t)) \bullet (\mathbf{P}^t(t) \times \mathbf{P}^{tt}(t)) = 0,$$

which is equivalent to

$$(\mathbf{P}^t(t) \times \mathbf{P}^{tt}(t))_x^2 + (\mathbf{P}^t(t) \times \mathbf{P}^{tt}(t))_y^2 + (\mathbf{P}^t(t) \times \mathbf{P}^{tt}(t))_z^2 = 0,$$

$$\text{or} \quad (y^t z^{tt} - z^t y^{tt})^2 + (z^t x^{tt} - x^t z^{tt})^2 + (x^t y^{tt} - y^t x^{tt})^2 = 0. \qquad (1.25)$$

This is the sum of three nonnegative quantities, so each must be zero. Since

$$\frac{dy}{dx} = \frac{dy}{dt} \Big/ \frac{dx}{dt} = \frac{y^t}{x^t},$$

we get

$$\frac{d^2 y}{dx^2} = \frac{d}{dt}\left(\frac{y^t}{x^t}\right)\frac{dt}{dx} = \frac{x^t y^{tt} - x^{tt} y^t}{(x^t)^3}.$$

Therefore, saying that the three quantities above are zero is the same as saying that

$$\frac{d^2 y}{dx^2} = \frac{d^2 x}{dz^2} = \frac{d^2 z}{dy^2} = 0.$$

Equation (1.25) can be used to show that a two-dimensional parametric cubic can have at most two inflection points. We denote a general PC by

$$\mathbf{P}(t) = \mathbf{a}t^3 + \mathbf{b}t^2 + \mathbf{c}t + \mathbf{d} = (a_x, a_y)t^3 + (b_x, b_y)t^2 + (c_x, c_y)t + (d_x, d_y),$$

which implies $x^t = 3a_x t^2 + 2b_x t + c_x$ and $x^{tt} = 6a_x t + b_x$, and similarly for y^t and y^{tt}. Using this notation, we write Equation (1.25) explicitly (notice that for a two-dimensional PC, only the third part is nonzero) as

$$\begin{aligned} 0 &= x^t y^{tt} - y^t x^{tt} \\ &= (3a_x t^2 + 2b_x t + c_x)(6a_y t + b_y) - (3a_y t^2 + 2b_y t + c_y)(6a_x t + b_x) \\ &= 6(a_y b_x - a_x b_y)t^2 + 6(a_y c_x - a_x c_y)t + 2(b_y c_x - b_x c_y). \end{aligned}$$

This is a quadratic equation in t, so there can be at most two solutions.

1.7 Special and Degenerate Curves

Parametric curves may exhibit unusual behavior when their derivatives satisfy certain conditions. Such curves are referred to as special or degenerate. Here are four examples:

1. If the first derivative $\mathbf{P}^t(t)$ of a curve $\mathbf{P}(t)$ is zero for all values of t, then $\mathbf{P}(t)$ degenerates to the point $\mathbf{P}(0)$.

2. If $\mathbf{P}^t(t) \neq 0$ and $\mathbf{P}^t(t) \times \mathbf{P}^{tt}(t) = 0$ (i.e., the tangent vector points in the direction of the acceleration vector), then $\mathbf{P}(t)$ is a straight line.

3. If $\mathbf{P}^t(t) \times \mathbf{P}^{tt}(t) \neq 0$ and $|\mathbf{P}^t(t)\,\mathbf{P}^{tt}(t)\,\mathbf{P}^{ttt}(t)| = 0$, then $\mathbf{P}(t)$ is a plane curve. (The notation $|\mathbf{a}\,\mathbf{b}\,\mathbf{c}|$ refers to the determinant whose three columns are $\mathbf{a}$, $\mathbf{b}$, and $\mathbf{c}$.)

4. Finally, if both $\mathbf{P}^t(t) \times \mathbf{P}^{tt}(t)$ and $|\mathbf{P}^t(t)\,\mathbf{P}^{tt}(t)\,\mathbf{P}^{ttt}(t)|$ are nonzero, the curve $\mathbf{P}(t)$ is nonplanar (i.e., it is a space curve).

1.8 Basic Concepts of Surfaces

Section 1.3 mentions the explicit, implicit, and parametric representations of curves. Surfaces can also be represented in these three ways. The explicit representation of a surface is $z = f(x, y)$ and the implicit representation is $F(x, y, z) = 0$ (Figure C.3). In practice, however, the parametric representation is used almost exclusively, for the same reasons that parametric curves are so important.

A simple, intuitive way to grasp the concept of a parametric surface is to visualize it as a set of curves. Figure 1.17a shows a single curve and Figure 1.17b shows how it is duplicated several times to create a family of identical curves. The brain finds it natural to interpret such a family as a surface. If we denote the curve by $\mathbf{P}(u)$, we can denote each of its copies in the family by $\mathbf{P}_i(u)$, where i is an integer index.

Taking this idea a step further, a solid surface is obtained by creating infinitely many copies of the curve and placing them next to each other without any gaps in between. It makes sense to replace the integer index i of each curve by a real (continuous) index w. The solid version of the surface of Figure 1.17b can therefore be denoted by $\mathbf{P}_w(u)$, where varying u moves us along a curve and varying w moves us from curve to curve in steps that can be arbitrarily small.

The next step is to obtain a general surface by varying the shape of the curves so they are not identical (Figure 1.17c). The shape of a curve should therefore depend on w, which suggests a notation such as $\mathbf{P}(u, w)$ for the surface. The shape of each curve depends on both u and w but in a special way. Each of the two parameters moves us along a different direction on the surface, so we can talk about the u direction and the w direction (Figure 1.17d).

The general form of a parametric surface is $\mathbf{P}(u, w) = (f_1(u, w), f_2(u, w), f_3(u, w))$. The surface depends on two parameters, u and w, that vary independently in some interval $[a, b]$ (normally, but not always, limited to $[0, 1]$). For each pair (u, w), the expression above produces the three coordinates of a point on the surface.

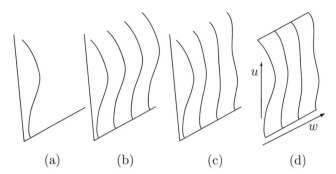

Figure 1.17: A Surface as a Family of Curves.

◇ **Exercise 1.27:** A curve can be either two-dimensional or three-dimensional. A surface, however, exists only in three dimensions, and each surface point has three coordinates. Why is it that the expression for the surface depends on two, and not on three, parameters? We would expect the surface to be of the form $\mathbf{P}(u, v, w)$, a function of three parameters. What's the explanation?

A simple example of a parametric surface is

$$\mathbf{P}(u, w) = [0.5(1 - u)w + u, w, (1 - u)(1 - w)] \tag{1.26}$$

[this is also Equation (2.11)]. Such a surface is called *bilinear* since it is linear in both parameters. We use this example to discuss the concept of a surface patch and to show how a wire-frame surface can be displayed.

1.8.1 A Surface Patch

The expression $\mathbf{P}(u, 0.2)$ (where w is held fixed and u varies) depends on just one parameter and is therefore a curve on the surface. The four curves $\mathbf{P}(u, 0)$, $\mathbf{P}(u, 1)$, $\mathbf{P}(0, w)$, and $\mathbf{P}(1, w)$ are of special interest. They are the *boundary curves* of the surface (Figure 1.18a). Since there are four such curves, our surface is a *patch* that has a (roughly) rectangular shape. Of special interest are the four quantities $\mathbf{P}(0, 0)$, $\mathbf{P}(0, 1)$, $\mathbf{P}(1, 0)$, and $\mathbf{P}(1, 1)$. They are the corner points of the surface patch and are sometimes denoted by $\mathbf{P}_{ij}$.

We say that the curve $\mathbf{P}(u, 0.2)$ lies on the surface in the u direction. It is an *isoparametric curve*. Similarly, any curve $\mathbf{P}(u_0, w)$ where u_0 is fixed, lies in the w direction and is an isoparametric curve. These are the two main directions on a rectangular surface patch.

Two more special curves, the *surface diagonals*, are $\mathbf{P}(u, 1 - u)$ and $\mathbf{P}(u, u)$. The former goes from $\mathbf{P}_{01}$ to $\mathbf{P}_{10}$ and the latter goes from $\mathbf{P}_{00}$ to $\mathbf{P}_{11}$.

A large surface is obtained by constructing a number of patches and connecting them. The method used to construct the patch should allow for smooth connection of patches.

◇ **Exercise 1.28:** Compute the corner points, boundary curves, and diagonals of the bilinear surface patch of Equation (1.26).

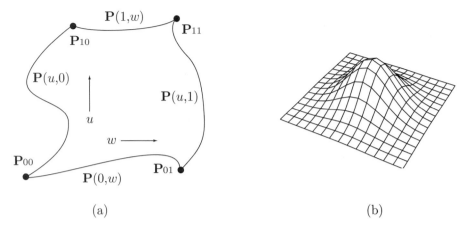

(a)   (b)

Figure 1.18: (a) A Surface Patch. (b) A Wire Frame.

⋄ **Exercise 1.29:** Calculate the corner points and boundary curves of the surface patch

$$\mathbf{P}(u, w) = \big((c - a)u + a, (d - b)w + b, 0\big),$$

where a, b, c, and d are given constants and the parameters u and w vary independently in the range $[0, 1]$. What kind of a surface is this?

1.8.2 Displaying a Surface Patch

A surface patch can be displayed either as a wire frame (Figure 1.18b) or as a solid surface. The pseudo-code of Figure 1.19 shows how to display a surface patch as a wire frame. The code consists of two similar loops—one drawing the curves in the w direction and the other drawing the curves in the u direction. The first loop varies u from 0 to 1 in steps of 0.2, thereby drawing six curves. Each of the six is drawn by varying w in small steps (0.01 in the example). The second loop is similar and draws six curves in the u direction.

Procedure `SurfacePoint` receives the current values of u and w, and calculates the coordinates (x, y, z) of one surface point. Procedure `PersProj` uses these coordinates to calculate the screen coordinates (xs, ys) of a pixel (it projects the three-dimensional pixel on the two-dimensional screen using perspective projection). Finally, procedure `Pixel` actually displays the pixel in the desired color. Better results are obtained by eliminating those parts of the surface that are hidden by other parts, but this topic is outside the scope of this book.

To display a solid surface, the *normal vector* of the surface (Section 1.13) has to be calculated at every point and a shading algorithm applied to compute the amount of light reflected from the point. Most texts on computer graphics discuss shading models and algorithms.

```
for u:=0 to 1 step 0.2 do          for w:=0 to 1 step 0.2 do
  begin                              begin
  for w:=0 to 1 step 0.01 do         for u:=0 to 1 step 0.01 do
    begin                              begin
    SurfacePoint(u,w,x,y,z);           SurfacePoint(u,w,x,y,z);
    PersProj(x,y,z,xs,ys);             PersProj(x,y,z,xs,ys);
    Pixel(xs,ys,color)                 Pixel(xs,ys,color)
    end;                               end;
  end;                               end;
```

Figure 1.19: Procedure for a Wire-Frame Surface.

1.9 The Cartesian Product

The concept of blending was introduced in Section 1.2. This is an important concept that is used in many curve and surface algorithms. This section shows how blending can be used in surface design. We start with two parametric curves $\mathbf{Q}(u) = \sum_{i=1}^{n} f_i(u)\mathbf{Q}_i$ and $\mathbf{R}(w) = \sum_{i=1}^{m} g_i(w)\mathbf{R}_i$ where $\mathbf{Q}_i$ and $\mathbf{R}_i$ can be points or vectors. Now examine the function

$$\mathbf{P}(u,w) = \sum_{i=1}^{n}\sum_{j=1}^{m} f_i(u)g_j(w)\mathbf{P}_{ij} = \sum_{i=1}^{n}\sum_{j=1}^{m} h_{ij}(u,w)\mathbf{P}_{ij}, \qquad (1.27)$$

where $h_{ij}(u,w) = f_i(u)g_j(w)$. The function $\mathbf{P}(u,w)$ describes a surface, since it is a function of the two independent parameters u and w. For any value of the pair (u,w), the function computes a weighted sum of the quantities $\mathbf{P}_{ij}$. These quantities—which are normally points, but can also be vectors—are triplets, so $\mathbf{P}(u,w)$ returns a triplet (x,y,z) that are the three-dimensional coordinates of a point on the surface. When u and w vary over their ranges independently, $\mathbf{P}(u,w)$ computes all the three-dimensional points of a surface patch.

> I don't blend in at a family picnic.
> —Batman in *Batman Forever*, 1995.

The technique of blending quantities $\mathbf{P}_{ij}$ into a surface by means of weights taken from two curves is called the *Cartesian product*, although the terms *tensor product* and *cross-product* are also sometimes used. The quantities $\mathbf{P}_{ij}$ can be points, tangent vectors, or second derivatives. Equation (1.27) can also be written in the compact form

$$\mathbf{P}(u,w) = \big(f_1(u), \ldots, f_n(u)\big) \begin{pmatrix} \mathbf{P}_{11} & \mathbf{P}_{12} & \cdots & \mathbf{P}_{1m} \\ \vdots & \vdots & & \vdots \\ \mathbf{P}_{n1} & \mathbf{P}_{n2} & \cdots & \mathbf{P}_{nm} \end{pmatrix} \begin{pmatrix} g_1(w) \\ \vdots \\ g_m(w) \end{pmatrix}. \qquad (1.28)$$

Notice that it uses a matrix whose elements are nonscalar quantities (triplets). Even more important, Equation (1.27), combined with the isotropic principle (Section 1.1), tells us that if all $\mathbf{P}_{ij}$ are points, then the surface $\mathbf{P}(u,w)$ is independent of the particular coordinate axes used if $\sum_{ij} h_{ij}(u, w) = 1$. If the two original curves $\mathbf{Q}(u)$ and $\mathbf{R}(w)$ are isotropic, then it's easy to see that the surface is also isotropic because

$$\sum_{ij} h_{ij}(u, w) = \sum_i \sum_j f_i g_j = \left(\sum_j g_j\right)\left(\sum_i f_i\right) = 1.$$

The following two examples illustrate the importance of the Cartesian product. The first example applies this technique to derive the equation of the bilinear surface (Section 2.3) from that of a straight segment. The parametric representation of the line segment from $\mathbf{P}_0$ to $\mathbf{P}_1$ is Equation (2.1)

$$\mathbf{P}(t) = (1 - t)\mathbf{P}_0 + t\mathbf{P}_1 = \mathbf{P}_0 + (\mathbf{P}_1 - \mathbf{P}_0)t$$

$$= [1 - t, t]\begin{bmatrix} \mathbf{P}_0 \\ \mathbf{P}_1 \end{bmatrix} = [B_{10}(t), B_{11}(t)]\begin{bmatrix} \mathbf{P}_0 \\ \mathbf{P}_1 \end{bmatrix}, \tag{1.29}$$

where $B_{1i}(t)$ are the Bernstein polynomials of degree 1 [Equation (6.5)]. The Cartesian product of Equation (1.29) with itself is

$$\mathbf{P}(u, w) = [B_{10}(u), B_{11}(u)]\begin{bmatrix} \mathbf{P}_{00} & \mathbf{P}_{01} \\ \mathbf{P}_{10} & \mathbf{P}_{11} \end{bmatrix}\begin{bmatrix} B_{10}(w) \\ B_{11}(w) \end{bmatrix}$$

$$= [1 - u, u]\begin{bmatrix} \mathbf{P}_{00} & \mathbf{P}_{01} \\ \mathbf{P}_{10} & \mathbf{P}_{11} \end{bmatrix}\begin{bmatrix} 1 - w \\ w \end{bmatrix}$$

$$= \mathbf{P}_{00}(1 - u)(1 - w) + \mathbf{P}_{01}(1 - u)w + \mathbf{P}_{10}u(1 - w) + \mathbf{P}_{11}uw,$$

and this is the parametric expression of the bilinear surface patch, Equation (2.8).

The second example starts with the parametric cubic polynomial that passes through four given points. This curve is derived from first principles in Section 3.1 and is given by Equation (3.6), duplicated here

$$\mathbf{P}(t) = (t^3, t^2, t, 1)\begin{bmatrix} -4.5 & 13.5 & -13.5 & 4.5 \\ 9.0 & -22.5 & 18 & -4.5 \\ -5.5 & 9.0 & -4.5 & 1.0 \\ 1.0 & 0 & 0 & 0 \end{bmatrix}\begin{bmatrix} \mathbf{P}_1 \\ \mathbf{P}_2 \\ \mathbf{P}_3 \\ \mathbf{P}_4 \end{bmatrix}$$

$$= (t^3, t^2, t, 1)\mathbf{N}\begin{bmatrix} \mathbf{P}_1 \\ \mathbf{P}_2 \\ \mathbf{P}_3 \\ \mathbf{P}_4 \end{bmatrix}. \tag{3.6}$$

The principle of Cartesian product is now applied to multiply this curve by itself in order to obtain a bicubic surface patch that passes through 16 given points. The result is obtained immediately

$$\mathbf{P}(u, w) = (u^3, u^2, u, 1)\mathbf{N}\begin{bmatrix} \mathbf{P}_{33} & \mathbf{P}_{32} & \mathbf{P}_{31} & \mathbf{P}_{30} \\ \mathbf{P}_{23} & \mathbf{P}_{22} & \mathbf{P}_{21} & \mathbf{P}_{20} \\ \mathbf{P}_{13} & \mathbf{P}_{12} & \mathbf{P}_{11} & \mathbf{P}_{10} \\ \mathbf{P}_{03} & \mathbf{P}_{02} & \mathbf{P}_{01} & \mathbf{P}_{00} \end{bmatrix}\mathbf{N}^T\begin{bmatrix} w^3 \\ w^2 \\ w \\ 1 \end{bmatrix}. \tag{1.30}$$

Note that this result is also obtained in Section 3.6.1 [Equation (3.27)], where it is derived from first principles and requires the solution of a system of 16 equations. Cartesian product is obviously a useful, simple, and elegant method to easily derive the expressions of many types of surfaces.

1.10 Connecting Surface Patches

Often, a complex surface is constructed of individual patches that have to be connected smoothly, which is why this short section examines the conditions required for the smooth connection of two rectangular patches. Figure 1.20 illustrates two patches $\mathbf{P}(u, w)$ and $\mathbf{Q}(u, w)$ connected along the w direction such that $\mathbf{P}(1, w) = \mathbf{Q}(0, w)$ for $0 \leq w \leq 1$. Specifically, the two corner points $\mathbf{Q}_{00}$ and $\mathbf{P}_{10}$ are identical and so are $\mathbf{Q}_{01}$ and $\mathbf{P}_{11}$. The two patches will connect smoothly if any of the following conditions are met:

1. $\mathbf{Q}^u(0, w) = \mathbf{P}^u(1, w)$ for $0 \leq w \leq 1$.
2. $\mathbf{Q}^u(0, w) = f(w)\mathbf{P}^u(1, w)$ for $0 \leq w \leq 1$ and a positive function $f(w)$.
3. $\mathbf{Q}^u(0, w) = f(w)\mathbf{P}^u(1, w) + g(w)\mathbf{P}^w(1, w)$ for $0 \leq w \leq 1$ and positive functions $f(w)$ and $g(w)$.

These conditions involve the three tangent vectors:

1. $\mathbf{Q}^u(0, w)$, the tangent in the u direction of patch $\mathbf{Q}$ at $u = 0$.
2. $\mathbf{P}^u(1, w)$, the tangent in the u direction of $\mathbf{P}$ at $u = 1$.
3. $\mathbf{P}^w(1, w)$, the tangent in the w direction of $\mathbf{P}$ at $u = 1$.

Condition 1 implies that tangents 1 and 2 are equal. Condition 2 implies that they point in the same direction but their sizes differ. Condition 3 means that tangent 1 does not point in the direction of tangent 2, but lies in the plane defined by tangents 2 and 3.

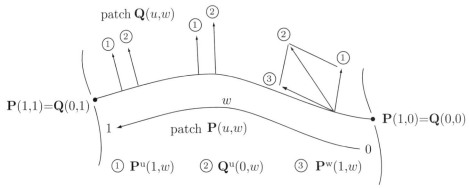

Figure 1.20: Tangent Vectors For Smooth Connection.

Note that condition 3 includes condition 2 (in the special case $g(w) = 0$) and condition 2 includes condition 1 (in the special case $f(w) = 1$).

1.11 Fast Computation of a Bicubic Patch

A complete rectangular surface patch is displayed as a wireframe by drawing two families of curves, in the u and w directions, as pointed out in Section 1.8.2. This section shows how to apply the technique of forward differences to the problem of fast computation of these curves. The material presented here is an extension of the ideas and methods presented in Section 1.5.1. We limit this discussion to a general bicubic surface patch, whose expression is

$$\mathbf{P}(u,w) = (u^3, u^2, u, 1) \begin{bmatrix} \mathbf{M}_{00} & \mathbf{M}_{01} & \mathbf{M}_{02} & \mathbf{M}_{03} \\ \mathbf{M}_{10} & \mathbf{M}_{11} & \mathbf{M}_{12} & \mathbf{M}_{13} \\ \mathbf{M}_{20} & \mathbf{M}_{21} & \mathbf{M}_{22} & \mathbf{M}_{23} \\ \mathbf{M}_{30} & \mathbf{M}_{31} & \mathbf{M}_{32} & \mathbf{M}_{33} \end{bmatrix} \begin{bmatrix} w^3 \\ w^2 \\ w \\ 1 \end{bmatrix}. \tag{1.31}$$

(Where matrix elements $\mathbf{M}_{ij}$ are derived from the 16 points $\mathbf{P}_{ij}$ and from the elements of matrix $\mathbf{N}$. Compare with Equation (3.21).)

For a fixed w, the surface $\mathbf{P}(u,w)$ reduces to a PC curve in the u direction $\mathbf{P}_w(u) = \mathbf{A}u^3 + \mathbf{B}u^2 + \mathbf{C}u + \mathbf{D}$. Each of the four coefficients is a cubic polynomial in w as follows:

$$\mathbf{A}(w) = \mathbf{M}_{00}w^3 + \mathbf{M}_{01}w^2 + \mathbf{M}_{02}w + \mathbf{M}_{03},$$
$$\mathbf{B}(w) = \mathbf{M}_{10}w^3 + \mathbf{M}_{11}w^2 + \mathbf{M}_{12}w + \mathbf{M}_{13},$$
$$\mathbf{C}(w) = \mathbf{M}_{20}w^3 + \mathbf{M}_{21}w^2 + \mathbf{M}_{22}w + \mathbf{M}_{23},$$
$$\mathbf{D}(w) = \mathbf{M}_{30}w^3 + \mathbf{M}_{31}w^2 + \mathbf{M}_{32}w + \mathbf{M}_{33}.$$

Applying the forward differences technique of Section 1.5.1, we can compute the n points $\mathbf{P}_w(0)$, $\mathbf{P}_w(\Delta)$, $\mathbf{P}_w(2\Delta), \ldots, \mathbf{P}_w([n-1]\Delta)$ [where $(n-1)\Delta = 1$] with three additions and three assignments for each point. This, however, requires that the four quantities $\mathbf{A}(w)$, $\mathbf{B}(w)$, $\mathbf{C}(w)$, and $\mathbf{D}(w)$ be computed first, which involves multiplications and exponentiations. Moreover, to display the entire surface patch we need to compute and display U curves $\mathbf{P}_w(u)$ for U values of w in the interval $[0,1]$. The natural solution is to apply forward differences to the computations of $\mathbf{A}(w)$, $\mathbf{B}(w)$, $\mathbf{C}(w)$, and $\mathbf{D}(w)$ for each value of w.

To compute $\mathbf{A}(w) = \mathbf{M}_{00}w^3 + \mathbf{M}_{01}w^2 + \mathbf{M}_{02}w + \mathbf{M}_{03}$ we compute the following

$$\mathbf{A}(0) = \mathbf{M}_{03}, \quad \mathbf{dA}(0) = \mathbf{M}_{00}\Delta^3 + \mathbf{M}_{01}\Delta^2 + \mathbf{M}_{02}\Delta, \quad \mathbf{ddA}(0) = 6\mathbf{M}_{00}\Delta^3 + 2\mathbf{M}_{01}\Delta^2,$$
$$\mathbf{dddA} = 6\mathbf{M}_{00}\Delta^3,$$
$$\mathbf{A}(\Delta) = \mathbf{A}(0) + \mathbf{dA}(0), \quad \mathbf{dA}(\Delta) = \mathbf{dA}(0) + \mathbf{ddA}(0), \quad \mathbf{ddA}(\Delta) = \mathbf{ddA}(0) + \mathbf{dddA},$$
$$\mathbf{A}([j+1]\Delta) = \mathbf{A}(j\Delta) + \mathbf{dA}(j\Delta),$$
$$\mathbf{dA}([j+1]\Delta) = \mathbf{dA}(j\Delta) + \mathbf{ddA}(j\Delta),$$
$$\mathbf{ddA}([j+1]\Delta) = \mathbf{ddA}(j\Delta) + \mathbf{dddA},$$

and similarly for $\mathbf{B}(w)$, $\mathbf{C}(w)$, and $\mathbf{D}(w)$. Each requires three additions and three assignments, for a total of 12 additions and 12 assignments.

Thus, a complete curve $\mathbf{P}(u, j\Delta)$ is drawn in the u direction on the surface in the following two steps:

1. Compute $\mathbf{A}(j\Delta)$ from $\mathbf{A}([j-1]\Delta)$, $\mathbf{dA}([j-1]\Delta)$, and $\mathbf{ddA}([j-1]\Delta)$ and similarly for $\mathbf{B}(j\Delta)$, $\mathbf{C}(j\Delta)$, and $\mathbf{D}(j\Delta)$, in 12 additions and 12 assignments.

2. Use these four quantities to compute the n points $\mathbf{P}(0, j\Delta)$, $\mathbf{P}(\Delta, j\Delta)$, $\mathbf{P}(2\Delta, j\Delta)$, up to $\mathbf{P}(1, j\Delta)$, in three additions and three assignments for each point.

The total number of simple operations required for drawing curve $\mathbf{P}(u, j\Delta)$ is therefore $12 + 12 + n(3 + 3) = 6n + 24$. If U such curves are drawn in the u direction, the total number of operations is $(6n + 24)U$.

To complete the wireframe, another family of W curves of the form $\mathbf{P}(i\Delta, w)$ should be computed and displayed. We assume that m points are computed for each curve, which brings the total number of operations for this family of curves to $(6m + 24)W$.

A PC curve $\mathbf{P}_u(w)$ in the w direction on the surface has the form $\mathbf{P}_u(w) = \mathbf{E}w^3 + \mathbf{F}w^2 + \mathbf{G}w + \mathbf{H}$, where each of the four coefficients is a cubic polynomial in u as follows:

$$\mathbf{E}(u) = \mathbf{M}_{00}u^3 + \mathbf{M}_{10}u^2 + \mathbf{M}_{20}u + \mathbf{M}_{30},$$
$$\mathbf{F}(u) = \mathbf{M}_{01}u^3 + \mathbf{M}_{11}u^2 + \mathbf{M}_{21}u + \mathbf{M}_{31},$$
$$\mathbf{G}(u) = \mathbf{M}_{02}u^3 + \mathbf{M}_{12}u^2 + \mathbf{M}_{22}u + \mathbf{M}_{32},$$
$$\mathbf{H}(u) = \mathbf{M}_{03}u^3 + \mathbf{M}_{13}u^2 + \mathbf{M}_{23}u + \mathbf{M}_{33}.$$

Thus, $\mathbf{E}$, $\mathbf{F}$, $\mathbf{G}$, and $\mathbf{H}$ are similar to $\mathbf{A}(w)$, $\mathbf{B}(w)$, $\mathbf{C}(w)$, and $\mathbf{D}(w)$, but are computed with the transpose of matrix $\mathbf{M}$.

A complete curve $\mathbf{P}(i\Delta, w)$ is drawn in the w direction on the surface in the following two steps:

1. Compute $\mathbf{E}(i\Delta)$, $\mathbf{F}(i\Delta)$, $\mathbf{G}(i\Delta)$, and $\mathbf{H}(i\Delta)$ from the corresponding quantities for $[i-1]\Delta$ in 12 additions and 12 assignments.

2. Use these four quantities to compute the m points $\mathbf{P}(i\Delta, 0)$, $\mathbf{P}(i\Delta, \Delta)$, $\mathbf{P}(i\Delta, 2\Delta)$, up to $\mathbf{P}(i\Delta, 1)$, in three additions and three assignments for each point.

The total number of simple operations required to compute the m points for curve $\mathbf{P}(i\Delta, w)$ is therefore $6m + 24$. If W such curves are drawn in the w direction, the total number of operations is $(6m + 24)W$.

Thus, it seems that the entire wireframe can be computed and drawn with $(6n + 24)U + (6m + 24)W$ operations. For $m = n$ and $U = W$ this becomes $2(6n + 24)U$. Typical values of these parameters may be $m = n = 100$ and $U = W = 15$, which results in $624 \times 30 = 18{,}720$ operations.

However, as Figure 1.21 illustrates, some of the points traversed by the curves of the two families are identical, so a sophisticated algorithm may identify them and store them in memory to eliminate double computations and thereby reduce the total number of operations. The figure shows seven curves in the w direction, with 13 points each (the white circles) and five curves in the u direction, consisting of 19 points each (the black circles). Thus, $n = 19$, $m = 13$, $W = 7$, and $U = 5$. The total number of points is $19 \times 5 + 13 \times 7 = 186$, and of these, 7×5, or about 19%, are identical (the $U \times W$ squares).

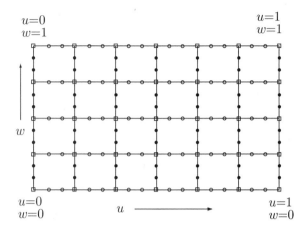

Figure 1.21: A Rectangular Wireframe With 186 Points.

1.12 Subdividing a Surface Patch

The surface subdivision method illustrated here is based on the approach employed in Section 1.5.2 to subdivide a curve. Hence, the reader is advised to read and understand Section 1.5.2 before tackling the material presented here.

Imagine a user trying to construct a surface patch with an interactive algorithm. The patch is based on quantities $\mathbf{P}_{ij}$ that are normally points (some of these quantities may be tangent vectors, but we'll refer to them as points), but the surface refuses to take the desired shape even after the points $\mathbf{P}_{ij}$ have been moved about, shuffled, and manipulated endlessly. This is a common case and it indicates that more points are needed. Just adding new points is a bad approach, because the extra points will modify the shape of the surface and will therefore require the designer to start afresh. A better solution is to add points in such a way that the new surface will have the same shape as the original one. A surface subdivision method takes a surface patch defined by n points $\mathbf{P}_{ij}$ and partitions it into several smaller patches such that together those patches have the same shape as the original surface, and each is defined by n points $\mathbf{Q}_{ij}$, each of which is computed from the original points.

We illustrate this approach to surface subdivision using the bicubic surface patch as an example. The general expression of such a patch is Equation (3.21), duplicated here

$$\mathbf{P}(u,w) = (u^3, u^2, u, 1)\mathbf{N} \begin{bmatrix} \mathbf{P}_{33} & \mathbf{P}_{32} & \mathbf{P}_{31} & \mathbf{P}_{30} \\ \mathbf{P}_{23} & \mathbf{P}_{22} & \mathbf{P}_{21} & \mathbf{P}_{20} \\ \mathbf{P}_{13} & \mathbf{P}_{12} & \mathbf{P}_{11} & \mathbf{P}_{10} \\ \mathbf{P}_{03} & \mathbf{P}_{02} & \mathbf{P}_{01} & \mathbf{P}_{00} \end{bmatrix} \mathbf{N}^T \begin{bmatrix} w^3 \\ w^2 \\ w \\ 1 \end{bmatrix} = \mathbf{UNPN}^T\mathbf{W}^T,$$

where both u and w vary independently over the interval $[0,1]$. We now select four numbers u_1, u_2, w_1, and w_2 that satisfy $0 \le u_1 < u_2 \le 1$ and $0 \le w_1 < w_2 \le 1$. The expression $\mathbf{P}(u,w)$ where u and w vary in the intervals $[u_1, u_2]$ and $[w_1, w_2]$, respectively, is a rectangle on this surface (Figure 1.22a).

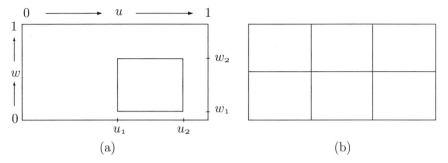

Figure 1.22: Rectangles On A Bicubic Surface Patch.

The next step is to substitute new parameters t and v for u and w, respectively, and express rectangle $\mathbf{P}(u, w)$ as $\mathbf{P}(t, v)$ where both t and v vary independently in $[0, 1]$. If the original rectangle is expressed as

$$\mathbf{P}(u, w) = \mathbf{UNPN}^T \mathbf{W}^T, \quad u_1 \le u \le u_2, \quad w_1 \le w \le w_2,$$

then after the substitutions its shape will be the same and its form will be

$$\mathbf{P}(t, v) = \mathbf{TNQN}^T \mathbf{V}^T, \text{ for } 0 \le t \le 1, \quad 0 \le v \le 1.$$

Both rectangles have the same shape, but $\mathbf{P}(t, v)$ is defined by means of new points $\mathbf{Q}_{ij}$, and the main task is to figure out how to compute the $\mathbf{Q}_{ij}$'s from the original points $\mathbf{P}_{ij}$ while preserving the shape.

Once this is clear, a surface patch can be divided into several rectangles, as in Figure 1.22b, and each expressed in terms of new points. Each new rectangle has the same shape as that part of the surface from which it came, but is defined by the same number of points as the entire original surface. Each rectangle can now be reshaped because of the extra points.

The parameter substitutions from u and w to t and v are the linear relations $t = (u - u_1)/(u_2 - u_1)$ and $v = (w - w_1)/(w_2 - w_1)$. These imply

$$u = (u_2 - u_1) \left[t + \frac{u_1}{u_2 - u_1} \right] \text{ and } w = (w_2 - w_1) \left[v + \frac{w_1}{w_2 - w_1} \right].$$

The rectangle is expressed by means of the new parameters in the form

$$\mathbf{P}(t, v)$$
$$= \left[(u_2 - u_1)^3 \left[t + \frac{u_1}{u_2 - u_1} \right]^3, (u_2 - u_1)^2 \left[t + \frac{u_1}{u_2 - u_1} \right]^2, (u_2 - u_1) \left[t + \frac{u_1}{u_2 - u_1} \right], 1 \right]$$

$$\times \mathbf{NPN}^T \begin{bmatrix} (w_2 - w_1)^3 \left[v + \frac{w_1}{w_2 - w_1} \right]^3 \\ (w_2 - w_1)^2 \left[v + \frac{w_1}{w_2 - w_1} \right]^2 \\ (w_2 - w_1) \left[v + \frac{w_1}{w_2 - w_1} \right] \\ 1 \end{bmatrix}$$

$$= [t^3, t^2, t, 1] \begin{bmatrix} (u_2 - u_1)^3 & 0 & 0 & 0 \\ 3u_1(u_2 - u_1)^2 & (u_2 - u_1)^2 & 0 & 0 \\ 3u_1^2(u_2 - u_1) & 2u_1(u_2 - u_1) & u_2 - u_1 & 0 \\ u_1^3 & u_1^2 & u_1 & 0 \end{bmatrix} \tag{1.32}$$

$$\times \mathbf{NPN}^T \begin{bmatrix} (w_2 - w_1)^3 & 3w_1(w_2 - w_1)^2 & 3w_1^2(w_2 - w_1) & w_1^3 \\ 0 & (w_2 - w_1)^2 & 2w_1(w_2 - w_1) & w_1^2 \\ 0 & 0 & w_2 - w_1 & w_1 \\ 0 & 0 & 0 & 1 \end{bmatrix} \begin{bmatrix} v^3 \\ v^2 \\ v \\ 1 \end{bmatrix}$$

$$= [t^3, t^2, t, 1] \mathbf{LNPN}^T \mathbf{R} [v^3, v^2, v, 1]^T$$

$$= [t^3, t^2, t, 1] \mathbf{NQN}^T [v^3, v^2, v, 1]^T,$$

where the new points $\mathbf{Q}$ are related to the original points by $\mathbf{Q} = \mathbf{N}^{-1} \mathbf{LNPN}^T \mathbf{R} (\mathbf{N}^T)^{-1}$.

To illustrate the application of matrices $\mathbf{L}$ and $\mathbf{R}$ of Equation (1.32), we apply them to the special case $u_1 = 0$, $u_2 = 1/2$, $w_1 = 1/2$, and $w_2 = 1$ to isolate the gray rectangle of Figure 1.23. The resulting matrices are

$$\mathbf{L} = \begin{pmatrix} 1/8 & 0 & 0 & 0 \\ 0 & 1/4 & 0 & 0 \\ 0 & 0 & 1/2 & 0 \\ 0 & 0 & 0 & 1 \end{pmatrix} \quad \mathbf{R} = \begin{pmatrix} 1/8 & 3/8 & 3/8 & 1/8 \\ 0 & 1/4 & 1/2 & 1/4 \\ 0 & 0 & 1/2 & 1/2 \\ 0 & 0 & 0 & 1 \end{pmatrix}.$$

These should be compared with matrices $\mathbf{L}$ and $\mathbf{R}$ of Equations (1.12) and (1.14), respectively.

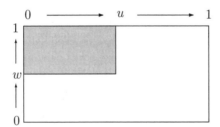

Figure 1.23: A Rectangle on a Surface Patch.

1.13 Surface Normals

The main aim of computer graphics is to display real-looking, solid surfaces. This is done by applying a shading algorithm to every pixel on the surface. Such algorithms may be very complex, but the main task of shading is to compute the amount of light reflected from every surface point. This requires the calculation of the normal to the surface at every point. The normal is the vector that's perpendicular to the surface at the point. It can be defined in two ways:

1. We imagine a flat plane touching the surface at the point (this is called the *osculating plane*). The normal is the vector that's perpendicular to this plane.

2. We calculate two tangent vectors to the surface at the point. The normal is the vector that's perpendicular to both tangents.

The following shows how to calculate the normal vectors for various types of surfaces.

■ The normal to the implicit surface $F(x, y, z) = 0$ at point (x_0, y_0, z_0) is the vector

$$\left(\frac{\partial F(x_0, y_0, z_0)}{\partial x}, \frac{\partial F(x_0, y_0, z_0)}{\partial y}, \frac{\partial F(x_0, y_0, z_0)}{\partial z} \right).$$

Example: The ellipsoid $x^2/a^2 + y^2/b^2 + z^2/c^2 - 1 = 0$. A partial derivative would be, for example, $\partial f/\partial x = 2x/a^2$, so the normal is

$$\left(\frac{2x}{a^2}, \frac{2y}{b^2}, \frac{2z}{c^2} \right) \quad \text{which is in the same direction as} \quad \left(\frac{x}{a^2}, \frac{y}{b^2}, \frac{z}{c^2} \right).$$

For example, the normal at point $(0, 0, -c)$ is $(0, 0, -c/c^2) = (0, 0, -1/c)$. This is a vector in the direction $(0, 0, -1)$.

◇ **Exercise 1.30:** What is the normal to the explicit surface $z = f(x, y)$ at point (x_0, y_0)?

> No money, no job, no rent. Hey, I'm back to normal.
>   —Mickey Rourke (as Henry Chinaski) in *Barfly*, 1987.

■ The normal to the parametric surface $\mathbf{P}(u, w)$ is calculated in two steps. In step 1, the two tangent vectors $\mathbf{U} = \partial \mathbf{P}(u, w)/\partial u$ and $\mathbf{V} = \partial \mathbf{P}(u, w)/\partial w$ are calculated. In step 2, the normal is calculated as their cross-product $\mathbf{U} \times \mathbf{V}$ (Equation (1.5), page 7).

■ The normal to a polygon in a polygonal surface (Section 2.2) can be calculated as shown for an implicit surface. The (implicit) plane equation is $F(x, y, z) = Ax + By + Cz + D = 0$, so the normal is $\left(\frac{\partial F}{\partial x}, \frac{\partial F}{\partial y}, \frac{\partial F}{\partial z} \right)$, which is simply (A, B, C). Another way of calculating the normal, especially suited for triangles, is to find two vectors on the surface and calculate their cross-product. Two suitable vectors are $\mathbf{U} = \mathbf{P}_1 - \mathbf{P}_2$ and $\mathbf{V} = \mathbf{P}_1 - \mathbf{P}_3$, where $\mathbf{P}_1$, $\mathbf{P}_2$, and $\mathbf{P}_3$ are the triangle's corners. Their cross product is

$$\mathbf{U} \times \mathbf{V} = (U_y V_z - U_z V_y, U_z V_x - U_x V_z, U_x V_y - U_y V_x).$$

Example: A polygon with vertices $(1, 1, -1)$, $(1, 1, 1)$ $(1, -1, 1)$, and $(1, -1, -1)$. All the vertices have $x = 1$, so they are on the $x = 1$ plane, which means that the normal should be a vector in the x direction. The calculation is straightforward:

$$\mathbf{U} = (1, 1, 1) - (1, 1, -1) = (0, 0, 2),$$
$$\mathbf{V} = (1, -1, 1) - (1, 1, -1) = (0, -2, 2),$$
$$\mathbf{U} \times \mathbf{V} = (0 - (-4), 0 - 0, 0 - 0) = (4, 0, 0).$$

This is a vector in the right direction.

◇ **Exercise 1.31:** What will happen if we calculate $\mathbf{U}$ as $(1,1,-1) - (1,1,1)$?

◇ **Exercise 1.32:** Find the normal to the pyramid face of Equation (Ans.4).

◇ **Exercise 1.33:** Find the normal to the cone of Equation (Ans.3).

◇ **Exercise 1.34:** Construct a cylinder as a sweep surface (Chapter 9) and find its normal vector. Assume that the cylinder is swept when the line from $(-a, 0, R)$ to $(a, 0, R)$ is rotated $360°$ about the x axis.

John's leaning against the window, probably trying to figure out what parametric equation generated the petals on that eight-foot-tall, carnivorous plant. He turns around to be introduced. "John Cantrell."
"Harvard Li. Didn't you get my e-mail?"
Harvard Li! Now Randy is starting to remember this guy. Founder of Harvard Computer Company, a medium-sized PC clone manufacturer in Taiwan.

Neal Stephenson, *Cryptonomicon* (2002)

2
Linear Interpolation

In order to achieve realism, the many algorithms and techniques employed in computer graphics have to construct mathematical models of curved surfaces, models that are based on curves. It seems that straight line segments and flat surface patches, which are simple geometric figures, cannot play an important role in achieving realism, yet they turn out to be useful in many instances. A smooth curve can be approximated by a set of short straight segments. A smooth, curved surface can similarly be approximated by a set of surface patches, each a small, flat polygon. Thus, this chapter discusses straight lines and flat surfaces that are defined by points. The application of these simple geometric figures to computer graphics is referred to as *linear interpolation*. The chapter also presents two types of surfaces, bilinear and lofted, that are curved, but are partly based on straight lines.

2.1 Straight Segments

We start with the parametric equation of a straight segment. Given any two points $\mathbf{A}$ and $\mathbf{C}$, the expression $\mathbf{A} + \alpha(\mathbf{C} - \mathbf{A})$ is the sum of a point and a vector, so it is a point (see page 2) that we can denote by $\mathbf{B}$. The vector $\mathbf{C} - \mathbf{A}$ points from $\mathbf{A}$ to $\mathbf{C}$, so adding it to $\mathbf{A}$ results in a point on the line connecting $\mathbf{A}$ to $\mathbf{C}$. Thus, we conclude that the three points $\mathbf{A}$, $\mathbf{B}$, and $\mathbf{C}$ are collinear. Note that the expression $\mathbf{B} = \mathbf{A} + \alpha(\mathbf{C} - \mathbf{A})$ can be written $\mathbf{B} = (1 - \alpha)\mathbf{A} + \alpha\mathbf{C}$, showing that $\mathbf{B}$ is a linear combination of $\mathbf{A}$ and $\mathbf{C}$ with barycentric weights. In general, any of three collinear points can be written as a linear combination of the other two. Such points are not independent.

We therefore conclude that given two arbitrary points $\mathbf{P}_0$ and $\mathbf{P}_1$, the parametric representation of the line segment from $\mathbf{P}_0$ to $\mathbf{P}_1$ is

$$\mathbf{P}(t) = (1 - t)\mathbf{P}_0 + t\mathbf{P}_1 = \mathbf{P}_0 + (\mathbf{P}_1 - \mathbf{P}_0)t = \mathbf{P}_0 + t\mathbf{d}, \quad \text{for} \quad 0 \leq t \leq 1. \quad (2.1)$$

The tangent vector of this line is the constant vector $\frac{d\mathbf{P}(t)}{dt} = \mathbf{P}_1 - \mathbf{P}_0 = \mathbf{d}$, the direction from $\mathbf{P}_0$ to $\mathbf{P}_1$.

If we think of $\mathbf{P}_i$ as the vector from the origin to point $\mathbf{P}_i$, then the figure on the right shows how the straight line is obtained as a linear, barycentric combination of the two vectors $\mathbf{P}_0$ and $\mathbf{P}_1$, with coefficients $(1 - t)$ and t. We can think of this combination as a vector that pivots from $\mathbf{P}_0$ to $\mathbf{P}_1$ while varying its magnitude, so its tip always stays on the line.

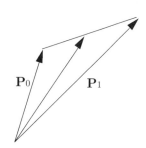

The expression $\mathbf{P}_0 + t\mathbf{d}$ is also useful. It describes the line as the sum of the point $\mathbf{P}_0$ and the vector $t\mathbf{d}$, a vector pointing from $\mathbf{P}_0$ to $\mathbf{P}_1$, whose magnitude depends on t. This representation is useful in cases where the direction of the line and one point on it are known. Notice that varying t in the interval $[-\infty, +\infty]$ constructs the infinite line that contains $\mathbf{P}_0$ and $\mathbf{P}_1$.

2.1.1 Distance of a Point From a Line

Given a line in parametric form $\mathbf{L}(t) = \mathbf{P}_0 + t\mathbf{v}$ (where $\mathbf{v}$ is a vector in the direction of the line) and a point $\mathbf{P}$, what is the distance between them? Assume that $\mathbf{Q}$ is the point on $\mathbf{L}(t)$ that's the closest to $\mathbf{P}$. Point $\mathbf{Q}$ can be expressed as $\mathbf{Q} = \mathbf{L}(t_0) = \mathbf{P}_0 + t_0\mathbf{v}$ for some t_0. The vector from $\mathbf{Q}$ to $\mathbf{P}$ is $\mathbf{P} - \mathbf{Q}$. Since $\mathbf{Q}$ is the nearest point to $\mathbf{P}$, this vector should be perpendicular to the line. Thus, we end up with the condition $(\mathbf{P} - \mathbf{Q}) \bullet \mathbf{v} = 0$ or $(\mathbf{P} - \mathbf{P}_0 - t_0\mathbf{v}) \bullet \mathbf{v} = 0$, which is satisfied by

$$ t_0 = \frac{(\mathbf{P} - \mathbf{P}_0) \bullet \mathbf{v}}{\mathbf{v} \bullet \mathbf{v}}. $$

Substituting this value of t_0 in the line equation gives

$$ \mathbf{Q} = \mathbf{P}_0 + \frac{(\mathbf{P} - \mathbf{P}_0) \bullet \mathbf{v}}{\mathbf{v} \bullet \mathbf{v}}\mathbf{v}. \tag{2.2} $$

The distance between $\mathbf{Q}$ and $\mathbf{P}$ is the magnitude of vector $\mathbf{P} - \mathbf{Q}$.

This method always works since vector $\mathbf{v}$ cannot be zero (otherwise there would be no line).

In the two-dimensional case, the line can be represented explicitly as $y = ax + b$ and the problem can be easily solved with just elementary trigonometry. Figure 2.1 shows a general point $\mathbf{P} = (P_x, P_y)$ at a distance d from a line $y = ax + b$. It is easy to see that the vertical distance e between the line and $\mathbf{P}$ is $|P_y - aP_x - b|$. We also know from trigonometry that

$$ 1 = \sin^2 \alpha + \cos^2 \alpha = \tan^2 \alpha \cos^2 \alpha + \cos^2 \alpha = \cos^2 \alpha(1 + \tan^2 \alpha), $$

implying

$$ \cos^2 \alpha = \frac{1}{1 + \tan^2 \alpha}. $$

We therefore get

$$d = e \cos \alpha = e\sqrt{\cos^2 \alpha} = \frac{e}{\sqrt{1 + \tan^2 \alpha}} = \frac{|P_y - aP_x - b|}{\sqrt{1 + a^2}}. \tag{2.3}$$

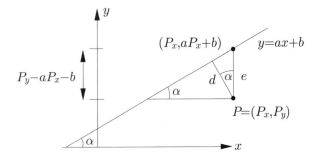

Figure 2.1: Distance Between $\mathbf{P}$ and $y = ax + b$.

⋄ **Exercise 2.1:** Many mathematics problems can be solved in more than one way and this problem is a good example. It is easy to solve by approaching it from different directions. Suggest some approaches to the solution.

> A man who boasts about never changing his views is a man who's decided always to travel in a straight line—the kind of idiot who believes in absolutes.
> —Honoré de Balzac, *Père Goriot*, 1834

2.1.2 Intersection of Lines

Here is a simple, fast algorithm for finding the intersection point(s) of two line segments. Assuming that the two segments $\mathbf{P}_1 + \alpha(\mathbf{P}_2 - \mathbf{P}_1)$ and $\mathbf{P}_3 + \beta(\mathbf{P}_4 - \mathbf{P}_3)$ are given [Equation (2.1)], their intersection point satisfies

$$\mathbf{P}_1 + \alpha(\mathbf{P}_2 - \mathbf{P}_1) = \mathbf{P}_3 + \beta(\mathbf{P}_4 - \mathbf{P}_3),$$

or

$$\alpha(\mathbf{P}_2 - \mathbf{P}_1) - \beta(\mathbf{P}_4 - \mathbf{P}_3) + (\mathbf{P}_1 - \mathbf{P}_3) = 0.$$

This can also be written $\alpha\mathbf{A} + \beta\mathbf{B} + \mathbf{C} = 0$, where $\mathbf{A} = \mathbf{P}_2 - \mathbf{P}_1$, $\mathbf{B} = \mathbf{P}_3 - \mathbf{P}_4$, and $\mathbf{C} = \mathbf{P}_1 - \mathbf{P}_3$. The solutions are

$$\alpha = \frac{B_y C_x - B_x C_y}{A_y B_x - A_x B_y}, \qquad \beta = \frac{A_x C_y - A_y C_x}{A_y B_x - A_x B_y}.$$

The calculation of $\mathbf{A}$, $\mathbf{B}$, and $\mathbf{C}$ requires six subtractions. The calculation of α and β requires three subtractions, six multiplications (since the denominators are identical), and two divisions.

Example: To calculate the intersection of the line segment from $\mathbf{P}_1 = (-1, 1)$ to $\mathbf{P}_2 = (1, -1)$ with the line segment from $\mathbf{P}_3 = (-1, -1)$ to $\mathbf{P}_4 = (1, 1)$, we first calculate

$$\mathbf{A} = \mathbf{P}_2 - \mathbf{P}_1 = (2, -2), \quad \mathbf{B} = \mathbf{P}_3 - \mathbf{P}_4 = (-2, -2), \quad \mathbf{C} = \mathbf{P}_1 - \mathbf{P}_3 = (0, 2).$$

Then calculate

$$\alpha = \frac{0 + 4}{4 + 4} = \frac{1}{2}, \quad \beta = \frac{4 - 0}{4 + 4} = \frac{1}{2}.$$

The lines intersect at their midpoints.

Example: The line segment from $\mathbf{P}_1 = (0, 0)$ to $\mathbf{P}_2 = (1, 0)$ and the line segment from $\mathbf{P}_3 = (2, 0)$ to $\mathbf{P}_4 = (2, 1)$ don't intersect. However, the calculation shows the values of α and β necessary for them to intersect,

$$\mathbf{A} = \mathbf{P}_2 - \mathbf{P}_1 = (1, 0), \quad \mathbf{B} = \mathbf{P}_3 - \mathbf{P}_4 = (0, -1), \quad \mathbf{C} = \mathbf{P}_1 - \mathbf{P}_3 = (-2, 0),$$

yields

$$\alpha = \frac{2 - 0}{0 + 1} = 2, \quad \beta = \frac{0 - 0}{0 + 1} = 0.$$

The lines would intersect at $\alpha = 2$ (i.e., if we extend the first segment to twice its length beyond $\mathbf{P}_2$) and $\beta = 0$ (i.e., point $\mathbf{P}_3$).

⋄ **Exercise 2.2:** How can we identify overlapping lines (i.e., the case of infinitely many intersection points) and parallel lines (no intersection points)? See Figure 2.2.

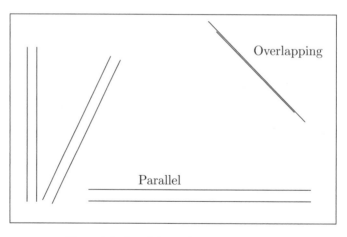

Figure 2.2: Parallel and Overlapped Lines.

2.2 Polygonal Surfaces

A polygonal surface consists of a number of flat faces, each a polygon. A polygon in such a surface is typically a triangle, because the three points of a triangle are always on the same plane. With higher-order polygons, the surface designer should make sure that all the corners of the polygon are on the same plane.

Each polygon is a collection of vertices (the points defining it) and edges (the lines connecting the points). Such a surface is easy to display, either as a wire frame or as a solid surface. In the former case, the edges of all the polygons should be displayed. In the latter case, all the points in a polygon are assigned the same color and brightness. They are all assumed to reflect the same amount of light, since the polygon is flat and has only one normal vector. As a result, a polygonal surface shaded this way appears angular and unnatural, but there is a simple method, known as Gouraud's algorithm [Gouraud 71], that smooths out the reflections from the individual polygons and makes the entire polygonal surface look curved.

Three methods are described for representing such a surface in memory:

1. Explicit polygons. Each polygon is represented as a list

$$\big((x_1, y_1, z_1), (x_2, y_2, z_2), \ldots, (x_n, y_n, z_n) \big)$$

of its vertices, and it is assumed that there is an edge from point 1 to point 2, from 2 to 3, and so on, and also an edge from point n to point 1.

This representation is simple but has two disadvantages:

I. A point may be shared by several polygons, so several copies have to be stored. If the user decides to modify the point, all its copies have to be located and updated. This is a minor problem, because an edge is rarely shared by more than two polygons.

II. An edge may also be shared by several polygons. When displaying the surface, such an edge will be displayed several times, slowing down the entire process.

2. Polygon definition by pointers. There is one list

$$\mathbf{V} = \big((x_1, y_1, z_1), (x_2, y_2, z_2), \ldots, (x_n, y_n, z_n) \big)$$

of all the vertices of the surface. A polygon is represented as a list of pointers, each pointing to a vertex in $\mathbf{V}$. Hence, $\mathbf{P} = (3, 5, 7, 10)$ implies that polygon $\mathbf{P}$ consists of vertices 3, 5, 7, and 10 in $\mathbf{V}$. Problem II still exists.

3. Explicit edges. List $\mathbf{V}$ is as before, and there is also an edge list

$$\mathbf{E} = \big((v_1, v_6, p_3), (v_5, v_7, p_1, p_3, p_6, p_8), \ldots \big).$$

Each element of $\mathbf{E}$ represents an edge. It contains two pointers to the vertices of the edge followed by pointers to all the polygons that share the edge. Each polygon is represented by a list of pointers to $\mathbf{E}$, for example, $\mathbf{P}_1 = (e_1, e_4, e_5)$. Problem II still exists, but it is minor.

2.2.1 Polygon Planarity

Given a polygon defined by points $\mathbf{P}_1$, $\mathbf{P}_2$, ..., $\mathbf{P}_n$, we use the scalar triple product [Equation (1.7)] to test for polygon planarity (i.e., to check whether all the polygon's vertices $\mathbf{P}_i$ are on the same plane). Such a test is necessary only if $n > 3$. We select $\mathbf{P}_1$ as the "pivot" point and calculate the $n - 1$ pivot vectors $\mathbf{v}_i = \mathbf{P}_i - \mathbf{P}_1$ for $i = 2, \dots, n$. Next, we calculate the $n - 3$ scalar triple products $\mathbf{v}_i \bullet (\mathbf{v}_2 \times \mathbf{v}_3)$ for $i = 4, \dots, n$. If any of these products are nonzero, the polygon is not planar. Note that limited accuracy on some computers may cause an otherwise null triple product to come out as a small floating-point number.

⋄ **Exercise 2.3:** Consider the polygon defined by the four points $\mathbf{P}_1 = (1, 0, 0)$, $\mathbf{P}_2 = (0, 1, 0)$, $\mathbf{P}_3 = (1, a, 1)$, and $\mathbf{P}_4 = (0, -a, 0)$. For what values of a will it be planar?

2.2.2 Plane Equations

A polygonal surface consists of flat polygons (often triangles). To calculate the normal to a polygon, we first need to know the polygon's equation. The implicit equation of a flat plane is $Ax + By + Cz + D = 0$. It seems that we need four equations in order to calculate the four unknown coefficients A, B, C, and D, but it turns out that three equations are enough. Assuming that the three points $\mathbf{P}_i = (x_i, y_i, z_i)$, $i = 1, 2, 3$, are given, we can write the four equations

$$
\begin{aligned}
Ax + By + Cz + D &= 0, \\
Ax_1 + By_1 + Cz_1 + D &= 0, \\
Ax_2 + By_2 + Cz_2 + D &= 0, \\
Ax_3 + By_3 + Cz_3 + D &= 0.
\end{aligned}
$$

The first equation is true for any point (x, y, z) on the plane. We cannot solve this system of four equations in four unknowns, but we know that it has a solution if and only if its determinant is zero. The expression below assumes this and also expands the determinant by its top row:

$$
\begin{aligned}
0 &= \begin{vmatrix} x & y & z & 1 \\ x_1 & y_1 & z_1 & 1 \\ x_2 & y_2 & z_2 & 1 \\ x_3 & y_3 & z_3 & 1 \end{vmatrix} \\
&= x \begin{vmatrix} y_1 & z_1 & 1 \\ y_2 & z_2 & 1 \\ y_3 & z_3 & 1 \end{vmatrix} - y \begin{vmatrix} x_1 & z_1 & 1 \\ x_2 & z_2 & 1 \\ x_3 & z_3 & 1 \end{vmatrix} + z \begin{vmatrix} x_1 & y_1 & 1 \\ x_2 & y_2 & 1 \\ x_3 & y_3 & 1 \end{vmatrix} - \begin{vmatrix} x_1 & y_1 & z_1 \\ x_2 & y_2 & z_2 \\ x_3 & y_3 & z_3 \end{vmatrix}.
\end{aligned}
$$

This expression is of the form $Ax + By + Cz + D = 0$ where

$$
A = \begin{vmatrix} y_1 & z_1 & 1 \\ y_2 & z_2 & 1 \\ y_3 & z_3 & 1 \end{vmatrix} \quad B = -\begin{vmatrix} x_1 & z_1 & 1 \\ x_2 & z_2 & 1 \\ x_3 & z_3 & 1 \end{vmatrix} \quad C = \begin{vmatrix} x_1 & y_1 & 1 \\ x_2 & y_2 & 1 \\ x_3 & y_3 & 1 \end{vmatrix} \quad D = -\begin{vmatrix} x_1 & y_1 & z_1 \\ x_2 & y_2 & z_2 \\ x_3 & y_3 & z_3 \end{vmatrix}.
$$

$$(2.4)$$

⋄ **Exercise 2.4:** Calculate the expression of the plane containing the z axis and passing through the point $(1, 1, 0)$.

⋄ **Exercise 2.5:** In the plane equation $Ax + By + Cz + D = 0$, if $D = 0$, then the plane passes through the origin. Assuming $D \neq 0$, we can write the same equation as $x/a + y/b + z/c = 1$, where $a = -D/A$, $b = -D/B$, and $c = -D/C$. What is the geometrical interpretation of a, b, and c?

We operate with nothing but things which do not exist, with lines, planes, bodies, atoms, divisible time, divisible space—how should explanation even be possible when we first make everything into an image, into our own image!

—Friedrich Nietzsche

In some practical situations, the normal to the plane as well as one point on the plane, are known. It is easy to derive the plane equation in such a case.

We assume that $\mathbf{N}$ is the (known) normal vector to the plane, $\mathbf{P}_1$ is a known point, and $\mathbf{P}$ is any point in the plane. The vector $\mathbf{P} - \mathbf{P}_1$ is perpendicular to $\mathbf{N}$, so their dot product $\mathbf{N} \bullet (\mathbf{P} - \mathbf{P}_1)$ equals zero. Since the dot product is associative, we can write $\mathbf{N} \bullet \mathbf{P} = \mathbf{N} \bullet \mathbf{P}_1$. The dot product $\mathbf{N} \bullet \mathbf{P}_1$ is just a number, to be denoted by s, so we obtain

$$\mathbf{N} \bullet \mathbf{P} = s \quad \text{or} \quad N_x x + N_y y + N_z z - s = 0. \tag{2.5}$$

Equation (2.5) can now be written as $Ax + By + Cz + D = 0$, where $A = N_x$, $B = N_y$, $C = N_z$, and $D = -s = -\mathbf{N} \bullet \mathbf{P}_1$. The three unknowns A, B, and C are therefore the components of the normal vector and D can be calculated from any known point $\mathbf{P}_1$ on the plane. The expression $\mathbf{N} \bullet \mathbf{P} = s$ is a useful equation of the plane and is used elsewhere in this book.

⋄ **Exercise 2.6:** Given $\mathbf{N} = (1, 1, 1)$ and $\mathbf{P}_1 = (1, 1, 1)$, calculate the plane equation.

Note that the direction of the normal in this case is unimportant. Substituting $(-A, -B, -C)$ for (A, B, C) would also change the sign of D, resulting in the same equation. However, the direction of the normal is important when the surface is to be shaded. To be used for the calculation of reflection, the normal has to point *outside* the surface. This has to be verified by the user, since the computer has no idea of the shape of the surface and the meaning of "inside" and "outside." In the case where a plane is defined by three points, the direction of the normal can be specified by arranging the three points (in the data structure in memory) in a certain order.

It is also easy to derive the equation of a plane when three points on the plane, $\mathbf{P}_1$, $\mathbf{P}_2$, and $\mathbf{P}_3$, are known. In order for the points to define a plane, they should not be collinear. We consider the vectors $\mathbf{r} = \mathbf{P}_2 - \mathbf{P}_1$ and $\mathbf{s} = \mathbf{P}_3 - \mathbf{P}_1$ a local coordinate system on the plane. Any point $\mathbf{P}$ on the plane can be expressed as a linear combination $\mathbf{P} = u\mathbf{r} + w\mathbf{s}$, where u and w are real numbers. Since $\mathbf{r}$ and $\mathbf{s}$ are local coordinates on the plane, the position of point $\mathbf{P}$ relative to the origin is expressed as (Figure 2.3)

$$\mathbf{P}(u, w) = \mathbf{P}_1 + u\mathbf{r} + w\mathbf{s}, \quad -\infty < u, w < \infty. \tag{2.6}$$

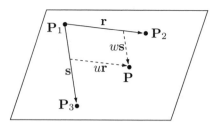

Figure 2.3: Three Points on a Plane.

◇ **Exercise 2.7:** Given the three points $\mathbf{P}_1 = (3,0,0)$, $\mathbf{P}_2 = (0,3,0)$, and $\mathbf{P}_3 = (0,0,3)$, write the equation of the plane defined by them.

2.2.3 Space Division

An infinite plane divides the entire three-dimensional space into two parts. We can call them "outside" and "inside" (or "above" and "below"), and define the outside direction as the direction pointed to by the normal. Using the plane equation, $\mathbf{N} \bullet \mathbf{P} = s$, it is possible to tell if a given point $\mathbf{P}_i$ lies inside, outside, or on the plane. All that's necessary is to examine the sign of the dot product $\mathbf{N} \bullet (\mathbf{P}_i - \mathbf{P})$, where $\mathbf{P}$ is any point on the plane, different from $\mathbf{P}_i$.

This dot product can also be written $|\mathbf{N}|\,|\mathbf{P}_i - \mathbf{P}|\cos\theta$, where θ is the angle between the normal $\mathbf{N}$ and the vector $\mathbf{P}_i - \mathbf{P}$. The sign of the dot product equals the sign of $\cos\theta$, and Figure 2.4a shows that for $-90° < \theta < 90°$, point $\mathbf{P}_i$ lies outside the plane, for $\theta = 90°$, point $\mathbf{P}_i$ lies on the plane, and for $\theta > 90°$, $\mathbf{P}_i$ lies inside the plane.

The regular division of the plane into congruent figures evoking an association in the observer with a familiar natural object is one of these hobbies or problems. ... I have embarked on this geometric problem again and again over the years, trying to throw light on different aspects each time. I cannot imagine what my life would be like if this problem had never occurred to me; one might say that I am head over heels in love with it, and I still don't know why.

—M. C. Escher

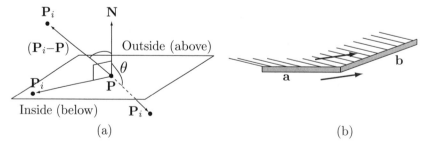

Figure 2.4: (a) Space Division. (b) Turning On a Polygon.

2.2.4 Turning Around on a Polygon

When moving along the edges of a polygon from vertex to vertex, we make a turn at each vertex. Sometimes, the "sense" of the turn (left or right) is important. However, the terms "left" and "right" are relative, depending on the location of the observer, and are therefore ambiguous. Consider Figure 2.4b. It shows two edges, **a** and **b**, of a "thick" polygon, with two arrows pointing from **a** to **b**. Imagine each arrow to be a bug crawling on the polygon. The bug on the top considers the turn from **a** to **b** a left turn, while the bug crawling on the bottom considers the same turn to be a "right" turn.

It is therefore preferable to define terms such as "positive turn" and "negative turn," that depend on the polygon and on the coordinate axes, but not on the position of any observer. To define these terms, consider the plane defined by the vectors **a** and **b** (if they are parallel, they don't define any plane, but then there is no sense talking about turning from **a** to **b**). The cross product **a** × **b** is a vector perpendicular to the plane. It can point in the direction of the normal **N** to the plane, or in the opposite direction. In the former case, we say that the turn from **a** to **b** is positive; in the latter case, the turn is said to be negative.

To calculate the sense of the turn, simply check the sign of the triple scalar product **N** • (**a** × **b**). A positive sign implies a positive turn.

◇ **Exercise 2.8:** Why?

2.2.5 Convex Polygons

Given a polygon, we select two arbitrary points on its edges and connect them with a straight line. If for any two such points the line is fully contained in the polygon, then the polygon is called convex. Another way to define a convex polygon is to say that a line can intersect such a polygon at only two points (unless the line is identical to one of the edges or it grazes the polygon at one point).

The sense of a turn (positive or negative) can also serve to define a convex polygon. When traveling from vertex to vertex in such a polygon all turns should have the same sense. They should all be positive or all negative. In contrast, when traveling along a concave polygon, both positive and negative turns must be made (Figure 2.5).

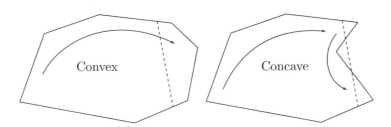

Figure 2.5: Convex and Concave Polygons.

We can think of a polygon as a set of points in two dimensions. The concept of a set of points, however, exists in any number of dimensions. A set of points is convex if it satisfies the definition regardless of the number of dimensions. One important concept associated with a set of points is the *convex hull* of the set. This is the set of "extreme" points that satisfies the following: the set obtained by connecting the points of the convex hull contains all the points of the set. (A simple, two-dimensional analogy is to consider the points nails driven into a board. A rubber band placed around all the nails and stretched will identify the points that constitute the convex hull.)

2.2.6 Line and Plane Intersection

Given a plane $\mathbf{N} \bullet \mathbf{P} = s$ and a line $\mathbf{P} = \mathbf{P}_1 + t\mathbf{d}$ [Equation (2.1)], it is easy to calculate their intersection point. We simply substitute the value of $\mathbf{P}$ in the plane equation to obtain $\mathbf{N} \bullet (\mathbf{P}_1 + t\mathbf{d}) = s$. This results in $t = (s - \mathbf{N} \bullet \mathbf{P}_1)/(\mathbf{N} \bullet \mathbf{d})$. Thus, we compute the value of t and substitute it in the line equation, to get the point of intersection. Such a process is important in ray tracing, an important rendering algorithm where the intersections of light rays and polygons are computed all the time.

⋄ **Exercise 2.9:** The intersection of a line parallel to a plane is either the entire line (if the line happens to be in the plane) or is empty. How do we distinguish these cases from the equation above?

2.2.7 Triangles

A polygonal surface is often constructed of triangles. A triangle is flat but finite, whereas the plane equation describes an infinite plane. We therefore need to modify this equation to describe only the area inside a given triangle

Given any three noncollinear points $\mathbf{P}_1$, $\mathbf{P}_2$, and $\mathbf{P}_3$ in three dimensions, we first derive the equation of the (infinite) plane defined by them. Following that, we limit ourselves to just that part of the plane that's inside the triangle. We start with the two vectors $(\mathbf{P}_2 - \mathbf{P}_1)$ and $(\mathbf{P}_3 - \mathbf{P}_1)$. They can serve as local coordinate axes on the plane (even though they are not normally perpendicular), with point $\mathbf{P}_1$ as the local origin. The linear combination $u(\mathbf{P}_2 - \mathbf{P}_1) + w(\mathbf{P}_3 - \mathbf{P}_1)$, where both u and w can take any real values, is a vector on the plane. To get the coordinates of an arbitrary point on the plane, we simply add point $\mathbf{P}_1$ to this linear combination (recall that the sum of a point and a vector is a point). The resulting plane equation is

$$\mathbf{P}_1 + u(\mathbf{P}_2 - \mathbf{P}_1) + w(\mathbf{P}_3 - \mathbf{P}_1) = \mathbf{P}_1(1 - u - w) + \mathbf{P}_2 u + \mathbf{P}_3 w. \qquad (2.7)$$

To limit the area covered to just the triangle whose corners are $\mathbf{P}_1$, $\mathbf{P}_2$, and $\mathbf{P}_3$, we note that Equation (2.7) yields

$$\mathbf{P}_1, \text{ when } u = 0 \text{ and } w = 0,$$
$$\mathbf{P}_2, \text{ when } u = 1 \text{ and } w = 0,$$
$$\mathbf{P}_3, \text{ when } u = 0 \text{ and } w = 1.$$

The entire triangle can therefore be obtained by varying u and w under the conditions $u \geq 0, w \geq 0$, and $u + w \leq 1$.

⋄ **Exercise 2.10:** Given the three points $\mathbf{P}_1 = (10, -5, 4)$, $\mathbf{P}_2 = (8, -4, 3.2)$, and $\mathbf{P}_3 = (8, 4, 3.2)$, derive the equation of the triangle defined by them.

> If triangles had a God, He'd have three sides.
> —Yiddish proverb

⋄ **Exercise 2.11:** Given the three points $\mathbf{P}_1 = (10, -5, 4)$, $\mathbf{P}_2 = (8, -4, 3.2)$, and $\mathbf{P}_3 = (12, -6, 4.8)$, calculate the triangle defined by them.

For more information, see [Triangles 04] or [Kimberling 94].

2.3 Bilinear Surfaces

A flat polygon is the simplest type of surface. The bilinear surface is the simplest nonflat (curved) surface because it is fully defined by means of its four corner points. It is discussed here because its four boundary curves are straight lines and because the coordinates of any point on this surface are derived by linear interpolations. Since this patch is completely defined by its four corner points, it cannot have a very complex shape. Nevertheless it may be highly curved. If the four corners are coplanar, the bilinear patch defined by them is flat.

Let the corner points be the four distinct points $\mathbf{P}_{00}$, $\mathbf{P}_{01}$, $\mathbf{P}_{10}$, and $\mathbf{P}_{11}$. The top and bottom boundary curves are straight lines and are easy to calculate (Figure 2.6). They are $\mathbf{P}(u,0) = (\mathbf{P}_{10} - \mathbf{P}_{00})u + \mathbf{P}_{00}$ and $\mathbf{P}(u,1) = (\mathbf{P}_{11} - \mathbf{P}_{01})u + \mathbf{P}_{01}$.

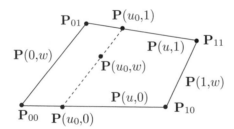

Figure 2.6: A Bilinear Surface.

To linearly interpolate between these boundary curves, we first calculate two corresponding points $\mathbf{P}(u_0, 0)$ and $\mathbf{P}(u_0, 1)$, one on each curve, then connect them with a straight line $\mathbf{P}(u_0, w)$. The two points are

$$\mathbf{P}(u_0, 0) = (\mathbf{P}_{10} - \mathbf{P}_{00})u_0 + \mathbf{P}_{00} \quad \text{and} \quad \mathbf{P}(u_0, 1) = (\mathbf{P}_{11} - \mathbf{P}_{01})u_0 + \mathbf{P}_{01},$$

and the straight segment connecting them is

$$\mathbf{P}(u_0, w) = (\mathbf{P}(u_0, 1) - \mathbf{P}(u_0, 0))\, w + \mathbf{P}(u_0, 0)$$
$$= \left[(\mathbf{P}_{11} - \mathbf{P}_{01})u_0 + \mathbf{P}_{01} - ((\mathbf{P}_{10} - \mathbf{P}_{00})u_0 + \mathbf{P}_{00})\right] w$$
$$+ (\mathbf{P}_{10} - \mathbf{P}_{00})u_0 + \mathbf{P}_{00}.$$

The expression for the entire surface is obtained when we release the parameter u from its fixed value u_0 and let it vary. The result is:

$$
\begin{aligned}
\mathbf{P}(u,w) &= \mathbf{P}_{00}(1-u)(1-w) + \mathbf{P}_{01}(1-u)w + \mathbf{P}_{10}u(1-w) + \mathbf{P}_{11}uw \\
&= \sum_{i=0}^{1}\sum_{j=0}^{1} B_{1i}(u)\mathbf{P}_{ij}B_{1j}(w), \\
&= [B_{10}(u), B_{11}(u)] \begin{bmatrix} \mathbf{P}_{00} & \mathbf{P}_{01} \\ \mathbf{P}_{10} & \mathbf{P}_{11} \end{bmatrix} \begin{bmatrix} B_{10}(w) \\ B_{11}(w) \end{bmatrix},
\end{aligned}
\tag{2.8}
$$

where the functions $B_{1i}(t)$ are the Bernstein polynomials of degree 1, introduced in Section 6.16. This implies that the bilinear surface is a special case of the rectangular Bézier surface, introduced in the same section. (The Bernstein polynomials crop up in unexpected places.) Mathematically, the bilinear surface is a hyperbolic paraboloid (see answer to exercise 2.12). Its parametric expression is linear in both u and w.

The expression $\mathbf{P}(t) = (1-t)\mathbf{P}_1 + t\mathbf{P}_2$ has already been introduced. This is the straight segment from point $\mathbf{P}_1$ to point $\mathbf{P}_2$ expressed as a blend (or a barycentric sum) of the points with the two weights $(1-t)$ and t. Since $B_{10}(t) = 1-t$ and $B_{11}(t) = t$, this expression can also be written in the form

$$
[B_{10}(t), B_{11}(t)] \begin{bmatrix} \mathbf{P}_1 \\ \mathbf{P}_2 \end{bmatrix}.
\tag{2.9}
$$

The reader should notice the similarity between Equations (2.8) and (2.9). The former expression is a direct extension of the latter and is a simple example of the technique of Cartesian product, discussed in Section 1.9, which is used to extend many curves to surfaces.

Figure 2.7 shows a bilinear surface together with the *Mathematica* code that produced it. The coordinates of the four corner points and the final, simplified expression of the surface are also included. The figure illustrates the bilinear nature of this surface. Every line in the u or in the w directions on this surface is straight, but the surface itself is curved.

Example: We select the four points $\mathbf{P}_{00} = (0,0,1)$, $\mathbf{P}_{10} = (1,0,0)$, $\mathbf{P}_{01} = (1,1,1)$, and $\mathbf{P}_{11} = (0,1,0)$ (Figure 2.7) and apply Equation (2.8). The resulting surface patch is

$$
\begin{aligned}
P(u,w) &= (0,0,1)(1-u)(1-w) + (1,1,1)(1-u)w + (1,0,0)u(1-w) + (0,1,0)uw \\
&= \big(u + w - 2uw, w, 1 - u\big).
\end{aligned}
\tag{2.10}
$$

It is easy to check the expression by substituting $u = 0, 1$ and $w = 0, 1$, which reduces the expression to the four corner points. The tangent vectors can easily be calculated. They are

$$
\frac{\partial \mathbf{P}(u,w)}{\partial u} = (1 - 2w, 0, -1), \quad \frac{\partial \mathbf{P}(u,w)}{\partial w} = (1 - 2u, 1, 0).
$$

The first vector lies in the xz plane, and the second lies in the xy plane.

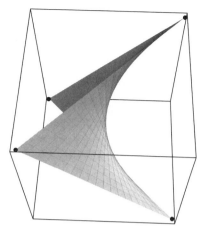

```
(* a bilinear surface patch *)
Clear[bilinear,pnts,u,w];
<<:Graphics:ParametricPlot3D.m;
pnts=ReadList["Points",{Number,Number,Number}, RecordLists->True];
bilinear[u_,w_]:=pnts[[1,1]](1-u)(1-w)+pnts[[1,2]]u(1-w) \
 +pnts[[2,1]]w(1-u)+pnts[[2,2]]u w;
Simplify[bilinear[u,w]]
g1=Graphics3D[{AbsolutePointSize[5], Table[Point[pnts[[i,j]]],{i,1,2},{j,1,2}]}];
g2=ParametricPlot3D[bilinear[u,w],{u,0,1,.05},{w,0,1,.05}, Compiled->False,
 DisplayFunction->Identity];
Show[g1,g2, ViewPoint->{0.063, -1.734, 2.905}];
{{0, 0, 1}, {1, 1, 1}, {1, 0, 0}, {0, 1, 0}}
{u + w - 2 u w, u, 1 - w}
```

Figure 2.7: A Bilinear Surface.

Example: The four points $\mathbf{P}_{00} = (0,0,1)$, $\mathbf{P}_{10} = (1,0,0)$, $\mathbf{P}_{01} = (0.5,1,0)$, and $\mathbf{P}_{11} = (1,1,0)$ are selected and Equation (2.8) is applied to them. The resulting surface patch is (Figure 2.8)

$$
\begin{aligned}
P(u,w) &= (0,0,1)(1-u)(1-w) + (0.5,1,0)(1-u)w + (1,0,0)u(1-w) + (1,1,0)uw \\
&= \big(0.5(1-u)w + u, w, (1-u)(1-w)\big).
\end{aligned}
\tag{2.11}
$$

Note that the y coordinate is simply w. This means that points with the same w value, such as $\mathbf{P}(0.1, w)$ and $\mathbf{P}(0.5, w)$ have the same y coordinate and are therefore located on the same horizontal line. Also, the z coordinate is a simple function of u and w, varying from 1 (when $u = w = 0$) to 0 as we move toward $u = 1$ or $w = 1$.

The boundary curves are very easy to calculate from Equation (2.11). Here are two of them

$$
\mathbf{P}(0, w) = (0.5w, w, 1 - w), \qquad \mathbf{P}(u, 1) = (0.5(1 - u) + u, 1, 0).
$$

The tangent vectors can also be obtained from Equation (2.11)

$$
\frac{\partial \mathbf{P}(u,w)}{\partial u} = (-0.5w + 1, 0, w - 1), \qquad \frac{\partial \mathbf{P}(u,w)}{\partial w} = (0.5(1 - u), 1, u - 1).
\tag{2.12}
$$

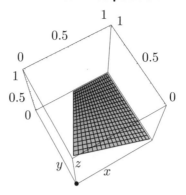

```
(* Another bilinear surface example *)
ParametricPlot3D[{0.5(1-u)w+u,w,(1-u)(1-w)}, {u,0,1},{w,0,1}, Compiled->False,
ViewPoint->{-0.846, -1.464, 3.997}, DefaultFont->{"cmr10", 10}];
```

Figure 2.8: A Bilinear Surface.

The first is a vector in the xz plane, while the second is a vector in the $y = 1$ plane. The following two tangent values are especially simple: $\frac{\partial \mathbf{P}(u,1)}{\partial u} = (0.5, 0, 0)$ and $\frac{\partial \mathbf{P}(1,w)}{\partial w} = (0, 1, 0)$. The first is a vector in the x direction and the second is a vector in the y direction.

Finally, we compute the *normal vector* to the surface. This vector is normal to the surface at any point, so it is perpendicular to the two tangent vectors $\partial \mathbf{P}(u, w)/\partial u$ and $\partial \mathbf{P}(u, w)/\partial w$ and is therefore the cross-product [Equation (1.5)] of these vectors. The calculation is straightforward:

$$\mathbf{N}(u, w) = \frac{\partial \mathbf{P}}{\partial u} \times \frac{\partial \mathbf{P}}{\partial w} = (1 - w, 0.5(1 - u), 1 - 0.5w). \tag{2.13}$$

There are two ways of satisfying ourselves that Equation (2.13) is the correct expression for the normal:

1. It is easy to prove, by directly calculating the dot products, that the normal vector of Equation (2.13) is perpendicular to both tangents of Equation (2.12).

2. A closer look at the coordinates of our points shows that three of them have a z coordinate of zero and only $\mathbf{P}_{00}$ has $z = 1$. This means that the surface approaches a flat xy surface as one moves away from point $\mathbf{P}_{00}$. It also means that the normal should approach the z direction when u and w move away from zero, and it should move away from that direction when u and w approach zero. It is, in fact, easy to confirm the following limits:

$$\lim_{u,w \to 1} \mathbf{N}(u, w) = (0, 0, 0.5), \qquad \lim_{u,w \to 0} \mathbf{N}(u, w) = (1, 0.5, 1).$$

⋄ **Exercise 2.12:** (1) Calculate the bilinear surface for the points $(0, 0, 0)$, $(1, 0, 0)$, $(0, 1, 0)$, and $(1, 1, 1)$. (2) Guess the explicit representation $z = F(x, y)$ of this surface. (3) What curve results from the intersection of this surface with the plane $z = k$ (parallel to the

xy plane). (4) What curve results from the intersection of this surface with a plane containing the z axis?

Example: This is the third example of a bilinear surface. The four points $\mathbf{P}_{00} = (0, 0, 1)$, $\mathbf{P}_{10} = (1, 0, 0)$, and $\mathbf{P}_{01} = \mathbf{P}_{11} = (0, 1, 0)$ create a *triangular surface patch* (Figure 2.9) because two of them are identical. The surface expression is

$$P(u, w) = (0, 0, 1)(1-u)(1-w) + (0, 1, 0)(1-u)w + (1, 0, 0)u(1-w) + (0, 1, 0)uw$$
$$= \big(u(1 - w), w, (1 - u)(1 - w)\big).$$

Notice that the boundary curve $\mathbf{P}(u, 1)$ degenerates to the single point $(0, 1, 0)$, i.e., it does not depend on u.

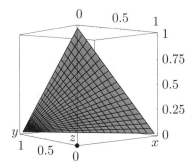

```
(* A Triangular bilinear surface example *)
ParametricPlot3D[{u(1-w),w,(1-u)(1-w)}, {u,0,1},{w,0,1}, Compiled->False,
ViewPoint->{-2.673, -3.418, 0.046}, DefaultFont->{"cmr10", 10}];
```

Figure 2.9: A Triangular Bilinear Surface.

⋄ **Exercise 2.13:** Calculate the tangent vectors and the normal vector of this surface.

⋄ **Exercise 2.14:** Given the two points $\mathbf{P}_{00} = (-1, -1, 0)$ and $\mathbf{P}_{10} = (1, -1, 0)$, consider them the endpoints of a straight segment $\mathbf{L}_1$.

(1) Construct the endpoints of the three straight segments $\mathbf{L}_2$, $\mathbf{L}_3$, and $\mathbf{L}_4$. Each should be translated one unit above its predecessor on the y axis and should be rotated $60°$ about the y axis, as shown in Figure 2.10. Denote the four pairs of endpoints by $\mathbf{P}_{00}\mathbf{P}_{10}$, $\mathbf{P}_{01}\mathbf{P}_{11}$, $\mathbf{P}_{02}\mathbf{P}_{12}$ and $\mathbf{P}_{03}\mathbf{P}_{13}$.

(2) Calculate the three bilinear surface patches

$$\mathbf{P}_1(u,w) = \mathbf{P}_{00}(1-u)(1-w) + \mathbf{P}_{01}(1-u)w + \mathbf{P}_{10}u(1-w) + \mathbf{P}_{11}uw,$$
$$\mathbf{P}_2(u,w) = \mathbf{P}_{01}(1-u)(1-w) + \mathbf{P}_{02}(1-u)w + \mathbf{P}_{11}u(1-w) + \mathbf{P}_{12}uw,$$
$$\mathbf{P}_3(u,w) = \mathbf{P}_{02}(1-u)(1-w) + \mathbf{P}_{03}(1-u)w + \mathbf{P}_{12}u(1-w) + \mathbf{P}_{13}uw.$$

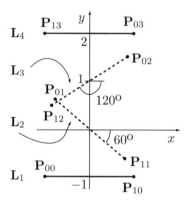

Figure 2.10: Four Straight Segments for Exercise 2.14.

2.4 Lofted Surfaces

This kind of surface patch is curved, but it belongs in this chapter because it is linear in one direction. It is bounded by two arbitrary curves [that we denote by $\mathbf{P}(u,0)$ and $\mathbf{P}(u,1)$] and by two straight segments $\mathbf{P}(0,w)$ and $\mathbf{P}(1,w)$ connecting them. Surface lines in the w direction are therefore straight, whereas each line in the u direction is a blend of $\mathbf{P}(u,0)$ and $\mathbf{P}(u,1)$. The blend of the two curves is simply $(1-w)\mathbf{P}(u,0) + w\mathbf{P}(u,1)$, and this blend, which is linear in w, constitutes the expression of the surface

$$\mathbf{P}(u,w) = (1-w)\mathbf{P}(u,0) + w\mathbf{P}(u,1). \tag{2.14}$$

This expression is linear in w, implying straight lines in the w direction. Moving in the u direction, we travel on a curve whose shape depends on the value of w. For $w_0 \approx 0$, the curve $\mathbf{P}(u,w_0)$ is close to the boundary curve $\mathbf{P}(u,0)$. For $w_0 \approx 1$, it is close to the boundary curve $\mathbf{P}(u,1)$. For $w_0 = 0.5$, it is $0.5\mathbf{P}(u,0) + 0.5\mathbf{P}(u,1)$, an equal mixture of the two.

Note that this kind of surface is fully defined by specifying the two boundary curves. The four corner points are implicit in these curves. These surfaces are sometimes called *ruled*, because straight lines are an important part of their description. This is also the reason why this type of surface is sometimes defined as follows: a surface is a lofted surface if and only if through every point on it there is a straight line that lies completely on the surface.

This definition implies that any cylinder is a lofted surface, but a little thinking shows that even a bilinear surface is lofted.

Example: We start with the six points $\mathbf{P}_1 = (-1, 0, 0)$, $\mathbf{P}_2 = (0, -1, 0)$, $\mathbf{P}_3 = (1, 0, 0)$, $\mathbf{P}_4 = (-1, 0, 1)$, $\mathbf{P}_5 = (0, -1, 1)$, and $\mathbf{P}_6 = (1, 0, 1)$. Because of the special coordinates of the points (and because of the way we will compute the boundary curves), the surface is easy to visualize (Figure 2.11). This helps to intuitively make sense of the expressions for the tangent vectors and the normal. Note especially that the left and right edges of the surface are in the xz plane, whereas we will see that all the other lines in the w direction have a small negative y component.

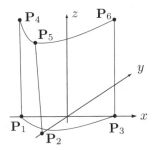

Figure 2.11: A Lofted Surface.

We proceed in six steps as follows:

1. As the top boundary curve, $\mathbf{P}(u, 1)$, we select the quadratic polynomial passing through the top three points $\mathbf{P}_4$, $\mathbf{P}_5$, and $\mathbf{P}_6$. There is only one such curve and it has the form $\mathbf{P}(u, 1) = \mathbf{A} + \mathbf{B}u + \mathbf{C}u^2$, where the coefficients $\mathbf{A}$, $\mathbf{B}$, and $\mathbf{C}$ have to be calculated. We use the fact that the curve passes through the three points to set up the three equations $\mathbf{P}(0, 1) = \mathbf{P}_4$, $\mathbf{P}(0.5, 1) = \mathbf{P}_5$, and $\mathbf{P}(1, 1) = \mathbf{P}_6$, that are written explicitly as

$$\mathbf{A} + \mathbf{B} \times 0 + \mathbf{C} \times 0^2 = (-1, 0, 1),$$
$$\mathbf{A} + \mathbf{B} \times 0.5 + \mathbf{C} \times 0.5^2 = (0, -1, 1),$$
$$\mathbf{A} + \mathbf{B} \times 1 + \mathbf{C} \times 1^2 = (1, 0, 1).$$

These are easy to solve and result in $\mathbf{A} = (-1, 0, 1)$, $\mathbf{B} = (2, -4, 0)$, and $\mathbf{C} = (0, 4, 0)$. The top boundary curve is therefore $\mathbf{P}(u, 1) = \big(2u - 1, 4u(u - 1), 1\big)$.

2. As the bottom boundary curve, we select the quadratic Bézier curve [Equation (6.6)] defined by the three points $\mathbf{P}_1$, $\mathbf{P}_2$, and $\mathbf{P}_3$. The curve is

$$
\begin{aligned}
\mathbf{P}(u, 0) &= \sum_{i=0}^{2} B_{2i}(u) \mathbf{P}_{i+1} \\
&= (1 - u)^2(-1, 0, 0) + 2u(1 - u)(0, -1, 0) + u^2(1, 0, 0) \\
&= \big(2u - 1, -2u(1 - u), 0\big).
\end{aligned}
$$

3. The expression of the surface is immediately obtained

$$\mathbf{P}(u, w) = \mathbf{P}(u, 0)(1 - w) + \mathbf{P}(u, 1)w = \big(2u - 1, 2u(u - 1)(1 + w), w\big).$$

(Notice that it does not pass through $\mathbf{P}_2$.)

4. The two tangent vectors are also easy to compute

$$\frac{\partial \mathbf{P}}{\partial u} = \big(2, 2(2u-1)(1+w), 0\big), \quad \frac{\partial \mathbf{P}}{\partial w} = \big(0, 2u(u-1), 1\big).$$

5. The normal, as usual, is the cross-product of the tangents and is given by $\mathbf{N}(u, w) = \big(2(2u-1)(1+w), -2, 4u(u-1)\big)$.

6. The most important feature of this example is the ease with which the expressions of the tangents and the normal can be visualized. This is possible because of the simple shape and orientation of the surface (again, see Figure 2.11). The reader should examine the expressions and make sure the following points are clear:

■ The two boundary curves are very similar. One difference between them is, of course, the x and z coordinates. However, the *only* important difference is in the y coordinate. Both curves are quadratic polynomials in u, but although $\mathbf{P}(u, 1)$ passes through the three top points, $\mathbf{P}(u, 0)$ passes only through the first and last points.

■ The tangent in the u direction, $\partial \mathbf{P}/\partial u$, features $z = 0$; it is a vector in the xy plane. At the bottom of the surface, where $w = 0$, it changes direction from $(2, -2, 0)$ (when $u = 0$) to $(2, 2, 0)$ (when $u = 1$), both $45°$ directions in the xy plane. However, at the top, where $w = 1$, the tangent changes direction from $(2, -4, 0)$ to $(2, 4, 0)$, both $63°$ directions. This is because the top boundary curve goes deeper in the y direction.

■ The tangent in the w direction, $\partial \mathbf{P}/\partial w$ features $x = 0$; it is a vector in the yz plane. Its z coordinate is a constant 1, and its y coordinate varies from 0 (on the left, where $u = 0$), to -0.5 (in the middle, where $u = 0.5$), and back to 0 (on the right, where $u = 1$). On the left and right edges of the surface, this vector is therefore vertical $(0, 0, 1)$. In the middle, it is $(0, -0.5, 1)$, making a negative half-step in y for each step in z.

■ The normal vector features $y = -2$ with a small z component. It therefore points mostly in the negative y direction, and a little in x. At the bottom $(w = 0)$, it varies from $(-2, -2, 0)$, to $(0, -2, -1)$,* and ends in $(2, -2, 0)$. At the top $(w = 1)$, it varies from $(-4, -2, 0)$, to $(0, -2, -1)$, and ends in $(4, -2, 0)$. The top boundary curve is deeper, causing the tangent to be more in the y direction and the normal to be more in the x direction, than on the bottom boundary curve.

⋄ **Exercise 2.15:** (a) Given the two three-dimensional points $\mathbf{P}_1 = (-1, -1, 0)$ and $\mathbf{P}_2 = (1, -1, 0)$, calculate the straight line from $\mathbf{P}_1$ to $\mathbf{P}_2$. This will become the bottom boundary curve of a lofted surface.

(b) Given the three three-dimensional points $\mathbf{P}_4 = (-1, 1, 0)$, $\mathbf{P}_5 = (0, 1, 1)$, and $\mathbf{P}_6 = (1, 1, 0)$, calculate the quadratic polynomial $\mathbf{P}(t) = \mathbf{A}t^2 + \mathbf{B}t + \mathbf{C}$ that passes through them. This will become the top boundary curve of the surface.

(c) Calculate the expression of the lofted surface patch and the coordinates of its center point $\mathbf{P}(0.5, 0.5)$.

---

* it has a small z component, reflecting the fact that the surface is not completely vertical at $u = 0.5$.

2.4.1 A Double Helix

This example illustrates how the well-known double helix can be derived as a lofted surface. The two-dimensional parametric curve $(\cos t, \sin t)$ is, of course, a circle (of radius one unit, centered on the origin). As a result, the three-dimensional curve $(\cos t, \sin t, t)$ is a helix spiraling around the z axis upward from the origin. The similar curve $(\cos(t + \pi), \sin(t + \pi), t)$ is another helix, at a 180° phase difference with the first. We consider these the two boundary curves of a lofted surface and create the entire surface as a linear interpolation of the two curves. Hence,

$$\mathbf{P}(u, w) = (\cos u, \sin u, u)(1 - w) + (\cos(u + \pi), \sin(u + \pi), u)w,$$

where $0 \leq w \leq 1$, and u can vary in any range. The two curves form a double helix, so the surface looks like a twisted ribbon. Figure 2.12 shows such a surface, together with the code that generated it.

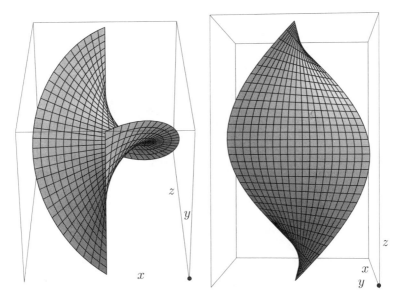

```
Clear[loftedSurf]; (* double helix as a lofted surface *)
<<:Graphics:ParametricPlot3D.m;
loftedSurf:={Cos[u],Sin[u],u}(1-w)+{Cos[u+Pi],Sin[u+Pi],u}w;
ParametricPlot3D[loftedSurf, {u,0,Pi,.1},{w,0,1}, Compiled->False,
Ticks->False, ViewPoint->{-2.640, -0.129, 0.007}]
```

Figure 2.12: The Double Helix as a Lofted Surface.

◇ **Exercise 2.16:** Calculate the expression of a cone as a lofted surface. Assume that the vertex of the cone is located at the origin, and the base is a circle of radius R, centered on the z axis and located on the plane $z = H$.

◇ **Exercise 2.17:** Derive the expression for a square pyramid where each face is a lofted surface. Assume that the base is a square, $2a$ units on a side, centered about the origin on the xy plane. The top is point $(0, 0, H)$.

2.4.2 A Cusp

Given the two curves $\mathbf{P}_1(u) = (8, 4, 0)u^3 - (12, 9, 0)u^2 + (6, 6, 0)u + (-1, 0, 0)$ and $\mathbf{P}_2(u) = (2u - 1, 4u(u - 1), 1)$, the lofted surface defined by them is easy to calculate. Notice that the curves pass through the points $\mathbf{P}_1(0) = (-1, 0, 0)$, $\mathbf{P}_1(0.5) = (0, 5/4, 0)$, $\mathbf{P}_1(1) = (1, 1, 0)$, $\mathbf{P}_2(0) = (-1, 0, 1)$, $\mathbf{P}_2(0.5) = (0, -1, 1)$, and $\mathbf{P}_2(1) = (1, 0, 1)$, which makes it easy to visualize the surface (Figure 2.13). The tangent vectors of the two curves are

$$\mathbf{P}_1^u(u) = (24, 12, 0)u^2 - (24, 18, 0)u + (6, 6, 0), \quad \mathbf{P}_2^u(u) = (2, 8u - 4, 0).$$

Notice that $\mathbf{P}_1^u(0.5)$ equals $(0, 0, 0)$, which implies that $\mathbf{P}_1(u)$ has a cusp at $u = 0.5$. The lofted surface defined by the two curves is

$$\mathbf{P}(u, w) = \big(4u^2(2u - 3)(1 - w) - 4uw + 6u - 1, u^2(4u - 9)(1 - w) + 4u^2w - 10uw + 6u, w\big).$$

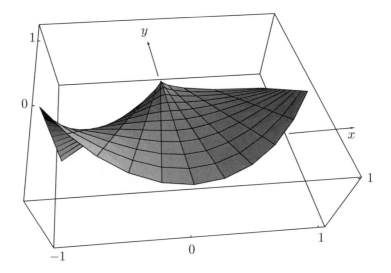

```
(* Another lofted surface example *)
<<:Graphics:ParametricPlot3D.m
Clear[ls];
ls=Simplify[{8u^3-12u^2+6u-1,4u^3-9u^2+6u,0}(1-w)+{2u-1,4u(u-1),1}w];
ParametricPlot3D[ls, {u,0,1,.1},{w,0,1,.1}, Compiled->False,
  ViewPoint->{-0.139, -1.179, 1.475}, DefaultFont->{"cmr10", 10},
  AspectRatio->Automatic, Ticks->{{0,1},{0,1},{0,1}}];
```

Figure 2.13: A Lofted Surface Patch.

Now, look Gwen, y'know if we're gonna keep living together in this loft, we're gonna have to have some rules.

—Leah Remini (as Terri Reynolds) in *Fired Up* (1997)

◇ **Exercise 2.18:** Calculate the tangent vector of this surface in the u direction, and compute its value at the cusp.

LERP, a quasi-acronym for Linear Interpolation, used as a verb or noun for the operation. "Bresenham's algorithm lerps incrementally between the two endpoints of the line."

The New Hacker's Dictionary version 4.2.2, a.k.a., The Jargon File

3
Polynomial Interpolation

Definition: A polynomial of degree n in x is the function

$$P_n(x) = \sum_{i=0}^{n} a_i x^i = a_0 + a_1 x + a_2 x^2 + \cdots + a_n x^n,$$

where a_i are the coefficients of the polynomial (in our case, they are real numbers). Note that there are $n + 1$ coefficients.

Calculating a polynomial involves additions, multiplications, and exponentiations, but there are two methods that greatly simplify this calculation. They are the following:

1. Horner's rule. A degree-3 polynomial can be written in the form

$$P(x) = \big((a_3 x + a_2)x + a_1\big)x + a_0,$$

thereby eliminating all exponentiations.

2. Forward differences. This is one of Newton's many contributions to mathematics and it is described in some detail in Section 1.5.1. Only the first step requires multiplications. All other steps are performed with additions and assignments only.

Given a set of points, it is possible to construct a polynomial that when plotted passes through the points. When fully computed and displayed, such a polynomial becomes a curve that's referred to as a *polynomial interpolation* of the points. The first part of this chapter discusses methods for polynomial interpolation and shows their limitations. The second part extends the discussion to a two-dimensional grid of points, and shows how to compute a two-parameter polynomial that passes through the points. When fully computed and displayed, such a polynomial becomes a surface. The methods described here apply the algebra of polynomials to the geometry of curves and surfaces, but this application is limited, because high-degree polynomials tend to oscillate. Section 1.5, and especially Exercise 1.20 show why this is so. Still, there are cases where high-degree polynomials are useful.

This chapter starts with a simple example where four points are given and a cubic polynomial that passes through them is derived from first principles. Following this, the Lagrange and Newton polynomial interpolation methods are introduced. The chapter continues with a description of several simple surface algorithms based on polynomials. It concludes with the Coons and Gordon surfaces, which also employ polynomials.

3.1 Four Points

Four points (two-dimensional or three-dimensional) $\mathbf{P}_1$, $\mathbf{P}_2$, $\mathbf{P}_3$, and $\mathbf{P}_4$ are given. We are looking for a PC curve that passes through these points and has the form

$$\mathbf{P}(t) = \mathbf{a}t^3 + \mathbf{b}t^2 + \mathbf{c}t + \mathbf{d} = (t^3, t^2, t, 1)(\mathbf{a}, \mathbf{b}, \mathbf{c}, \mathbf{d})^T = \mathbf{T}(t)\mathbf{A} \quad \text{for} \quad 0 \le t \le 1, \quad (3.1)$$

where each of the four coefficients $\mathbf{a}$, $\mathbf{b}$, $\mathbf{c}$, and $\mathbf{d}$ is a pair (or a triplet), $\mathbf{T}(t)$ is the row vector $(t^3, t^2, t, 1)$, and $\mathbf{A}$ is the column vector $(\mathbf{a}, \mathbf{b}, \mathbf{c}, \mathbf{d})^T$. The only unknowns are $\mathbf{a}$, $\mathbf{b}$, $\mathbf{c}$, and $\mathbf{d}$.

Since the four points can be located anywhere, we cannot assume anything about their positions and we make the general assumption that $\mathbf{P}_1$ and $\mathbf{P}_4$ are the two end-points $\mathbf{P}(0)$ and $\mathbf{P}(1)$ of the curve, and that $\mathbf{P}_2$ and $\mathbf{P}_3$ are the two interior points $\mathbf{P}(1/3)$ and $\mathbf{P}(2/3)$. (Having no information about the locations of the points, the best we can do is to use equi-distant values of the parameter t.) We therefore write the four equations $\mathbf{P}(0) = \mathbf{P}_1$, $\mathbf{P}(1/3) = \mathbf{P}_2$, $\mathbf{P}(2/3) = \mathbf{P}_3$, and $\mathbf{P}(1) = \mathbf{P}_4$, or explicitly

$$\begin{aligned}
\mathbf{a}(0)^3 + \mathbf{b}(0)^2 + \mathbf{c}(0) + \mathbf{d} &= \mathbf{P}_1, \\
\mathbf{a}(1/3)^3 + \mathbf{b}(1/3)^2 + \mathbf{c}(1/3) + \mathbf{d} &= \mathbf{P}_2, \\
\mathbf{a}(2/3)^3 + \mathbf{b}(2/3)^2 + \mathbf{c}(2/3) + \mathbf{d} &= \mathbf{P}_3, \\
\mathbf{a}(1)^3 + \mathbf{b}(1)^2 + \mathbf{c}(1) + \mathbf{d} &= \mathbf{P}_4.
\end{aligned} \quad (3.2)$$

The solutions of this system of equations are

$$\begin{aligned}
\mathbf{a} &= -(9/2)\mathbf{P}_1 + (27/2)\mathbf{P}_2 - (27/2)\mathbf{P}_3 + (9/2)\mathbf{P}_4, \\
\mathbf{b} &= 9\mathbf{P}_1 - (45/2)\mathbf{P}_2 + 18\mathbf{P}_3 - (9/2)\mathbf{P}_4, \\
\mathbf{c} &= -(11/2)\mathbf{P}_1 + 9\mathbf{P}_2 - (9/2)\mathbf{P}_3 + \mathbf{P}_4, \\
\mathbf{d} &= \mathbf{P}_1.
\end{aligned} \quad (3.3)$$

Substituting these solutions into Equation (3.1) gives

$$\begin{aligned}
\mathbf{P}(t) = &\left(-(9/2)\mathbf{P}_1 + (27/2)\mathbf{P}_2 - (27/2)\mathbf{P}_3 + (9/2)\mathbf{P}_4\right)t^3 \\
&+ \left(9\mathbf{P}_1 - (45/2)\mathbf{P}_2 + 18\mathbf{P}_3 - (9/2)\mathbf{P}_4\right)t^2 \\
&+ \left(-(11/2)\mathbf{P}_1 + 9\mathbf{P}_2 - (9/2)\mathbf{P}_3 + \mathbf{P}_4\right)t + \mathbf{P}_1.
\end{aligned}$$

After rearranging, this becomes

$$
\begin{aligned}
\mathbf{P}(t) =& (-4.5t^3 + 9t^2 - 5.5t + 1)\mathbf{P}_1 + (13.5t^3 - 22.5t^2 + 9t)\mathbf{P}_2 \\
& + (-13.5t^3 + 18t^2 - 4.5t)\mathbf{P}_3 + (4.5t^3 - 4.5t^2 + t)\mathbf{P}_4 \\
=& G_1(t)\mathbf{P}_1 + G_2(t)\mathbf{P}_2 + G_3(t)\mathbf{P}_3 + G_4(t)\mathbf{P}_4 \\
=& \mathbf{G}(t)\mathbf{P},
\end{aligned}
\tag{3.4}
$$

where the four functions $G_i(t)$ are cubic polynomials in t

$$
\begin{aligned}
G_1(t) = (-4.5t^3 + 9t^2 - 5.5t + 1), \quad & G_3(t) = (-13.5t^3 + 18t^2 - 4.5t), \\
G_2(t) = (13.5t^3 - 22.5t^2 + 9t), \quad & G_4(t) = (4.5t^3 - 4.5t^2 + t),
\end{aligned}
\tag{3.5}
$$

$\mathbf{P}$ is the column $(\mathbf{P}_1, \mathbf{P}_2, \mathbf{P}_3, \mathbf{P}_4)^T$ and $\mathbf{G}(t)$ is the row $(G_1(t), G_2(t), G_3(t), G_4(t))$ (see also Exercise 3.8 for a different approach to this polynomial).

The functions $G_i(t)$ are called blending functions because they represent any point on the curve as a blend of the four given points. Note that they are barycentric (they should be, since they blend points, and this is shown in the next paragraph). We can also write

$$
G_1(t) = (t^3, t^2, t, 1)(-4.5, 9, -5.5, 1)^T
$$

and similarly for $G_2(t)$, $G_3(t)$, and $G_4(t)$. The curve can now be expressed as

$$
\mathbf{P}(t) = \mathbf{G}(t)\mathbf{P} = (t^3, t^2, t, 1)
\begin{bmatrix}
-4.5 & 13.5 & -13.5 & 4.5 \\
9.0 & -22.5 & 18 & -4.5 \\
-5.5 & 9.0 & -4.5 & 1.0 \\
1.0 & 0 & 0 & 0
\end{bmatrix}
\begin{bmatrix}
\mathbf{P}_1 \\
\mathbf{P}_2 \\
\mathbf{P}_3 \\
\mathbf{P}_4
\end{bmatrix}
= \mathbf{T}(t)\,\mathbf{N}\,\mathbf{P}. \tag{3.6}
$$

Matrix $\mathbf{N}$ is called the basis matrix and $\mathbf{P}$ is the geometry vector. Equation (3.1) tells us that $\mathbf{P}(t) = \mathbf{T}(t)\,\mathbf{A}$, so we conclude that $\mathbf{A} = \mathbf{N}\mathbf{P}$.

The four functions $G_i(t)$ are barycentric because of the nature of Equation (3.2), not because of the special choice of the four t values. To see why this is so, we write Equation (3.2) for four different, arbitrary values t_1, t_2, t_3, and t_4 (they have to be different, otherwise two or more equations would be contradictory).

$$
\begin{aligned}
at_1^3 + bt_1^2 + ct_1 + d &= P_1, \\
at_2^3 + bt_2^2 + ct_2 + d &= P_2, \\
at_3^3 + bt_3^2 + ct_3 + d &= P_3, \\
at_4^3 + bt_4^2 + ct_4 + d &= P_4,
\end{aligned}
\tag{3.7}
$$

(where we treat the four values P_i as numbers, not points, and as a result, a, b, c, and d are also numbers). The solutions are of the form

$$
\begin{aligned}
a &= c_{11}P_1 + c_{12}P_2 + c_{13}P_3 + c_{14}P_4, \\
b &= c_{21}P_1 + c_{22}P_2 + c_{23}P_3 + c_{24}P_4, \\
c &= c_{31}P_1 + c_{32}P_2 + c_{33}P_3 + c_{34}P_4, \\
d &= c_{41}P_1 + c_{42}P_2 + c_{43}P_3 + c_{44}P_4.
\end{aligned}
\tag{3.8}
$$

Comparing Equation (3.8) to Equations (3.3) and (3.5) shows that the four functions $G_i(t)$ can be expressed in terms of the c_{ij} in the form

$$G_i(t) = (c_{1i}t^3 + c_{2i}t^2 + c_{3i}t + c_{4i}). \tag{3.9}$$

The point is that the 16 coefficients c_{ij} do not depend on the four values P_i. They are the same for any choice of the P_i. As a special case, we now select $P_1 = P_2 = P_3 = P_4 = 1$ which reduces Equation (3.8) to

$$at_1^3 + bt_1^2 + ct_1 + d = 1, \quad at_2^3 + bt_2^2 + ct_2 + d = 1,$$
$$at_3^3 + bt_3^2 + ct_3 + d = 1, \quad at_4^3 + bt_4^2 + ct_4 + d = 1.$$

Because the four values t_i are arbitrary, the four equations above can be written as the single equation $at^3 + bt^2 + ct + d = 1$, that holds for any t. Its solutions must therefore be $a = b = c = 0$ and $d = 1$.

Thus, we conclude that when all four values P_i are 1, a must be zero. In general, $a = c_{11}P_1 + c_{12}P_2 + c_{13}P_3 + c_{14}P_4$, which implies that $c_{11} + c_{12} + c_{13} + c_{14}$ must be zero. Similar arguments show that $c_{21} + c_{22} + c_{23} + c_{24} = 0$, $c_{31} + c_{32} + c_{33} + c_{34} = 0$, and $c_{41} + c_{42} + c_{43} + c_{44} = 1$. These relations, combined with Equation (3.9), show that the four $G_i(t)$ are barycentric.

To calculate the curve, we only need to calculate the four quantities **a**, **b**, **c**, and **d** (that constitute vector **A**), and write Equation (3.1) using the numerical values of **a**, **b**, **c**, and **d**.

Example: (This example is in two dimensions, each of the four points $\mathbf{P}_i$ along with the four coefficients $\mathbf{a}, \mathbf{b}, \mathbf{c}, \mathbf{d}$ form a pair. For three-dimensional curves the method is the same except that triplets are used instead of pairs.) Given the four two-dimensional points $\mathbf{P}_1 = (0,0)$, $\mathbf{P}_2 = (1,0)$, $\mathbf{P}_3 = (1,1)$, and $\mathbf{P}_4 = (0,1)$, we set up the equation

$$\begin{pmatrix} \mathbf{a} \\ \mathbf{b} \\ \mathbf{c} \\ \mathbf{d} \end{pmatrix} = \mathbf{A} = \mathbf{NP} = \begin{pmatrix} -4.5 & 13.5 & -13.5 & 4.5 \\ 9.0 & -22.5 & 18 & -4.5 \\ -5.5 & 9.0 & -4.5 & 1.0 \\ 1.0 & 0 & 0 & 0 \end{pmatrix} \begin{pmatrix} (0,0) \\ (1,0) \\ (1,1) \\ (0,1) \end{pmatrix}.$$

Its solutions are

$$\mathbf{a} = -4.5(0,0) + 13.5(1,0) - 13.5(1,1) + 4.5(0,1) = (0,-9),$$
$$\mathbf{b} = 19(0,0) - 22.5(1,0) + 18(1,1) - 4.5(0,1) = (-4.5, 13.5),$$
$$\mathbf{c} = -5.5(0,0) + 9(1,0) - 4.5(1,1) + 1(0,1) = (4.5, -3.5),$$
$$\mathbf{d} = 1(0,0) - 0(1,0) + 0(1,1) - 0(0,1) = (0,0).$$

So the curve $\mathbf{P}(t)$ that passes through the given points is

$$\mathbf{P}(t) = \mathbf{T}(t)\,\mathbf{A} = (0,-9)t^3 + (-4.5, 13.5)t^2 + (4.5, -3.5)t.$$

It is now easy to calculate and verify that $\mathbf{P}(0) = (0,0) = \mathbf{P}_1$, and

$$\mathbf{P}(1/3) = (0,-9)(1/27) + (-4.5, 13.5)(1/9) + (4.5, -3.5)(1/3) = (1,0) = \mathbf{P}_2,$$
$$\mathbf{P}(1) = (0,-9)1^3 + (-4.5, 13.5)1^2 + (4.5, -3.5)1 = (0,1) = \mathbf{P}_4.$$

◇ **Exercise 3.1:** Calculate $\mathbf{P}(2/3)$ and verify that it equals $\mathbf{P}_3$.

◇ **Exercise 3.2:** Imagine the circular arc of radius 1 in the first quadrant (a quarter circle). Write the coordinates of the four points that are equally spaced on this arc. Use the coordinates to calculate a PC approximating this arc. Calculate point $\mathbf{P}(1/2)$. How far does it deviate from the midpoint of the true quarter circle?

◇ **Exercise 3.3:** Calculate the PC that passes through the four points $\mathbf{P}_1$ through $\mathbf{P}_4$ assuming that only the three relative coordinates $\Delta_1 = \mathbf{P}_2 - \mathbf{P}_1$, $\Delta_2 = \mathbf{P}_3 - \mathbf{P}_2$, and $\Delta_3 = \mathbf{P}_4 - \mathbf{P}_3$ are given. Show a numeric example.

The main advantage of this method is its simplicity. Given the four points, it is easy to calculate the PC that passes through them. This, however, is also the reason for the downside of the method. It produces only *one* PC that passes through four given points. If that PC does not have the required shape, there is nothing the user can do. This simple curve method is not interactive.

Even though this method is not very useful for curve drawing, it may be useful for interpolation. Given two points $\mathbf{P}_1$ and $\mathbf{P}_2$, we know that the point midway between them is their average, $(\mathbf{P}_1 + \mathbf{P}_2)/2$. A natural question is: given four points $\mathbf{P}_1$ through $\mathbf{P}_4$, what point is located midway between them? We can answer this question by calculating the average, $(\mathbf{P}_1 + \mathbf{P}_2 + \mathbf{P}_3 + \mathbf{P}_4)/4$, but this weighted sum assigns the same weight to each of the four points. If we want to assign more weight to the interior points $\mathbf{P}_2$ and $\mathbf{P}_3$, we can calculate the PC that passes through the points and compute $\mathbf{P}(0.5)$ from Equation (3.6). The result is

$$\mathbf{P}(0.5) = -0.0625\mathbf{P}_1 + 0.5625\mathbf{P}_2 + 0.5625\mathbf{P}_3 - 0.0625\mathbf{P}_4.$$

This is a weighted sum that assigns more weight to the interior points. Notice that the weights are barycentric. Exercise 3.13 provides a hint as to why the two extreme weights are negative. This method can be extended to a two-dimensional grid of points (Section 3.6.1).

> A precisian professor had the habit of saying: "...quartic polynomial $ax^4 + bx^3 + cx^2 + dx + e$, where e need not be the base of the natural logarithms."
> —J. E. Littlewood, *A Mathematician's Miscellany*

◇ **Exercise 3.4:** The preceding method makes sense if the four points are (approximately) equally spaced along the curve. If they are not, the following approach may be taken. Instead of using 1/3 and 2/3 as the intermediate values, the user may specify values α and β, both in the interval $(0, 1)$, such that $\mathbf{P}_2 = \mathbf{P}(\alpha)$ and $\mathbf{P}_3 = \mathbf{P}(\beta)$. Generalize Equation (3.6) such that it depends on α and β.

3.2 The Lagrange Polynomial

The preceding section shows how a cubic interpolating polynomial can be derived for a set of four given points. This section discusses the Lagrange polynomial, a general approach to the problem of polynomial interpolation.

Given the $n + 1$ data points $\mathbf{P}_0 = (x_0, y_0)$, $\mathbf{P}_1 = (x_1, y_1), \ldots, \mathbf{P}_n = (x_n, y_n)$, the problem is to find a function $y = f(x)$ that will pass through all of them. We first try an expression of the form $y = \sum_{i=0}^{n} y_i L_i^n(x)$. This is a weighted sum of the individual y_i coordinates where the weights depend on the x_i coordinates. This sum will pass through the points if

$$L_i^n(x) = \begin{cases} 1, & x = x_i, \\ 0, & \text{otherwise.} \end{cases}$$

A good mathematician can easily guess that such functions are given by

$$L_i^n(x) = \frac{\Pi_{j \neq i}(x - x_j)}{\Pi_{j \neq i}(x_i - x_j)} = \frac{(x - x_0)(x - x_1) \cdots (x - x_{i-1})(x - x_{i+1})(x - x_n)}{(x_i - x_0) \cdots (x_i - x_{i-1})(x_i - x_{i+1}) \cdots (x_i - x_n)}.$$

(Note that $(x - x_i)$ is missing from the numerator and $(x_i - x_i)$ is missing from the denominator.) The function $y = \sum_{i=0}^{n} y_i L_i^n(x)$ is called the Lagrange polynomial because it was originally developed by Lagrange [Lagrange 77] and it is a polynomial of degree n. It is denoted by LP.

Horner's rule and the method of forward differences make polynomials very desirable to use. In practice, however, polynomials are used in parametric form as illustrated in Section 1.5, since any explicit function $y = f(x)$ is limited in the shapes of curves it can generate (note that the explicit form $y = \sum_{i=0}^{n} y_i L_i^n(x)$ of the LP cannot be calculated if two of the $n + 1$ given data points have the same x coordinate).

The LP has two properties that make it impractical for interactive curve design, it is of a high degree and it is unique.

1. Writing $P_n(x) = 0$ creates an equation of degree n in x. It has n solutions (some may be complex numbers), so when plotted as a curve it intercepts the x axis n times. For large n, such a curve may be loose because it tends to oscillate wildly. In practice, we normally prefer tight curves.

2. It is easy to show that the LP is unique (see below). There are infinitely many curves that pass through any given set of points and the one we are looking for may not be the LP. Any useful, practical mathematical method for curve design should make it easy for the designer to change the shape of the curve by varying the values of parameters.

It's easy to show that there is only one polynomial of degree n that passes through any given set of $n + 1$ points.

A root of the polynomial $P_n(x)$ is a value x_r such that $P_n(x_r) = 0$. A polynomial $P_n(x)$ can have at most n distinct roots (unless it is the zero polynomial). Suppose that there is another polynomial $Q_n(x)$ that passes through the same $n + 1$ data points. At the points, we would have $P_n(x_i) = Q_n(x_i) = y_i$ or $(P_n - Q_n)(x_i) = 0$. The difference $(P_n - Q_n)$ is a polynomial whose degree must be $\leq n$, so it cannot have more than n distinct roots. On the other hand, this difference is 0 at the $n + 1$ data points, so it has

$n+1$ roots. We conclude that it must be the zero polynomial, which implies that $P_n(x)$ and $Q_n(x)$ are identical.

This uniqueness theorem can also be employed to show that the Lagrange weights $L_i^n(x)$ are barycentric. Given a function $f(x)$, select $n+1$ distinct values x_0 through x_n, and consider the $n+1$ *support* points $(x_0, f(x_0))$ through $(x_n, f(x_n))$. The uniqueness theorem states that there is a unique polynomial $p(x)$ of degree n or less that passes through the points, i.e., $p(x_k) = f(x_k)$ for $k = 0, 1, \ldots, n$. We say that this polynomial interpolates the points. Now consider the constant function $f(x) \equiv 1$. The Lagrange polynomial that interpolates its points is

$$\mathrm{LP}(x) = \sum_{i=0}^{n} y_i L_i^n(x) = \sum_{i=0}^{n} 1 \times L_i^n(x) = \sum_{i=0}^{n} L_i^n(x).$$

On the other hand, $\mathrm{LP}(x)$ must be identical to 1, because $\mathrm{LP}(x_k) = f(x_k)$ and $f(x_k) = 1$ for any point x_k. Thus, we conclude that $\sum_{i=0}^{n} L_i^n(x) = 1$ for any x.

Because of these two properties, we conclude that a practical curve design method should be based on polynomials of low degree and should depend on parameters that control the shape of the curve. Such methods are discussed in the chapters that follow. Still, polynomial interpolation may be useful in special situations, which is why it is discussed in the remainder of this chapter.

⋄ **Exercise 3.5:** Calculate the LP between the two points $\mathbf{P}_0 = (x_0, y_0)$ and $\mathbf{P}_1 = (x_1, y_1)$. What kind of a curve is it?

> I have another method not yet communicated... a convenient, rapid and general solution of this problem, *To draw a geometrical curve which shall pass through any number of given points...* These things are done at once geometrically with no calculation intervening... Though at first glance it looks unmanageable, yet the matter turns out otherwise. For it ranks among the most beautiful of all that I could wish to solve.
> (Isaac Newton in a letter to Henry Oldenburg, October 24, 1676, quoted in [Turnbull 59], vol. II, p 188.)
>
> —James Gleick, *Isaac Newton* (2003).

The LP can also be expressed in parametric form. Given the $n+1$ data points $\mathbf{P}_0, \mathbf{P}_1, \ldots, \mathbf{P}_n$, we need to construct a polynomial $\mathbf{P}(t)$ that passes through all of them, such that $\mathbf{P}(t_0) = \mathbf{P}_0, \mathbf{P}(t_1) = \mathbf{P}_1, \ldots, \mathbf{P}(t_n) = \mathbf{P}_n$, where $t_0 = 0$, $t_n = 1$, and t_1 through t_{n-1} are certain values between 0 and 1 (the t_i are called *knot* values). The LP has the form $\mathbf{P}(t) = \sum_{i=0}^{n} \mathbf{P}_i L_i^n(t)$. This is a weighted sum of the individual points where the weights (or basis functions) are given by

$$L_i^n(t) = \frac{\Pi_{j \neq i}^{n}(t - t_j)}{\Pi_{j \neq i}^{n}(t_i - t_j)}. \tag{3.10}$$

Note that $\sum_{i=0}^{n} L_i^n(t) = 1$, so these weights are barycentric.

⋄ **Exercise 3.6:** Calculate the parametric LP between the two general points $\mathbf{P}_0$ and $\mathbf{P}_1$.

⋄ **Exercise 3.7:** Calculate the parametric LP for the three points $\mathbf{P}_0 = (0,0)$, $\mathbf{P}_1 = (0,1)$, and $\mathbf{P}_2 = (1,1)$.

⋄ **Exercise 3.8:** Calculate the parametric LP for the four equally-spaced points $\mathbf{P}_1$, $\mathbf{P}_2$, $\mathbf{P}_3$, and $\mathbf{P}_4$ and show that it is identical to the interpolating PC given by Equation (3.4).

The parametric LP is also mentioned on page 109, in connection with Gordon surfaces.

The LP has another disadvantage. If the resulting curve is not satisfactory, the user may want to fine-tune it by adding one more point. However, all the basis functions $L_i^n(t)$ will have to be recalculated in such a case, since they also depend on the points, not only on the knot values. This disadvantage makes the LP slow to use in practice, which is why the Newton polynomial (Section 3.3) is sometimes used instead.

3.2.1 The Quadratic Lagrange Polynomial

Equation (3.10) can easily be employed to obtain the Lagrange polynomial for three points $\mathbf{P}_0$, $\mathbf{P}_1$, and $\mathbf{P}_2$. The weights in this case are

$$L_0^2(t) = \frac{\prod_{j\neq0}^{2}(t - t_j)}{\prod_{j\neq0}^{2}(t_0 - t_j)} = \frac{(t - t_1)(t - t_2)}{(t_0 - t_1)(t_0 - t_2)},$$

$$L_1^2(t) = \frac{\prod_{j\neq1}^{2}(t - t_j)}{\prod_{j\neq1}^{2}(t_1 - t_j)} = \frac{(t - t_0)(t - t_2)}{(t_1 - t_0)(t_1 - t_2)}, \qquad (3.11)$$

$$L_2^2(t) = \frac{\prod_{j\neq2}^{2}(t - t_j)}{\prod_{j\neq2}^{2}(t_2 - t_j)} = \frac{(t - t_0)(t - t_1)}{(t_2 - t_0)(t_2 - t_1)},$$

and the polynomial $\mathbf{P}_2(t) = \sum_{i=0}^{2} \mathbf{P}_i L_i^2(t)$ is easy to calculate once the values of t_0, t_1, and t_2 have been determined.

The Uniform Quadratic Lagrange Polynomial is obtained when $t_0 = 0$, $t_1 = 1$, and $t_2 = 2$. (See discussion of uniform and nonuniform parametric curves in Section 1.4.1.) Equation (3.11) yields

$$\mathbf{P}_{2u}(t) = \frac{t^2 - 3t + 2}{2}\mathbf{P}_0 - (t^2 - 2t)\mathbf{P}_1 + \frac{t^2 - t}{2}\mathbf{P}_2$$

$$= (t^2, t, 1) \begin{pmatrix} 1/2 & -1 & 1/2 \\ -3/2 & 2 & -1/2 \\ 1 & 0 & 0 \end{pmatrix} \begin{pmatrix} \mathbf{P}_0 \\ \mathbf{P}_1 \\ \mathbf{P}_2 \end{pmatrix}. \qquad (3.12)$$

The sums of three rows of the matrix of Equation (3.12) are (from top to bottom) 0, 0, and 1, showing that the three basis functions are barycentric, as they should be.

The Nonuniform Quadratic Lagrange Polynomial is obtained when $t_0 = 0$, $t_1 = t_0 + \Delta_0 = \Delta_0$, and $t_2 = t_1 + \Delta_1 = \Delta_0 + \Delta_1$ for some positive Δ_0 and Δ_1. Equation (3.11) gives

$$L_0^2(t) = \frac{(t - \Delta_0)(t - \Delta_0 - \Delta_1)}{(-\Delta_0)(-\Delta_0 - \Delta_1)}, \, L_1^2(t) = \frac{(t - 0)(t - \Delta_0 - \Delta_1)}{\Delta_0(-\Delta_1)}, \, L_2^2(t) = \frac{(t - 0)(t - \Delta_0)}{(\Delta_0 + \Delta_1)\Delta_1},$$

and the nonuniform polynomial is

$$\mathbf{P}_{2nu}(t) = (t^2, t, 1) \begin{bmatrix} \dfrac{1}{\Delta_0(\Delta_0 + \Delta_1)} & -\dfrac{1}{\Delta_0\Delta_1} & \dfrac{1}{(\Delta_0 + \Delta_1)\Delta_1} \\ \dfrac{-1}{\Delta_0 + \Delta_1} - \dfrac{1}{\Delta_0} & \dfrac{1}{\Delta_0} + \dfrac{1}{\Delta_1} & -\dfrac{1}{\Delta_1} + \dfrac{1}{\Delta_0 + \Delta_1} \\ 1 & 0 & 0 \end{bmatrix} \begin{bmatrix} \mathbf{P}_0 \\ \mathbf{P}_1 \\ \mathbf{P}_2 \end{bmatrix}. \quad (3.13)$$

For $\Delta_0 = \Delta_1 = 1$, Equation (3.13) reduces to the uniform polynomial, Equation (3.12). For $\Delta_0 = \Delta_1 = 1/2$, the parameter t varies in the "standard" range $[0, 1]$ and Equation (3.13) becomes

$$\mathbf{P}_{2std}(t) = (t^2, t, 1) \begin{pmatrix} 2 & -4 & 2 \\ -3 & 4 & -1 \\ 1 & 0 & 0 \end{pmatrix} \begin{pmatrix} \mathbf{P}_0 \\ \mathbf{P}_1 \\ \mathbf{P}_2 \end{pmatrix}. \quad (3.14)$$

(Notice that the three rows again sum to 0, 0, and 1, to produce three barycentric basis functions.) In most cases, Δ_0 and Δ_1 should be set to the chord lengths $|\mathbf{P}_1 - \mathbf{P}_0|$ and $|\mathbf{P}_2 - \mathbf{P}_1|$, respectively.

⋄ **Exercise 3.9:** Use Cartesian product to generalize Equation (3.14) to a surface patch that passes through nine given points.

    Example: The three points $\mathbf{P}_0 = (1, 0)$, $\mathbf{P}_1 = (1.3, .5)$, and $\mathbf{P}_2 = (4, 0)$ are given. The uniform LP is obtained when $\Delta_0 = \Delta_1 = 1$ and it equals

$$\mathbf{P}_{2u}(t) = \left(1 - 0.9t + 1.2t^2, 0.5(2 - t)t\right).$$

Many nonuniform polynomials are possible. We select the one that's obtained when the Δ values are the chord lengths between the points. In our case, they are $\Delta_0 = |\mathbf{P}_1 - \mathbf{P}_0| \approx 0.583$ and $\Delta_1 = |\mathbf{P}_2 - \mathbf{P}_1| \approx 2.75$. This polynomial is

$$\mathbf{P}_{2nu}(t) = (1 + 0.433t + 0.14t^2, 1.04t - 0.312t^2).$$

These uniform and nonuniform polynomials are shown in Figure 3.1. The figure illustrates how the nonuniform curve based on the chord lengths between the points is tighter (features smaller overall curvature). Such a curve is generally considered a better interpolation of the three points.

    Figure 3.2 shows three examples of nonuniform Lagrange polynomials that pass through the three points $\mathbf{P}_0 = (1, 1)$, $\mathbf{P}_1 = (2, 2)$, and $\mathbf{P}_2 = (4, 0)$. The value of Δ_0 is 1.414, the chord length between $\mathbf{P}_0$ and $\mathbf{P}_1$. The chord length between $\mathbf{P}_1$ and $\mathbf{P}_2$ is 2.83 and Δ_1 is first assigned this value, then half this value, and finally twice it. The three

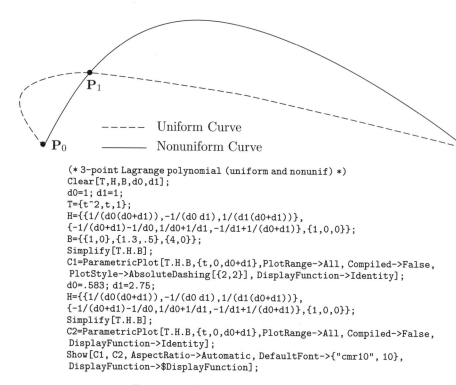

```
(* 3-point Lagrange polynomial (uniform and nonunif) *)
Clear[T,H,B,d0,d1];
d0=1; d1=1;
T={t^2,t,1};
H={{1/(d0(d0+d1)),-1/(d0 d1),1/(d1(d0+d1))},
{-1/(d0+d1)-1/d0,1/d0+1/d1,-1/d1+1/(d0+d1)},{1,0,0}};
B={{1,0},{1.3,.5},{4,0}};
Simplify[T.H.B];
C1=ParametricPlot[T.H.B,{t,0,d0+d1},PlotRange->All, Compiled->False,
 PlotStyle->AbsoluteDashing[{2,2}], DisplayFunction->Identity];
d0=.583; d1=2.75;
H={{1/(d0(d0+d1)),-1/(d0 d1),1/(d1(d0+d1))},
{-1/(d0+d1)-1/d0,1/d0+1/d1,-1/d1+1/(d0+d1)},{1,0,0}};
Simplify[T.H.B];
C2=ParametricPlot[T.H.B,{t,0,d0+d1},PlotRange->All, Compiled->False,
 DisplayFunction->Identity];
Show[C1, C2, AspectRatio->Automatic, DefaultFont->{"cmr10", 10},
 DisplayFunction->$DisplayFunction];
```

Figure 3.1: Three-Point Lagrange Polynomials.

resulting curves illustrate how the Lagrange polynomial can be reshaped by modifying the Δ_i parameters. The three polynomials in this case are

$$(1 + 0.354231t + 0.249634t^2, 1 + 1.76716t - 0.749608t^2),$$
$$(1 + 0.70738t - 0.000117766t^2, 1 + 1.1783t - 0.333159t^2),$$
$$(1 + 0.777945t - 0.0500221t^2, 1 + 0.919208t - 0.149925t^2).$$

3.2.2 The Cubic Lagrange Polynomial

Equation (3.10) is now applied to the cubic Lagrange polynomial that interpolates the four points $\mathbf{P}_0$, $\mathbf{P}_1$, $\mathbf{P}_2$, and $\mathbf{P}_3$. The weights in this case are

$$L_0^3(t) = \frac{\prod_{j\neq0}^3(t - t_j)}{\prod_{j\neq0}^3(t_0 - t_j)} = \frac{(t - t_1)(t - t_2)(t - t_3)}{(t_0 - t_1)(t_0 - t_2)(t_0 - t_3)},$$

$$L_1^3(t) = \frac{\prod_{j\neq1}^3(t - t_j)}{\prod_{j\neq1}^3(t_1 - t_j)} = \frac{(t - t_0)(t - t_2)(t - t_3)}{(t_1 - t_0)(t_1 - t_2)(t_1 - t_3)},$$

$$L_2^3(t) = \frac{\prod_{j\neq2}^3(t - t_j)}{\prod_{j\neq2}^3(t_2 - t_j)} = \frac{(t - t_0)(t - t_1)(t - t_3)}{(t_2 - t_0)(t_2 - t_1)(t_2 - t_3)},$$

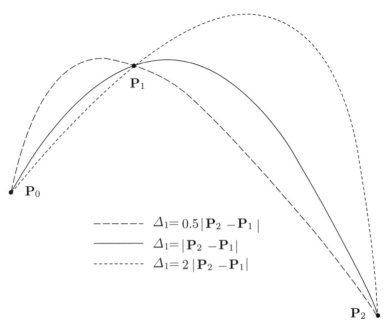

$$\Delta_1 = 0.5\,|\mathbf{P}_2 - \mathbf{P}_1|$$
$$\Delta_1 = |\mathbf{P}_2 - \mathbf{P}_1|$$
$$\Delta_1 = 2\,|\mathbf{P}_2 - \mathbf{P}_1|$$

```
(* 3-point Lagrange polynomial (3 examples of nonuniform) *)
Clear[T,H,B,d0,d1,C1,C2,C3];
d0=1.414; d1=1.415; (* d1=0.5|P2-P1| *)
T={t^2,t,1};
H={{1/(d0(d0+d1)),-1/(d0 d1),1/(d1(d0+d1))},
{-1/(d0+d1)-1/d0,1/d0+1/d1,-1/d1+1/(d0+d1)},{1,0,0}};
B={{1,1},{2,2},{4,0}};
Simplify[T.H.B]
C1=ParametricPlot[T.H.B,{t,0,d0+d1},PlotRange->All, Compiled->False,
DisplayFunction->Identity];
d1=2.83; (* d1=|P2-P1| *)
H={{1/(d0(d0+d1)),-1/(d0 d1),1/(d1(d0+d1))},
{-1/(d0+d1)-1/d0,1/d0+1/d1,-1/d1+1/(d0+d1)},{1,0,0}};
Simplify[T.H.B]
C2=ParametricPlot[T.H.B,{t,0,d0+d1},PlotRange->All, Compiled->False,
DisplayFunction->Identity];
d1=5.66; (* d1=2|P2-P1| *)
H={{1/(d0(d0+d1)),-1/(d0 d1),1/(d1(d0+d1))},
{-1/(d0+d1)-1/d0,1/d0+1/d1,-1/d1+1/(d0+d1)},{1,0,0}};
Simplify[T.H.B]
C3=ParametricPlot[T.H.B,{t,0,d0+d1},PlotRange->All, Compiled->False,
DisplayFunction->Identity];
Show[C1,C2,C3, AspectRatio->Automatic, DefaultFont->{"cmr10", 10},
DisplayFunction->$DisplayFunction];
(* (1/24,-1/8)t^3+(-1/3,3/4)t^2+(1,-1)t *)
```

Figure 3.2: Three-Point Nonuniform Lagrange Polynomials.

$$L_3^3(t) = \frac{\prod_{j\neq 3}^3 (t - t_j)}{\prod_{j\neq 3}^3 (t_3 - t_j)} = \frac{(t - t_0)(t - t_1)(t - t_2)}{(t_3 - t_0)(t_3 - t_1)(t_3 - t_2)}, \tag{3.15}$$

and the polynomial $\mathbf{P}_3(t) = \sum_{i=0}^3 \mathbf{P}_i L_i^3(t)$ is easy to calculate once the values of t_0, t_1, t_2, and t_3 have been determined.

The Nonuniform Cubic Lagrange Polynomial is obtained when $t_0 = 0$, $t_1 = t_0 + \Delta_0 = \Delta_0$, $t_2 = t_1 + \Delta_1 = \Delta_0 + \Delta_1$, and $t_3 = t_2 + \Delta_2 = \Delta_0 + \Delta_1 + \Delta_2$ for positive Δ_i. The expression for the polynomial is

$$\mathbf{P}_{3nu}(t) = (t^3, t^2, t, 1)\mathbf{Q}\begin{pmatrix} \mathbf{P}_0 \\ \mathbf{P}_1 \\ \mathbf{P}_2 \\ \mathbf{P}_3 \end{pmatrix}, \tag{3.16}$$

where $\mathbf{Q}$ is the matrix

$$\mathbf{Q} = \begin{pmatrix}
\frac{1}{(-\Delta_0)(-\Delta_0-\Delta_1)(-\Delta_0-\Delta_1-\Delta_2)} & \frac{1}{\Delta_0(-\Delta_1)(-\Delta_1-\Delta_2)} \\[2mm]
-\frac{3\Delta_0+2\Delta_1+\Delta_2}{(-\Delta_0)(-\Delta_0-\Delta_1)(-\Delta_0-\Delta_1-\Delta_2)} & -\frac{2\Delta_0+2\Delta_1+\Delta_2}{\Delta_0(-\Delta_1)(-\Delta_1-\Delta_2)} \\[2mm]
\frac{\Delta_0(\Delta_0+\Delta_1)+(\Delta_0+\Delta_1)(\Delta_0+\Delta_1+\Delta_2)+(\Delta_0+\Delta_1+\Delta_2)\Delta_0}{(-\Delta_0)(-\Delta_0-\Delta_1)(-\Delta_0-\Delta_1-\Delta_2)} & \frac{(\Delta_0+\Delta_1)(\Delta_0+\Delta_1+\Delta_2)}{\Delta_0(-\Delta_1)(-\Delta_1-\Delta_2)} \\[2mm]
-\frac{\Delta_0(\Delta_0+\Delta_1)(\Delta_0+\Delta_1+\Delta_2)}{(-\Delta_0)(-\Delta_0-\Delta_1)(-\Delta_0-\Delta_1-\Delta_2)} & 0
\end{pmatrix}$$

$$\begin{pmatrix}
\frac{1}{(\Delta_0+\Delta_1)\Delta_1(-\Delta_2)} & \frac{1}{(\Delta_0+\Delta_1+\Delta_2)(\Delta_1+\Delta_2)\Delta_2} \\[2mm]
-\frac{2\Delta_0+\Delta_1+\Delta_2}{(\Delta_0+\Delta_1)\Delta_1(-\Delta_2)} & -\frac{2\Delta_0+\Delta_1}{(\Delta_0+\Delta_1+\Delta_2)(\Delta_1+\Delta_2)\Delta_2} \\[2mm]
\frac{\Delta_0(\Delta_0+\Delta_1+\Delta_2)}{(\Delta_0+\Delta_1)\Delta_1(-\Delta_2)} & \frac{\Delta_0(\Delta_0+\Delta_1)}{(\Delta_0+\Delta_1+\Delta_2)(\Delta_1+\Delta_2)\Delta_2} \\[2mm]
0 & 0
\end{pmatrix}.$$

The Uniform Cubic Lagrange Polynomial. We construct the "standard" case, where t varies from 0 to 1. This implies $t_0 = 0$, $t_1 = 1/3$, $t_2 = 2/3$, and $t_3 = 1$. Equation (3.16) reduces to

$$\mathbf{P}_{3u}(t) = (t^3, t^2, t, 1)\begin{pmatrix} -9/2 & 27/2 & -27/2 & 9/2 \\ 9 & -45/2 & 18 & -9/2 \\ -11/2 & 9 & -9/2 & 1 \\ 1 & 0 & 0 & 0 \end{pmatrix}\begin{pmatrix} \mathbf{P}_0 \\ \mathbf{P}_1 \\ \mathbf{P}_2 \\ \mathbf{P}_3 \end{pmatrix}. \tag{3.17}$$

Figure 3.3 shows the quadratic and cubic Lagrange basis functions. It is easy to see that there are values of t (indicated by arrows) for which one of the basis functions is 1 and the others are zeros. This is how the curve (which is a weighted sum of the functions) passes through a point. The functions add up to 1, but most climb above 1 and are negative in certain regions. In the nonuniform case, the particular choice of the various Δ_i reshapes the basis functions in such a way that a function still retains its basic shape, but its areas above and below the t axis may increase or decrease significantly. Those willing to experiment can copy Matrix $\mathbf{Q}$ of Equation (3.16) into

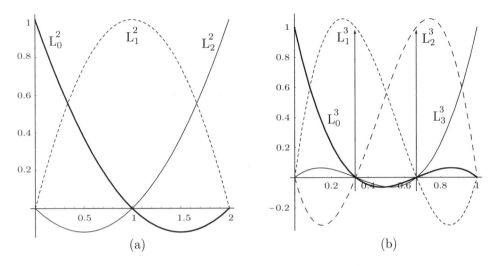

```
(* Plot quadratic and cubic Lagrange basis functions *)
lagq={t^2,t,1}.{{1/2,-1,1/2},{-3/2,2,-1/2},{1,0,0}};
Plot[{lagq[[1]],lagq[[2]],lagq[[3]]},{t,0,2},
  PlotRange->All, AspectRatio->Automatic, DefaultFont->{"cmr10", 10}];
lagc={t^3,t^2,t,1}.{{-9/2,27/2,-27/2,9/2},{9,-45/2,18,-9/2},
  2{-11/2,9,-9/2,1},{1,0,0,0}}
Plot[{lagc[[1]],lagc[[2]],lagc[[3]],lagc[[4]]},{t,0,1},
  PlotRange->All, AspectRatio->Automatic, DefaultFont->{"cmr10", 10}];
```

Figure 3.3: (a) Quadratic and (b) Cubic Lagrange Basis Functions.

appropriate mathematical software and use code similar to that of Figure 3.3 to plot the basis functions for various values of Δ_i.

It should be noted that the basis functions of the Bézier curve (Section 6.2) are more intuitive and provide easier control of the shape of the curve, which is why Lagrange interpolation is not popular and is used in special cases only.

3.2.3 Barycentric Lagrange Interpolation

Given the $n+1$ data points $\mathbf{P}_0 = (x_0, y_0)$ through $\mathbf{P}_n = (x_n, y_n)$, the explicit (nonparametric) Lagrange polynomial that interpolates them is $\text{LP}(x) = \sum_{i=0}^{n} y_i L_i^n(x)$, where

$$L_i^n(x) = \frac{\Pi_{j \neq i}^n (x - x_j)}{\Pi_{j \neq i}^n (x_i - x_j)} = \frac{(x - x_0)(x - x_1) \cdots (x - x_{i-1})(x - x_{i+1})(x - x_n)}{(x_i - x_0) \cdots (x_i - x_{i-1})(x_i - x_{i+1}) \cdots (x_i - x_n)}.$$

This representation of the Lagrange polynomial has the following disadvantages:

1. The denominator of $L_i^n(x)$ requires n subtractions and $n-1$ multiplications, for a total of $O(n)$ operations. The denominators of the $n+1$ weights therefore require $O(n^2)$ operations. The numerators also require $O(n^2)$ operations, but have to be recomputed for each value of x.

2. Adding a new point $\mathbf{P}_{n+1}$ requires the computation of a new weight $L_{n+1}^{n+1}(x)$

and a recomputation of all the original weights $L_i^n(x)$, because

$$L_i^{n+1}(x) = L_i^n(x) \frac{x - x_{n+1}}{x_i - x_{n+1}}, \quad \text{for} \quad i = 0, 1, \dots, n.$$

3. The computations are numerically unstable. A small change in any of the data points may cause a large change in $\text{LP}(x)$.

Numerical analysts have long believed that these reasons make the Newton polynomial (Section 3.3) more attractive for practical work. However, recent research has resulted in a new, barycentric form of the LP, that makes Lagrange interpolation more attractive. This section is based on [Berrut and Trefethen 04].

The barycentric form of the LP is

$$
\begin{aligned}
\text{LP}(x) &= \sum_{i=0}^{n} y_i L_i^n(x) = \sum_{i=0}^{n} y_i \frac{\Pi_{j \neq i}^n (x - x_j)}{\Pi_{j \neq i}^n (x_i - x_j)} \\
&= \sum_{i=0}^{n} y_i \frac{w_i}{x - x_i} \left[\Pi_{j=0}^n (x - x_j) \right] = \Pi_{j=0}^n (x - x_j) \sum_{i=0}^{n} y_i \frac{w_i}{x - x_i} \\
&= L(x) \sum_{i=0}^{n} y_i \frac{w_i}{x - x_i},
\end{aligned}
\tag{3.18}
$$

where

$$w_i = \frac{1}{\Pi_{j \neq i}^n (x_i - x_j)}, \quad \text{for} \quad i = 0, 1, \dots, n.$$

Each weight w_i requires $O(n)$ operations, for a total of $O(n^2)$, but these weights no longer depend on x and consequently have to be computed just once! The only quantity that depends on x is $L(x)$ and it requires only $O(n)$ operations. Also, when a new point is added, the only operations required are (1) divide each w_i by $(x_i - x_{n+1})$ and (2) compute w_{n+1}. These require $O(n)$ steps.

A better form of Equation (3.18), one that's more numerically stable, is obtained when we consider the case $y_i = 1$. If all the data points are of the form $(x_i, 1)$, then the interpolating LP should satisfy $\text{LP}(x) \equiv 1$, which brings Equation (3.18) to the form

$$1 = L(x) \sum_{i=0}^{n} \frac{w_i}{x - x_i}, \tag{3.19}$$

We can now divide Equation (3.18) by Equation (3.19) to obtain

$$\text{LP}(x) = \left[\sum_{i=0}^{n} y_i \frac{w_i}{x - x_i} \right] \bigg/ \left[\sum_{j=0}^{n} \frac{w_j}{x - x_j} \right]. \tag{3.20}$$

The weights of Equation (3.20) are

$$\frac{\dfrac{w_i}{x - x_i}}{\sum_{j=0}^{n} w_j / (x - x_j)}, \quad i = 0, 1, \dots, n,$$

and it's easy to see that they are barycentric. Also, any common factors in the weights can now be cancelled out. For example, it can be shown that in the case of data points that are uniformly distributed in the interval $[-1, +1]$

$$\mathbf{P}_0 = (-1, y_0), \quad \mathbf{P}_1 = (-1 + h, y_1), \quad \mathbf{P}_2 = (-1 + 2h, y_2), \ldots, \mathbf{P}_n = (+1, y_n)$$

(where $h = 2/n$), the weights become $w_i = (-1)^{n-i} \binom{n}{i} / (h^n n!)$. The common factors are those that do not depend on i. When they are cancelled out, the weights become the simple expressions

$$w_i = (-1)^i \binom{n}{i}.$$

(This is also true for points that are equidistant in any interval $[a, b]$. Incidentally, it can be shown that the case of equidistant data points is ill conditioned and the LP, in any form, can change its value wildly in response to even small changes in the data points.)

3.3 The Newton Polynomial

The Newton polynomial offers an alternative approach to the problem of polynomial interpolation. The final interpolating polynomial is identical to the LP, but the derivation is different. It allows the user to easily add more points and thereby provide fine control over the shape of the curve. We again assume that $n + 1$ data points $\mathbf{P}_0, \mathbf{P}_1, \ldots, \mathbf{P}_n$ are given and are assigned knot values

$$t_0 = 0 < t_1 < \cdots < t_{n-1} < t_n = 1.$$

We are looking for a curve expressed by the degree-n parametric polynomial

$$\mathbf{P}(t) = \sum_{i=0}^{n} N_i(t) \mathbf{A}_i,$$

where the basis functions $N_i(t)$ depend only on the knot values and not on the data points. Only the (unknown) coefficients $\mathbf{A}_i$ depend on the points. This definition (originally proposed by Newton) is useful because each coefficient $\mathbf{A}_i$ depends only on points $\mathbf{P}_0$ through $\mathbf{P}_i$. If the user decides to add a point $\mathbf{P}_{n+1}$, only one coefficient, $\mathbf{A}_{n+1}$, and one basis function, $N_{n+1}(t)$, need be recomputed.

The definition of the basis functions is

$$N_0(t) = 1 \quad \text{and} \quad N_i(t) = (t - t_0)(t - t_1) \cdots (t - t_{i-1}), \quad \text{for} \quad i = 1, \ldots, n.$$

To calculate the unknown coefficients, we write the equations

$$\mathbf{P}_0 = \mathbf{P}(t_0) = \mathbf{A}_0,$$
$$\mathbf{P}_1 = \mathbf{P}(t_1) = \mathbf{A}_0 + \mathbf{A}_1(t_1 - t_0),$$
$$\mathbf{P}_2 = \mathbf{P}(t_2) = \mathbf{A}_0 + \mathbf{A}_1(t_2 - t_0) + \mathbf{A}_2(t_2 - t_0)(t_2 - t_1),$$
$$\vdots$$
$$\mathbf{P}_n = \mathbf{P}(t_n) = \mathbf{A}_0 + \cdots.$$

These equations don't have to be solved simultaneously. Each can easily be solved after all its predecessors have been solved. The solutions are

$$\mathbf{A}_0 = \mathbf{P}_0,$$

$$\mathbf{A}_1 = \frac{\mathbf{P}_1 - \mathbf{P}_0}{t_1 - t_0},$$

$$\mathbf{A}_2 = \frac{\mathbf{P}_2 - \mathbf{P}_0 - \dfrac{(\mathbf{P}_1 - \mathbf{P}_0)(t_2 - t_0)}{t_1 - t_0}}{(t_2 - t_0)(t_2 - t_1)} = \frac{\dfrac{\mathbf{P}_2 - \mathbf{P}_1}{t_2 - t_1} - \dfrac{\mathbf{P}_1 - \mathbf{P}_0}{t_1 - t_0}}{t_2 - t_0}.$$

This obviously gets very complicated quickly, so we use the method of *divided differences* to express all the solutions in compact notation. The divided difference of the knots $t_i t_k$ is denoted $[t_i t_k]$ and is defined as

$$[t_i t_k] \stackrel{\text{def}}{=} \frac{\mathbf{P}_i - \mathbf{P}_k}{t_i - t_k}.$$

The solutions can now be expressed as

$$\mathbf{A}_0 = \mathbf{P}_0,$$

$$\mathbf{A}_1 = \frac{\mathbf{P}_1 - \mathbf{P}_0}{t_1 - t_0} = [t_1 t_0],$$

$$\mathbf{A}_2 = [t_2 t_1 t_0] = \frac{[t_2 t_1] - [t_1 t_0]}{t_2 - t_0},$$

$$\mathbf{A}_3 = [t_3 t_2 t_1 t_0] = \frac{[t_3 t_2 t_1] - [t_2 t_1 t_0]}{t_3 - t_0},$$

$$\vdots$$

$$\mathbf{A}_n = [t_n \ldots t_1 t_0] = \frac{[t_n \ldots t_1] - [t_{n-1} \ldots t_0]}{t_n - t_0}.$$

◇ **Exercise 3.10:** Given the same points and knot values as in Exercise 3.7, calculate the Newton polynomial that passes through the points.

◇ **Exercise 3.11:** The tangent vector to a curve $\mathbf{P}(t)$ is the derivative $\frac{d\mathbf{P}(t)}{dt}$, which we denote by $\mathbf{P}^t(t)$. Calculate the tangent vectors to the curve of Exercises 3.7 and 3.10 at the three points. Also calculate the slopes of the curve at the points.

3.4 Polynomial Surfaces

The polynomial $y = \sum a_i x^i$ is the explicit representation of a curve. Similarly, the parametric polynomial $\mathbf{P}(t) = \sum t^i \mathbf{P}_i$ and also $\mathbf{P}(t) = \sum a_i(t)\mathbf{P}_i$ (where $a_i(t)$ is a polynomial in t) are parametric representations of curves. These expressions can be extended to polynomials in two variables, which represent surfaces. Thus, the double polynomial $z = \sum_i \sum_j a_{ij} x^i y^j$ is the explicit representation of a surface patch, because it yields a z value for any pair of coordinates (x, y). Similarly, the double parametric polynomial $\mathbf{P}(u, w) = \sum_i \sum_j u^i w^j \mathbf{P}_{ij}$ is the parametric representation of a surface patch. For the cubic case (polynomials of degree 3), such a double polynomial can be expressed compactly in matrix notation

$$
\mathbf{P}(u, w) = [u^3, u^2, u, 1]\mathbf{N}
\begin{bmatrix}
\mathbf{P}_{33} & \mathbf{P}_{32} & \mathbf{P}_{31} & \mathbf{P}_{30} \\
\mathbf{P}_{23} & \mathbf{P}_{22} & \mathbf{P}_{21} & \mathbf{P}_{20} \\
\mathbf{P}_{13} & \mathbf{P}_{12} & \mathbf{P}_{11} & \mathbf{P}_{10} \\
\mathbf{P}_{03} & \mathbf{P}_{02} & \mathbf{P}_{01} & \mathbf{P}_{00}
\end{bmatrix}
\mathbf{N}^T
\begin{bmatrix}
w^3 \\
w^2 \\
w \\
1
\end{bmatrix}.
\tag{3.21}
$$

The corresponding surface patch is accordingly referred to as bicubic.

3.5 The Biquadratic Surface Patch

This section introduces the biquadratic surface patch and constructs this simple surface as a Cartesian product. Given the two quadratic (degree 2) polynomials

$$
\mathbf{Q}(u) = \sum_{i=0}^{2} f_i(u)\mathbf{Q}_i \quad \text{and} \quad \mathbf{R}(w) = \sum_{j=0}^{2} g_j(w)\mathbf{R}_j
$$

the biquadratic surface immediately follows from the principle of Cartesian product

$$
\mathbf{P}(u, w) = \sum_{i=0}^{2} \sum_{j=0}^{2} f_i(u)g_j(w)\mathbf{P}_{ij}.
\tag{3.22}
$$

Different constructions are possible depending on the geometric meaning of the nine quantities $\mathbf{P}_{ij}$. The following section presents such a construction and Section 4.10 discusses another approach, based on points, tangent vectors, and twist vectors.

3.5.1 Nine Points

Equation (3.14), duplicated below, gives the quadratic standard Lagrange polynomial that interpolates three given points:

$$
\mathbf{P}_{2std}(t) = (t^2, t, 1)
\begin{pmatrix}
2 & -4 & 2 \\
-3 & 4 & -1 \\
1 & 0 & 0
\end{pmatrix}
\begin{pmatrix}
\mathbf{P}_0 \\
\mathbf{P}_1 \\
\mathbf{P}_2
\end{pmatrix}.
\tag{3.14}
$$

Cartesian product yields the corresponding biquadratic surface

$$
\mathbf{P}(u, w) = (u^2, u, 1) \begin{pmatrix} 2 & -4 & 2 \\ -3 & 4 & -1 \\ 1 & 0 & 0 \end{pmatrix} \begin{pmatrix} \mathbf{P}_{22} & \mathbf{P}_{21} & \mathbf{P}_{20} \\ \mathbf{P}_{12} & \mathbf{P}_{11} & \mathbf{P}_{10} \\ \mathbf{P}_{02} & \mathbf{P}_{01} & \mathbf{P}_{00} \end{pmatrix}
$$
$$
\times \begin{pmatrix} 2 & -4 & 2 \\ -3 & 4 & -1 \\ 1 & 0 & 0 \end{pmatrix}^T \begin{pmatrix} w^2 \\ w \\ 1 \end{pmatrix},
$$

(3.23)

where the nine quantities $\mathbf{P}_{ij}$ are points defining this surface patch. They should be roughly equally spaced over the surface.

Example: Given the nine points of Figure 3.4a, we compute and draw the biquadratic surface patch defined by them. The surface is shown in Figure 3.4b. The code is also listed.

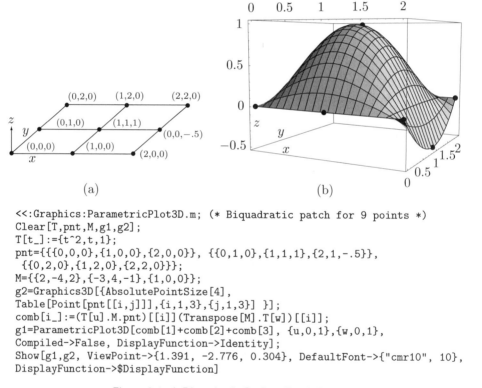

(a)   (b)

```
<<:Graphics:ParametricPlot3D.m; (* Biquadratic patch for 9 points *)
Clear[T,pnt,M,g1,g2];
T[t_]:={t^2,t,1};
pnt={{{0,0,0},{1,0,0},{2,0,0}}, {{0,1,0},{1,1,1},{2,1,-.5}},
 {{0,2,0},{1,2,0},{2,2,0}}};
M={{2,-4,2},{-3,4,-1},{1,0,0}};
g2=Graphics3D[{AbsolutePointSize[4],
Table[Point[pnt[[i,j]]],{i,1,3},{j,1,3}] }];
comb[i_]:=(T[u].M.pnt)[[i]](Transpose[M].T[w])[[i]];
g1=ParametricPlot3D[comb[1]+comb[2]+comb[3], {u,0,1},{w,0,1},
Compiled->False, DisplayFunction->Identity];
Show[g1,g2, ViewPoint->{1.391, -2.776, 0.304}, DefaultFont->{"cmr10", 10},
DisplayFunction->$DisplayFunction]
```

Figure 3.4: A Biquadratic Surface Patch Example.

It is also possible to construct similar biquadratic surfaces from the expressions for the uniform and nonuniform quadratic Lagrange polynomials, Equations (3.12) and (3.13).

◇ **Exercise 3.12:** The geometry vector of Equation (3.14) has point $\mathbf{P}_0$ at the top, but the geometry matrix of Equation (3.23) has point $\mathbf{P}_{00}$ at its bottom-right instead of its top-left corner. Why is that?

3.6 The Bicubic Surface Patch

The parametric cubic (PC) curve, Equation (3.1), is useful, since it can be used when either four points, or two points and two tangent vectors, are known. The latter approach is the topic of Chapter 4. The PC curve can easily be extended to a bicubic surface patch by means of the Cartesian product.

A PC curve has the form $\mathbf{P}(t) = \sum_{i=0}^{3} \mathbf{a}_i t^i$. Two such curves, $\mathbf{P}(u)$ and $\mathbf{P}(w)$, can be combined to form the Cartesian product surface patch

$$
\begin{aligned}
\mathbf{P}(u, w) \\
= \sum_{i=0}^{3} \sum_{j=0}^{3} \mathbf{a}_{ij} u^i w^j \\
= \mathbf{a}_{33} u^3 w^3 + \mathbf{a}_{32} u^3 w^2 + \mathbf{a}_{31} u^3 w + \mathbf{a}_{30} u^3 + \mathbf{a}_{23} u^2 w^3 + \mathbf{a}_{22} u^2 w^2 + \mathbf{a}_{21} u^2 w + \mathbf{a}_{20} u^2 \\
+ \mathbf{a}_{13} u w^3 + \mathbf{a}_{12} u w^2 + \mathbf{a}_{11} u w + \mathbf{a}_{10} u + \mathbf{a}_{03} w^3 + \mathbf{a}_{02} w^2 + \mathbf{a}_{01} w + \mathbf{a}_{00} \qquad (3.24)
\end{aligned}
$$

$$
= (u^3, u^2, u, 1) \begin{pmatrix} \mathbf{a}_{33} & \mathbf{a}_{32} & \mathbf{a}_{31} & \mathbf{a}_{30} \\ \mathbf{a}_{23} & \mathbf{a}_{22} & \mathbf{a}_{21} & \mathbf{a}_{20} \\ \mathbf{a}_{13} & \mathbf{a}_{12} & \mathbf{a}_{11} & \mathbf{a}_{10} \\ \mathbf{a}_{03} & \mathbf{a}_{02} & \mathbf{a}_{01} & \mathbf{a}_{00} \end{pmatrix} \begin{pmatrix} w^3 \\ w^2 \\ w \\ 1 \end{pmatrix}, \quad \text{where } 0 \le u, w \le 1. \qquad (3.25)
$$

This is a double cubic polynomial (hence the name *bicubic*) with 16 terms, where each of the 16 coefficients $\mathbf{a}_{ij}$ is a triplet [compare with Equation (3.21)]. When w is set to a fixed value w_0, Equation (3.25) becomes $\mathbf{P}(u, w_0)$, which is a PC curve. The same is true for $\mathbf{P}(u_0, w)$. The conclusion is that curves that lie on this surface in the u or in the w directions are parametric cubics. The four boundary curves are consequently also PC curves.

Notice that the shape and location of the surface depend on all 16 coefficients. Any change in any of them produces a different surface patch. Equation (3.25) is the algebraic representation of the bicubic patch. In order to use it in practice, the 16 unknown coefficients have to be expressed in terms of known geometrical quantities, such as points, tangent vectors, or second derivatives.

Two types of bicubic surfaces are discussed here. The first is based on 16 data points and the second is constructed from four known curves. A third type—defined by four data points, eight tangent vectors, and four twist vectors—is the topic of Section 4.9.

> Milo... glanced curiously at the strange circular room, where sixteen tiny arched windows corresponded exactly to the sixteen points of the compass. Around the entire circumference were numbers from zero to three hundred and sixty, marking the degrees of the circle, and on the floor, walls, tables, chairs, desks, cabinets, and ceiling were labels showing their heights, widths, depths, and distances to and from each other.
>
> —Norton Juster, *The Phantom Tollbooth.*

3.6.1 Sixteen Points

We start with the sixteen given points

$$
\begin{array}{cccc}
\mathbf{P}_{03} & \mathbf{P}_{13} & \mathbf{P}_{23} & \mathbf{P}_{33} \\
\mathbf{P}_{02} & \mathbf{P}_{12} & \mathbf{P}_{22} & \mathbf{P}_{32} \\
\mathbf{P}_{01} & \mathbf{P}_{11} & \mathbf{P}_{21} & \mathbf{P}_{31} \\
\mathbf{P}_{00} & \mathbf{P}_{10} & \mathbf{P}_{20} & \mathbf{P}_{30}.
\end{array}
$$

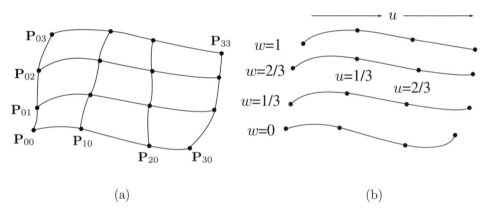

(a)   (b)

Figure 3.5: (a) Sixteen Points. (b) Four Curves.

We assume that the points are (roughly) equally spaced on the rectangular surface patch as shown in Figure 3.5a. We know that the bicubic surface has the form

$$
\mathbf{P}(u, w) = \sum_{i=0}^{3} \sum_{j=0}^{3} \mathbf{a}_{ij} u^i w^j, \tag{3.26}
$$

where each of the 16 coefficients $\mathbf{a}_{ij}$ is a triplet. To calculate the 16 unknown coefficients, we write 16 equations, each based on one of the given points

$$
\begin{array}{llll}
\mathbf{P}(0,0) = \mathbf{P}_{00}, & \mathbf{P}(0,1/3) = \mathbf{P}_{01}, & \mathbf{P}(0,2/3) = \mathbf{P}_{02}, & \mathbf{P}(0,1) = \mathbf{P}_{03}, \\
\mathbf{P}(1/3,0) = \mathbf{P}_{10}, & \mathbf{P}(1/3,1/3) = \mathbf{P}_{11}, & \mathbf{P}(1/3,2/3) = \mathbf{P}_{12}, & \mathbf{P}(1/3,1) = \mathbf{P}_{13}, \\
\mathbf{P}(2/3,0) = \mathbf{P}_{20}, & \mathbf{P}(2/3,1/3) = \mathbf{P}_{21}, & \mathbf{P}(2/3,2/3) = \mathbf{P}_{22}, & \mathbf{P}(2/3,1) = \mathbf{P}_{23}, \\
\mathbf{P}(1,0) = \mathbf{P}_{30}, & \mathbf{P}(1,1/3) = \mathbf{P}_{31}, & \mathbf{P}(1,2/3) = \mathbf{P}_{32}, & \mathbf{P}(1,1) = \mathbf{P}_{33}.
\end{array}
$$

After solving, the final expression for the surface patch becomes

$$\mathbf{P}(u,w) = (u^3, u^2, u, 1)\mathbf{N}\begin{pmatrix} \mathbf{P}_{00} & \mathbf{P}_{10} & \mathbf{P}_{20} & \mathbf{P}_{30} \\ \mathbf{P}_{01} & \mathbf{P}_{11} & \mathbf{P}_{21} & \mathbf{P}_{31} \\ \mathbf{P}_{02} & \mathbf{P}_{12} & \mathbf{P}_{22} & \mathbf{P}_{32} \\ \mathbf{P}_{03} & \mathbf{P}_{13} & \mathbf{P}_{23} & \mathbf{P}_{33} \end{pmatrix}\mathbf{N}^T\begin{pmatrix} w^3 \\ w^2 \\ w \\ 1 \end{pmatrix}, \tag{3.27}$$

where

$$\mathbf{N} = \begin{pmatrix} -4.5 & 13.5 & -13.5 & 4.5 \\ 9.0 & -22.5 & 18 & -4.5 \\ -5.5 & 9.0 & -4.5 & 1.0 \\ 1.0 & 0 & 0 & 0 \end{pmatrix}.$$

is the basis matrix used to blend four points in a PC [Equation (3.6)]. As mentioned, this type of surface patch has only limited use because it cannot have a very complex shape. A larger surface, made up of a number of such patches, can be constructed, but it is difficult to connect the individual patches smoothly.

(This type of surface is also derived in Section 1.9 as a Cartesian product.)

Example: Given the 16 points listed in Figure 3.6, we compute and plot the bicubic surface patch defined by them. The figure shows two views of this surface.

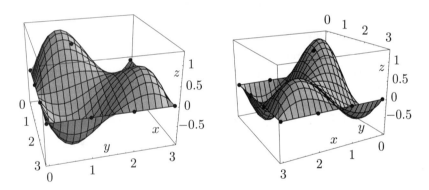

```
<<:Graphics:ParametricPlot3D.m; (* BiCubic patch for 16 points *)
Clear[T,pnt,M,g1,g2];
T[t_]:={t^3,t^2,t,1};
pnt={{{0,0,0},{1,0,0},{2,0,0},{3,0,0}}, {{0,1,0},{1,1,1},{2,1,-.5},{3,1,0}},
 {{0,2,-.5},{1,2,0},{2,2,.5},{3,2,0}},{{0,3,0},{1,3,0},{2,3,0},{3,3,0}}};
M={{-4.5,13.5,-13.5,4.5},{9,-22.5,18,-4.5},{-5.5,9,-4.5,1},{1,0,0,0}};
g2=Graphics3D[{AbsolutePointSize[3],
Table[Point[pnt[[i,j]]],{i,1,4},{j,1,4}] }];
comb[i_]:=(T[u].M.pnt)[[i]](Transpose[M].T[w])[[i]];
g1=ParametricPlot3D[comb[1]+comb[2]+comb[3]+comb[4], {u,0,1},{w,0,1},
Compiled->False, DisplayFunction->Identity];
Show[g1,g2, ViewPoint->{2.752, -0.750, 1.265}, DefaultFont->{"cmr10", 10},
(*  ViewPoint->{1.413, 2.605, 0.974}  for alt view *)
DisplayFunction->$DisplayFunction]
```

Figure 3.6: A Bicubic Surface Patch Example.

Even though this type of surface has limited use in graphics, it can be used for two-dimensional bicubic polynomial interpolation of points and numbers. Given a set of three-dimensional points arranged in a two-dimensional grid, the problem is to compute a weighted sum of the points and employ it to predict the value of a new point at the center of the grid. It makes sense to assign more weights to points that are closer to the center, and a natural way to achieve this is to calculate the surface patch $\mathbf{P}(u, w)$ that passes through all the points in the grid and use the value $\mathbf{P}(0.5, 0.5)$ as the interpolated value at the center of the grid.

The MLP image compression method [Salomon 04] is an example of the use of this approach. The problem is to interpolate the values of a group of 4×4 pixels in an image in order to predict the value of a pixel at the center of this group. The simple solution is to calculate the surface patch defined by the 16 pixels and to use the surface point $\mathbf{P}(0.5, 0.5)$ as the interpolated value of the pixel at the center of the group. Substituting $u = 0.5$ and $w = 0.5$ in Equation (3.27) produces

$$
\begin{aligned}
\mathbf{P}&(0.5, 0.5) \\
&= 0.00390625\mathbf{P}_{00} - 0.0351563\mathbf{P}_{01} - 0.0351563\mathbf{P}_{02} + 0.00390625\mathbf{P}_{03} \\
&\quad - 0.0351563\mathbf{P}_{10} + 0.316406\mathbf{P}_{11} + 0.316406\mathbf{P}_{12} - 0.0351563\mathbf{P}_{13} \\
&\quad - 0.0351563\mathbf{P}_{20} + 0.316406\mathbf{P}_{21} + 0.316406\mathbf{P}_{22} - 0.0351563\mathbf{P}_{23} \\
&\quad + 0.00390625\mathbf{P}_{30} - 0.0351563\mathbf{P}_{31} - 0.0351563\mathbf{P}_{32} + 0.00390625\mathbf{P}_{33}.
\end{aligned}
$$

The 16 coefficients are the ones used by MLP.

⋄ **Exercise 3.13:** The center point of the surface is calculated as a weighted sum of the 16 equally-spaced data points (this technique is known as bicubic interpolation). It makes sense to assign small weights to points located away from the center, but our result assigns *negative* weights to eight of the 16 points. Explain the meaning of negative weights and show what role they play in interpolating the center of the surface.

Readers who find it tedious to follow the details above should compare the way two-dimensional bicubic polynomial interpolation is presented here to the way it is discussed by [Press and Flannery 88]; the following quotation is from their page 125: "...the formulas that obtain the c's from the function and derivative values are just a complicated linear transformation, with coefficients which, having been determined once, in the mists of numerical history, can be tabulated and forgotten."

Seated at his disorderly desk, caressed by a counterpane of drifting tobacco haze, he would pore over the manuscript, crossing out, interpolating, re-arguing, and then referring to volumes on his shelves.

—Christopher Morley, *The Haunted Bookshop* (1919).

3.6.2 Four Curves

A variant of the previous method starts with four curves (any curves, not just PCs), $\mathbf{P}_0(u)$, $\mathbf{P}_1(u)$, $\mathbf{P}_2(u)$, and $\mathbf{P}_3(u)$, roughly parallel, all going in the u direction (Figure 3.5b). It is possible to select four points $\mathbf{P}_i(0)$, $\mathbf{P}_i(1/3)$, $\mathbf{P}_i(2/3)$, and $\mathbf{P}_i(1)$ on each curve $\mathbf{P}_i(u)$, for a total of 16 points. The surface patch can then easily be constructed from Equation (3.27).

Example: The surface of Figure 3.7 is defined by the following four curves (shown in the diagram in an inset). All go along the x axis, at different y values, and are sine curves (with different phases) along the z axis.

$$\mathbf{P}_0(u) = (u, 0, \sin(\pi u)), \quad \mathbf{P}_1(u) = (u, 1 + u/10, \sin(\pi(u + 0.1))),$$
$$\mathbf{P}_2(u) = (u, 2, \sin(\pi(u + 0.2))), \quad \mathbf{P}_3(u) = (u, 3 + u/10, \sin(\pi(u + 0.3))),$$

The *Mathematica* code of Figure 3.7 shows how matrix `basis` is created with the 16 points

$$\begin{pmatrix} \mathbf{P}_0(0) & \mathbf{P}_0(.33) & \mathbf{P}_0(.67) & \mathbf{P}_0(1) \\ \mathbf{P}_1(0) & \mathbf{P}_1(.33) & \mathbf{P}_1(.67) & \mathbf{P}_1(1) \\ \mathbf{P}_2(0) & \mathbf{P}_2(.33) & \mathbf{P}_2(.67) & \mathbf{P}_2(1) \\ \mathbf{P}_3(0) & \mathbf{P}_3(.33) & \mathbf{P}_3(.67) & \mathbf{P}_3(1) \end{pmatrix}.$$

3.7 Coons Surfaces

This type of surface is based on the pioneering work of Steven Anson Coons at MIT in the 1960s. His efforts are summarized in [Coons 64] and [Coons 67].

We start with the linear Coons surface, which is a generalization of lofted surfaces. This type of surface patch is defined by its four boundary curves. All four boundary curves are given, and none has to be a straight line. Naturally, the boundary curves have to meet at the corner points, so these points are implicitly known.

Coons decided to search for an expression $\mathbf{P}(u, w)$ of the surface that satisfies (1) it is symmetric in u and w and (2) it is an interpolation of $\mathbf{P}(u, 0)$ and $\mathbf{P}(u, 1)$ in one direction and of $\mathbf{P}(0, w)$ and $\mathbf{P}(1, w)$ in the other direction. He found a surprisingly simple, two-step solution.

The first step is to construct two lofted surfaces from the two sets of opposite boundary curves. They are $\mathbf{P}_a(u, w) = \mathbf{P}(0, w)(1 - u) + \mathbf{P}(1, w)u$ and $\mathbf{P}_b(u, w) = \mathbf{P}(u, 0)(1 - w) + \mathbf{P}(u, 1)w$.

The second step is to tentatively attempt to create the final surface $\mathbf{P}(u, w)$ as the sum $\mathbf{P}_a(u, w) + \mathbf{P}_b(u, w)$. It is clear that this is not the expression we are looking for because it does not converge to the right curves at the boundaries. For $u = 0$, for example, we want $\mathbf{P}(u, w)$ to converge to boundary curve $\mathbf{P}(0, w)$. The sum above, however, converges to $\mathbf{P}(0, w) + \mathbf{P}(0, 0)(1 - w) + \mathbf{P}(0, 1)w$. We therefore have to subtract $\mathbf{P}(0, 0)(1 - w) + \mathbf{P}(0, 1)w$. Similarly, for $u = 1$, the sum converges to $\mathbf{P}(1, w) + \mathbf{P}(1, 0)(1 - w) + \mathbf{P}(1, 1)w$, so we have to subtract $\mathbf{P}(1, 0)(1 - w) + \mathbf{P}(1, 1)w$. For $w = 0$, we have to subtract $\mathbf{P}(0, 0)(1 - u) + \mathbf{P}(1, 0)u$, and for $w = 1$, we should subtract $\mathbf{P}(0, 1)(1 - u) + \mathbf{P}(1, 1)u$.

Note that the expressions $\mathbf{P}(0, 0)$, $\mathbf{P}(0, 1)$, $\mathbf{P}(1, 0)$, and $\mathbf{P}(1, 1)$ are simply the four corner points. A better notation for them may be $\mathbf{P}_{00}$, $\mathbf{P}_{01}$, $\mathbf{P}_{10}$, and $\mathbf{P}_{11}$.

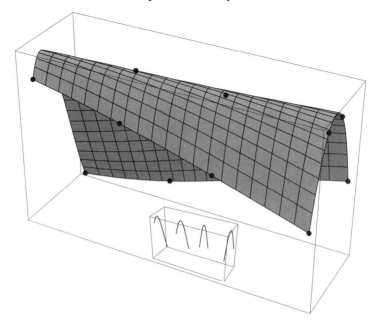

```
Clear[p0,p1,p2,p3,basis,fourP,g0,g1,g2,g3,g4,g5];
p0[u_]:={u,0,Sin[Pi u]}; p1[u_]:={u,1+u/10,Sin[Pi(u+.1)]};
p2[u_]:={u,2,Sin[Pi(u+.2)]}; p3[u_]:={u,3+u/10,Sin[Pi(u+.3)]};
(* matrix 'basis' has dimensions 4x4x3 *)
basis:={{p0[0],p0[.33],p0[.67],p0[1]},{p1[0],p1[.33],p1[.67],p1[1]},
{p2[0],p2[.33],p2[.67],p2[1]},{p3[0],p3[.33],p3[.67],p3[1]}};
fourP:= (* basis matrix for a 4-point curve *)
{{-4.5,13.5,-13.5,4.5},{9,-22.5,18,-4.5},{-5.5,9,-4.5,1},{1,0,0,0}};
prt[i_]:= (* extracts component i from the 3rd dimen of 'basis' *)
basis[[Range[1,4],Range[1,4],i]];
coord[i_]:= (* calc. the 3 parametric components of the surface *)
{u^3,u^2,u,1}.fourP.prt[i].Transpose[fourP].{w^3,w^2,w,1};
g0=ParametricPlot3D[p0[u], {u,0,1}]
g1=ParametricPlot3D[p1[u], {u,0,1}]
g2=ParametricPlot3D[p2[u], {u,0,1}]
g3=ParametricPlot3D[p3[u], {u,0,1}]
g4=Graphics3D[{AbsolutePointSize[4],
Table[Point[basis[[i,j]]],{i,1,4},{j,1,4}]}];
g5=ParametricPlot3D[{coord[1],coord[2],coord[3]},
{u,0,1,.05},{w,0,1,.05}, DisplayFunction->Identity];
Show[g0,g1,g2,g3, ViewPoint->{-2.576, -1.365, 1.718},
 Ticks->False, DisplayFunction->$DisplayFunction]
Show[g4,g5, ViewPoint->{-2.576, -1.365, 1.718},
 DisplayFunction->$DisplayFunction]
```

Figure 3.7: A Four-Curve Surface.

Today, this type of surface is known as the linear Coons surface. Its expression is $\mathbf{P}(u, w) = \mathbf{P}_a(u, w) + \mathbf{P}_b(u, w) - \mathbf{P}_{ab}(u, w)$, where

$$\mathbf{P}_{ab}(u, w) = \mathbf{P}_{00}(1 - u)(1 - w) + \mathbf{P}_{01}(1 - u)w + \mathbf{P}_{10}u(1 - w) + \mathbf{P}_{11}uw.$$

Note that $\mathbf{P}_a$ and $\mathbf{P}_b$ are lofted surfaces, whereas $\mathbf{P}_{ab}$ is a bilinear surface. The final expression is

$$
\begin{aligned}
\mathbf{P}(u, w) &= \mathbf{P}_a(u, w) + \mathbf{P}_b(u, w) - \mathbf{P}_{ab}(u, w) \\
&= (1 - u, u) \begin{pmatrix} \mathbf{P}(0, w) \\ \mathbf{P}(1, w) \end{pmatrix} + (1 - w, w) \begin{pmatrix} \mathbf{P}(u, 0) \\ \mathbf{P}(u, 1) \end{pmatrix} \\
&\quad - (1 - u, u) \begin{pmatrix} \mathbf{P}_{00} & \mathbf{P}_{01} \\ \mathbf{P}_{10} & \mathbf{P}_{11} \end{pmatrix} \begin{pmatrix} 1 - w \\ w \end{pmatrix} \qquad (3.28) \\
&= (1 - u, u, 1) \begin{pmatrix} -\mathbf{P}_{00} & -\mathbf{P}_{01} & \mathbf{P}(0, w) \\ -\mathbf{P}_{10} & -\mathbf{P}_{11} & \mathbf{P}(1, w) \\ \mathbf{P}(u, 0) & \mathbf{P}(u, 1) & (0, 0, 0) \end{pmatrix} \begin{pmatrix} 1 - w \\ w \\ 1 \end{pmatrix}. \qquad (3.29)
\end{aligned}
$$

Equation (3.28) is more useful than Equation (3.29) since it shows how the surface is defined in terms of the two barycentric pairs $(1 - u, u)$ and $(1 - w, w)$. They are the blending functions of the linear Coons surface. It turns out that many pairs of barycentric functions $\big(f_1(u), f_2(u)\big)$ and $\big(g_1(w), g_2(w)\big)$ can serve as blending functions, out of which more general Coons surfaces can be constructed. All that the blending functions have to satisfy is

$$
\begin{aligned}
f_1(0) = 1, \quad & f_1(1) = 0, \quad f_2(0) = 0, \quad f_2(1) = 1, \quad f_1(u) + f_2(u) = 1, \\
g_1(0) = 1, \quad & g_1(1) = 0, \quad g_2(0) = 0, \quad g_2(1) = 1, \quad g_1(w) + g_2(w) = 1.
\end{aligned}
\qquad (3.30)
$$

Example: We select the four (nonpolynomial) boundary curves

$$
\begin{aligned}
\mathbf{P}_{u0} = (u, 0, \sin(\pi u)), \quad & \mathbf{P}_{u1} = (u, 1, \sin(\pi u)), \\
\mathbf{P}_{0w} = (0, w, \sin(\pi w)), \quad & \mathbf{P}_{1w} = (1, w, \sin(\pi w)).
\end{aligned}
$$

Each is one-half of a sine wave. The first two proceed along the x axis, and the other two go along the y axis. They meet at the four corner points $\mathbf{P}_{00} = (0, 0, 0)$, $\mathbf{P}_{01} = (0, 1, 0)$, $\mathbf{P}_{10} = (1, 0, 0)$, and $\mathbf{P}_{11} = (1, 1, 0)$. The surface and the *Mathematica* code that produced it are shown in Figure 3.8. Note the `Simplify` command, which displays the final, simplified expression of the surface `{u, w, Sin[Pi u] + Sin[Pi w]}`.

Example: Given the four corner points $\mathbf{P}_{00} = (-1, -1, 0)$, $\mathbf{P}_{01} = (-1, 1, 0)$, $\mathbf{P}_{10} = (1, -1, 0)$, and $\mathbf{P}_{11} = (1, 1, 0)$ (notice that they lie on the xy plane), we calculate the four boundary curves of a linear Coons surface patch as follows:

1. We select boundary curve $\mathbf{P}(0, w)$ as the straight line from $\mathbf{P}_{00}$ to $\mathbf{P}_{01}$:

$$\mathbf{P}(0, w) = \mathbf{P}_{00}(1 - w) + \mathbf{P}_{01}w = (-1, 2w - 1, 0).$$

```
<<:Graphics:ParametricPlot3D.m;
Clear[p00,p01,p10,p11,pu0,pu1,p0w,p1w];
p00:={0,0,0}; p01:={0,1,0};
p10:={1,0,0}; p11:={1,1,0};
pu0:={u,0,Sin[Pi u]};
pu1:={u,1,Sin[Pi u]};
p0w:={0,w,Sin[Pi w]};
p1w:={1,w,Sin[Pi w]};
Simplify[
{1-u,u}.{p0w,p1w}+{1-w,w}.{pu0,pu1}
-p00(1-u)(1-w)-p01(1-u)w
-p10(1-w)u-p11 u w]
ParametricPlot3D[%,
{u,0,1,.2},{w,0,1,.2},
PlotRange->All,
AspectRatio->Automatic,
RenderAll->False,
Ticks->{{1},{0,1},{0,1}},
Prolog->AbsoluteThickness[.4]]
```

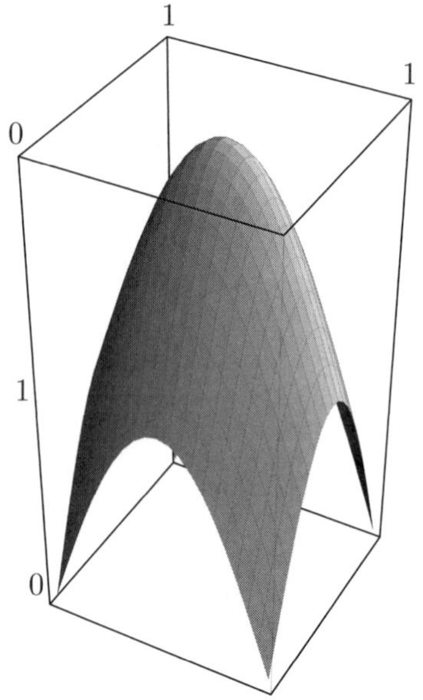

Figure 3.8: A Coons Surface.

2. We place the two points $(1, -0.5, 0.5)$ and $(1, 0.5, -0.5)$ between $\mathbf{P}_{10}$ and $\mathbf{P}_{11}$ and calculate boundary curve $\mathbf{P}(1, w)$ as the cubic Lagrange polynomial [Equation (3.17)] determined by these four points

$$\mathbf{P}(1,w) = \frac{1}{2}(w^3, w^2, w, 1) \begin{bmatrix} -9 & -27 & 27 & 9 \\ 18 & -45 & 36 & -9 \\ -11 & 18 & -9 & 2 \\ 2 & 0 & 0 & 0 \end{bmatrix} \begin{bmatrix} (1,-1,0) \\ (1,-0.5,0.5) \\ (1,0.5,-0.5) \\ (1,1,0) \end{bmatrix}$$

$$= \left(1, (-4 - w + 27w^2 - 18w^3)/4, 27(w - 3w^2 + 2w^3)/4\right).$$

3. The single point $(0, -1, -0.5)$ is placed between points $\mathbf{P}_{00}$ and $\mathbf{P}_{10}$ and boundary curve $\mathbf{P}(u, 0)$ is calculated as the quadratic Lagrange polynomial [Equation (3.14)] determined by these three points:

$$\mathbf{P}(u,0) = (u^2, u, 1) \begin{bmatrix} 2 & -4 & 2 \\ -3 & 4 & -1 \\ 1 & 0 & 0 \end{bmatrix} \begin{bmatrix} (-1,-1,0) \\ (0,-1,-.5) \\ (1,-1,0) \end{bmatrix} = (2u - 1, -1, 2u^2 - 2u).$$

4. Similarly, a new point $(0, 1, .5)$ is placed between points $\mathbf{P}_{01}$ and $\mathbf{P}_{11}$, and boundary curve $\mathbf{P}(u, 1)$ is calculated as the quadratic Lagrange polynomial determined

by these three points:

$$\mathbf{P}(u,1) = (u^2, u, 1)\begin{bmatrix} 2 & -4 & 2 \\ -3 & 4 & -1 \\ 1 & 0 & 0 \end{bmatrix}\begin{bmatrix} (-1,1,0) \\ (0,1,.5) \\ (1,1,0) \end{bmatrix} = (2u-1, 1, -2u^2 + 2u).$$

The four boundary curves and the four corner points now become the linear Coons surface patch given by Equation (3.28):

$$\mathbf{P}(u,w) = (1-u, u, 1)\begin{bmatrix} -(-1,-1,0) & -(-1,1,0) \\ -(1,-1,0) & -(1,1,0) \\ (2u-1,-1,2u^2-2u) & (2u-1,1,-2u^2+2u) \\ \end{bmatrix}$$

$$\left.\begin{matrix} (-1,2w-1,0) \\ (1,(-4-w+27w^2-18w^3)/4, 27(w-3w^2+2w^3)/4) \\ 0 \end{matrix}\right]\begin{bmatrix} 1-w \\ w \\ 1 \end{bmatrix}.$$

This is simplified with the help of appropriate software and becomes

$$\begin{aligned} \mathbf{P}(u,w) = \big(&-1+2u+(1-u)(1-w)-u(1-w)+(-1+2u)(1-w) \\ &+ (1-u)w - uw + (-1+2u)w, \\ &-1 + (1-u)(1-w) + u(1-w) + 2w - (1-u)w \\ &- uw + (1-u)(-1+2w) + u(-4-w+27w^2-18w^3)/4, \\ &(-2u+2u^2)(1-w) + (2u-2u^2)w + 27u(w-3w^2+2w^3)/4\big). \end{aligned}$$

The surface patch and the eight points involved are shown in Figure 3.9.

3.7.1 Translational Surfaces

Given two curves $\mathbf{P}(u,0)$ and $\mathbf{P}(0,w)$ that intersect at a point

$$\mathbf{P}(u,0)|_{u=0} = \mathbf{P}(0,w)|_{w=0} \stackrel{\text{def}}{=} \mathbf{P}_{00},$$

it is easy to construct the surface patch created by sliding one of the curves, say, $\mathbf{P}(u,0)$, along the other one (Figure 3.10).

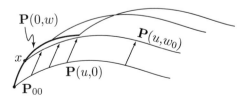

Figure 3.10: A Translational Surface.

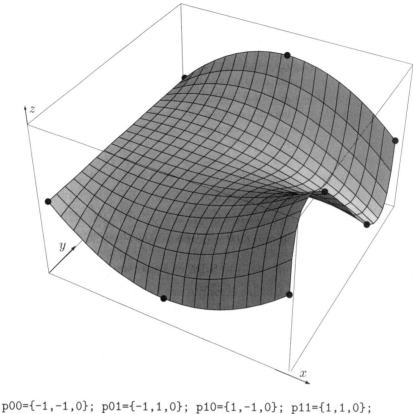

```
p00={-1,-1,0}; p01={-1,1,0}; p10={1,-1,0}; p11={1,1,0};
pnts={p00,p01,p10,p11,{1,-1/2,1/2},{1,1/2,-1/2},
 {0,-1,-1/2},{0,1,1/2}};
p0w[w_]:={-1,2w-1,0};
p1w[w_]:={1,(-4-w+27w^2-18w^3)/4,27(w-3w^2+2w^3)/4};
pu0[u_]:={2u-1,-1,2u^2-2u};
pu1[u_]:={2u-1,1,-2u^2+2u};
p[u_,w_]:=(1-u)p0w[w]+u p1w[w]+(1-w)pu0[u]+w pu1[u] \
 -p00(1-u)(1-w)-p01(1-u)
w-p10 u(1-w)-p11 u w;
g1=Graphics3D[{AbsolutePointSize[5], Table[Point[pnts[[i]]],
{i,1,8}]}];
g2=ParametricPlot3D[p[u,w], {u,0,1},{w,0,1}, Compiled->False,
Ticks->{{-1,1},{-1,1},{-1,1}}, DisplayFunction->Identity];
Show[g1,g2]
```

Figure 3.9: A Coons Surface Patch and Code.

We fix w at a certain value w_0 and compute the vector from the intersection point $\mathbf{P}_{00}$ to point $\mathbf{P}(0, w_0)$ (marked with an x in the figure). This vector is the difference $\mathbf{P}(0, w_0) - \mathbf{P}_{00}$, implying that any point on the curve $\mathbf{P}(u, w_0)$ can be obtained by adding this vector to the corresponding point on curve $\mathbf{P}(u, 0)$. The entire curve $\mathbf{P}(u, w_0)$ is therefore constructed as the sum $\mathbf{P}(u, 0) + [\mathbf{P}(0, w_0) - \mathbf{P}_{00}]$ for $0 \leq u \leq 1$. The resulting *translational surface* $\mathbf{P}(u, w)$ is obtained when w is released and is varied in the interval $[0, 1]$

$$\mathbf{P}(u, w) = \mathbf{P}(u, 0) + \mathbf{P}(0, w) - \mathbf{P}_{00}.$$

There is an interesting relation between the linear Coons surface and translational surfaces. The Coons patch is constructed from four intersecting curves. Consider a pair of such curves that intersect at a corner $\mathbf{P}_{ij}$ of the Coons patch. We can employ this pair and the corner to construct a translational surface $\mathbf{P}_{ij}(u, w)$. Once we construct the four translational surfaces for the four corners of the Coons patch, they can be used to express the entire Coons linear surface patch by a special version of Equation (3.29)

$$(1 - u, u) \begin{bmatrix} \mathbf{P}_{00}(u, w) & \mathbf{P}_{01}(u, w) \\ \mathbf{P}_{10}(u, w) & \mathbf{P}_{11}(u, w) \end{bmatrix} \begin{bmatrix} 1 - w \\ w \end{bmatrix}.$$

This version expresses the Coons surface patch as a weighted combination of four translational surfaces

3.7.2 Higher-Degree Coons Surfaces

One possible pair of blending functions is the cubic Hermite polynomials, functions $F_1(t)$ and $F_2(t)$ of Equation (4.6)

$$\begin{aligned} H_{3,0}(t) &= B_{3,0}(t) + B_{3,1}(t) = (1 - t)^3 + 3t(1 - t)^2 = 1 + 2t^3 - 3t^2, \\ H_{3,3}(t) &= B_{3,2}(t) + B_{3,3}(t) = 3t^2(1 - t) + t^3 = 3t^2 - 2t^3, \end{aligned} \tag{3.31}$$

where $B_{n,i}(t)$ are the Bernstein polynomials, Equation (6.5). The sum $H_{3,0}(t) + H_{3,3}(t)$ is identically 1 (because the Bernstein polynomials are barycentric), so these functions can be used to construct the *bicubic Coons surface*. Its expression is

$$\begin{aligned} \mathbf{P}(u, w) &= (H_{3,0}(u), H_{3,3}(u), 1) \begin{bmatrix} -\mathbf{P}_{00} & -\mathbf{P}_{01} & \mathbf{P}(0, w) \\ -\mathbf{P}_{10} & -\mathbf{P}_{11} & \mathbf{P}(1, w) \\ \mathbf{P}(u, 0) & \mathbf{P}(u, 1) & 0 \end{bmatrix} \begin{bmatrix} H_{3,0}(w) \\ H_{3,3}(w) \\ 1 \end{bmatrix} \\ &= (1 + 2u^3 - 3u^2, 3u^2 - 2u^3, 1) \begin{bmatrix} -\mathbf{P}_{00} & -\mathbf{P}_{01} & \mathbf{P}(0, w) \\ -\mathbf{P}_{10} & -\mathbf{P}_{11} & \mathbf{P}(1, w) \\ \mathbf{P}(u, 0) & \mathbf{P}(u, 1) & (0, 0, 0) \end{bmatrix} \begin{bmatrix} 1 + 2w^3 - 3w^2 \\ 3w^2 - 2w^3 \\ 1 \end{bmatrix}. \end{aligned} \tag{3.32}$$

One advantage of the bicubic Coons surface patch is that it is especially easy to connect smoothly to other patches of the same type. This is because its blending functions satisfy

$$\left.\frac{dH_{3,0}(t)}{dt}\right|_{t=0} = 0, \quad \left.\frac{dH_{3,0}(t)}{dt}\right|_{t=1} = 0, \quad \left.\frac{dH_{3,3}(t)}{dt}\right|_{t=0} = 0, \quad \left.\frac{dH_{3,3}(t)}{dt}\right|_{t=1} = 0. \tag{3.33}$$

Figure 3.11 shows two bicubic Coons surface patches, $\mathbf{P}(u,w)$ and $\mathbf{Q}(u,w)$, connected along their boundary curves $\mathbf{P}(u,1)$ and $\mathbf{Q}(u,0)$, respectively. The condition for patch connection is, of course, $\mathbf{P}(u,1) = \mathbf{Q}(u,0)$. The condition for smooth connection is

$$\left.\frac{\partial \mathbf{P}(u,w)}{\partial w}\right|_{w=1} = \left.\frac{\partial \mathbf{Q}(u,w)}{\partial w}\right|_{w=o} \tag{3.34}$$

(but see Section 1.10 for other, less restrictive conditions).

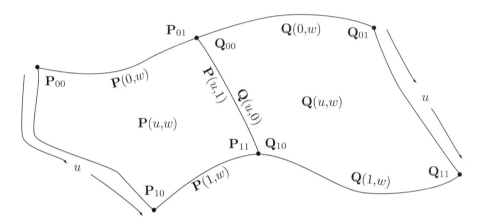

Figure 3.11: Smooth Connection of Bicubic Coons Surface Patches.

The partial derivatives of $\mathbf{P}(u,w)$ are easy to calculate from Equation (3.32). They are

$$\left.\frac{\partial \mathbf{P}(u,w)}{\partial w}\right|_{w=1} = H_{3,0}(u)\left.\frac{d\mathbf{P}(0,w)}{dw}\right|_{w=1} + H_{3,3}(u)\left.\frac{d\mathbf{P}(1,w)}{dw}\right|_{w=1},$$

$$\left.\frac{\partial \mathbf{Q}(u,w)}{\partial w}\right|_{w=0} = H_{3,0}(u)\left.\frac{d\mathbf{Q}(0,w)}{dw}\right|_{w=0} + H_{3,3}(u)\left.\frac{d\mathbf{Q}(1,w)}{dw}\right|_{w=0}. \tag{3.35}$$

[All other terms vanish because the blending functions satisfy Equation (3.33).] The condition for smooth connection, Equation (3.34), is therefore satisfied if

$$\left.\frac{d\mathbf{P}(0,w)}{dw}\right|_{w=1} = \left.\frac{d\mathbf{Q}(0,w)}{dw}\right|_{w=0} \quad \text{and} \quad \left.\frac{d\mathbf{P}(1,w)}{dw}\right|_{w=1} = \left.\frac{d\mathbf{Q}(1,w)}{dw}\right|_{w=0},$$

or, expressed in words, if the two boundary curves $\mathbf{P}(0,w)$ and $\mathbf{Q}(0,w)$ on the $u=0$ side of the patch connect smoothly, and the same for the two boundary curves $\mathbf{P}(1,w)$ and $\mathbf{Q}(1,w)$ on the $u=1$ side of the patch.

The reader should now find it easy to appreciate the advantage of the degree-5 Hermite blending functions [functions $F_1(t)$ and $F_2(t)$ of Equation (4.17)]

$$H_{5,0}(t) = B_{5,0}(t) + B_{5,1}(t) + B_{5,2}(t) = 1 - 10t^3 + 15t^4 - 6t^5,$$
$$H_{5,5}(t) = B_{5,3}(t) + B_{5,4}(t) + B_{5,5}(t) = 10t^3 - 15t^4 + 6t^5. \tag{3.36}$$

They are based on the Bernstein polynomials $B_{5,i}(t)$ hence they satisfy the conditions of Equation (3.30). They further have the additional property that their first *and* second derivatives are zero for $t = 0$ and for $t = 1$. The degree-5 Coons surface constructed by them is

$$\mathbf{P}_5(u, w) = \big(H_{5,0}(u), H_{5,5}(u), 1\big) \begin{bmatrix} -\mathbf{P}_{00} & -\mathbf{P}_{01} & \mathbf{P}(0, w) \\ -\mathbf{P}_{10} & -\mathbf{P}_{11} & \mathbf{P}(1, w) \\ \mathbf{P}(u, 0) & \mathbf{P}(u, 1) & 0 \end{bmatrix} \begin{bmatrix} H_{5,0}(w) \\ H_{5,5}(w) \\ 1 \end{bmatrix} . \quad (3.37)$$

Adjacent patches of this type of surface are easy to connect with G^2 continuity. All that's necessary is to have two pairs of boundary curves $\mathbf{P}(0, w)$, $\mathbf{Q}(0, w)$ and $\mathbf{P}(1, w)$, $\mathbf{Q}(1, w)$, where the two curves of each pair connect with G^2 continuity.

3.7.3 The Tangent Matching Coons Surface

The original aim of Coons was to construct a surface patch where all four boundary curves are specified by the user. Such patches are easy to compute and the conditions for connecting them smoothly are simple. It is possible to extend the original ideas of Coons to a surface patch where the user specifies the four boundary curves and also four functions that describe how (in what direction) this surface approaches its boundaries. Figure 3.12 illustrates the meaning of this statement. It shows a rectangular surface patch with some curves of the form $\mathbf{P}(u, w_i)$. Each of these curves goes from boundary curve $\mathbf{P}(0, w)$ to the opposite boundary curve $\mathbf{P}(1, w)$ by varying its parameter u from 0 to 1. Each has a different value of w_i. When such a curve reaches its end, it is moving in a certain, well-defined direction shown in the diagram. The end tangent vectors of these curves are different and we can imagine a function that yields these tangents as we move along the boundary curve $\mathbf{P}(1, w)$, varying w from 0 to 1. A good name for such a function is $\mathbf{P}_u(1, w)$, where the subscript u indicates that this tangent of the surface is in the u direction, the index 1 indicates the tangent at the end ($u = 1$), and the w indicates that this tangent vector is a function of w.

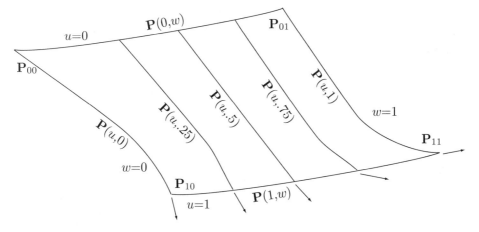

Figure 3.12: Tangent Matching in a Coons Surface.

There are four such functions, namely $\mathbf{P}_u(0, w)$, $\mathbf{P}_u(1, w)$, $\mathbf{P}_w(u, 0)$, and $\mathbf{P}_w(u, 1)$. Assuming that the user provides these functions, as well as the four boundary curves, our task is to obtain an expression $\mathbf{P}(u, w)$ for the surface that will satisfy the following:

1. When we substitute 0 or 1 for u and w in $\mathbf{P}(u, w)$, we get the four given corner points and the four given boundary curves. This condition can be expressed as the eight constraints

$$\mathbf{P}(0,0) = \mathbf{P}_{00}, \quad \mathbf{P}(0,1) = \mathbf{P}_{01}, \quad \mathbf{P}(1,0) = \mathbf{P}_{10}, \quad \mathbf{P}(1,1) = \mathbf{P}_{11},$$
$$\mathbf{P}(0, w), \quad \mathbf{P}(1, w), \quad \mathbf{P}(u, 0), \text{ and } \mathbf{P}(u, 1) \quad \text{are the given boundary curves.}$$

2. When we substitute 0 or 1 for u and w in the partial first derivatives of $\mathbf{P}(u, w)$, we get the four given tangent functions and their values at the four corner points. This condition can be expressed as the 12 constraints

$$\left.\frac{\partial \mathbf{P}(u, w)}{\partial u}\right|_{u=0} = \mathbf{P}_u(0, w), \qquad \left.\frac{\partial \mathbf{P}(u, w)}{\partial u}\right|_{u=1} = \mathbf{P}_u(1, w),$$

$$\left.\frac{\partial \mathbf{P}(u, w)}{\partial w}\right|_{w=0} = \mathbf{P}_w(u, 0), \qquad \left.\frac{\partial \mathbf{P}(u, w)}{\partial w}\right|_{w=1} = \mathbf{P}_w(u, 1),$$

$$\left.\frac{\partial \mathbf{P}(u, w)}{\partial u}\right|_{u=0,w=0} = \mathbf{P}_u(0, 0), \qquad \left.\frac{\partial \mathbf{P}(u, w)}{\partial u}\right|_{u=0,w=1} = \mathbf{P}_u(0, 1),$$

$$\left.\frac{\partial \mathbf{P}(u, w)}{\partial u}\right|_{u=1,w=0} = \mathbf{P}_u(1, 0), \qquad \left.\frac{\partial \mathbf{P}(u, w)}{\partial u}\right|_{u=1,w=1} = \mathbf{P}_u(1, 1),$$

$$\left.\frac{\partial \mathbf{P}(u, w)}{\partial w}\right|_{u=0,w=0} = \mathbf{P}_w(0, 0), \qquad \left.\frac{\partial \mathbf{P}(u, w)}{\partial w}\right|_{u=0,w=1} = \mathbf{P}_w(0, 1).$$

$$\left.\frac{\partial \mathbf{P}(u, w)}{\partial w}\right|_{u=1,w=0} = \mathbf{P}_w(1, 0), \qquad \left.\frac{\partial \mathbf{P}(u, w)}{\partial w}\right|_{u=1,w=1} = \mathbf{P}_w(1, 1).$$

3. When we substitute 0 or 1 for u and w in the partial second derivatives of $\mathbf{P}(u, w)$, we get the four first derivatives of the given tangent functions at the four corner points. This condition can be expressed as the four constraints

$$\left.\frac{\partial^2 \mathbf{P}(u, w)}{\partial u \partial w}\right|_{u=0,w=0} = \left.\frac{d\mathbf{P}_u(0, w)}{dw}\right|_{w=0} = \left.\frac{d\mathbf{P}_u(u, 0)}{du}\right|_{u=0} \stackrel{\text{def}}{=} \mathbf{P}_{uw}(0, 0),$$

$$\left.\frac{\partial^2 \mathbf{P}(u, w)}{\partial u \partial w}\right|_{u=0,w=1} = \left.\frac{d\mathbf{P}_u(0, w)}{dw}\right|_{w=1} = \left.\frac{d\mathbf{P}_u(u, 1)}{du}\right|_{u=0} \stackrel{\text{def}}{=} \mathbf{P}_{uw}(0, 1),$$

$$\left.\frac{\partial^2 \mathbf{P}(u, w)}{\partial u \partial w}\right|_{u=1,w=0} = \left.\frac{d\mathbf{P}_u(1, w)}{dw}\right|_{w=0} = \left.\frac{d\mathbf{P}_u(u, 0)}{du}\right|_{u=1} \stackrel{\text{def}}{=} \mathbf{P}_{uw}(1, 0),$$

$$\left.\frac{\partial^2 \mathbf{P}(u, w)}{\partial u \partial w}\right|_{u=1,w=1} = \left.\frac{d\mathbf{P}_u(1, w)}{dw}\right|_{w=1} = \left.\frac{d\mathbf{P}_u(u, 1)}{du}\right|_{u=1} \stackrel{\text{def}}{=} \mathbf{P}_{uw}(1, 1).$$

This is a total of 24 constraints. A derivation of this type of surface can be found in [Beach 91]. Here, we only quote the final result

$$\mathbf{P}(u, w) = \big(B_0(u), B_1(u), C_0(u), C_1(u), 1\big)\mathbf{M} \begin{bmatrix} B_0(w) \\ B_1(w) \\ C_0(w) \\ C_1(w) \\ 1 \end{bmatrix}, \qquad (3.38)$$

where $\mathbf{M}$ is the 5×5 matrix

$$\mathbf{M} = \begin{bmatrix} -\mathbf{P}_{00} & -\mathbf{P}_{01} & -\mathbf{P}_w(0,0) & -\mathbf{P}_w(0,1) & \mathbf{P}(0,w) \\ -\mathbf{P}_{10} & -\mathbf{P}_{11} & -\mathbf{P}_w(1,0) & -\mathbf{P}_w(1,1) & \mathbf{P}(1,w) \\ -\mathbf{P}_u(0,0) & -\mathbf{P}_u(0,1) & -\mathbf{P}_{uw}(0,0) & -\mathbf{P}_{uw}(0,1) & \mathbf{P}_u(0,w) \\ -\mathbf{P}_u(1,0) & -\mathbf{P}_u(1,1) & -\mathbf{P}_{uw}(1,0) & -\mathbf{P}_{uw}(1,1) & \mathbf{P}_u(1,w) \\ \mathbf{P}(u,0) & \mathbf{P}(u,1) & \mathbf{P}_w(u,0) & \mathbf{P}_w(u,1) & (0,0,0) \end{bmatrix}. \qquad (3.39)$$

The two blending functions $B_0(t)$ and $B_1(t)$ can be any functions satisfying conditions (3.30) and (3.33). Examples are the pairs $H_{3,0}(t)$, $H_{3,3}(t)$ and $H_{5,0}(t)$, $H_{5,5}(t)$ of Equations (3.31) and (3.36). The two blending functions $C_0(t)$ and $C_1(t)$ should satisfy

$$C_0(0) = 0, \quad C_0(1) = 0, \quad C_0'(0) = 1, \quad C_0'(1) = 0,$$
$$C_1(0) = 0, \quad C_1(1) = 0, \quad C_1'(0) = 0, \quad C_1'(1) = 1.$$

One choice is the pair $C_0(t) = t - 2t^2 + t^3$ and $C_1(t) = -t^2 + t^3$.

Such a surface patch is difficult to specify. The user has to input the four boundary curves and four tangent functions, a total of eight functions. The user then has to calculate the coordinates of the four corner points and the other 12 quantities required by the matrix of Equation (3.39). The advantage of this type of surface is that once fully specified, such a surface patch is easy to connect smoothly to other patches of the same type since the tangents along the boundaries are fully specified by the user.

3.7.4 The Triangular Coons Surface

A triangular surface patch is bounded by three boundary curves and has three corner points. Such surface patches are handy in situations like the one depicted in Figure 3.15, where a triangular Coons patch is used to smoothly connect two perpendicular lofted surface patches. Section 6.23 discusses the triangular Bézier surface patch which is commonly used in practice. Our approach to constructing the triangular Coons surface is to merge two of the four corner points and explore the behavior of the resulting surface patch. We arbitrarily decide to set $\mathbf{P}_{01} = \mathbf{P}_{11}$, which reduces the boundary curve $\mathbf{P}(u,1)$ to a single point (Figure 3.13). The expression of this triangular surface patch is

$$\mathbf{P}(u, w) = \big(B_0(u), B_1(u), 1\big) \begin{pmatrix} -\mathbf{P}_{00} & -\mathbf{P}_{11} & \mathbf{P}(0,w) \\ -\mathbf{P}_{10} & -\mathbf{P}_{11} & \mathbf{P}(1,w) \\ \mathbf{P}(u,0) & \mathbf{P}_{11} & (0,0,0) \end{pmatrix} \begin{pmatrix} B_0(w) \\ B_1(w) \\ 1 \end{pmatrix}, \qquad (3.40)$$

where the blending functions $B_0(t)$, $B_1(t)$ can be the pair $H_{3,0}$ and $H_{3,3}$, or the pair $H_{5,0}$ and $H_{5,5}$, or any other pair of blending functions satisfying Equations (3.30) and (3.33).

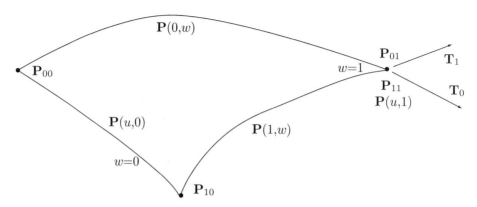

Figure 3.13: A Triangular Coons Surface Patch.

The tangent vector of the surface along the degenerate boundary curve $\mathbf{P}(u,1)$ is given by Equation (3.35):

$$\left.\frac{\partial\mathbf{P}(u,w)}{\partial w}\right|_{w=1} = B_0(u)\left.\frac{d\mathbf{P}(0,w)}{dw}\right|_{w=1} + B_1(u)\left.\frac{d\mathbf{P}(1,w)}{dw}\right|_{w=1}. \tag{3.41}$$

Thus, this tangent vector is a linear combination of the two tangents

$$\mathbf{T}_0 \stackrel{\text{def}}{=} \left.\frac{d\mathbf{P}(0,w)}{dw}\right|_{w=1} \quad \text{and} \quad \mathbf{T}_1 \stackrel{\text{def}}{=} \left.\frac{d\mathbf{P}(1,w)}{dw}\right|_{w=1},$$

and therefore lies in the plane defined by them. As u varies from 0 to 1, this tangent vector swings from $\mathbf{T}_0$ to $\mathbf{T}_1$ while the curve $\mathbf{P}(u,1)$ stays at the common point $\mathbf{P}_{01} = \mathbf{P}_{11}$. Once this behavior is grasped, the reader should be able to accept the following statement: The triangular patch will be well behaved in the vicinity of the common point if this tangent vector does not reverse its movement while swinging from $\mathbf{T}_0$ to $\mathbf{T}_1$. If it starts moving toward $\mathbf{T}_1$, then reverses and goes back toward $\mathbf{T}_0$, then reverses again, the surface may have a *fold* close to the common point. To guarantee this smooth behavior of the tangent vector, the blending functions $B_0(t)$ and $B_1(t)$ must satisfy one more condition, namely $B_0(t)$ should be monotonically decreasing in t and $B_1(t)$ should be monotonically increasing in t. The two sets of blending functions $H_{3,0}$, $H_{3,3}$ and $H_{5,0}$, $H_{5,5}$ satisfy this condition and can therefore be used to construct triangular Coons surface patches.

Example: Given the three corners $\mathbf{P}_{00} = (0,0,0)$, $\mathbf{P}_{10} = (2,0,0)$, and $\mathbf{P}_{01} = \mathbf{P}_{11} = (1,1,0)$, we compute and plot the triangular Coons surface patch defined by them. The first step is to compute the three boundary curves. We assume that the "bottom" boundary curve $\mathbf{P}(u,0)$ goes from $\mathbf{P}_{00}$ through $(1,0,-1)$ to $\mathbf{P}_{10}$. We similarly require that the "left" boundary curve $\mathbf{P}(0,w)$ goes from $\mathbf{P}_{00}$ through $(0.5,0.5,1)$ to $\mathbf{P}_{01}$ and the "right" boundary curve $\mathbf{P}(1,w)$ goes from $\mathbf{P}_{10}$ through $(1.5,0.5,1)$ to $\mathbf{P}_{11}$. All three curves are computed as standard quadratic Lagrange polynomials from Equation (3.14).

They become

$$\mathbf{P}(u,0) = (2u, 0, 4u(u-1)),$$
$$\mathbf{P}(0,w) = (w, w, 4w(1-w)),$$
$$\mathbf{P}(1,w) = (2-w, w, 4w(w-1)).$$

Figure 3.14 shows two views of this surface and illustrates the downside of this type of surface. The technique of drawing a surface patch as a wireframe with two families of curves works well for rectangular surface patches but is unsuitable for triangular patches. The figure shows how one family of curves converges to the double corner point, thereby making the wireframe look unusually dense in the vicinity of the point. Section 6.23 presents a better approach to the display of a triangular surface patch as a wireframe.

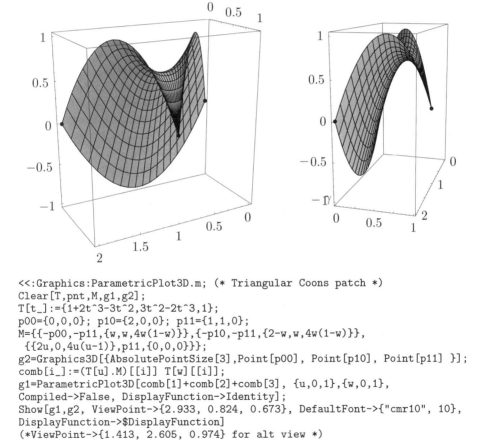

```
<<:Graphics:ParametricPlot3D.m; (* Triangular Coons patch *)
Clear[T,pnt,M,g1,g2];
T[t_]:={1+2t^3-3t^2,3t^2-2t^3,1};
p00={0,0,0}; p10={2,0,0}; p11={1,1,0};
M={{-p00,-p11,{w,w,4w(1-w)}},{-p10,-p11,{2-w,w,4w(1-w)}},
  {{2u,0,4u(u-1)},p11,{0,0,0}}};
g2=Graphics3D[{AbsolutePointSize[3],Point[p00], Point[p10], Point[p11] }];
comb[i_]:=(T[u].M)[[i]] T[w][[i]];
g1=ParametricPlot3D[comb[1]+comb[2]+comb[3], {u,0,1},{w,0,1},
Compiled->False, DisplayFunction->Identity];
Show[g1,g2, ViewPoint->{2.933, 0.824, 0.673}, DefaultFont->{"cmr10", 10},
DisplayFunction->$DisplayFunction]
(*ViewPoint->{1.413, 2.605, 0.974} for alt view *)
```

Figure 3.14: A Triangular Coons Surface Patch Example.

⋄ **Exercise 3.14:** What happens if the blending functions of the triangular Coons surface patch do not satisfy the condition of Equation (3.33)?

"Now, don't worry, my pet," Mrs. Whatsit said cheerfully. "We took care of that before we left. Your mother has had enough to worry her with you and Charles to cope with, and not knowing about your father, without our adding to her anxieties. We took a time wrinkle as well as a space wrinkle. It's very easy to do if you just know how."

—Madeleine L'Engle, A Wrinkle in Time (1962).

◇ **Exercise 3.15:** Given the four points $\mathbf{P}_{00} = (0, 0, 1)$, $\mathbf{P}_{10} = (1, 0, 0)$, $\mathbf{P}_{01} = (0.5, 1, 0)$, and $\mathbf{P}_{11} = (1, 1, 0)$, calculate the Coons surface defined by them, assuming straight lines as boundary curves. What type of a surface is this?

3.7.5 Summarizing Example

The surface shown in Figure 3.15 consists of four (intentionally separated) patches. A flat bilinear patch $\mathbf{B}$ at the top, two lofted patches $\mathbf{L}$ and $\mathbf{F}$ on both sides, and a triangular Coons patch $\mathbf{C}$ filling up the corner.

The bilinear patch is especially simple since it is defined by its four corner points. Its expression is

$$
\begin{aligned}
\mathbf{B}(u, w) &= (0, 1/2, 1)(1 - u)(1 - w) + (1, 1/2, 1)(1 - u)w \\
&+ (0, 3/2, 1)(1 - w)u + (1, 3/2, 1)uw \\
&= (w, 1/2 + u, 1).
\end{aligned}
$$

The calculation of lofted patch $\mathbf{L}$ starts with the two boundary curves $\mathbf{L}(u, 0)$ and $\mathbf{L}(u, 1)$. Each is calculated using Hermite interpolation (Chapter 4) since its extreme tangents, as well as its endpoints, are easy to figure out from the diagram. The boundary curves are

$$
\mathbf{L}(u, 0) = (u^3, u^2, u, 1)\mathbf{H}\big((0, 0, 0), (0, 1/2, 1), (0, 0, 1), (0, 1, 0)\big)^T,
$$
$$
\mathbf{L}(u, 1) = (u^3, u^2, u, 1)\mathbf{H}\big((1, 0, 0), (1, 1/2, 1), (0, 0, 1), (0, 1, 0)\big)^T,
$$

where $\mathbf{H}$ is the Hermite basis matrix, Equation (4.7). Surface patch L is thus

$$
\mathbf{L}(u, w) = \mathbf{L}(u, 0)(1 - w) + \mathbf{L}(u, 1)w = (w, u^2/2, u + u^2 - u^3).
$$

Lofted patch $\mathbf{F}$ is calculated similarly. Its boundary curves are

$$
\mathbf{F}(u, 0) = (u^3, u^2, u, 1)\mathbf{H}\big((3/2, 1/2, 0), (1, 1/2, 1), (0, 0, 1), (-1, 0, 0)\big)^T,
$$
$$
\mathbf{F}(u, 1) = (u^3, u^2, u, 1)\mathbf{H}\big((3/2, 3/2, 0), (1, 3/2, 1), (0, 0, 1), (-1, 0, 0)\big)^T,
$$

and the patch itself is

$$
\mathbf{F}(u, w) = \mathbf{F}(u, 0)(1 - w) + \mathbf{F}(u, 1)w = \big((3 - u^2)/2, 1/2 + w, u + u^2 - u^3\big).
$$

The triangular Coons surface $\mathbf{C}$ has corner points $\mathbf{C}_{00} = (1, 0, 0)$, $\mathbf{C}_{10} = (3/2, 1/2, 0)$, and $\mathbf{C}_{01} = \mathbf{C}_{11} = (1, 1/2, 1)$. Its bottom boundary curve is

$$
\mathbf{C}(u, 0) = (u^3, u^2, u, 1)\mathbf{H}\big((1, 0, 0), (3/2, 1/2, 0), (1, 0, 0), (0, 1, 0)\big)^T,
$$

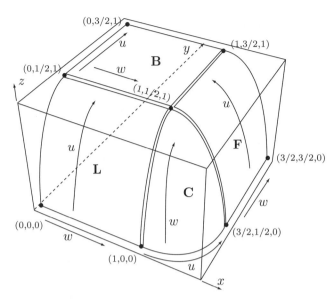

```
b[u_,w_]:={0,1/2,1}(1-u)(1-w)+{1,1/2,1}(1-u)w
 +{0,3/2,1}(1-w)u+{1,3/2,1}u w;
H={{2,-2,1,1},{-3,3,-2,-1},{0,0,1,0},{1,0,0,0}};
lu0={u^3,u^2,u,1}.H.{{0,0,0},{0,1/2,1},{0,0,1},{0,1,0}};
lu1={u^3,u^2,u,1}.H.{{1,0,0},{1,1/2,1},{0,0,1},{0,1,0}};
l[u_,w_]:=lu0(1-w)+lu1 w;
fu0={u^3,u^2,u,1}.H.{{3/2,1/2,0},{1,1/2,1},{0,0,1},{-1,0,0}};
fu1={u^3,u^2,u,1}.H.{{3/2,3/2,0},{1,3/2,1},{0,0,1},{-1,0,0}};
f[u_,w_]:=fu0(1-w)+fu1 w;
cu0={u^3,u^2,u,1}.H.{{1,0,0},{3/2,1/2,0},{1,0,0},{0,1,0}};
cu1={1,1/2,1};
c0w={w^3,w^2,w,1}.H.{{1,0,0},{1,1/2,1},{0,0,1},{0,1,0}};
c1w={w^3,w^2,w,1}.H.{{3/2,1/2,0},{1,1/2,1},{0,0,1},{-1,0,0}};
c[u_,w_]:=(1-u)c0w+u c1w+(1-w)cu0+w cu1 \
 -(1-u)(1-w){1,0,0}-u(1-w){3/2,1/2,0}-w(1-u)cu1- u w cu1;
g1=ParametricPlot3D[b[u,w], {u,0,1},{w,0,1}]
g2=ParametricPlot3D[l[u,w], {u,0,1},{w,0,1}]
g3=ParametricPlot3D[f[u,w], {u,0,1},{w,0,1}]
g4=ParametricPlot3D[c[u,w], {u,0,1},{w,0,1}]
Show[g1,g2,g3,g4]
```

Figure 3.15: Bilinear, Lofted, and Coons Surface Patches.

and its top boundary curve $\mathbf{C}(u, 1)$ is the multiple point $\mathbf{C}_{01} = \mathbf{C}_{11}$. The two boundary curves in the w direction are

$$\mathbf{C}(0, w) = (w^3, w^2, w, 1)\mathbf{H}\big((1, 0, 0), (3/1, 1/2, 1), (0, 0, 1), (0, 1, 0)\big)^T,$$
$$\mathbf{C}(1, w) = (w^3, w^2, w, 1)\mathbf{H}\big((3/1, 1/2, 0), (1, 1/2, 1), (0, 0, 1), (-1, 0, 0)\big)^T,$$

and the surface patch itself equals

$$\begin{aligned}
\mathbf{C}(u, w) &= (1 - u)\mathbf{C}(0, w) + u\mathbf{C}(1, w) + (1 - w)\mathbf{C}(u, 0) + w\mathbf{C}(u, 1) \\
&\quad - (1 - u)(1 - w)1, 0, 0 - u(1 - w)3/2, 1/2, 0 - w(1 - u)\mathbf{C}_{11} - uw\mathbf{C}_{11} \\
&= ((2 + u^2(-1 + w) - u(-2 + w + w^2))/2, \\
&\quad (-u^2(-1 + w) - u(-1 + w)w + w^2)/2, w + w^2 - w^3).
\end{aligned}$$

3.8 Gordon Surfaces

The Gordon surface is a generalization of Coons surfaces. A linear Coons surface is fully defined by means of four boundary curves, so its shape cannot be too complex. A Gordon surface (Figure 3.16) is defined by means of two families of curves, one in each of the u and w directions. It can have very complex shapes and is a good candidate for use in applications where realism is important.

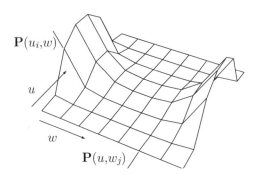

Figure 3.16: A Gordon Surface.

We denote the curves by $\mathbf{P}(u_i, w)$, where $i = 0, \ldots, m$, and $\mathbf{P}(u, w_j)$, $j = 0, \ldots, n$. The main idea is to find an expression for a surface $\mathbf{P}_a(u, w)$ that interpolates the first family of curves, add it to a similar expression for a surface $\mathbf{P}_b(u, w)$ that interpolates the second family of curves, and subtract a surface $\mathbf{P}_{ab}(u, w)$ that represents multiple contributions from $\mathbf{P}_a$ and $\mathbf{P}_b$.

The first surface, $\mathbf{P}_a(u, w)$, should interpolate the family of $m + 1$ curves $\mathbf{P}(u_i, w)$. When moving on this surface in the u direction (fixed w), we want to intersect all $m + 1$ curves. For a given, fixed w, we therefore need to find a curve that will pass through the $m + 1$ points $\mathbf{P}(u_i, w)$. A natural (albeit not the only) candidate for such a curve is our old acquaintance the Lagrange polynomial (Section 3.2). We write it as $\mathbf{P}_a(u, w) = \sum_{i=o}^{m} \mathbf{P}(u_i, w) L_i^m(u)$, and it is valid for any value of w. Similarly, we can write the second surface as the Lagrange polynomial $\mathbf{P}_b(u, w) = \sum_{j=o}^{n} \mathbf{P}(u, w_j) L_j^n(w)$.

The surface representing multiple contributions is similar to the bilinear part of Equation (3.28). It is

$$\mathbf{P}_{ab}(u, w) = \sum_{i=o}^{m} \sum_{j=o}^{n} \mathbf{P}(u_i, w_j) L_i^m(u) L_j^n(w),$$

and the final expression of the Gordon surface is $\mathbf{P}(u, w) = \mathbf{P}_a(u, w) + \mathbf{P}_b(u, w) - \mathbf{P}_{ab}(u, w)$. Note that the $(m + 1) \times (n + 1)$ points $\mathbf{P}(u_i, w_j)$ should be located on *both* curves. For such a surface to make sense, the curves have to intersect.

A friend comes to you and asks if a particular polynomial $p(x)$ of degree 25 in $F_2[x]$ is irreducible. The friend explains that she has tried dividing $p(x)$ by every polynomial in $F_2[x]$ of degree from 1 to 18 and has found that $p(x)$ is not divisible by any of them. She is getting tired of doing all these divisions and wonders if there's an easier way to check whether or not $p(x)$ is irreducible. You surprise your friend with the statement that she need not do any more work: $p(x)$ is indeed irreducible!

—John Palmieri, *Introduction to Modern Algebra for Teachers*

4
Hermite Interpolation

The curve and surface methods of the preceding chapters are based on points. Using polynomials, it is easy to construct a parametric curve segment (or surface patch) that passes through a given one-dimensional array or two-dimensional grid of points.

The downside of these methods is that they are not interactive. If the resulting curve or surface isn't the one the designer wants, the only way to modify it is to add points. Moving the points is not an option because the curve has to pass through the original data points. Adding points provides some control over the shape of the curve, but slows down the computations.

A practical, useful curve/surface design algorithm should be interactive. It should provide user-controlled parameters that modify the shape of the curve in a predictable, intuitive way. The Hermite interpolation approach, the topic of this chapter, is such a method.

Hermite interpolation is based on two points $\mathbf{P}_1$ and $\mathbf{P}_2$ and two tangent vectors $\mathbf{P}_1^t$ and $\mathbf{P}_2^t$. It computes a curve segment that starts at $\mathbf{P}_1$, going in direction $\mathbf{P}_1^t$ and ends at $\mathbf{P}_2$ moving in direction $\mathbf{P}_2^t$. Before delving into the details, the reader may find it useful to peruse Figure 4.1 where several such curves are shown, with their endpoints and extreme tangent vectors.

It is obvious that a single Hermite segment can take on many different shapes. It can even have a cusp and can develop a loop. A complete curve, however, normally requires several segments connected with C^0, C^1, or C^2 continuities, as illustrated in Section 1.4.2. Spline methods for constructing such a curve are discussed in Chapter 5.

The method is called Hermite interpolation after Charles Hermite who developed it and derived its blending functions in the 1870s, as part of his work on approximation and interpolation. He was not concerned with the computation of curves and surfaces (and was actually known to hate geometry), and developed his method as a way to interpolate any mathematical quantity from an initial value to a final value given the rates of change of the quantity at the start and at the end.

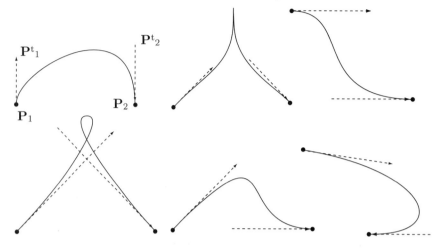

Figure 4.1: Various Hermite Curve Segments.

[Hermite] had a kind of positive hatred of geometry and once curiously reproached me with having made a geometrical memoir.

—Jacques Hadamard.

4.1 Interactive Control

Hermite interpolation has an important advantage; it is interactive. If a Hermite curve segment has a wrong shape, the user can edit it by modifying the tangent vectors.

⋄ **Exercise 4.1:** In the case of a four-point PC, we can change the shape of the curve by moving the points. Why then is the four-point method considered noninteractive?

Figure 4.1 illustrates how the shape of the curve depends on the directions of the tangent vectors. Figure 4.2 shows how the curve can be edited by modifying the magnitudes of those vectors. The figure shows three curves that start in a 45° direction and end up going vertically down. The effect illustrated here is simple. As the magnitude of the start tangent increases, the curve continues longer in the original direction. This behavior implies that short tangents produce a curve that changes its direction early and starts moving straight toward the final point. Such a curve is close to a straight segment, so we conclude that a long tangent results in a loose curve and a short tangent produces a tight curve (see also exercise 4.7).

The reason the magnitudes, and not just the directions, of the tangents affect the shape of the curve is that the three-dimensional Hermite segment is a PC and calculating a PC involves four coefficients, each a triplet, for a total of 12 unknown numbers. The two endpoints supply six known quantities and the two tangents should supply the remaining six. However, if we consider only the direction of a vector and not its magnitude, then the vectors $(1, 0.5, 0.3)$, $(2, 1, 0.6)$, and $(4, 2, 1.2)$ are all equal. In such a case, only two of the three vector components are independent and two vectors supply only four independent quantities.

Figure 4.2: Effects of Varying the Tangent's Magnitude.

◇ **Exercise 4.2:** Discuss this claim in detail.

> A sketch tells as much in a glance as a dozen pages of print.
> —Ivan Turgenev, *Fathers and Sons* (1862).

4.2 The Hermite Curve Segment

The Hermite curve segment is easy to derive. It is a PC curve (a degree-3 polynomial in t) with four coefficients that depend on the two points and two tangents. The basic equation of a PC curve is Equation (3.1) duplicated here

$$\mathbf{P}(t) = \mathbf{a}t^3 + \mathbf{b}t^2 + \mathbf{c}t + \mathbf{d} = (t^3, t^2, t, 1)(\mathbf{a}, \mathbf{b}, \mathbf{c}, \mathbf{d})^T = \mathbf{T}(t)\mathbf{A}. \tag{3.1}$$

This is the *algebraic representation* of the curve, in which the four coefficients are still unknown. Once these coefficients are expressed in terms of the known quantities, which are geometric, the curve will be expressed geometrically.

The tangent vector to a curve $\mathbf{P}(t)$ is the derivative $d\mathbf{P}(t)/dt$, which we denote by $\mathbf{P}^t(t)$. The tangent vector of a PC curve is therefore

$$\mathbf{P}^t(t) = 3\mathbf{a}t^2 + 2\mathbf{b}t + \mathbf{c}. \tag{4.1}$$

We denote the two given points by $\mathbf{P}_1$ and $\mathbf{P}_2$ and the two given tangents by $\mathbf{P}_1^t$ and $\mathbf{P}_2^t$. The four quantities are now used to calculate the *geometric representation* of the PC by writing equations that relate the four unknown coefficients $\mathbf{a}$, $\mathbf{b}$, $\mathbf{c}$, and $\mathbf{d}$ to the four known ones, $\mathbf{P}_1$, $\mathbf{P}_2$, $\mathbf{P}_1^t$, and $\mathbf{P}_2^t$. The equations are $\mathbf{P}(0) = \mathbf{P}_1$, $\mathbf{P}(1) = \mathbf{P}_2$, $\mathbf{P}^t(0) = \mathbf{P}_1^t$, and $\mathbf{P}^t(1) = \mathbf{P}_2^t$ [compare with Equations (3.2)]. Their explicit forms are

$$\begin{aligned}
\mathbf{a}{\cdot}0^3 + \mathbf{b}{\cdot}0^2 + \mathbf{c}{\cdot}0 + \mathbf{d} &= \mathbf{P}_1, \\
\mathbf{a}{\cdot}1^3 + \mathbf{b}{\cdot}1^2 + \mathbf{c}{\cdot}1 + \mathbf{d} &= \mathbf{P}_2, \\
3\mathbf{a}{\cdot}0^2 + 2\mathbf{b}{\cdot}0 + \mathbf{c} &= \mathbf{P}_1^t, \\
3\mathbf{a}{\cdot}1^2 + 2\mathbf{b}{\cdot}1 + \mathbf{c} &= \mathbf{P}_2^t.
\end{aligned} \tag{4.2}$$

They are easy to solve and the solutions are

$$\mathbf{a} = 2\mathbf{P}_1 - 2\mathbf{P}_2 + \mathbf{P}_1^t + \mathbf{P}_2^t, \quad \mathbf{b} = -3\mathbf{P}_1 + 3\mathbf{P}_2 - 2\mathbf{P}_1^t - \mathbf{P}_2^t, \quad \mathbf{c} = \mathbf{P}_1^t, \quad \mathbf{d} = \mathbf{P}_1. \tag{4.3}$$

Substituting these solutions into Equation (3.1) gives

$$\mathbf{P}(t) = (2\mathbf{P}_1 - 2\mathbf{P}_2 + \mathbf{P}_1^i + \mathbf{P}_2^i)t^3 + (-3\mathbf{P}_1 + 3\mathbf{P}_2 - 2\mathbf{P}_1^i - \mathbf{P}_2^i)t^2 + \mathbf{P}_1^t t + \mathbf{P}_1, \quad (4.4)$$

which, after rearranging, becomes

$$
\begin{aligned}
\mathbf{P}(t) &= (2t^3 - 3t^2 + 1)\mathbf{P}_1 + (-2t^3 + 3t^2)\mathbf{P}_2 + (t^3 - 2t^2 + t)\mathbf{P}_1^t + (t^3 - t^2)\mathbf{P}_2^t \\
&= F_1(t)\mathbf{P}_1 + F_2(t)\mathbf{P}_2 + F_3(t)\mathbf{P}_1^t + F_4(t)\mathbf{P}_2^t \\
&= (F_1(t), F_2(t), F_3(t), F_4(t))(\mathbf{P}_1, \mathbf{P}_2, \mathbf{P}_1^t, \mathbf{P}_2^t)^T \\
&= \mathbf{F}(t)\mathbf{B}, \hspace{6cm} (4.5)
\end{aligned}
$$

where

$$
\begin{aligned}
F_1(t) = (2t^3 - 3t^2 + 1), \quad F_2(t) = (-2t^3 + 3t^2) = 1 - F_1(t), \\
F_3(t) = (t^3 - 2t^2 + t), \quad F_4(t) = (t^3 - t^2),
\end{aligned} \hspace{2cm} (4.6)
$$

$\mathbf{B}$ is the column $(\mathbf{P}_1, \mathbf{P}_2, \mathbf{P}_1^t, \mathbf{P}_2^t)^T$, and $\mathbf{F}(t)$ is the row $(F_1(t), F_2(t), F_3(t), F_4(t))$. Equations (4.4) and (4.5) are the geometric representation of the Hermite PC segment.

Functions $F_i(t)$ are the Hermite blending functions. They create any point on the curve as a blend of the four given quantities. They are shown in Figure 4.3. Note that $F_1(t) + F_2(t) \equiv 1$. These two functions blend points, not tangent vectors, and should therefore be barycentric. We can also write $F_1(t) = (t^3, t^2, t, 1)(2, -3, 0, 1)^T$ and similarly for $F_2(t)$, $F_3(t)$, and $F_4(t)$. In matrix notation this becomes

$$
\mathbf{F}(t) = (t^3, t^2, t, 1)
\begin{pmatrix}
2 & -2 & 1 & 1 \\
-3 & 3 & -2 & -1 \\
0 & 0 & 1 & 0 \\
1 & 0 & 0 & 0
\end{pmatrix}
= \mathbf{T}(t)\,\mathbf{H}.
$$

The curve can now be written

$$
\mathbf{P}(t) = \mathbf{F}(t)\mathbf{B} = \mathbf{T}(t)\,\mathbf{H}\,\mathbf{B} = (t^3, t^2, t, 1)
\begin{pmatrix}
2 & -2 & 1 & 1 \\
-3 & 3 & -2 & -1 \\
0 & 0 & 1 & 0 \\
1 & 0 & 0 & 0
\end{pmatrix}
\begin{pmatrix}
\mathbf{P}_1 \\
\mathbf{P}_2 \\
\mathbf{P}_1^t \\
\mathbf{P}_2^t
\end{pmatrix}. \hspace{1cm} (4.7)
$$

Equation (3.1) tells us that $\mathbf{P}(t) = \mathbf{T}(t)\,\mathbf{A}$, which implies $\mathbf{A} = \mathbf{H}\,\mathbf{B}$. Matrix $\mathbf{H}$ is called the Hermite basis matrix.

The following is *Mathematica* code to display a single Hermite curve segment.

```
Clear[T,H,B]; (* Hermite Interpolation *)
T={t^3,t^2,t,1};
H={{2,-2,1,1},{-3,3,-2,-1},{0,0,1,0},{1,0,0,0}};
B={{0,0},{2,1},{1,1},{1,0}};
ParametricPlot[T.H.B,{t,0,1},PlotRange->All]
```

⋄ **Exercise 4.3:** Express the midpoint $\mathbf{P}(0.5)$ of a Hermite segment in terms of the two endpoints and two tangent vectors. Draw a diagram to illustrate the geometric interpretation of the result.

4.2.1 Hermite Blending Functions

The four Hermite blending functions of Equation (4.6) are illustrated graphically in Figure 4.3. An analysis of these functions is essential for a thorough understanding of the Hermite interpolation method.

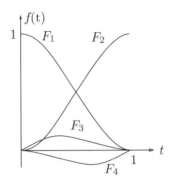

Figure 4.3: Hermite Weight Functions

Function $F_1(t)$ is the weight assigned to the start point $\mathbf{P}_1$. It goes down from its maximum $F_1(0) = 1$ to $F_1(1) = 0$. This shows why for small values of t the curve is close to $\mathbf{P}_1$ and why $\mathbf{P}_1$ has little or no influence on the curve for large values of t. The opposite is true for $F_2(t)$, the weight of the endpoint $\mathbf{P}_2$. Function $F_3(t)$ is a bit trickier. It starts at zero, has a maximum at $t = 1/3$, then drops slowly back to zero. This behavior is interpreted as follows:

1. For small values of t, function $F_3(t)$ has almost no effect. The curve stays close to $\mathbf{P}_1$ regardless of the extreme tangents or anything else.

2. For t values around $1/3$, weight $F_3(t)$ exerts some influence on the curve. For these t values, weight $F_4(t)$ is small, and the curve is (approximately) the sum of (1) point $F_1(t)\mathbf{P}_1$ (large contribution), (2) point $F_2(t)\mathbf{P}_2$ (small contribution), and (3) vector $F_3(t)\mathbf{P}_1^t$. The sum of a point $\mathbf{P} = (x, y)$ and a vector $\mathbf{v} = (v_x, v_y)$ is a point located at $(x + v_x, y + v_y)$, which is how weight $F_3(t)$ "pulls" the curve in the direction of tangent vector $\mathbf{P}_1^t$.

3. For large t values, function $F_3(t)$ again has almost no effect. The curve moves closer to $\mathbf{P}_2$ because weight $F_2(t)$ becomes dominant.

Function $F_4(t)$ is interpreted in a similar way. It has almost no effect for small and for large values of t. Its maximum (actually, minimum, because it is negative) occurs at $t = 2/3$, so it affects the curve only in this region. For t values close to $2/3$, the curve is the sum of point $F_2(t)\mathbf{P}_2$ (large contribution), point $F_1(t)\mathbf{P}_1$ (small contribution), and vector $-|F_4(t)|\mathbf{P}_2^t$. Because $F_4(t)$ is negative, this sum is equivalent to $(x - v_x, y - v_y)$, which is why the curve approaches endpoint $\mathbf{P}_2$ while moving in direction $\mathbf{P}_2^t$.

Another important feature of the Hermite weight functions is that $F_1(t)$ and $F_2(t)$ are barycentric. They have to be, since they blend two points, and a detailed look at the four equations (4.2) explains why they are. The first of these equations is simply $\mathbf{d} = \mathbf{P}_1$, which reduces the second one to $\mathbf{a}+\mathbf{b}+\mathbf{c}+\mathbf{d} = \mathbf{P}_2$ or $\mathbf{a}+\mathbf{b}+\mathbf{c} = \mathbf{P}_2-\mathbf{P}_1$. The third equation solves $\mathbf{c}$, and the fourth equation, combined with the second equation, is finally used to compute $\mathbf{a}$ and $\mathbf{b}$. All this implies that $\mathbf{a}$ and $\mathbf{b}$ have the form $\mathbf{a} = \alpha(\mathbf{P}_2 - \mathbf{P}_1) + \cdots$, $\mathbf{b} = \beta(\mathbf{P}_2 - \mathbf{P}_1) + \cdots$. The final PC therefore has the form

$$\mathbf{P}(t) = \mathbf{a}t^3 + \mathbf{b}t^2 + \mathbf{c}t + \mathbf{d} = (\alpha\mathbf{P}_2 - \alpha\mathbf{P}_1 + \cdots)t^3 + (\beta\mathbf{P}_2 - \beta\mathbf{P}_1 + \cdots)t^2 + (\cdots)t + \mathbf{P}_1,$$

where the ellipsis represent parts that depend only on the tangent vectors, not on the endpoints. When this is rearranged, the result is

$$\mathbf{P}(t) = (-\alpha t^3 - \beta t^2 + 1)\mathbf{P}_1 + (\alpha t^3 + \beta t^2)\mathbf{P}_2 + (\cdots)\mathbf{P}_1^t + (\cdots)\mathbf{P}_2^t,$$

which is why the coefficients of $\mathbf{P}_1$ and $\mathbf{P}_2$ add up to unity.

4.2.2 Hermite Derivatives

The concept of blending can be applied to the calculation of the derivatives of a curve, not just to the curve itself. One way to calculate $\mathbf{P}^t(t)$ is to differentiate $\mathbf{T}(t) = (t^3, t^2, t, 1)$. The result is

$$\mathbf{P}^t(t) = \mathbf{T}^t(t)\mathbf{H}\mathbf{B} = (3t^2, 2t, 1, 0)\mathbf{H}\mathbf{B}.$$

A more general method is to use the relation $\mathbf{P}(t) = \mathbf{F}(t)\mathbf{B}$, which implies

$$\mathbf{P}^t(t) = \mathbf{F}^t(t)\mathbf{B} = \big(F_1^t(t), F_2^t(t), F_3^t(t), F_4^t(t)\big)\mathbf{B}.$$

The individual derivatives $F_i^t(t)$ can be obtained from Equation (4.6). The results can be expressed as

$$\mathbf{P}^t(t) = (t^3, t^2, t, 1)\begin{bmatrix} 0 & 0 & 0 & 0 \\ 6 & -6 & 3 & 3 \\ -6 & 6 & -4 & -2 \\ 0 & 0 & 1 & 0 \end{bmatrix}\begin{bmatrix} \mathbf{P}_1 \\ \mathbf{P}_2 \\ \mathbf{P}_1^t \\ \mathbf{P}_2^t \end{bmatrix} = \mathbf{T}(t)\mathbf{H}_t\mathbf{B}. \qquad (4.8)$$

Similarly, the second derivatives of the Hermite segment can be expressed as

$$\mathbf{P}^{tt}(t) = (t^3, t^2, t, 1)\begin{bmatrix} 0 & 0 & 0 & 0 \\ 0 & 0 & 0 & 0 \\ 12 & -12 & 6 & 6 \\ -6 & 6 & -4 & -2 \end{bmatrix}\begin{bmatrix} \mathbf{P}_1 \\ \mathbf{P}_2 \\ \mathbf{P}_1^t \\ \mathbf{P}_2^t \end{bmatrix} = \mathbf{T}(t)\mathbf{H}_{tt}\mathbf{B}. \qquad (4.9)$$

These expressions make it easy to calculate the first and second derivatives at any point on a Hermite segment. Similar expressions can be derived for any other curves that are based on the blending of geometrical quantities.

⋄ **Exercise 4.4:** What is H_{ttt}?

Example: The two two-dimensional points $\mathbf{P}_1 = (0,0)$ and $\mathbf{P}_2 = (1,0)$ and the two tangents $\mathbf{P}_1^t = (1,1)$ and $\mathbf{P}_2^t = (0,-1)$ are given. The segment should therefore start at the origin, going in a 45° direction, and end at point $(1,0)$, going straight down. The calculation of $\mathbf{P}(t)$ is straightforward:

$$\mathbf{P}(t) = \mathbf{T}(t)\,\mathbf{A} = \mathbf{T}(t)\,\mathbf{H}\,\mathbf{B}$$

$$= (t^3, t^2, t, 1) \begin{bmatrix} 2 & -2 & 1 & 1 \\ -3 & 3 & -2 & -1 \\ 0 & 0 & 1 & 0 \\ 1 & 0 & 0 & 0 \end{bmatrix} \begin{bmatrix} (0,0) \\ (1,0) \\ (1,1) \\ (0,-1) \end{bmatrix}$$

$$= (t^3, t^2, t, 1) \begin{bmatrix} 2(0,0) - 2(1,0) + 1(1,1) + 1(0,-1) \\ -3(0,0) + 3(1,0) - 2(1,1) - 1(0,-1) \\ 0(0,0) + 0(1,0) + 1(1,1) + 0(0,-1) \\ 1(0,0) + 0(1,0) + 0(1,1) + 0(0,-1) \end{bmatrix}$$

$$= (t^3, t^2, t, 1) \begin{bmatrix} (-1,0) \\ (1,-1) \\ (1,1) \\ (0,0) \end{bmatrix}$$

$$= (-1,0)t^3 + (1,-1)t^2 + (1,1)t. \tag{4.10}$$

⋄ **Exercise 4.5:** Use Equation (4.10) to show that the segment really passes through points $(0,0)$ and $(1,0)$. Calculate the tangent vectors and use them to show that the segment really starts and ends in the right directions.

⋄ **Exercise 4.6:** Repeat the example above with $\mathbf{P}_1^t = (2,2)$. The new curve segment should go through the same points, in the same directions. However, it should continue longer in the original 45° direction, since the size of the new tangent is $\sqrt{2^2 + 2^2} = 2\sqrt{2}$, twice as long as the previous one, which is $\sqrt{1^2 + 1^2} = \sqrt{2}$.

⋄ **Exercise 4.7:** Calculate the Hermite curve for two given points $\mathbf{P}_1$ and $\mathbf{P}_2$ assuming that the tangent vectors at the two points are zero (indeterminate). What kind of a curve is this?

⋄ **Exercise 4.8:** Use the Hermite method to calculate PC segments for the cases where the known quantities are as follows:

1. The three tangent vectors at the start, middle, and end of the segment.

2. The two interior points $\mathbf{P}(1/3)$ and $\mathbf{P}(2/3)$, and the two extreme tangent vectors $\mathbf{P}^t(0)$ and $\mathbf{P}^t(1)$.

3. The two extreme points $\mathbf{P}(0)$ and $\mathbf{P}(1)$, and the two interior tangent vectors $\mathbf{P}^t(1/3)$ and $\mathbf{P}^t(2/3)$ (this is similar to case 2, so it's easy).

Example: Given the two three-dimensional points $\mathbf{P}_1 = (0,0,0)$ and $\mathbf{P}_2 = (1,1,1)$ and the two tangent vectors $\mathbf{P}_1^t = (1,0,0)$ and $\mathbf{P}_2^t = (0,1,0)$, the curve segment is the simple cubic polynomial shown in Figure 4.4

$$\mathbf{P}(t) = (t^3, t^2, t, 1) \begin{bmatrix} 2 & -2 & 1 & 1 \\ -3 & 3 & -2 & -1 \\ 0 & 0 & 1 & 0 \\ 1 & 0 & 0 & 0 \end{bmatrix} \begin{bmatrix} (0,0,0) \\ (1,1,1) \\ (1,0,0) \\ (0,1,0) \end{bmatrix}$$

$$= (-t^3 + t^2 + t, -t^3 + 2t^2, -2t^3 + 3t^2). \tag{4.11}$$

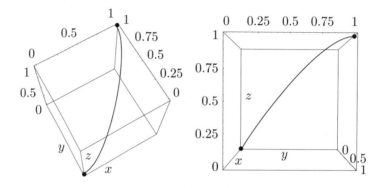

```
<<:Graphics:ParametricPlot3D.m; (* Hermite 3D example *)
Clear[T,H,B];
T={t^3,t^2,t,1};
H={{2,-2,1,1},{-3,3,-2,-1},{0,0,1,0},{1,0,0,0}};
B={{0,0,0},{1,1,1},{1,0,0},{0,1,0}};
ParametricPlot3D[T.H.B,{t,0,1}, Compiled->False,
 ViewPoint->{-0.846, -1.464, 3.997}, DefaultFont->{"cmr10", 10}];
(* ViewPoint->{3.119, -0.019, 0.054} alt view *)
```

Figure 4.4: A Hermite Curve Segment in Space.

> I'm retired—goodbye tension, hello pension!
> —Anonymous.

4.2.3 Hermite Segments With Tension

This section shows how to create a Hermite curve segment under tension by employing a nonuniform Hermite segment. Such a segment is obtained when the parameter t varies in the interval $[0, \Delta]$, where Δ can be any real positive number. The derivation of this case is similar to the uniform case. Equation (4.2) becomes

$$\mathbf{a}\cdot 0^3 + \mathbf{b}\cdot 0^2 + \mathbf{c}\cdot 0 + \mathbf{d} = \mathbf{P}_1,$$
$$\mathbf{a}\Delta^3 + \mathbf{b}\Delta^2 + \mathbf{c}\Delta + \mathbf{d} = \mathbf{P}_2,$$
$$3\mathbf{a}\cdot 0^2 + 2\mathbf{b}\cdot 0 + \mathbf{c} = \mathbf{P}_1^t,$$
$$3\mathbf{a}\Delta^2 + 2\mathbf{b}\Delta + \mathbf{c} = \mathbf{P}_2^t,$$

with solutions

$$\mathbf{a} = \frac{2(\mathbf{P}_1 - \mathbf{P}_2)}{\Delta^3} + \frac{\mathbf{P}_1^t + \mathbf{P}_2^t}{\Delta^2},$$

$$\mathbf{b} = \frac{3(\mathbf{P}_2 - \mathbf{P}_1)}{\Delta^2} - \frac{2\mathbf{P}_1^t}{\Delta} - \frac{\mathbf{P}_2^t}{\Delta},$$

$$\mathbf{c} = \mathbf{P}_1^t,$$

$$\mathbf{d} = \mathbf{P}_1.$$

The curve segment can now be expressed, similar to Equation (4.7), in the form

$$\mathbf{P}_{nu}(t) = (t^3, t^2, t, 1) \begin{pmatrix} \frac{2}{\Delta^3} & \frac{-2}{\Delta^3} & \frac{1}{\Delta^2} & \frac{1}{\Delta^2} \\ \frac{-3}{\Delta^2} & \frac{3}{\Delta^2} & \frac{-2}{\Delta} & \frac{-1}{\Delta} \\ 0 & 0 & 1 & 0 \\ 1 & 0 & 0 & 0 \end{pmatrix} \begin{pmatrix} \mathbf{P}_1 \\ \mathbf{P}_2 \\ \mathbf{P}_1^t \\ \mathbf{P}_2^t \end{pmatrix} = \mathbf{T}(t)\mathbf{H}_{nu}\mathbf{B}. \tag{4.12}$$

It is easy to verify that matrix $\mathbf{H}_{nu}$ reduces to $\mathbf{H}$ for $\Delta = 1$. Figure 4.5 shows a typical nonuniform Hermite segment drawn three times for $\Delta = 0.5, 1$, and 2. Careful examination of the three curves shows that increasing the value of Δ causes the curve segment to continue longer in its initial and final directions; it has the same effect as increasing the magnitudes of the tangent vectors of the uniform Hermite segment. Once this is grasped, the reader should not be surprised to learn that the nonuniform curve of Equation (4.12) can also be expressed as

$$\mathbf{P}_{nu}(t) = (t^3, t^2, t, 1) \begin{pmatrix} 2 & -2 & 1 & 1 \\ -3 & 3 & -2 & -1 \\ 0 & 0 & 1 & 0 \\ 1 & 0 & 0 & 0 \end{pmatrix} \begin{pmatrix} \mathbf{P}_1 \\ \mathbf{P}_2 \\ \Delta\mathbf{P}_1^t \\ \Delta\mathbf{P}_2^t \end{pmatrix}. \tag{4.13}$$

This shows that the nonuniform Hermite curve segment is a special case of the uniform curve. Any nonuniform Hermite curve can also be obtained as a uniform Hermite curve by adjusting the magnitudes of the tangent vectors. However, varying the magnitudes of both tangent vectors has an important geometric interpretation, it changes the *tension* of the curve segment. Imagine that the two endpoints are nails driven into the page and the curve segment is a rubber string. When the string is pulled at both sides, its shape approaches a straight line. Figure 4.5 shows how decreasing Δ results in a curve with higher tension, so instead of working with nonuniform Hermite segments, we can consider Δ a tension parameter. Practical curve methods that create a spline curve out of individual Hermite segments can add a tension parameter to the spline, thereby making the method more interactive. An example is the cardinal splines method (Section 5.4).

4.2.4 PC Conic Approximations

Hermite interpolation can be applied to compute (approximate) conic sections (see Appendix A for more on conics). Given three points $\mathbf{P}_0$, $\mathbf{P}_1$, and $\mathbf{P}_2$ and a scalar α, we construct the 4-tuple

$$(\mathbf{P}_0, \mathbf{P}_2, 4\alpha(\mathbf{P}_1 - \mathbf{P}_0), 4\alpha(\mathbf{P}_2 - \mathbf{P}_1)), \quad \text{where } 0 \leq \alpha \leq 1, \tag{4.14}$$

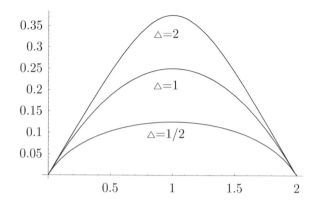

```
Clear[T,H,B]; (* Nonuniform Hermite segments *)
T={t^3,t^2,t,1};
H={{2,-2,1,1},{-3,3,-2,-1},{0,0,1,0},{1,0,0,0}};
B[delta_]:={{0,0},{2,0},delta{2,1},delta{2,-1}};
g1=ParametricPlot[T.H.B[0.5],{t,0,1},Compiled->False,
 DisplayFunction->Identity];
g2=ParametricPlot[T.H.B[1],{t,0,1},Compiled->False,
 DisplayFunction->Identity];
g3=ParametricPlot[T.H.B[1.5],{t,0,1},Compiled->False,
 DisplayFunction->Identity];
Show[g1,g2,g3, DisplayFunction->$DisplayFunction, DefaultFont->{"cmr10", 10}]
```

Figure 4.5: Three Nonuniform Hermite Segments.

to become our two points and two extreme tangent vectors and compute a segment that approximates a conic section. We obtain an ellipse when $0 \le \alpha < 0.5$, a parabola when $\alpha = 0.5$, and a hyperbola when $0.5 < \alpha \le 1$ (see below for a circle).

The tangent vectors at the two ends are $\mathbf{P}^t(0) = 4\alpha(\mathbf{P}_1 - \mathbf{P}_0)$ and $\mathbf{P}^t(1) = 4\alpha(\mathbf{P}_2 - \mathbf{P}_1)$ (note their directions). The tangent vector halfway is $\mathbf{P}^t(0.5) = (1.5 - \alpha)(\mathbf{P}_2 - \mathbf{P}_0)$. It is parallel to the vector $\mathbf{P}_2 - \mathbf{P}_0$.

The case of the parabola is especially useful and is explicitly shown here. Substituting $\alpha = 0.5$ in Equation (4.14) and applying Equation (4.7) yields the Hermite segment

$$
\mathbf{P}(t) = (t^3, t^2, t, 1) \begin{bmatrix} 2 & -2 & 1 & 1 \\ -3 & 3 & -2 & -1 \\ 0 & 0 & 1 & 0 \\ 1 & 0 & 0 & 0 \end{bmatrix} \begin{bmatrix} \mathbf{P}_0 \\ \mathbf{P}_2 \\ 2(\mathbf{P}_1 - \mathbf{P}_0) \\ 2(\mathbf{P}_2 - \mathbf{P}_1) \end{bmatrix}
$$
$$
= (1 - t)^2 \mathbf{P}_0 + 2t(1 - t)\mathbf{P}_1 + t^2 \mathbf{P}_2.
$$

This is the parabola produced in Exercises 4.9 and 6.2.

◇ **Exercise 4.9:** We know that any three points $\mathbf{P}_0$, $\mathbf{P}_1$, and $\mathbf{P}_2$ define a unique parabola (i.e., a triangle defines a parabola). Use Hermite interpolation to calculate the parabola from $\mathbf{P}_0$ to $\mathbf{P}_2$ whose start and end tangents go in the directions from $\mathbf{P}_0$ to $\mathbf{P}_1$ and from $\mathbf{P}_1$ to $\mathbf{P}_2$, respectively.

Hermite interpolation provides a simple way to construct approximate circles and circular arcs. Figure 4.6a shows how this method is employed to construct a circular arc of unit radius about the origin. We assume that an arc spanning an angle 2θ is needed and we place its two endpoints $\mathbf{P}_1$ and $\mathbf{P}_2$ at locations $(\cos\theta, -\sin\theta)$ and $(\cos\theta, \sin\theta)$, respectively. This arc is symmetric about the x axis, but we later show how to rotate it to have an arbitrary arc. Since a circle is always perpendicular to its radius, we select as our start and end tangents two vectors that are perpendicular to $\mathbf{P}_1$ and $\mathbf{P}_2$. They are $\mathbf{P}_1^t = a(\sin\theta, \cos\theta)$ and $\mathbf{P}_2^t = a(-\sin\theta, \cos\theta)$, where a is a parameter to be determined. The Hermite curve segment defined by these points and vectors is, as usual,

$$
\begin{aligned}
\mathbf{P}(t) = (t^3, t^2, t, 1) &\begin{bmatrix} 2 & -2 & 1 & 1 \\ -3 & 3 & -2 & -1 \\ 0 & 0 & 1 & 0 \\ 1 & 0 & 0 & 0 \end{bmatrix} \begin{bmatrix} (\cos\theta, -\sin\theta) \\ (\cos\theta, \sin\theta) \\ a(\sin\theta, \cos\theta) \\ a(-\sin\theta, \cos\theta) \end{bmatrix} \\
&= (2t^3 - 3t^2 + 1)(\cos\theta, -\sin\theta) + (-2t^3 + 3t^2)(\cos\theta, \sin\theta) \\
&\quad + (t^3 - 2t^2 + t)a(\sin\theta, \cos\theta) + (t^3 - t^2)a(-\sin\theta, \cos\theta).
\end{aligned} \tag{4.15}
$$

We need an equation in order to determine a and we obtain it by requiring that the curve segment passes through the circular arc at its center, i.e., $\mathbf{P}(0.5) = (1,0)$. This produces the equation

$$
\begin{aligned}
(1,0) = \mathbf{P}(0.5) &= \left(\frac{2}{8} - \frac{3}{4} + 1\right)(\cos\theta, -\sin\theta) + \left(-\frac{2}{8} + \frac{3}{4}\right)(\cos\theta, \sin\theta) \\
&\quad + \left(\frac{1}{8} - \frac{2}{4} + \frac{1}{2}\right)a(\sin\theta, \cos\theta) + \left(\frac{1}{8} - \frac{1}{4}\right)a(-\sin\theta, \cos\theta) \\
&= \frac{1}{8}(8\cos\theta + 2a\sin\theta, 0),
\end{aligned}
$$

whose solution is

$$
a = \frac{4(1 - \cos\theta)}{\sin\theta}.
$$

The curve can now be written in the form

$$
\mathbf{P}(t) = (t^3, t^2, t, 1) \begin{bmatrix} 2 & -2 & 1 & 1 \\ -3 & 3 & -2 & -1 \\ 0 & 0 & 1 & 0 \\ 1 & 0 & 0 & 0 \end{bmatrix} \begin{bmatrix} (\cos\theta, -\sin\theta) \\ (\cos\theta, \sin\theta) \\ \left(4(1 - \cos\theta), \frac{4(1-\cos\theta)}{\tan\theta}\right) \\ \left(-4(1 - \cos\theta), \frac{4(1-\cos\theta)}{\tan\theta}\right) \end{bmatrix}.
$$

This curve provides an excellent approximation to a circular arc, even for angles θ as large as $90°$.

◇ **Exercise 4.10:** Write Equation (4.15) for $\theta = 90°$; calculate $\mathbf{P}(0.25)$ and the deviation of the curve from a true circle at this point.

In general, an arc with a unit radius is not symmetric about the x axis but may look as in Figure 4.6b, where $\mathbf{P}_1$ and $\mathbf{P}_2$ are any points at a distance of one unit from

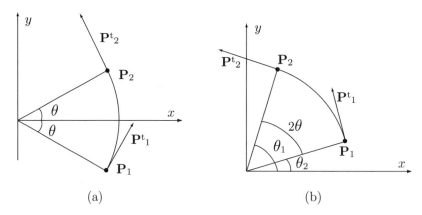

Figure 4.6: Hermite Segment and a Circular Arc.

the origin. All that's necessary to calculate the arc from Equation (4.15) is the value of θ (where 2θ is the angle between $\mathbf{P}_1$ and $\mathbf{P}_2$) and this can be calculated numerically from the two points using the relations

$$\theta = (\theta_1 - \theta_2)/2, \quad \cos\theta_1 = \mathbf{P}_1 \bullet (1,0), \quad \cos\theta_2 = \mathbf{P}_2 \bullet (1,0),$$
$$\cos(2\theta) = \cos(\theta_1 - \theta_2) = \cos\theta_1 \cos\theta_2 + \sin\theta_1 \sin\theta_2,$$
$$\cos\theta = \pm\sqrt{[1 + \cos(2\theta)]/2}, \quad \sin\theta = \sqrt{1 - \cos^2\theta}.$$

4.3 Degree-5 Hermite Interpolation

It is possible to extend the basic idea of Hermite interpolation to polynomials of higher degree. Naturally, more data is needed in order to calculate such a polynomial, and this data is provided by the user, normally in the form of higher-order derivatives of the curve. If the user specifies the two endpoints, the two extreme tangent vectors, and the two extreme second derivatives, the software can use these six data items to calculate the six coefficients of a fifth-degree polynomial that interpolates the two points. In general, if the two endpoints and the first k pairs of derivatives at the extreme points are known (a total of $2k + 2$ items), they can be used to calculate an interpolating polynomial of degree $2k + 1$. These higher-degree polynomials are not as useful as the cubic, but the fifth-degree polynomial is shown here, as a demonstration of the power of Hermite interpolation (see also Section 5.3).

Given two endpoints $\mathbf{P}_1$ and $\mathbf{P}_2$, the values of two tangent vectors $\mathbf{P}_1^t$ and $\mathbf{P}_2^t$, and of two second derivatives $\mathbf{P}_1^{tt}$ and $\mathbf{P}_2^{tt}$, we can calculate the polynomial

$$\mathbf{P}(t) = \mathbf{a}t^5 + \mathbf{b}t^4 + \mathbf{c}t^3 + \mathbf{d}t^2 + \mathbf{e}t + \mathbf{f} \tag{4.16}$$

by writing the six equations

$$\mathbf{P}(0) = \mathbf{a}t^5 + \mathbf{b}t^4 + \mathbf{c}t^3 + \mathbf{d}t^2 + \mathbf{e}t + \mathbf{f}|_0 = \mathbf{f} = \mathbf{P}_1,$$

$$\mathbf{P}(1) = \mathbf{a}t^5 + \mathbf{b}t^4 + \mathbf{c}t^3 + \mathbf{d}t^2 + \mathbf{e}t + \mathbf{f}|_1 = \mathbf{a} + \mathbf{b} + \mathbf{c} + \mathbf{d} + \mathbf{e} + \mathbf{f} = \mathbf{P}_2,$$

$$\mathbf{P}^t(0) = 5\mathbf{a}t^4 + 4\mathbf{b}t^3 + 3\mathbf{c}t^2 + 2\mathbf{d}t + \mathbf{e}|_0 = \mathbf{e} = \mathbf{P}_1^t,$$

$$\mathbf{P}^t(1) = 5\mathbf{a}t^4 + 4\mathbf{b}t^3 + 3\mathbf{c}t^2 + 2\mathbf{d}t + \mathbf{e}|_1 = 5\mathbf{a} + 4\mathbf{b} + 3\mathbf{c} + 2\mathbf{d} + \mathbf{e} = \mathbf{P}_2^t,$$

$$\mathbf{P}^{tt}(0) = 20\mathbf{a}t^3 + 12\mathbf{b}t^2 + 6\mathbf{c}t + 2\mathbf{d}|_0 = 2\mathbf{d} = \mathbf{P}_1^{tt},$$

$$\mathbf{P}^{tt}(1) = 20\mathbf{a}t^3 + 12\mathbf{b}t^2 + 6\mathbf{c}t + 2\mathbf{d}|_1 = 20\mathbf{a} + 12\mathbf{b} + 6\mathbf{c} + 2\mathbf{d} = \mathbf{P}_2^{tt}.$$

Solving for the six unknown coefficients yields the degree-5 Hermite interpolating polynomial

$$\mathbf{P}(t) = F_1(t)\mathbf{P}_1 + F_2(t)\mathbf{P}_2 + F_3(t)\mathbf{P}_1^t + F_4(t)\mathbf{P}_2^t + F_5(t)\mathbf{P}_1^{tt} + F_6(t)\mathbf{P}_2^{tt}$$

$$= (-6t^5 + 15t^4 - 10t^3 + 1)\mathbf{P}_1 + (6t^5 - 15t^4 + 10t^3)\mathbf{P}_2$$

$$+ (-3t^5 + 8t^4 - 6t^3 + t)\mathbf{P}_1^t + (-3t^5 + 7t^4 - 4t^3)\mathbf{P}_2^t$$

$$+ \left(-(1/2)t^5 + (3/2)t^4 - (3/2)t^3 + (1/2)t^2\right)\mathbf{P}_1^{tt} + \left((1/2)t^5 - t^4 + (1/2)t^3\right)\mathbf{P}_2^{tt}$$

$$= (t^5, t^4, t^3, t^2, t, 1)
\begin{bmatrix}
-6 & 6 & -3 & -3 & -1/2 & 1/2 \\
15 & -15 & 8 & 7 & 3/2 & -1 \\
-10 & 10 & -6 & -4 & -3/2 & 1/2 \\
0 & 0 & 0 & 0 & 1/2 & 0 \\
0 & 0 & 1 & 0 & 0 & 0 \\
1 & 0 & 0 & 0 & 0 & 0
\end{bmatrix}
\begin{bmatrix}
\mathbf{P}_1 \\
\mathbf{P}_2 \\
\mathbf{P}_1^t \\
\mathbf{P}_2^t \\
\mathbf{P}_1^{tt} \\
\mathbf{P}_2^{tt}
\end{bmatrix}. \tag{4.17}$$

4.4 Controlling the Hermite Segment

The Hermite method is interactive. In general, the points cannot be moved, but the tangent vectors can be varied. Even if their directions cannot be changed, their magnitudes normally are not fixed by the user and can be modified to edit the shape of the curve segment.

The simple experiment of this section illustrates the amount of editing and controlling that can be achieved just by varying the magnitudes of the tangents. We start with the Hermite segment defined by the two endpoints $\mathbf{P}_1 = (0,0)$ and $\mathbf{P}_2 = (2,1)$ and by the two tangent vectors $\mathbf{P}^t(0) = (1,1)$ and $\mathbf{P}^t(1) = (1,0)$. The curve starts in the $45°$ direction and ends in a horizontal direction. The curve is easy to calculate. Its expression is

$$\mathbf{P}(t) = (t^3, t^2, t, 1)
\begin{bmatrix}
2 & -2 & 1 & 1 \\
-3 & 3 & -2 & -1 \\
0 & 0 & 1 & 0 \\
1 & 0 & 0 & 0
\end{bmatrix}
\begin{bmatrix}
(0,0) \\
(2,1) \\
(1,1) \\
(1,0)
\end{bmatrix} = -(2,1)t^3 + (3,1)t^2 + (1,1)t. \tag{4.18}$$

Suppose that the user wants to raise the curve a bit, but also keep the same start and end directions and endpoints. The only way to edit the curve is to change the magnitudes of the tangents.

To keep the same directions, the new tangent vectors should have the form (a, a) and $(b, 0)$, where a and b are two new parameters that have to be computed. To raise the curve, we go through the following steps:

1. Calculate the midpoint of the curve. This is $\mathbf{P}(0.5) = (1, 5/8)$.

2. Decide by how much to raise it. Let's say we decide to raise the midpoint to $(1, 1)$.

3. Construct a new curve $\mathbf{Q}(t)$, based on the tangents (a, a) and $(b, 0)$.

4. Require that the new curve pass through $(1, 1)$ as its midpoint and determine a and b from this requirement.

The general form of the new curve is

$$\mathbf{Q}(t) = (t^3, t^2, t, 1) \begin{bmatrix} 2 & -2 & 1 & 1 \\ -3 & 3 & -2 & -1 \\ 0 & 0 & 1 & 0 \\ 1 & 0 & 0 & 0 \end{bmatrix} \begin{bmatrix} (0,0) \\ (2,1) \\ (a,a) \\ (b,0) \end{bmatrix}$$

$$= (a+b-4, a-2)t^3 + (-2a-b+6, 3-2a)t^2 + (a,a)t. \qquad (4.19)$$

The requirement $\mathbf{Q}(0.5) = (1, 1)$ can now be written

$$(a+b-4, a-2)/8 + (-2a-b+6, 3-2a)/4 + (a, a)/2 = (1, 1),$$

which yields the two equations $a+b-4+2(-2a-b+6)+4a = 8$ and $a-2+2(3-2a)+4a = 8$. The solutions are $a = b = 4$, so the new curve has the form

$$\mathbf{Q}(t) = (4, 2)t^3 - (6, 5)t^2 + (4, 4)t. \qquad (4.20)$$

A simple check verifies that this curve really starts at $(0, 0)$, ends at $(2, 1)$, has the extreme tangents $(4, 4)$ and $(4, 0)$, and passes midway through $(1, 1)$.

Raising the midpoint from $(1, 5/8)$ to $(1, 1)$ has completely changed the curve (Equations (4.18) and (4.20) are different). The new curve starts going in the same $45°$ direction, then starts going up, reaches point $(1, 1)$, starts going down, and still has "time" to arrive at point $(2, 1)$ moving horizontally. An interesting question is: How much can we raise the midpoint? If we raise it from $(1, 5/8)$ to, say, $(1, 100)$, would the curve be able to change directions, climb up, pass through the new midpoint, dive down, and still approach $(2, 1)$ moving horizontally?

To check this, let's assume that we raise the midpoint from $(1, 5/8)$ to $(1, 5/8 + \alpha)$, where α is a real number. The curve is constrained by $\mathbf{Q}(0.5) = (1, 5/8 + \alpha)$, which yields the equation

$$(a+b-4, a-2)/8 + (-2a-b+6, 3-2a)/4 + (a, a)/2 = (1, 5/8 + \alpha).$$

The solutions are $a = b = 1 + 8\alpha$. This means that α can vary without limit. When α is positive, the curve is pulled up. Negative values of α push the curve down. The value $\alpha = -1/8$ is special. It implies $a = b = 0$ and results in the curve $\mathbf{Q}(t) = (6t^2 - 4t^3, 3t^2 - 2t^3)$. The parameter substitution $u = 3t^2 - 2t^3$ yields $\mathbf{Q}(u) = (2u, u)$. This curve is the straight line from $(0,0)$ to $(2,1)$. Its midpoint is $(1, 1/2)$.

⋄ **Exercise 4.11:** Values $\alpha < -1/8$ result in negative a and b. Can they still be used in Equation (4.19)?

⋄ **Exercise 4.12:** How can we coerce the curve of Equation (4.19) to have point $(1, 0)$ as its midpoint?

Note: Raising the curve is done by increasing the size of the tangent vectors. This forces the curve to continue longer in the initial and final directions. This is also the reason why too much raising causes undesirable effects. Figure 4.7 shows the original curve ($\alpha = 0$) and the effects of increasing α. For $\alpha = 0.4$, the curve is raised and still has a reasonable shape. However, for larger values of α, the curve gets tight, develops a *cusp* (a kink), then starts looping on itself. It is easy to see that when $\alpha = 5/8$, the tangent vector becomes indefinite at the midpoint ($t = 0.5$). To show this, we differentiate the curve of Equation (4.19) to obtain the tangent

$$\mathbf{Q}^t(t) = 3(a + b - 4, a - 2)t^2 + 2(-2a - b + 6, 3 - 2a)t + (a, a).$$

From $a = b = 1 + 8\alpha$, we get

$$\mathbf{Q}^t(t) = (48\alpha - 6, 24\alpha - 3)t^2 + (6 - 48\alpha, 2 - 32\alpha)t + (1 + 8\alpha, 1 + 8\alpha).$$

For $\alpha = 5/8$, this reduces to $\mathbf{Q}^t(t) = (24, 12)t^2 - (24, 18)t + (6, 6)$, so $\mathbf{Q}^t(0.5) = (0, 0)$.

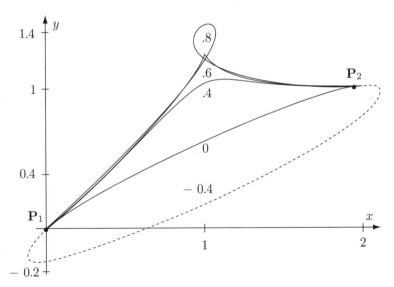

Figure 4.7: Effects of Changing α.

⬦ **Exercise 4.13:** Given the two endpoints $\mathbf{P}_1 = (0,0)$ and $\mathbf{P}_2 = (1,0)$ and the two tangent vectors $\mathbf{P}_1^t = \alpha(\cos\theta, \sin\theta)$ and $\mathbf{P}_1^t = \alpha(\cos\theta, -\sin\theta)$ (Figure 4.8), calculate the value of α for which the Hermite segment from $\mathbf{P}_1$ to $\mathbf{P}_2$ has a cusp.

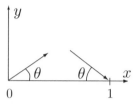

Figure 4.8: Tangents for Exercise 4.13.

The following problem may sometimes occur in practice. Given two endpoints $\mathbf{P}_1$ and $\mathbf{P}_2$, two *unit* tangent vectors $\mathbf{T}1$ and $\mathbf{T}2$, and a third point $\mathbf{P}_3$, find scale factors α and β such that the Hermite segment $\mathbf{P}(t)$ defined by points $\mathbf{P}_1$ and $\mathbf{P}_2$ and tangents $\alpha\mathbf{T}1$ and $\beta\mathbf{T}2$, respectively, will pass through $\mathbf{P}_3$. Also find the value t_0 for which $\mathbf{P}(t_0) = \mathbf{P}_3$.

We start with Equation (4.5), which in our case becomes

$$\mathbf{P}_3 = F_1(t_0)\mathbf{P}_1 + F_2(t_0)\mathbf{P}_2 + F_3(t_0)\alpha\mathbf{T}1 + F_4(t_0)\beta\mathbf{T}2,$$

where the $F_i(t)$ are given by Equation (4.6). Since $F_1(t) + F_2(t) \equiv 1$ we can write

$$\mathbf{P}_3 - \mathbf{P}_1 = F_2(t_0)(\mathbf{P}_2 - \mathbf{P}_1) + \alpha F_3(t_0)\mathbf{T}1 + \beta F_4(t_0)\mathbf{T}2.$$

This can now be written as the three scalar equations

$$\begin{aligned}
x_3 - x_1 &= F_2(t_0)(x_2 - x_1) + \alpha F_3(t_0)T1_x + \beta F_4(t_0)T2_x, \\
y_3 - y_1 &= F_2(t_0)(y_2 - y_1) + \alpha F_3(t_0)T1_y + \beta F_4(t_0)T2_y, \\
z_3 - z_1 &= F_2(t_0)(z_2 - z_1) + \alpha F_3(t_0)T1_z + \beta F_4(t_0)T2_z.
\end{aligned} \tag{4.21}$$

This is a system of three equations in the three unknowns α, β, and t_0. In principle, it should have a unique solution, but solving it is awkward since t_0 is included in the $F_i(t_0)$ functions, which are degree-3 polynomials in t_0. The first step is to isolate the two products $\alpha F_3(t_0)$ and $\beta F_4(t_0)$ in the first two equations. This yields

$$\begin{pmatrix} \alpha F_3(t_0) \\ \beta F_4(t_0) \end{pmatrix} = \begin{pmatrix} T1_x & T2_x \\ T1_y & T2_y \end{pmatrix}^{-1} \left[\begin{pmatrix} x_3 - x_1 \\ y_3 - y_1 \end{pmatrix} - \begin{pmatrix} x_2 - x_1 \\ y_2 - y_1 \end{pmatrix} F_2(t_0) \right].$$

This result is used in step two to eliminate $\alpha F_3(t_0)$ and $\beta F_4(t_0)$ from the third equation:

$$\begin{aligned}
z_3 - z_1 &= F_2(t_0)(z_2 - z_1) + (T1_z, T2_z)\begin{pmatrix} \alpha F_3(t_0) \\ \beta F_4(t_0) \end{pmatrix} \\
&= F_2(t_0)(z_2 - z_1) \\
&\quad + (T1_z, T2_z)\begin{pmatrix} T1_x & T2_x \\ T1_y & T2_y \end{pmatrix}^{-1} \left[\begin{pmatrix} x_3 - x_1 \\ y_3 - y_1 \end{pmatrix} - \begin{pmatrix} x_2 - x_1 \\ y_2 - y_1 \end{pmatrix} F_2(t_0) \right].
\end{aligned}$$

We now have an equation with the single unknown t_0. Step three is to simplify the result above by using the value $F_2(t_0) = -2t_0^3 + 3t_0^2$:

$$\begin{vmatrix} x_2 - x_1 & y_2 - y_1 & z_2 - z_1 \\ T1_x & T1_y & T1_z \\ T2_x & T2_y & T2_z \end{vmatrix} (-2t_0^3 + 3t_0^2) = \begin{vmatrix} x_3 - x_1 & y_3 - y_1 & z_3 - z_1 \\ T1_x & T1_y & T1_z \\ T2_x & T2_y & T2_z \end{vmatrix}. \quad (4.22)$$

Step four is to solve Equation (4.22) for t_0. Once t_0 is known, α and β can be computed from the other equations. Equation (4.22), however, is cubic in t_0, so it may have to be solved numerically and it may have between zero and three real solutions t_0. Any acceptable solution t_0 must be a real number in the range $[0, 1]$ and must result in positive α and β.

This, of course, is a slow, tedious approach and should only be used as a last resort, when nothing else works.

4.5 Truncating and Segmenting

Surfaces and solid objects are constructed of curves. When surfaces are joined, clipped, or intersected, there is sometimes a need to truncate curves. In general, the problem of truncating a curve starts with a parametric curve $\mathbf{P}(t)$ and the two values t_i and t_j. A new curve $\mathbf{Q}(T)$ needs be determined, that is identical to the segment $\mathbf{P}(t_i) \rightarrow \mathbf{P}(t_j)$ (Figure 4.9a) when T varies from 0 to 1. The discussion in this section is limited to Hermite segments. The endpoints of the new curve are $\mathbf{Q}(0) = \mathbf{P}(t_i)$ and $\mathbf{Q}(1) = \mathbf{P}(t_j)$. To understand how the two extreme tangent vectors of $\mathbf{Q}(T)$ are calculated, we first need to discuss reparametrization of parametric curves.

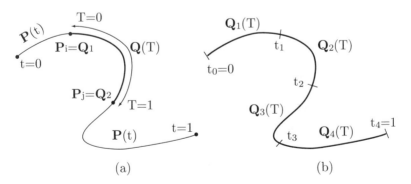

Figure 4.9: Truncating and Segmenting.

Reparametrization is the case where a new parameter $T(t)$ is substituted for the original parameter t. Notice that $T(t)$ is a function of t. One example of reparametrization is reversing the direction of a curve. It is easy to see that when t varies from 0 to 1, the simple function $T = 1 - t$ varies from 1 to 0. The two curves $\mathbf{P}(t)$ and $\mathbf{P}(1 - t)$ have the same shape and location but move in opposite directions. Another example of reparametrization is a curve $\mathbf{P}(t)$ with a parameter $0 \le t \le 1$ being transformed to a curve $\mathbf{Q}(T)$ with a parameter $a \le T \le b$ (Section 5.1.6 has an example). The simplest relation between T and t is linear, i.e., $T = at + b$. We can make two observations about this relation as follows:

1. At two different points i and j along the curve, the parameters are related by $T_i = at_i + b$ and $T_j = at_j + b$, respectively. Subtracting yields $T_j - T_i = a(t_j - t_i)$, so $a = (T_j - T_i)/(t_j - t_i)$.
   2. $T = at + b$ gives $dT = a\, dt$.

These two observations can be combined to produce the expression

$$\frac{dt}{dT} = \frac{1}{a} = \frac{t_j - t_i}{T_j - T_i}. \tag{4.23}$$

Equation (4.23) is used to calculate the extreme tangent vectors of our new curve $\mathbf{Q}(T)$. Since it goes from point $\mathbf{P}(t_i)$ (where $T = 0$) to point $\mathbf{P}(t_j)$ (where $T = 1$), we have $T_j - T_i = 1$. The tangent vectors of $\mathbf{Q}(T)$ are therefore

$$\mathbf{Q}^T(T) = \frac{d\mathbf{Q}(T)}{dT} = \frac{d\mathbf{P}(t)}{dt}\frac{dt}{dT} = \mathbf{P}^t(t) \cdot (t_j - t_i).$$

The two extreme tangents are $\mathbf{Q}^T(0) = (t_j - t_i)\mathbf{P}^t(t_i)$ and $\mathbf{Q}^T(1) = (t_j - t_i)\mathbf{P}^t(t_j)$. The new curve can now be calculated by

$$\mathbf{Q}(T) = (T^3, T^2, T, 1)\mathbf{H} \begin{bmatrix} \mathbf{P}(t_i) \\ \mathbf{P}(t_j) \\ (t_j - t_i)\mathbf{P}^t(t_i) \\ (t_j - t_i)\mathbf{P}^t(t_j) \end{bmatrix}, \tag{4.24}$$

where $\mathbf{H}$ is the Hermite matrix, Equation (4.7).

◇ **Exercise 4.14:** Compute the PC segment $\mathbf{Q}(T)$ that results from truncating $\mathbf{P}(t) = (-1, 0)t^3 + (1, -1)t^2 + (1, 1)t$ [Equation (4.10)] from $t_i = 0.25$ to $t_j = 0.75$.

Segmenting a curve is the problem of calculating several truncations. Assume that we are given values $0 = t_0 < t_1 < t_2 < \cdots < t_n = 1$, and we want to break a given curve $\mathbf{P}(t)$ into n segments such that segment i will go from point $\mathbf{P}(t_{i-1})$ to point $\mathbf{P}(t_i)$ (Figure 4.9b). Equation (4.24) gives segment i as

$$\mathbf{Q}_i(T) = (T^3, T^2, T, 1)\mathbf{H} \begin{bmatrix} \mathbf{P}(t_{i-1}) \\ \mathbf{P}(t_i) \\ (t_i - t_{i-1})\mathbf{P}^t(t_{i-1}) \\ (t_i - t_{i-1})\mathbf{P}^t(t_i) \end{bmatrix}.$$

4.5.1 Special and Degenerate Hermite Segments

The following special cases result in Hermite curve segments that are either especially simple (degenerate) or especially interesting

■   The case $\mathbf{P}_1 = \mathbf{P}_2$ and $\mathbf{P}_1^t = \mathbf{P}_2^t = (0, 0)$. Equation (4.4) yields $\mathbf{P}(t) = \mathbf{P}_1$; the curve degenerates to a point.

■ The case $\mathbf{P}_1^t = \mathbf{P}_2^t = \mathbf{P}_2 - \mathbf{P}_1$. The two tangents point in the same direction, from $\mathbf{P}_1$ to $\mathbf{P}_2$. Equation (4.4) yields

$$\mathbf{P}(t) = \big(2\mathbf{P}_1 - 2\mathbf{P}_2 + 2(\mathbf{P}_2 - \mathbf{P}_1)\big)t^3 + \big(-3\mathbf{P}_1 + 3\mathbf{P}_2 - 3(\mathbf{P}_2 - \mathbf{P}_1)\big)t^2$$
$$+ (\mathbf{P}_2 - \mathbf{P}_1)t + \mathbf{P}_1$$
$$= (\mathbf{P}_2 - \mathbf{P}_1)t + \mathbf{P}_1. \tag{4.25}$$

The curve reduces to a straight segment.

■ The case $\mathbf{P}_1 = \mathbf{P}_2$. Equation (4.4) yields $\mathbf{P}(t) = (\mathbf{P}_1^t + \mathbf{P}_2^t)t^3 + (-2\mathbf{P}_1^t - \mathbf{P}_2^t)t^2 + \mathbf{P}_1^t t + \mathbf{P}_1$. It is easy to see that this curve satisfies $\mathbf{P}(0) = \mathbf{P}(1)$. It is closed (but is not a circle).

■ The case $\mathbf{P}_1^t = \mathbf{P}_2^t = (x_2 - x_1, y_2 - y_1, 0)$. Equation (4.4) yields

$$\mathbf{P}(t) = \big(2\mathbf{P}_1 - 2\mathbf{P}_2 + 2(x_2 - x_1, y_2 - y_1, 0)\big)t^3$$
$$+ \big(-3\mathbf{P}_1 + 3\mathbf{P}_2 - 3(x_2 - x_1, y_2 - y_1, 0)\big)t^2$$
$$+ (x_2 - x_1, y_2 - y_1, 0)t + (x_1, y_1, z_1)$$
$$= \big(x_1 + (x_2 - x_1)t, y_1 + (y_2 - y_1)t, z_1 + (z_2 - z_1)(3t^2 - 2t^3)\big).$$

The x and y coordinates of this curve are linear functions of t, so its tangent vector has the form $(\alpha, \beta, z(t))$. Its x and y components are constants, so it always points in the same plane. Thus, the curve is planar.

4.6 Hermite Straight Segments

Equation (4.25) shows that the Hermite segment can sometimes degenerate into a straight segment. This section describes variations on Hermite straight segments. Specifically, we look in detail at the case where the two extreme tangent vectors point in the same direction, from $\mathbf{P}_1$ to $\mathbf{P}_2$, but have different magnitudes. We denote them by $\mathbf{P}_1^t = \alpha(\mathbf{P}_2 - \mathbf{P}_1)$ and $\mathbf{P}_2^t = \beta(\mathbf{P}_2 - \mathbf{P}_1)$, where α and β can be any real numbers. Equation (4.25) is obtained in the special case $\alpha = \beta = 1$.

The Hermite segment is expressed as $\mathbf{P}(t) = \mathbf{F}(t)\mathbf{B}$, where the four $F_i(t)$ functions are given by Equation (4.6), and $\mathbf{B}$ is the geometry vector, which, in our case, has the form

$$\mathbf{B} = \big(\mathbf{P}_1, \mathbf{P}_2, \alpha(\mathbf{P}_2 - \mathbf{P}_1), \beta(\mathbf{P}_2 - \mathbf{P}_1)\big)^T.$$

This can be written (since $F_1(t) + F_2(t) \equiv 1$) in the form

$$\mathbf{P}(t) = F_1(t)\mathbf{P}_1 + F_2(t)\mathbf{P}_2 + F_3(t)\alpha(\mathbf{P}_2 - \mathbf{P}_1) + F_4(t)\beta(\mathbf{P}_2 - \mathbf{P}_1)$$
$$= \mathbf{P}_1 + (F_2(t) + \alpha F_3(t) + \beta F_4(t))(\mathbf{P}_2 - \mathbf{P}_1)$$
$$= \mathbf{P}_1 + \big((1 - 2t^3 + 3t^2) + \alpha(t^3 - 2t^2 + t) + \beta(t^3 - t^2)\big)(\mathbf{P}_2 - \mathbf{P}_1)$$
$$= \mathbf{P}_1 + \big((\alpha + \beta - 2)t^3 - (2\alpha + \beta - 3)t^2 + \alpha t\big)(\mathbf{P}_2 - \mathbf{P}_1). \tag{4.26}$$

This has the form $\mathbf{P}(t) = \mathbf{P}_1 + G(t)(\mathbf{P}_2 - \mathbf{P}_1)$, which shows that all the points of $\mathbf{P}(t)$ lie on the straight line that passes through $\mathbf{P}_1$ and has the tangent vector $(\mathbf{P}_2 - \mathbf{P}_1)$. The precise form of $\mathbf{P}(t)$ depends on the values and signs of α and β. The remainder of this section analyzes several cases in detail. The remaining cases can be analyzed similarly. See also Exercise 6.7.

Case 1 is when $\alpha = \beta = 1$, which leads to Equation (4.25), a straight segment from $\mathbf{P}_1$ to $\mathbf{P}_2$.

Case 2 is when $\alpha = \beta = 0$. Equation (4.26) reduces in this case to

$$\mathbf{P}(t) = \mathbf{P}_1 + (-2t^3 + 3t^2)(\mathbf{P}_2 - \mathbf{P}_1), \qquad (4.27)$$

or $\mathbf{P}(T) = \mathbf{P}_1 + T(\mathbf{P}_2 - \mathbf{P}_1)$, where $T = -2t^3 + 3t^2$. This also is a straight segment from $\mathbf{P}_1$ to $\mathbf{P}_2$ but moving at a variable speed. It accelerates up to point $\mathbf{P}(0.5)$, then decelerates.

◇ **Exercise 4.15:** Explain why this is so.

Case 3 is when $\alpha = \beta = -1$. Equation (4.26) becomes in this case

$$\mathbf{P}(t) = \mathbf{P}_1 + (-4t^3 + 6t^2 - t)(\mathbf{P}_2 - \mathbf{P}_1), \qquad (4.28)$$

which is the curve shown in Figure 4.10a. It consists of three straight segments, but we can also think of it as a straight line that goes from $\mathbf{P}_1$ *backward* to a certain point $\mathbf{P}(i)$, then reverses direction, passes points $\mathbf{P}_1$ and $\mathbf{P}_2$, stops at point $\mathbf{P}(j)$, reverses direction again, and ends at $\mathbf{P}_2$. We can calculate i and j by calculating the tangent of Equation (4.28) and equating it to zero. The tangent vector is $\mathbf{P}^t(t) = (-12t^2 + 12t - 1)(\mathbf{P}_2 - \mathbf{P}_1)$ and the roots of the quadratic equation $-12t^2 + 12t - 1 = 0$ are (approximately) 0.083 and 0.92.

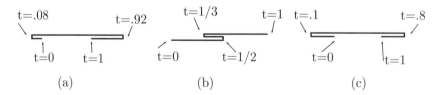

Figure 4.10: Straight Hermite Segments.

Case 4 is when $\alpha > 0$, $\beta > 0$. As an example, we try the values $\alpha = 2$ and $\beta = 4$. Equation (4.26) becomes in this case

$$\mathbf{P}(t) = \mathbf{P}_1 + (4t^3 - 5t^2 + 2t)(\mathbf{P}_2 - \mathbf{P}_1). \qquad (4.29)$$

This curve also consists of three straight segments (Figure 4.10b), but it behaves differently. It goes forward from $\mathbf{P}_1$ to a certain point $\mathbf{P}(i)$, then reverses direction, goes to point $\mathbf{P}(j)$, reverses direction again, and continues to $\mathbf{P}_2$. We can calculate i and j by calculating the tangent of Equation (4.29) and equating it to zero. The tangent vector is $\mathbf{P}^t(t) = (12t^2 - 10t + 2)(\mathbf{P}_2 - \mathbf{P}_1)$ and the roots of the quadratic equation $12t^2 - 10t + 2 = 0$ are $1/3$ and $1/2$.

Case 5 is when $\alpha < 0$, $\beta < 0$. As an example, we try the values $\alpha = -2$ and $\beta = -4$. Equation (4.26) becomes in this case

$$\mathbf{P}(t) = \mathbf{P}_1 + (-8t^3 + 11t^2 - 2t)(\mathbf{P}_2 - \mathbf{P}_1). \tag{4.30}$$

This curve again consists of three straight segments as in case 3, but points i and j are different (Figure 4.10c). The tangent of Equation (4.30) is $\mathbf{P}^t(t) = (-24t^2 + 22t - 2)(\mathbf{P}_2 - \mathbf{P}_1)$, and the roots of the quadratic equation $-24t^2 + 22t - 2 = 0$ are (approximately) 0.1 and 0.8.

Table 4.11 summarizes the nine possible cases of Equation (4.26).

Case	1	2	3	4	5	6	7	8	9
α	1	0	-1	> 0	< 0	> 0	< 0	≤ 0	≥ 0
β	1	0	-1	> 0	< 0	≤ 0	≥ 0	> 0	< 0

Table 4.11: Nine Cases of Straight Hermite Segments.

4.7 A Variant Hermite Segment

The Hermite method starts with four known quantities, two points and two tangents. These are used to set and solve four equations, so four unknowns can be calculated. A variation on this technique is the case where two points and just one tangent are given. These constitute only three quantities, so only three equations can be set and only three unknowns solved and determined. Thus, this variant curve can be only a quadratic (degree-2) polynomial. As usual, we denote the points by $\mathbf{P}_1$ and $\mathbf{P}_2$ and the tangent vector (which is assumed to be the start tangent, but can also be the end tangent) by $\mathbf{P}_1^t$. The quadratic polynomial is $\mathbf{P}(t) = \mathbf{a}t^2 + \mathbf{b}t + \mathbf{c}$, its tangent vector is $\mathbf{P}^t(t) = 2\mathbf{a}t + \mathbf{b}$, and we can immediately set up the three equations $\mathbf{P}(0) = \mathbf{P}_1$, $\mathbf{P}(1) = \mathbf{P}_2$, and $\mathbf{P}^t(0) = \mathbf{P}_1^t$ whose explicit forms are

$$\mathbf{a} \cdot 0^2 + \mathbf{b} \cdot 0 + \mathbf{c} = \mathbf{P}_1,$$
$$\mathbf{a} \cdot 1^2 + \mathbf{b} \cdot 1 + \mathbf{c} = \mathbf{P}_2, \tag{4.31}$$
$$2\mathbf{a} \cdot 0 + \mathbf{b} = \mathbf{P}_1^t.$$

The solutions are $\mathbf{c} = \mathbf{P}_1$, $\mathbf{b} = \mathbf{P}_1^t$, and $\mathbf{a} = \mathbf{P}_2 - \mathbf{b} - \mathbf{c} = \mathbf{P}_2 - \mathbf{P}_1 - \mathbf{P}_1^t$.

The quadratic polynomial is therefore

$$\begin{aligned}
\mathbf{P}(t) &= (\mathbf{P}_2 - \mathbf{P}_1 - \mathbf{P}_1^t)t^2 + \mathbf{P}_1^t t + \mathbf{P}_1 \\
&= (-t^2 + 1)\mathbf{P}_1 + t^2\mathbf{P}_2 + (-t^2 + t)\mathbf{P}_1^t \\
&= (t^2, t, 1) \begin{pmatrix} -1 & 1 & -1 \\ 0 & 0 & 1 \\ 1 & 0 & 0 \end{pmatrix} \begin{pmatrix} \mathbf{P}_1 \\ \mathbf{P}_2 \\ \mathbf{P}_1^t \end{pmatrix}.
\end{aligned} \tag{4.32}$$

Its tangent vector is $\mathbf{P}^t(t) = 2\mathbf{a}t + \mathbf{b} = 2(\mathbf{P}_2 - \mathbf{P}_1 - \mathbf{P}_1^t)t + \mathbf{P}_1^t$, which implies that the end tangent is

$$\mathbf{P}^t(1) = 2(\mathbf{P}_2 - \mathbf{P}_1) - \mathbf{P}_1^t. \tag{4.33}$$

Figure 4.12 shows the simple geometric interpretation of this.

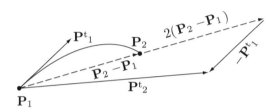

Figure 4.12: The Geometric Interpretation of the End Tangent.

⋄ **Exercise 4.16:** Derive the nonuniform version of this quadratic polynomial assuming that the parameter t varies from zero to some positive number Δ.

⋄ **Exercise 4.17:** Calculate a quadratic parametric polynomial $\mathbf{P}(t) = \mathbf{a}t^2 + \mathbf{b}t + \mathbf{c}$ assuming that only the two extreme tangent vectors $\mathbf{P}^t(0)$ and $\mathbf{P}^t(1)$ are given.

⋄ **Exercise 4.18:** Use your curve design skills to obtain the cubic polynomial equation of the curve segment $\mathbf{P}(t)$ defined by the following three conditions: (1) The two endpoints $\mathbf{P}_1$ and $\mathbf{P}_2$ are given, (2) the end tangent $\mathbf{P}_2^t$ is given, and (3) the start second derivative $\mathbf{P}^{tt}(0)$ is zero.

4.8 Ferguson Surfaces

A Ferguson surface patch [Ferguson 64] is an extension of the Hermite curve segment. The patch is specified by its four corner points $\mathbf{P}_{ij}$ and by two tangent vectors $\mathbf{P}_{ij}^u$ and $\mathbf{P}_{ij}^w$ in the u and w directions at each point; for a total of 12 three-dimensional quantities. Figure 4.13a,b illustrates the notation used. We start by deriving the expressions of the "bottom" and "top" boundary curves $\mathbf{P}(u, 0)$ and $\mathbf{P}(u, 1)$. Equation (4.5) yields

$$\mathbf{P}(u, 0) = F_1(u)\mathbf{P}_{00} + F_2(u)\mathbf{P}_{10} + F_3(u)\mathbf{P}_{00}^u + F_4(u)\mathbf{P}_{10}^u,$$
$$\mathbf{P}(u, 1) = F_1(u)\mathbf{P}_{01} + F_2(u)\mathbf{P}_{11} + F_3(u)\mathbf{P}_{01}^u + F_4(u)\mathbf{P}_{11}^u,$$

where functions $F_i(u)$ are given by Equation (4.6).

We now concentrate on the two tangent vectors $\mathbf{P}_{00}^w$ and $\mathbf{P}_{10}^w$. The points at the *tips* of those vectors are labeled $\mathbf{Q}_{00}$ and $\mathbf{Q}_{10}$, respectively and we derive the expression of the Hermite segment $\mathbf{Q}(u, 0)$ connecting these points by assuming that its tangents in the u direction are identical to those of boundary curve $\mathbf{P}(u, 0)$. Similarly, we denote the two points at the tips of tangents $\mathbf{P}_{01}^w$ and $\mathbf{P}_{11}^w$ by $\mathbf{Q}_{01}$ and $\mathbf{Q}_{11}$, respectively and derive the expression of the Hermite segment $\mathbf{Q}(u, 1)$ connecting them. The two segments are

$$\mathbf{Q}(u, 0) = F_1(u)\mathbf{Q}_{00} + F_2(u)\mathbf{Q}_{10} + F_3(u)\mathbf{P}_{00}^u + F_4(u)\mathbf{P}_{10}^u,$$
$$\mathbf{Q}(u, 1) = F_1(u)\mathbf{Q}_{01} + F_2(u)\mathbf{Q}_{11} + F_3(u)\mathbf{P}_{01}^u + F_4(u)\mathbf{P}_{11}^u.$$

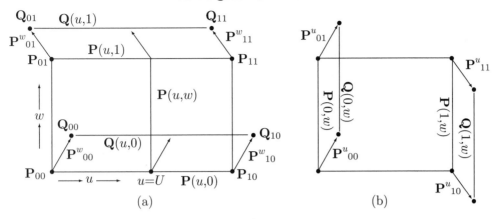

Figure 4.13: Ferguson Surface Patches.

Once the two curves $\mathbf{P}(u,0)$ and $\mathbf{Q}(u,0)$ are known, we can express the tangent vector $\mathbf{P}^w_{u0}$ in the w direction for any u as the difference

$$\mathbf{P}^w_{u0} = \mathbf{Q}(u,0) - \mathbf{P}(u,0) = F_1(u)[\mathbf{Q}_{00} - \mathbf{P}_{00}] + F_2(u)[\mathbf{Q}_{10} - \mathbf{P}_{10}] = F_1(u)\mathbf{P}^w_{00} + F_2(u)\mathbf{P}^w_{10},$$

and similarly

$$\mathbf{P}^w_{u1} = \mathbf{Q}(u,1) - \mathbf{P}(u,1) = F_1(u)\mathbf{P}^w_{01} + F_2(u)\mathbf{P}^w_{11}.$$

We now fix u at a certain value U and examine point $\mathbf{P}(U,0)$ on boundary curve $\mathbf{P}(u,0)$ and point $\mathbf{P}(U,1)$ on boundary curve $\mathbf{P}(u,1)$. The tangent in the w direction at point $\mathbf{P}(U,0)$ is the difference of points $\mathbf{Q}(U,0) - \mathbf{P}(U,0)$ and the tangent in the w direction at point $\mathbf{P}(U,1)$ is the difference of points $\mathbf{Q}(U,1) - \mathbf{P}(U,1)$. Once the two points $\mathbf{P}(U,0)$ and $\mathbf{P}(U,1)$ and the two tangents $\mathbf{Q}(U,0) - \mathbf{P}(U,0)$ and $\mathbf{Q}(U,1) - \mathbf{P}(U,1)$ are known, we can easily construct the Hermite segment defined by them. When u is released, this segment becomes the expression of the entire surface patch. The expression is

$$
\begin{aligned}
\mathbf{P}(u,w) &= F_1(w)\mathbf{P}(u,0) + F_2(w)\mathbf{P}(u,1) + F_3(w)\mathbf{P}^w(u,0) + F_4(w)\mathbf{P}^w(u,1) \\
&= F_1(w)\mathbf{P}(u,0) + F_2(w)\mathbf{P}(u,1) + F_3(w)\big[F_1(u)\mathbf{P}^w_{00} + F_2(u)\mathbf{P}^w_{10}\big] \\
&\quad + F_4(u)\big[F_1(u)\mathbf{P}^w_{01} + F_2(u)\mathbf{P}^w_{11}\big] \\
&= F_1(w)\big[F_1(u)\mathbf{P}_{00} + F_2(u)\mathbf{P}_{10} + F_3(u)\mathbf{P}^u_{00} + F_4(u)\mathbf{P}^u_{10}\big] \\
&\quad + F_2(w)\big[F_1(u)\mathbf{P}_{01} + F_2(u)\mathbf{P}_{11} + F_3(u)\mathbf{P}^u_{01} + F_4(u)\mathbf{P}^u_{11}\big] \\
&\quad + F_3(w)\big[F_1(u)\mathbf{P}^w_{00} + F_2(u)\mathbf{P}^w_{10}\big] + F_4(w)\big[F_1(u)\mathbf{P}^w_{01} + F_2(u)\mathbf{P}^w_{11}\big] \\
&= \big(F_1(u), F_2(u), F_3(u), F_4(u)\big)
\begin{bmatrix}
\mathbf{P}_{00} & \mathbf{P}_{01} & \mathbf{P}^w_{00} & \mathbf{P}^w_{01} \\
\mathbf{P}_{10} & \mathbf{P}_{11} & \mathbf{P}^w_{10} & \mathbf{P}^w_{11} \\
\mathbf{P}^u_{00} & \mathbf{P}^u_{01} & 0 & 0 \\
\mathbf{P}^u_{10} & \mathbf{P}^u_{11} & 0 & 0
\end{bmatrix}
\begin{bmatrix}
F_1(w) \\
F_2(w) \\
F_3(w) \\
F_4(w)
\end{bmatrix}. (4.34)
\end{aligned}
$$

Notice that even though we started with the two boundary curves $\mathbf{P}(u,0)$ and $\mathbf{P}(u,1)$ the final expression, Equation (4.34), is symmetric in u and w. It can also be derived by starting with the two boundary curves $\mathbf{P}(0, w)$ and $\mathbf{P}(1, w)$ (Figure 4.13b) and going through a similar process.

Notice that the Ferguson surface is very similar to the bicubic Hermite patch of Section 4.9, but is less flexible because it has zeros instead of the more general twist vectors.

The Ferguson surface patch is easy to connect smoothly with other patches of the same type. Given a set of points arranged roughly in a two-dimensional grid, with two tangent vectors for each point, as in Figure 4.14, Equation (4.34) can be applied to each set of four points and eight tangents to construct a surface patch and the patches will connect smoothly because the end tangents of a patch are the start tangents of the next patch.

As an example, Figure 4.15 shows two patches, one based on corner points $\mathbf{P}_{00}$, $\mathbf{P}_{01}$, $\mathbf{P}_{10}$, and $\mathbf{P}_{11}$, and the other based on $\mathbf{P}_{10}$, $\mathbf{P}_{11}$, $\mathbf{P}_{20}$, and $\mathbf{P}_{21}$. The 12 tangent vectors (two per point) are shown in the code with the figure. It's easy to see how the two patches (intentionally slightly separated in the figure) are connected smoothly.

4.9 Bicubic Hermite Patch

The spline methods covered in Chapter 5 are based on Hermite curve segments, which suggests that Hermite interpolation is useful. The Ferguson surface patch of Section 4.8 is an attempt to extend the technique of Hermite interpolation to surface patches. This section describes a more general extension. A single Hermite segment is a cubic polynomial, so we expect the Hermite surface patch, which is an extension of the Hermite curve segment, to be a bicubic surface. Its expression should be given by Equation (3.27), where matrix $\mathbf{H}$ [Equation (4.7)] should be substituted for $\mathbf{N}$, and the 16 quantities should be points and tangent vectors.

The basic idea is to ask the user to specify the four boundary curves as Hermite segments. Thus, the user should specify two points and two tangent vectors for each curve, for a total of eight points and eight tangents. For the four curves to form a surface, they have to meet at the four corners, so the eight points are reduced to four points. Four points and eight tangents provide 12 of the 16 quantities needed to construct the surface. Four more quantities are needed in order to calculate the 16 unknowns of Equation (3.26), and they are selected as the second derivatives of the surface at the corner points. They are called *twist vectors*.

To calculate the surface, 16 equations are written, expressing the way we require the surface to behave. For example, we want $\mathbf{P}(u, w)$ to approach the corner point $\mathbf{P}_{01}$ when $u \to 0$ and $w \to 1$. We also want $\mathbf{P}(0, w)$ to equal the PC between points $\mathbf{P}_{00}$ and $\mathbf{P}_{01}$. The equations are obtained from the 16 terms of Equation (3.24)

$$\mathbf{P}_{00} = \mathbf{a}_{00},$$
$$\mathbf{P}_{10} = \mathbf{a}_{30} + \mathbf{a}_{20} + \mathbf{a}_{10} + \mathbf{a}_{00},$$
$$\mathbf{P}_{01} = \mathbf{a}_{03} + \mathbf{a}_{02} + \mathbf{a}_{01} + \mathbf{a}_{00},$$
$$\mathbf{P}_{11} = \mathbf{a}_{33} + \mathbf{a}_{32} + \mathbf{a}_{31} + \mathbf{a}_{30} + \mathbf{a}_{23} + \mathbf{a}_{22} + \mathbf{a}_{21} + \mathbf{a}_{20}$$
$$+ \mathbf{a}_{13} + \mathbf{a}_{12} + \mathbf{a}_{11} + \mathbf{a}_{10} + \mathbf{a}_{03} + \mathbf{a}_{02} + \mathbf{a}_{01} + \mathbf{a}_{00},$$

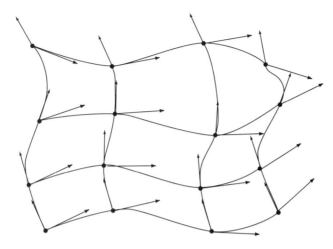

Figure 4.14: A Grid for a Ferguson Surface.

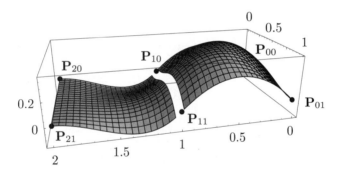

```
<<:Graphics:ParametricPlot3D.m; (* Two Ferguson patches *)
F1[t_]:=2t^3-3t^2+1; F2[t_]:=-2t^3+3t^2;
F3[t_]:=t^3-2t^2+t; F4[t_]:=t^3-t^2;
F[t_]:={F1[t],F2[t],F3[t],F4[t]};
p00={0,0,0}; p01={0,1,0}; pu00={1,0,1}; pw00={0,1,1}; pu01={1,0,1}; pw01={0,1,0};
p10={1,0,0}; p11={1,1,0}; pu10={1,0,-1}; pw10={0,1,0}; pu11={1,0,-1}; pw11={0,1,-1};
p20={2,0,0}; p21={2,1,0}; pu20={1,0,0}; pw20={0,1,0}; pu21={1,0,0}; pw21={0,1,0};
H={{p00,p01,pw00,pw01},{p10,p11,pw10,pw11},
 {pu00,pu01,{0,0,0},{0,0,0}},{pu10,pu11,{0,0,0},{0,0,0}}};
prt[i_]:=H[[Range[1,4],Range[1,4],i]];
g1=ParametricPlot3D[{F[u].prt[1].F[w],F[u].prt[2].F[w],F[u].prt[3].F[w]},
{u,0,.98,.05},{w,0,1,.05}, DisplayFunction->Identity];

H={{p10,p11,pw10,pw11},{p20,p21,pw20,pw21},
 {pu10,pu11,{0,0,0},{0,0,0}},{pu20,pu21,{0,0,0},{0,0,0}}};
g2=ParametricPlot3D[{F[u].prt[1].F[w],F[u].prt[2].F[w],F[u].prt[3].F[w]},
{u,0.05,1,.05},{w,0,1,.05}, DisplayFunction->Identity];

g3=Graphics3D[{AbsolutePointSize[4],
Point[p00],Point[p01],Point[p10],Point[p11],Point[p20],Point[p21]}];

Show[g1,g2,g3, ViewPoint->{0.322, 1.342, 0.506},
DefaultFont->{"cmr10", 10}, DisplayFunction->$DisplayFunction]
```

Figure 4.15: Two Ferguson Surface Patches.

$$\mathbf{P}_{00}^u = \mathbf{a}_{10},$$

$$\mathbf{P}_{00}^w = \mathbf{a}_{01},$$

$$\mathbf{P}_{10}^u = 3\mathbf{a}_{30} + 2\mathbf{a}_{20} + \mathbf{a}_{10},$$

$$\mathbf{P}_{10}^w = \mathbf{a}_{31} + \mathbf{a}_{21} + \mathbf{a}_{11} + \mathbf{a}_{01},$$

$$\mathbf{P}_{01}^u = \mathbf{a}_{13} + \mathbf{a}_{12} + \mathbf{a}_{11} + \mathbf{a}_{10},$$

$$\mathbf{P}_{01}^w = 3\mathbf{a}_{03} + 2\mathbf{a}_{02} + \mathbf{a}_{01},$$

$$\mathbf{P}_{11}^u = 3\mathbf{a}_{33} + 3\mathbf{a}_{32} + 3\mathbf{a}_{31} + 3\mathbf{a}_{30} + 2\mathbf{a}_{23} + 2\mathbf{a}_{22} + 2\mathbf{a}_{21}$$
$$+ 2\mathbf{a}_{20} + \mathbf{a}_{13} + \mathbf{a}_{12} + \mathbf{a}_{11} + \mathbf{a}_{10},$$

$$\mathbf{P}_{11}^w = 3\mathbf{a}_{33} + 2\mathbf{a}_{32} + \mathbf{a}_{31} + 3\mathbf{a}_{23} + 2\mathbf{a}_{22} + \mathbf{a}_{21} + 3\mathbf{a}_{13}$$
$$+ 2\mathbf{a}_{12} + \mathbf{a}_{11} + 3\mathbf{a}_{03} + 2\mathbf{a}_{02} + \mathbf{a}_{01},$$

$$\mathbf{P}_{00}^{uw} = \mathbf{a}_{11},$$

$$\mathbf{P}_{10}^{uw} = 3\mathbf{a}_{31} + 2\mathbf{a}_{21} + \mathbf{a}_{11},$$

$$\mathbf{P}_{01}^{uw} = 3\mathbf{a}_{13} + 2\mathbf{a}_{12} + \mathbf{a}_{11},$$

$$\mathbf{P}_{11}^{uw} = 9\mathbf{a}_{33} + 6\mathbf{a}_{32} + 3\mathbf{a}_{31} + 6\mathbf{a}_{23} + 4\mathbf{a}_{22}$$
$$+ 2\mathbf{a}_{21} + 3\mathbf{a}_{13} + 2\mathbf{a}_{12} + \mathbf{a}_{11}.$$

The solutions express the 16 coefficients $\mathbf{a}_{ij}$ in terms of the four corner points, eight tangent vectors, and four twist vectors:

$$\mathbf{a}_{01} = \mathbf{P}_{00}^w,$$

$$\mathbf{a}_{02} = -2\mathbf{P}_{00}^w - \mathbf{P}_{01}^w - 3\mathbf{P}_{00} + 3\mathbf{P}_{01},$$

$$\mathbf{a}_{03} = \mathbf{P}_{00}^w + \mathbf{P}_{01}^w + 2\mathbf{P}_{00} - 2\mathbf{P}_{01},$$

$$\mathbf{a}_{10} = \mathbf{P}_{00}^u,$$

$$\mathbf{a}_{11} = \mathbf{P}_{00}^{uw},$$

$$\mathbf{a}_{12} = -2\mathbf{P}_{00}^{uw} - \mathbf{P}_{01}^{uw} - 3\mathbf{P}_{00}^u + 3\mathbf{P}_{01}^u,$$

$$\mathbf{a}_{13} = \mathbf{P}_{00}^{uw} + \mathbf{P}_{01}^{uw} + 2\mathbf{P}_{00}^u - 2\mathbf{P}_{01}^u,$$

$$\mathbf{a}_{20} = -2\mathbf{P}_{00}^u - \mathbf{P}_{10}^u - 3\mathbf{P}_{00} + 3\mathbf{P}_{10},$$

$$\mathbf{a}_{21} = -2\mathbf{P}_{00}^{uw} - \mathbf{P}_{10}^{uw} - 3\mathbf{P}_{00}^w + 3\mathbf{P}_{10}^w,$$

$$\mathbf{a}_{22} = 4\mathbf{P}_{00}^{uw} + 2\mathbf{P}_{01}^{uw} + 2\mathbf{P}_{10}^{uw} + \mathbf{P}_{11}^{uw} + 6\mathbf{P}_{00}^u - 6\mathbf{P}_{01}^u + 3\mathbf{P}_{10}^u - 3\mathbf{P}_{11}^u + 6\mathbf{P}_{00}^w$$
$$+ 3\mathbf{P}_{01}^w - 6\mathbf{P}_{10}^w - 3\mathbf{P}_{11}^w + 9\mathbf{P}_{00} - 9\mathbf{P}_{01} - 9\mathbf{P}_{10} + 9\mathbf{P}_{11},$$

$$\mathbf{a}_{23} = -2\mathbf{P}_{00}^{uw} - 2\mathbf{P}_{01}^{uw} - \mathbf{P}_{10}^{uw} - \mathbf{P}_{11}^{uw} - 4\mathbf{P}_{00}^u + 4\mathbf{P}_{01}^u - 2\mathbf{P}_{10}^u + 2\mathbf{P}_{11}^u - 3\mathbf{P}_{00}^w$$
$$- 3\mathbf{P}_{01}^w + 3\mathbf{P}_{10}^w + 3\mathbf{P}_{11}^w - 6\mathbf{P}_{00} + 6\mathbf{P}_{01} + 6\mathbf{P}_{10} - 6\mathbf{P}_{11},$$

$$\mathbf{a}_{30} = \mathbf{P}_{00}^u + \mathbf{P}_{10}^u + 2\mathbf{P}_{00} - 2\mathbf{P}_{10},$$

$$\mathbf{a}_{31} = \mathbf{P}_{00}^{uw} + \mathbf{P}_{10}^{uw} + 2\mathbf{P}_{00}^w - 2\mathbf{P}_{10}^w,$$

$$\mathbf{a}_{32} = -2\mathbf{P}_{00}^{uw} - \mathbf{P}_{01}^{uw} - 2\mathbf{P}_{10}^{uw} - \mathbf{P}_{11}^{uw} - 3\mathbf{P}_{00}^u + 3\mathbf{P}_{01}^u - 3\mathbf{P}_{10}^u + 3\mathbf{P}_{11}^u - 4\mathbf{P}_{00}^w$$
$$- 2\mathbf{P}_{01}^w + 4\mathbf{P}_{10}^w + 2\mathbf{P}_{11}^w - 6\mathbf{P}_{00} + 6\mathbf{P}_{01} + 6\mathbf{P}_{10} - 6\mathbf{P}_{11},$$

$$\mathbf{a}_{33} = \mathbf{P}_{00}^{uw} + \mathbf{P}_{01}^{uw} + \mathbf{P}_{10}^{uw} + \mathbf{P}_{11}^{uw} + 2\mathbf{P}_{00}^u - 2\mathbf{P}_{01}^u + 2\mathbf{P}_{10}^u - 2\mathbf{P}_{11}^u + 2\mathbf{P}_{00}^w + 2\mathbf{P}_{01}^w$$
$$- 2\mathbf{P}_{10}^w - 2\mathbf{P}_{11}^w + 4\mathbf{P}_{00} - 4\mathbf{P}_{01} - 4\mathbf{P}_{10} + 4\mathbf{P}_{11}.$$

When Equation (3.26) is written in terms of these values, it becomes the compact expression

$$
\mathbf{P}(u, w) = (u^3, u^2, u, 1)\mathbf{H}
\begin{bmatrix}
\mathbf{P}_{00} & \mathbf{P}_{01} & \mathbf{P}_{00}^w & \mathbf{P}_{01}^w \\
\mathbf{P}_{10} & \mathbf{P}_{11} & \mathbf{P}_{10}^w & \mathbf{P}_{11}^w \\
\mathbf{P}_{00}^u & \mathbf{P}_{01}^u & \mathbf{P}_{00}^{uw} & \mathbf{P}_{01}^{uw} \\
\mathbf{P}_{10}^u & \mathbf{P}_{11}^u & \mathbf{P}_{10}^{uw} & \mathbf{P}_{11}^{uw}
\end{bmatrix}
\mathbf{H}^T
\begin{bmatrix}
w^3 \\
w^2 \\
w \\
1
\end{bmatrix}
\tag{4.35}
$$
$$
= \mathbf{U}\mathbf{H}\mathbf{B}\mathbf{H}^T\mathbf{W}^T,
$$

where H is the Hermite matrix, Equation (4.7). The quantities $\mathbf{P}_{ij}^{uw}$ are the twist vectors. They are usually not known in advance but the next section describes a way to estimate them.

4.10 Biquadratic Hermite Patch

Section 4.7 discusses a variation on the Hermite segment where two points $\mathbf{P}_1$ and $\mathbf{P}_2$ and just one tangent vector $\mathbf{P}_1^t$ are known. The curve segment is given by Equation (4.32), duplicated here

$$
\begin{aligned}
\mathbf{P}(t) &= (\mathbf{P}_2 - \mathbf{P}_1 - \mathbf{P}_1^t)t^2 + \mathbf{P}_1^t t + \mathbf{P}_1 \\
&= (-t^2 + 1)\mathbf{P}_1 + t^2\mathbf{P}_2 + (-t^2 + t)\mathbf{P}_1^t \\
&= (t^2, t, 1)
\begin{pmatrix}
-1 & 1 & -1 \\
0 & 0 & 1 \\
1 & 0 & 0
\end{pmatrix}
\begin{pmatrix}
\mathbf{P}_1 \\
\mathbf{P}_2 \\
\mathbf{P}_1^t
\end{pmatrix}.
\end{aligned}
\tag{4.32}
$$

If we denote the curve segment by $\mathbf{P}(t) = \mathbf{a}t^2 + \mathbf{b}t + \mathbf{c}$, then its tangent vector has the form $\mathbf{P}^t(t) = 2\mathbf{a}t + \mathbf{b} = 2(\mathbf{P}_2 - \mathbf{P}_1 - \mathbf{P}_1^t)t + \mathbf{P}_1^t$, which implies that the end tangent is $\mathbf{P}^t(1) = 2(\mathbf{P}_2 - \mathbf{P}_1) - \mathbf{P}_1^t$. The biquadratic surface constructed as the Cartesian product of two such curves is given by

$$
\mathbf{P}(u, w) = (u^2, u, 1)
\begin{pmatrix}
-1 & 1 & -1 \\
0 & 0 & 1 \\
1 & 0 & 0
\end{pmatrix}
\begin{pmatrix}
\mathbf{Q}_{22} & \mathbf{Q}_{21} & \mathbf{Q}_{20} \\
\mathbf{Q}_{12} & \mathbf{Q}_{11} & \mathbf{Q}_{10} \\
\mathbf{Q}_{02} & \mathbf{Q}_{01} & \mathbf{Q}_{00}
\end{pmatrix}
\begin{pmatrix}
-1 & 0 & 1 \\
1 & 0 & 0 \\
-1 & 1 & 0
\end{pmatrix}
\begin{pmatrix}
w^2 \\
w \\
1
\end{pmatrix},
\tag{4.36}
$$

where the nine quantities $\mathbf{Q}_{ij}$ still have to be assigned geometric meaning. This is done by computing $\mathbf{P}(u, w)$ and its partial derivatives for certain values of the parameters. Simple experimentation yields

$$
\begin{aligned}
\mathbf{P}(0, 0) = \mathbf{Q}_{22}, \quad \mathbf{P}(0, 1) = \mathbf{Q}_{21}, \quad \mathbf{P}(1, 0) = \mathbf{Q}_{12}, \quad \mathbf{P}(1, 1) = \mathbf{Q}_{11}, \\
\mathbf{P}^u(0, 0) = \mathbf{Q}_{02}, \quad \mathbf{P}^u(0, 1) = \mathbf{Q}_{01}, \quad \mathbf{P}^w(0, 0) = \mathbf{Q}_{20}, \quad \mathbf{P}^w(1, 0) = \mathbf{Q}_{10}, \\
\mathbf{P}^{uw}(0, 0) = \mathbf{Q}_{00}.
\end{aligned}
$$

This shows that the surface can be expressed as

$$
\mathbf{P}(u, w) = (u^2, u, 1)
\begin{pmatrix}
-1 & 1 & -1 \\
0 & 0 & 1 \\
1 & 0 & 0
\end{pmatrix}
\begin{pmatrix}
\mathbf{P}(0, 0) & \mathbf{P}(0, 1) & \mathbf{P}^w(0, 0) \\
\mathbf{P}(1, 0) & \mathbf{P}(1, 1) & \mathbf{P}^w(1, 0) \\
\mathbf{P}^u(0, 0) & \mathbf{P}^u(0, 1) & \mathbf{P}^{uw}(0, 0)
\end{pmatrix}
$$

$$\times \begin{pmatrix} -1 & 0 & 1 \\ 1 & 0 & 0 \\ -1 & 1 & 0 \end{pmatrix} \begin{pmatrix} w^2 \\ w \\ 1 \end{pmatrix} \tag{4.37}$$

$$= (u^2, u, 1) \begin{pmatrix} -1 & 1 & -1 \\ 0 & 0 & 1 \\ 1 & 0 & 0 \end{pmatrix} \begin{pmatrix} \mathbf{P}_{00} & \mathbf{P}_{01} & \mathbf{P}_{00}^w \\ \mathbf{P}_{10} & \mathbf{P}_{11} & \mathbf{P}_{10}^w \\ \mathbf{P}_{00}^u & \mathbf{P}_{01}^u & \mathbf{P}_{00}^{uw} \end{pmatrix} \begin{pmatrix} -1 & 0 & 1 \\ 1 & 0 & 0 \\ -1 & 1 & 0 \end{pmatrix} \begin{pmatrix} w^2 \\ w \\ 1 \end{pmatrix}.$$

Thus, this type of surface is defined by the following nine quantities:

The four corner points $\mathbf{P}_{00}$, $\mathbf{P}_{01}$, $\mathbf{P}_{10}$, and $\mathbf{P}_{11}$.

The two tangents in the u direction at points $\mathbf{P}_{00}$ and $\mathbf{P}_{01}$.

The two tangents in the w direction at points $\mathbf{P}_{00}$ and $\mathbf{P}_{10}$.

The second derivative at point $\mathbf{P}_{00}$.

The first eight quantities have simple geometric meaning, but the second derivative, which is a *twist vector*, has no simple geometrical interpretation. It can simply be set to zero or it can be estimated. Several methods exist to estimate the twist vectors of biquadratic and bicubic surface patches. The simple method described here is useful when a larger surface is constructed out of several such patches. We start by looking at the twist vector of a bilinear surface. Differentiating Equation (2.8) twice, with respect to u and w, produces the simple, constant expression

$$\mathbf{P}^{uw}(u, w) = \mathbf{P}_{00} - \mathbf{P}_{01} - \mathbf{P}_{10} + \mathbf{P}_{11} = (\mathbf{P}_{00} - \mathbf{P}_{01}) + (\mathbf{P}_{11} - \mathbf{P}_{10}), \tag{4.38}$$

that's a vector and is also independent of both parameters. This expression is now employed to estimate the twist vectors of all the patches that constitute a biquadratic or a bicubic surface. Figure 4.16a is an idealized diagram of such a surface, showing some individual patches. The first step is to apply Equation (4.38) to calculate a vector $\mathbf{T}_i$ for patch i from the four corner points of the patch. Vectors $\mathbf{T}_i$ are then averaged to provide estimates for the four twist vectors of each patch.

The principle is as follows: A corner point $\mathbf{P}_i$ with one index i belongs to just one patch (patch i) and is one of the four corner points of the entire surface ($\mathbf{P}_1$, $\mathbf{P}_4$, $\mathbf{P}_9$, and $\mathbf{P}_c$ of Figure 4.16a). The twist vector estimated for such a point is $\mathbf{T}_i$, the vector previously calculated for patch i. A point $\mathbf{P}_{ij}$ with two indexes ij is common to two patches i and j and is located on the boundary of the entire surface (examples are $\mathbf{P}_{15}$ and $\mathbf{P}_{59}$). The twist vector estimated for such a point is the average $(\mathbf{T}_i + \mathbf{T}_j)/2$. A point $\mathbf{P}_{ijkl}$ with four indexes is common to four patches. The twist vector estimated for such a point is the average $(\mathbf{T}_i + \mathbf{T}_j + \mathbf{T}_k + \mathbf{T}_l)/4$.

This method works well as a first estimate. After the surface is drawn, the twist vectors determined by this method may have to be modified to bring the surface closer to its required shape.

Example: Compute twist vectors for the four patches shown in Figure 4.16b. The first step is to compute a second derivative vector $\mathbf{P}_i^{uw}$ from Equation (4.38) for each patch i.

$$\mathbf{P}_1^{uw} = [(0,0,0) - (1,1,1)] + [(2,1,-1) - (1,0,0)] = (0,0,-2),$$
$$\mathbf{P}_2^{uw} = [(1,0,0) - (2,1,-1)] + [(3,1,1) - (2,0,2)] = (0,0,0),$$

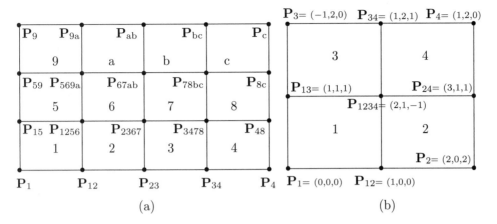

Figure 4.16: Estimating Twist Vectors.

$$\mathbf{P}_3^{uw} = [(1,1,1) - (-1,2,0)] + [(1,2,1) - (2,1,-1)] = (1,0,3),$$
$$\mathbf{P}_4^{uw} = [(2,1,-1) - (1,2,1)] + [(1,2,0) - (3,1,1)] = (-1,0,-3).$$

The second step is to compute a twist vector $\mathbf{T}_i$ for each of the nine points

$$\mathbf{T}_1 = \mathbf{P}_1^{uw} = (0,0,-2),$$
$$\mathbf{T}_{13} = [\mathbf{P}_1^{uw} + \mathbf{P}_3^{uw}]/2 = [(0,0,-2) + (1,0,3)]/2 = (.5,0,.5),$$
$$\mathbf{T}_3 = \mathbf{P}_3^{uw} = (1,0,3),$$
$$\mathbf{T}_{12} = [\mathbf{P}_1^{uw} + \mathbf{P}_2^{uw}]/2 = [(0,0,0) + (1,0,3)]/2 = (.5,0,1.5),$$
$$\mathbf{T}_{1234} = [\mathbf{P}_1^{uw} + \mathbf{P}_2^{uw} + \mathbf{P}_3^{uw} + \mathbf{P}_4^{uw}]/4$$
$$= [(0,0,-2) + (0,0,0) + (1,0,3) + (-1,0,-3)]/4 = (0,0,-.5),$$
$$\mathbf{T}_{34} = [\mathbf{P}_3^{uw} + \mathbf{P}_4^{uw}]/2 = [(1,0,3) + (-1,0,-3)]/2 = (0,0,0),$$
$$\mathbf{T}_2 = \mathbf{P}_2^{uw} = (0,0,0),$$
$$\mathbf{T}_{24} = [\mathbf{P}_2^{uw} + \mathbf{P}_4^{uw}]/2 = [(0,0,0) + (-1,0,-3)]/2 = (-.5,0,-1.5),$$
$$\mathbf{T}_4 = \mathbf{P}_4^{uw} = (-1,0,-3).$$

The last step is to compute one twist vector for each patch by averaging the four twist vectors of the four corners of the patch. For patch 1, the result is

$$[\mathbf{T}_1 + \mathbf{T}_{13} + \mathbf{T}_{1234} + \mathbf{T}_{12}]/4 = [(0,0,-2)+(.5,0,.5)+(0,0,-.5)+(.5,0,1.5)]/4 = (.25,0,-.125),$$

and similarly for the other three surface patches.

She could afterward calmly discuss with him such
blameless technicalities as hidden line algorithms and
buffer refresh times, cabinet versus cavalier projections
and Hermite versus Bézier parametric cubic curve forms.

John Updike, *Roger's Version* (1986)

5

Spline Interpolation

Given a set of points, it is easy to compute a polynomial that passes through the points. The LP of Section 3.2 is an example of such a polynomial. However, as the discussion in Section 1.5 (especially exercise 1.20) illustrates, a curve based on a high-degree polynomial may wiggle wildly and its shape may be far from what the user has in mind. In practical work we are normally interested in a smooth, tight curve that proceeds from point to point such that each segment between two points is a smooth arc. The spline approach to curve design, discussed in this chapter, constructs such a curve from individual segments, each a simple curve, generally a parametric cubic (PC). This chapter illustrates spline interpolation with three examples, cubic splines (Section 5.1), cardinal splines (Section 5.4), and Kochanek–Bartels splines (Section 5.6). Another important type, the B-spline, is the topic of Chapter 7. Other types of splines are known and are discussed in the scientific literature. A short history of splines can be found in [Schumaker 81] and [Farin 04].

   Definition: A spline is a set of polynomials of degree k that are smoothly connected at certain data points. At each data point, two polynomials connect, and their first derivatives (tangent vectors) have the same values. The definition also requires that all their derivatives up to the $(k-1)$st be the same at the point.

5.1 The Cubic Spline Curve

The cubic spline was originally introduced by James Ferguson in [Ferguson 64]. Given n data points that are numbered $\mathbf{P}_1$ through $\mathbf{P}_n$, there are infinitely many curves that pass through all the points in order of their numbers (Figure 5.1a), but the eye often tends to trace *one* imaginary smooth curve through the points, especially if the points are arranged in a familiar pattern. It is therefore useful to have an algorithm that does the same. Since the computer does not recognize familiar patterns the way humans do, such a method should be interactive, thereby allowing the user to create the desired curve.

The cubic spline method is such an algorithm. Given n data points, it constructs a smooth curve that passes through the points (see definition of data points in Section 1.3). The curve consists of $n-1$ individual Hermite segments that are smoothly connected at the $n-2$ interior points and that are easy to calculate and display. For the segments to meet at the interior points, their tangent vectors (first derivatives) must be the same at each interior point. An added feature of cubic splines is that their second derivatives are also the same at the interior points. The cubic spline method is interactive. The user can control the shape of the curve by varying the two extreme tangent vectors at the beginning and the end of the curve.

Given the n data points $\mathbf{P}_1$, $\mathbf{P}_2$, through $\mathbf{P}_n$, we look for $n-1$ parametric cubics $\mathbf{P}_1(t)$, $\mathbf{P}_2(t),\ldots,\mathbf{P}_{n-1}(t)$ such that $\mathbf{P}_k(t)$ is the polynomial segment from point $\mathbf{P}_k$ to point $\mathbf{P}_{k+1}$ (Figure 5.1b). The PCs will have to be smoothly connected at the $n-2$ interior points $\mathbf{P}_2$, $\mathbf{P}_3,\ldots,\mathbf{P}_{n-1}$, which means that their first derivatives will have to match at every interior point. The definition of a spline requires that their second derivatives match too. This requirement (the boundary condition of the cubic spline) is important because it provides the necessary equations and also results in a tight curve in the sense that once the curve is drawn, the eye can no longer detect the positions of the original data points.

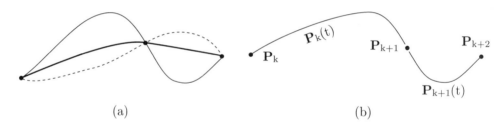

Figure 5.1: (a) Three Different Curves. (b) Two Segments.

The principle of cubic splines is to divide the set of n points into $n-1$ overlapping pairs of two points each and to fit a Hermite segment [Equations (4.4) and (4.5)] to each pair. The pairs are $(\mathbf{P}_1,\mathbf{P}_2)$, $(\mathbf{P}_2,\mathbf{P}_3)$, and so on, up to $(\mathbf{P}_{n-1},\mathbf{P}_n)$. Recall that a Hermite curve segment is specified by two points and two tangents. In our case, all the points are given, so the only unknowns are the tangent vectors. In order for segments $\mathbf{P}_k(t)$ and $\mathbf{P}_{k+1}(t)$ to connect smoothly at point $\mathbf{P}_{k+1}$, the end tangent of $\mathbf{P}_k(t)$ has to equal the start tangent of $\mathbf{P}_{k+1}(t)$. Thus, there is only one tangent vector per point, for a total of n unknowns.

The unknown tangent vectors are computed as the solutions of a system of n equations. The equations are derived from the requirement that the second derivatives of the individual segments match at every interior point. However, there are only $n-2$ interior points, so we can only have $n-2$ equations, enough to solve for only $n-2$ unknowns.

The key to resolving this shortage of equations is to ask the user to provide the software with the values of two tangent vectors (normally the first and last ones). Once this is done, the equations can easily be solved, yielding the remaining $n-2$ tangents.

This seems a strange way to solve equations, but it has the advantage of being interactive. If the resulting curve looks wrong, the user can repeat the calculation with two new tangent vectors. Before delving into the details, here is a summary of the steps involved.

1. The n data points are input into the program.
2. The user provides values (guesses or estimates) for two tangent vectors.
3. The program sets up $n - 2$ equations, with the remaining $n - 2$ tangent vectors as the unknowns, and solves them.
4. The program loops $n - 1$ times. In each iteration, it selects two adjacent points and their tangent vectors to calculate one Hermite segment.

We start with three adjacent points, $\mathbf{P}_k$, $\mathbf{P}_{k+1}$, and $\mathbf{P}_{k+2}$, of which $\mathbf{P}_{k+1}$ must be an interior point and the other two can be either interior or endpoints. Thus, k varies from 1 to $n - 2$. The Hermite segment from $\mathbf{P}_k$ to $\mathbf{P}_{k+1}$ is denoted by $\mathbf{P}_k(t)$, which implies that $\mathbf{P}_k(0) = \mathbf{P}_k$ and $\mathbf{P}_k(1) = \mathbf{P}_{k+1}$. The tangent vectors of $\mathbf{P}_k(t)$ at the endpoints are still unknown and are denoted by $\mathbf{P}_k^t$ and $\mathbf{P}_{k+1}^t$. The first step is to express segment $\mathbf{P}_k(t)$ geometrically, in terms of the two endpoints and the two tangents. Applying Equation (4.4) to our segment results in

$$
\begin{aligned}
\mathbf{P}_k(t) = \mathbf{P}_k + \mathbf{P}_k^t t + \left[3(\mathbf{P}_{k+1} - \mathbf{P}_k) - 2\mathbf{P}_k^t - \mathbf{P}_{k+1}^t\right] t^2 \\
+ \left[2(\mathbf{P}_k - \mathbf{P}_{k+1}) + \mathbf{P}_k^t + \mathbf{P}_{k+1}^t\right] t^3.
\end{aligned}
\tag{5.1}
$$

When the same equation is applied to the next segment $\mathbf{P}_{k+1}(t)$ (from $\mathbf{P}_{k+1}$ to $\mathbf{P}_{k+2}$), it becomes

$$
\begin{aligned}
\mathbf{P}_{k+1}(t) = \mathbf{P}_{k+1} + \mathbf{P}_{k+1}^t t + \left[3(\mathbf{P}_{k+2} - \mathbf{P}_{k+1}) - 2\mathbf{P}_{k+1}^t - \mathbf{P}_{k+2}^t\right] t^2 \\
+ \left[2(\mathbf{P}_{k+1} - \mathbf{P}_{k+2}) + \mathbf{P}_{k+1}^t + \mathbf{P}_{k+2}^t\right] t^3.
\end{aligned}
\tag{5.2}
$$

$\diamond$ **Exercise 5.1:** Where do we use the assumption that the first derivatives of segments $\mathbf{P}_k(t)$ and $\mathbf{P}_{k+1}(t)$ are equal at the interior point $\mathbf{P}_{k+1}$?

Next, we use the requirement that the second derivatives of the two segments be equal at the interior points. The second derivative $\mathbf{P}^{tt}(t)$ of a Hermite segment $\mathbf{P}(t)$ is obtained by differentiating Equation (4.1)

$$
\mathbf{P}^{tt}(t) = 6\mathbf{a}t + 2\mathbf{b}.
\tag{5.3}
$$

Equality of the second derivatives at the interior point $\mathbf{P}_{k+1}$ implies

$$
\mathbf{P}_k^{tt}(1) = \mathbf{P}_{k+1}^{tt}(0) \qquad \text{or} \qquad 6\mathbf{a}_k \times 1 + 2\mathbf{b}_k = 6\mathbf{a}_{k+1} \times 0 + 2\mathbf{b}_{k+1}.
\tag{5.4}
$$

Using the values of $\mathbf{a}$ and $\mathbf{b}$ from Equations (5.1) and (5.2), we get

$$
\begin{aligned}
6\left[2(\mathbf{P}_k - \mathbf{P}_{k+1}) + \mathbf{P}_k^t + \mathbf{P}_{k+1}^t\right] + 2\left[3(\mathbf{P}_{k+1} - \mathbf{P}_k) - 2\mathbf{P}_k^t - \mathbf{P}_{k+1}^t\right] \\
= 2\left[3(\mathbf{P}_{k+2} - \mathbf{P}_{k+1}) - 2\mathbf{P}_{k+1}^t - \mathbf{P}_{k+2}^t\right],
\end{aligned}
\tag{5.5}
$$

which, after simple algebraic manipulations, becomes

$$
\mathbf{P}_k^t + 4\mathbf{P}_{k+1}^t + \mathbf{P}_{k+2}^t = 3(\mathbf{P}_{k+2} - \mathbf{P}_k).
\tag{5.6}
$$

The three quantities on the left side of Equation (5.6) are unknown. The two quantities on the right side are known.

Equation (5.6) can be written $n - 2$ times for all the interior points $\mathbf{P}_{k+1} = \mathbf{P}_2, \mathbf{P}_3, \ldots, \mathbf{P}_{n-1}$ to obtain a system of $n - 2$ linear algebraic equations expressed in matrix form as

$$
n-2 \left\{ \underbrace{\begin{pmatrix} 1 & 4 & 1 & 0 & \cdots & 0 \\ 0 & 1 & 4 & 1 & \cdots & 0 \\ & & \ddots & \ddots & \ddots & \vdots \\ 0 & \cdots & \cdots & 1 & 4 & 1 \end{pmatrix}}_{n} \begin{pmatrix} \mathbf{P}_1^t \\ \mathbf{P}_2^t \\ \vdots \\ \mathbf{P}_n^t \end{pmatrix} = \begin{pmatrix} 3(\mathbf{P}_3 - \mathbf{P}_1) \\ 3(\mathbf{P}_4 - \mathbf{P}_2) \\ \vdots \\ 3(\mathbf{P}_n - \mathbf{P}_{n-2}) \end{pmatrix} \right. \tag{5.7}
$$

Equation (5.7) is a system of $n - 2$ equations in the n unknowns $\mathbf{P}_1^t, \mathbf{P}_2^t, \ldots, \mathbf{P}_n^t$. A practical approach to the solution is to let the user specify the values of the two extreme tangents $\mathbf{P}_1^t$ and $\mathbf{P}_n^t$. Once these values have been substituted in Equation (5.7), it's easy to solve it and obtain values for the remaining $n - 2$ tangents, $\mathbf{P}_2^t$ through $\mathbf{P}_{n-1}^t$. The n tangent vectors are now used to calculate the original coefficients $\mathbf{a}$, $\mathbf{b}$, $\mathbf{c}$, and $\mathbf{d}$ of each segment by means of Equations (4.3), (4.4), or (4.7), which should be written and solved $n - 1$ times, once for each segment of the spline.

The reader should notice that the matrix of coefficients of Equation (5.7) is tridiagonal and therefore diagonally dominant and thus nonsingular. This means that the system of equations can always be solved and that it has a unique solution. (Matrices and their properties are discussed in texts on linear algebra.)

This approach to solving Equation (5.7) is called the *clamped* end condition. Its advantage is that the user can vary the shape of the curve by entering new values for $\mathbf{P}_1^t$ and $\mathbf{P}_n^t$ and recalculating. This allows for interactive design, where each step brings the curve closer to the desired shape. Figure 5.1a is an example of three cubic splines that pass through the same points and differ only in $\mathbf{P}_1^t$ and $\mathbf{P}_n^t$. It illustrates how the shape of the entire curve can be radically changed by modifying the two extreme tangents.

It is possible to let the user specify any two tangent vectors, not just the two extreme ones. However, varying the two extreme tangents is a natural way to edit and reshape the curve in practical applications.

Tension control. Section 4.2.3 shows how to control the tension of a Hermite segment by varying the magnitudes of the tangent vectors. Since a cubic spline is based on Hermite segments, its tension can also be controlled in the same way. The user may input a tension parameter s and the software simply multiplies every tangent vector by s. Small values of s correspond to high tension, so a user-friendly algorithm inputs a parameter T in the interval $[0, 1]$ and multiplies each tangent vector by $s = \alpha(1 - T)$ for some predetermined α. Large values of T (close to 1) correspond to small s and therefore to high tension, while small values of T correspond to s close to α. This makes T a natural tension parameter. Section 5.4 has the similar relation $T = 1 - 2s$, which makes more sense for cardinal splines.

The downside of the cubic spline is the following:

1. There is no local control. Modifying the extreme tangent vectors changes Equation (5.7) and results in a different set of n tangent vectors. The entire curve is modified!

2. Equation (5.7) is a system of n equations that, for large values of n, may be too slow to solve.

| Picnic Blues (anagram of Cubic Spline.) |

5.1.1 Example

Given the four points $\mathbf{P}_1 = (0,0)$, $\mathbf{P}_2 = (1,0)$, $\mathbf{P}_3 = (1,1)$, and $\mathbf{P}_4 = (0,1)$, we are looking for three Hermite segments $\mathbf{P}_1(t)$, $\mathbf{P}_2(t)$, and $\mathbf{P}_3(t)$ that will connect smoothly at the two interior points $\mathbf{P}_2$ and $\mathbf{P}_3$ and will constitute the spline. We further select an initial direction $\mathbf{P}_1^t = (1,-1)$ and a final direction $\mathbf{P}_4^t = (-1,-1)$. Figure 5.2 shows the points, the two extreme tangent vectors, and the resulting curve.

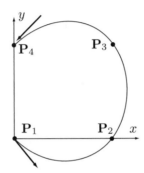

Figure 5.2: A Cubic Spline Example.

We first write Equation (5.7) for our special case ($n = 4$)

$$\begin{pmatrix} 1 & 4 & 1 & 0 \\ 0 & 1 & 4 & 1 \end{pmatrix} \begin{pmatrix} (1,-1) \\ \mathbf{P}_2^t \\ \mathbf{P}_3^t \\ (-1,-1) \end{pmatrix} = \begin{pmatrix} 3[(1,1)-(0,0)] \\ 3[(0,1)-(1,0)] \end{pmatrix} = \begin{pmatrix} (3,3) \\ (-3,3) \end{pmatrix},$$

or
$$(1,-1) + 4\mathbf{P}_2^t + \mathbf{P}_3^t = (3,3),$$
$$\mathbf{P}_2^t + 4\mathbf{P}_3^t + (-1,-1) = (-3,3).$$

This is a system of two equations in two unknowns. It is easy to solve and the solutions are $\mathbf{P}_2^t = (\frac{2}{3}, \frac{4}{5})$ and $\mathbf{P}_3^t = (-\frac{2}{3}, \frac{4}{5})$.

We now write Equation (4.7) three times, for the three spline segments. For the first segment, Equation (4.7) becomes

$$\mathbf{P}_1(t) = (t^3, t^2, t, 1) \begin{pmatrix} 2 & -2 & 1 & 1 \\ -3 & 3 & -2 & -1 \\ 0 & 0 & 1 & 0 \\ 1 & 0 & 0 & 0 \end{pmatrix} \begin{pmatrix} (0,0) \\ (1,0) \\ (1,-1) \\ (\frac{2}{3}, \frac{4}{5}) \end{pmatrix}$$
$$= (-\frac{1}{3}, -\frac{1}{5})t^3 + (\frac{1}{3}, \frac{6}{5})t^2 + (1,-1)t.$$

The second segment is calculated in a similar way:

$$\mathbf{P}_2(t) = (t^3, t^2, t, 1) \begin{pmatrix} 2 & -2 & 1 & 1 \\ -3 & 3 & -2 & -1 \\ 0 & 0 & 1 & 0 \\ 1 & 0 & 0 & 0 \end{pmatrix} \begin{pmatrix} (1,0) \\ (1,1) \\ (\frac{2}{3},\frac{4}{5}) \\ (-\frac{2}{3},\frac{4}{5}) \end{pmatrix}$$

$$= (0, -\tfrac{2}{5})t^3 + (-\tfrac{2}{3}, \tfrac{3}{5})t^2 + (\tfrac{2}{3}, \tfrac{4}{5})t + (1, 0).$$

Finally, we write, for the third segment,

$$\mathbf{P}_3(t) = (t^3, t^2, t, 1) \begin{pmatrix} 2 & -2 & 1 & 1 \\ -3 & 3 & -2 & -1 \\ 0 & 0 & 1 & 0 \\ 1 & 0 & 0 & 0 \end{pmatrix} \begin{pmatrix} (1,1) \\ (0,1) \\ (-\frac{2}{3},\frac{4}{5}) \\ (-1,-1) \end{pmatrix}$$

$$= (\tfrac{1}{3}, -\tfrac{1}{5})t^3 - (\tfrac{2}{3}, \tfrac{3}{5})t^2 + (-\tfrac{2}{3}, \tfrac{4}{5})t + (1, 1),$$

which completes the example.

◇ **Exercise 5.2:** Check to make sure that the three polynomial segments really connect at the two interior points. What are the tangent vectors at the points?

◇ **Exercise 5.3:** Redo the example of this section with an indefinite initial direction $\mathbf{P}_1^t = (0,0)$. What does it mean for a curve to start going in an indefinite direction?

5.1.2 Relaxed Cubic Splines

The original approach to the cubic spline curve is for the user to specify the two extreme tangent vectors. This approach is known as the *clamped* end condition. It is possible to have different end conditions, and the one described in this section is based on the simple idea of setting the two extreme second derivatives of the curve, $\mathbf{P}_1^{tt}(0)$ and $\mathbf{P}_{n-1}^{tt}(1)$, to zero. If we think of the second derivative as the acceleration of the curve (see the particle paradigm of Section 1.3), then this end condition implies constant speeds and therefore small curvatures at both ends of the curve. This is why this end condition is called *relaxed.*

It is easy to calculate the relaxed cubic spline. The second derivative of the parametric cubic $\mathbf{P}(t)$ is $\mathbf{P}^{tt}(t) = 6\mathbf{a}t + 2\mathbf{b}$ [Equation (5.3)]. The end condition $\mathbf{P}_1^{tt}(0) = 0$ implies $2\mathbf{b}_1 = 0$ or, from Equation (4.3)

$$-3\mathbf{P}_1 + 3\mathbf{P}_2 - 2\mathbf{P}_1^t - \mathbf{P}_2^t = 0, \quad \text{which yields} \quad \mathbf{P}_1^t = \tfrac{3}{2}(\mathbf{P}_2 - \mathbf{P}_1) - \tfrac{1}{2}\mathbf{P}_2^t. \tag{5.8}$$

The other end condition, $\mathbf{P}_{n-1}^{tt}(1) = 0$, implies $6\mathbf{a}_{n-1} + 2\mathbf{b}_{n-1} = 0$ or, from Equation (4.3)

$$6\left(2\mathbf{P}_{n-1} - 2\mathbf{P}_n + \mathbf{P}_{n-1}^t + \mathbf{P}_n^t\right) + 2\left(-3\mathbf{P}_{n-1} + 3\mathbf{P}_n - 2\mathbf{P}_{n-1}^t - \mathbf{P}_n^t\right) = 0,$$

or

$$\mathbf{P}_n^t = \tfrac{3}{2}(\mathbf{P}_n - \mathbf{P}_{n-1}) - \tfrac{1}{2}\mathbf{P}_{n-1}^t. \tag{5.9}$$

Substituting Eqs. (5.8) and (5.9) in Equation (5.7) results in

$$
n-2 \left\{
\underbrace{\begin{bmatrix}
1 & 4 & 1 & 0 & \cdots & 0 \\
0 & 1 & 4 & 1 & \cdots & 0 \\
 & & \ddots & \ddots & & \vdots \\
0 & \cdots & 1 & 4 & 1 & 0 \\
0 & \cdots & \cdots & 1 & 4 & 1
\end{bmatrix}}_{n}
\begin{bmatrix}
\frac{3}{2}(\mathbf{P}_2 - \mathbf{P}_1) - \frac{1}{2}\mathbf{P}_2^t \\
\mathbf{P}_2^t \\
\vdots \\
\mathbf{P}_{n-1}^t \\
\frac{3}{2}(\mathbf{P}_n - \mathbf{P}_{n-1}) - \frac{1}{2}\mathbf{P}_{n-1}^t
\end{bmatrix}
\right.
\tag{5.10}
$$

$$
= \begin{bmatrix}
3(\mathbf{P}_3 - \mathbf{P}_1) \\
3(\mathbf{P}_4 - \mathbf{P}_2) \\
\vdots \\
3(\mathbf{P}_{n-1} - \mathbf{P}_{n-3}) \\
3(\mathbf{P}_n - \mathbf{P}_{n-2})
\end{bmatrix}.
$$

This is a system of $n-2$ equations in the $n-2$ unknowns $\mathbf{P}_2^t, \mathbf{P}_3^t, \ldots, \mathbf{P}_{n-1}^t$. Calculating the relaxed cubic spline is done in the following steps:

1. Set up Equation (5.10) and solve it to obtain the $n-2$ interior tangent vectors.
2. Use $\mathbf{P}_2^t$ to calculate $\mathbf{P}_1^t$ from Equation (5.8). Similarly, use $\mathbf{P}_{n-1}^t$ to calculate $\mathbf{P}_n^t$ from Equation (5.9).
3. Now that the values of all n tangent vectors are known, write and solve Equation (4.4) or (4.7) $n-1$ times, each time calculating one spline segment.

The clamped cubic spline is interactive. The curve can be modified by varying the two extreme tangent vectors. The relaxed cubic spline, on the other hand, is not interactive. The only way to edit or modify it is to move the points or add points. The points, however, are data points that may be dictated by the problem on hand or that may be given by a user or a client, so it may not always be possible to move them.

Example: We use the same four points $\mathbf{P}_1 = (0,0)$, $\mathbf{P}_2 = (1,0)$, $\mathbf{P}_3 = (1,1)$, and $\mathbf{P}_4 = (0,1)$ of Section 5.1.1. The first step is to set up Equation (5.10) and solve it to obtain the two interior tangent vectors $\mathbf{P}_2^t$ and $\mathbf{P}_3^t$.

$$
\begin{pmatrix}
1 & 4 & 1 & 0 \\
0 & 1 & 4 & 1
\end{pmatrix}
\begin{pmatrix}
(\frac{3}{2},0) - \frac{1}{2}\mathbf{P}_2^t \\
\mathbf{P}_2^t \\
\mathbf{P}_3^t \\
(-\frac{3}{2},0) - \frac{1}{2}\mathbf{P}_3^t
\end{pmatrix}
= \begin{pmatrix}
(3,3) \\
(-3,3)
\end{pmatrix}.
$$

The solutions are

$$
\mathbf{P}_2^t = \left(\frac{3}{5}, \frac{2}{3}\right), \quad \mathbf{P}_3^t = \left(-\frac{3}{5}, \frac{2}{3}\right).
$$

The second step is to calculate $\mathbf{P}_1^t$ and $\mathbf{P}_4^t$

$$
\mathbf{P}_1^t = \frac{3}{2}(\mathbf{P}_2 - \mathbf{P}_1) - \frac{1}{2}\mathbf{P}_2^t = \left(\frac{3}{2},0\right) - \frac{1}{2}\left(\frac{3}{5}, \frac{2}{3}\right) = \left(\frac{6}{5}, -\frac{1}{3}\right),
$$

$$
\mathbf{P}_4^t = \frac{3}{2}(\mathbf{P}_4 - \mathbf{P}_3) - \frac{1}{2}\mathbf{P}_3^t = \left(-\frac{3}{2},0\right) - \frac{1}{2}\left(-\frac{3}{5}, \frac{2}{3}\right) = \left(-\frac{6}{5}, -\frac{1}{3}\right).
$$

Now that the values of all four tangent vectors are known, the last step is to write and solve Equation (4.4) or (4.7) three times to calculate each of the three segments of our example curve.

For the first segment, Equation (4.7) becomes

$$\mathbf{P}_1(t) = (t^3, t^2, t, 1) \begin{pmatrix} 2 & -2 & 1 & 1 \\ -3 & 3 & -2 & -1 \\ 0 & 0 & 1 & 0 \\ 1 & 0 & 0 & 0 \end{pmatrix} \begin{pmatrix} (0,0) \\ (1,0) \\ (\frac{6}{5}, -\frac{1}{3}) \\ (\frac{3}{5}, \frac{2}{3}) \end{pmatrix}$$

$$= (-\tfrac{1}{5}, \tfrac{1}{3})t^3 + (\tfrac{6}{5}, -\tfrac{1}{3})t.$$

For the second segment, Equation (4.7) becomes

$$\mathbf{P}_2(t) = (t^3, t^2, t, 1) \begin{pmatrix} 2 & -2 & 1 & 1 \\ -3 & 3 & -2 & -1 \\ 0 & 0 & 1 & 0 \\ 1 & 0 & 0 & 0 \end{pmatrix} \begin{pmatrix} (1,0) \\ (1,1) \\ (\frac{3}{5}, \frac{2}{3}) \\ (-\frac{3}{5}, \frac{2}{3}) \end{pmatrix}$$

$$= (0, -\tfrac{2}{3})t^3 + (-\tfrac{3}{5}, 1)t^2 + (\tfrac{3}{5}, \tfrac{2}{3})t + (1,0).$$

⋄ **Exercise 5.4:** Compute the third Hermite segment.

5.1.3 Cyclic Cubic Splines

The *cyclic* end condition is ideal for a closed cubic spline (Section 5.1.5) and also for a periodic cubic spline (Section 5.1.4). The condition is that the tangent vectors be equal at the two extremes of the curve (i.e., $\mathbf{P}_1^t = \mathbf{P}_n^t$) and the same for the second derivatives $\mathbf{P}_1^{tt} = \mathbf{P}_n^{tt}$. Notice that the curve doesn't have to be closed, i.e., a segment from $\mathbf{P}_n$ to $\mathbf{P}_1$ is not required.

Applying Equation (4.1) to the first condition yields

$$\mathbf{P}_1^t(0) = \mathbf{P}_{n-1}^t(1)$$

or

$$3\mathbf{a}_1 t^2 + 2\mathbf{b}_1 t + \mathbf{c}_1 |_{t=0} = 3\mathbf{a}_{n-1} t^2 + 2\mathbf{b}_{n-1} t + \mathbf{c}_{n-1} |_{t=1}$$

or

$$\mathbf{c}_1 = 3\mathbf{a}_{n-1} + 2\mathbf{b}_{n-1} + \mathbf{c}_{n-1}. \tag{5.11}$$

Applying Equation (5.3) to the second condition yields

$$\mathbf{P}_1^{tt}(0) = \mathbf{P}_{n-1}^{tt}(1)$$

or

$$6\mathbf{a}_1 t + 2\mathbf{b}_1 |_{t=0} = 6\mathbf{a}_{n-1} t + 2\mathbf{b}_{n-1} |_{t=1},$$

or

$$2\mathbf{b}_1 = 6\mathbf{a}_{n-1} + 2\mathbf{b}_{n-1}. \tag{5.12}$$

Subtracting Equations (5.11) and (5.12) yields $c_1 - 2b_1 = -3a_{n-1} + c_{n-1}$ or, from Equation (4.3)

$$\mathbf{P}_1^t - 2[-3\mathbf{P}_1 + 3\mathbf{P}_2 - 2\mathbf{P}_1^t - \mathbf{P}_2^t] = -3[2\mathbf{P}_{n-1} - 2\mathbf{P}_n + \mathbf{P}_{n-1}^t + \mathbf{P}_n^t] + \mathbf{P}_{n-1}^t.$$

This can be written

$$\mathbf{P}_1^t + 4\mathbf{P}_1^t + 3\mathbf{P}_n^t = 6(\mathbf{P}_2 - \mathbf{P}_1 + \mathbf{P}_n - \mathbf{P}_{n-1}) - (\mathbf{P}_2^t + \mathbf{P}_{n-1}^t).$$

Using the end condition $\mathbf{P}_1^t = \mathbf{P}_n^t$, we get

$$\mathbf{P}_1^t = \mathbf{P}_n^t = \tfrac{3}{4}(\mathbf{P}_2 - \mathbf{P}_1 + \mathbf{P}_n - \mathbf{P}_{n-1}) - \tfrac{1}{4}(\mathbf{P}_2^t + \mathbf{P}_{n-1}^t). \tag{5.13}$$

Substituting Equation (5.13) in Equation (5.7) results in

$$
n-2 \left\{
\underbrace{
\begin{bmatrix}
1 & 4 & 1 & 0 & \cdots & 0 \\
0 & 1 & 4 & 1 & \cdots & 0 \\
 & & \ddots & \ddots & & \vdots \\
0 & \cdots & 1 & 4 & 1 & 0 \\
0 & \cdots & \cdots & 1 & 4 & 1
\end{bmatrix}
}_{n}
\begin{bmatrix}
\tfrac{3}{4}(\mathbf{P}_2 - \mathbf{P}_1 + \mathbf{P}_n - \mathbf{P}_{n-1}) - \\
-\tfrac{1}{4}(\mathbf{P}_2^t + \mathbf{P}_{n-1}^t) \\
\mathbf{P}_2^t \\
\vdots \\
\mathbf{P}_{n-1}^t \\
\tfrac{3}{4}(\mathbf{P}_2 - \mathbf{P}_1 + \mathbf{P}_n - \mathbf{P}_{n-1}) - \\
-\tfrac{1}{4}(\mathbf{P}_2^t + \mathbf{P}_{n-1}^t)
\end{bmatrix}
\right.
$$

$$
=
\begin{bmatrix}
3(\mathbf{P}_3 - \mathbf{P}_1) \\
3(\mathbf{P}_4 - \mathbf{P}_2) \\
\vdots \\
3(\mathbf{P}_{n-1} - \mathbf{P}_{n-3}) \\
3(\mathbf{P}_n - \mathbf{P}_{n-2})
\end{bmatrix}, \tag{5.14}
$$

which is a system of $n - 2$ equations in the $n - 2$ unknowns $\mathbf{P}_2^t, \mathbf{P}_3^t, \ldots, \mathbf{P}_{n-1}^t$. Notice that in the case of a closed curve, these equations are somehow simplified because the two extreme points $\mathbf{P}_1$ and $\mathbf{P}_n$ are identical. Calculating the cyclic cubic spline is done in the following steps:

1. Set up Equation (5.14) and solve it to obtain the $n - 2$ interior tangent vectors.
2. Use $\mathbf{P}_2^t$ and $\mathbf{P}_{n-1}^t$ to calculate $\mathbf{P}_1^t$ and $\mathbf{P}_n^t$ from Equation (5.13).
3. Now that the values of all n tangent vectors are known, write and solve Equation (4.4) or (4.7) $n - 1$ times, each time calculating one spline segment.

Example: We select the five points $\mathbf{P}_1 = \mathbf{P}_5 = (0, -1)$, $\mathbf{P}_2 = (1, 0)$, $\mathbf{P}_3 = (0, 1)$, and $\mathbf{P}_4 = (-1, 0)$ and calculate the cubic spline with the cyclic end condition for these points. Notice that the curve is closed since $\mathbf{P}_1 = \mathbf{P}_5$. Also, since the points are symmetric about the origin, we can expect the resulting four PC segments to be similar. We start with Equation (5.14)

$$
\begin{bmatrix}
1 & 4 & 1 & 0 & 0 \\
0 & 1 & 4 & 1 & 0 \\
0 & 0 & 1 & 4 & 1
\end{bmatrix}
\begin{bmatrix}
\tfrac{3}{4}(\mathbf{P}_2 - \mathbf{P}_1 + \mathbf{P}_5 - \mathbf{P}_4) - \tfrac{1}{4}(\mathbf{P}_2^t + \mathbf{P}_4^t) \\
\mathbf{P}_2^t \\
\mathbf{P}_3^t \\
\mathbf{P}_4^t \\
\tfrac{3}{4}(\mathbf{P}_2 - \mathbf{P}_1 + \mathbf{P}_5 - \mathbf{P}_4) - \tfrac{1}{4}(\mathbf{P}_2^t + \mathbf{P}_4^t)
\end{bmatrix}
=
\begin{bmatrix}
3(\mathbf{P}_3 - \mathbf{P}_1) \\
3(\mathbf{P}_4 - \mathbf{P}_2) \\
3(\mathbf{P}_5 - \mathbf{P}_3)
\end{bmatrix},
$$

which is solved to yield $\mathbf{P}_2^t = (0, 3/2)$, $\mathbf{P}_3^t = (-3/2, 0)$, and $\mathbf{P}_4^t = (0, -3/2)$. These values are used to solve Equation (5.13)

$$\mathbf{P}_1^t = \mathbf{P}_5^t = \tfrac{3}{4}\left(\mathbf{P}_2 - \mathbf{P}_1 + \mathbf{P}_5 - \mathbf{P}_4\right) - \tfrac{1}{4}\left(\mathbf{P}_2^t + \mathbf{P}_4^t\right),$$

which gives $\mathbf{P}_1^t = \mathbf{P}_5^t = (3/2, 0)$. The four segments can now be calculated in the usual way. For the first segment, Equation (4.7) becomes

$$\mathbf{P}_1(t) = (t^3, t^2, t, 1) \begin{pmatrix} 2 & -2 & 1 & 1 \\ -3 & 3 & -2 & -1 \\ 0 & 0 & 1 & 0 \\ 1 & 0 & 0 & 0 \end{pmatrix} \begin{pmatrix} (0, -1) \\ (1, 0) \\ (\tfrac{3}{2}, 0) \\ (0, \tfrac{3}{2}) \end{pmatrix}$$

$$= -(\tfrac{1}{2}, \tfrac{1}{2})t^3 + (0, \tfrac{3}{2})t^2 + (\tfrac{3}{2}, 0)t + (0, -1).$$

For the second segment, Equation (4.7) becomes

$$\mathbf{P}_2(t) = (t^3, t^2, t, 1) \begin{pmatrix} 2 & -2 & 1 & 1 \\ -3 & 3 & -2 & -1 \\ 0 & 0 & 1 & 0 \\ 1 & 0 & 0 & 0 \end{pmatrix} \begin{pmatrix} (1, 0) \\ (0, 1) \\ (0, \tfrac{3}{2}) \\ (-\tfrac{3}{2}, 0) \end{pmatrix}$$

$$= (\tfrac{1}{2}, -\tfrac{1}{2})t^3 + (-\tfrac{3}{2}, 0)t^2 + (0, \tfrac{3}{2})t + (1, 0).$$

⋄ **Exercise 5.5:** Compute the third and fourth Hermite segments.

Notice how the symmetry of the problem causes the coefficients of $\mathbf{P}_1(t)$ and $\mathbf{P}_3(t)$ to have opposite signs, and the same for the coefficients of $\mathbf{P}_2(t)$ and $\mathbf{P}_4(t)$.

It is also possible to have an *anticyclic* end condition for the cubic spline. It requires that the two extreme tangent vectors have the same magnitudes but opposite directions $\mathbf{P}_1^t = -\mathbf{P}_n^t$ and the same condition for the second derivatives $\mathbf{P}_1^{tt} = -\mathbf{P}_n^{tt}$. Such an end condition makes sense for curves such as the cross section of a vase or any other surface of revolution.

Following steps similar to the ones for the cyclic case, we get for the anticyclic end condition

$$\mathbf{P}_1^t = -\mathbf{P}_n^t = \frac{3}{4}\left(\mathbf{P}_2 - \mathbf{P}_1 - \mathbf{P}_n + \mathbf{P}_{n-1}\right) - \frac{1}{4}\left(\mathbf{P}_2^t - \mathbf{P}_{n-1}^t\right). \qquad (5.15)$$

⋄ **Exercise 5.6:** Given the three points $\mathbf{P}_1 = (-1, 0)$, $\mathbf{P}_2 = (0, 1)$, and $\mathbf{P}_3 = (1, 0)$, calculate the anticyclic cubic spline for them and compare it to the clamped cubic spline for the same points.

5.1.4 Periodic Cubic Splines

A periodic function $f(x)$ is one that repeats itself. If p is the period of the function, then $f(x + p) = f(x)$ for any x. A two-dimensional cubic spline is periodic if it has the same extreme tangent vectors (i.e., if it starts and ends going in the same direction) and if its two extreme points $\mathbf{P}(0)$ and $\mathbf{P}(1)$ have the same y coordinate. If the curve satisfies these conditions, then we can place consecutive copies of it side by side and the result would look like a single periodic curve.

The case of a three-dimensional periodic cubic spline is less clear. It seems that the two extreme points can be any points (they don't have to have the same y or z coordinates or any other relationship), so the condition for periodicity is that the curve will have the same start and end tangents, i.e., it will be cyclic.

Example: Exercise 1.15 shows that the parametric expression $(\cos t, \sin t, t)$ describes a helix (see also Section 2.4.1 for a double helix). Modifying this expression to $\mathbf{P}(t) = (0.05t + \cos t, \sin t, .1t)$ creates a helix that moves in the x direction as it climbs up in the z direction. Figure 5.3 shows its behavior. This curve starts at $\mathbf{P}(0) = (1, 0, 0)$ and ends at $\mathbf{P}(10\pi) = (0.5\pi + 1, 0, \pi)$. There is no special relation between the start and end points, but the curve is periodic since both its start and end tangents equal $\mathbf{P}^t(0) = \mathbf{P}^t(10\pi) = (0.05, 1, 0.1)$. We can construct another period of this curve by copying it, moving the copy parallel to itself, and placing it such that the start point of the copy is at the end point of the original curve.

Notice that it is possible to make the start and end points even more unrelated by, for example, tilting the helix also in the y direction as it climbs up in the z direction. This kind of effect is achieved by an expression such as

$$\mathbf{P}(t) = (0.05t + \cos t, -0.05t^2 + \sin t, 0.1t).$$

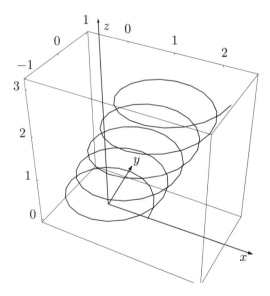

```
(* tilted helix as a periodic curve *)
ParametricPlot3D[{.05t+Cos[t],Sin[t],.1t}, {t,0,10Pi}, Compiled->False,
   Ticks->{{-1,0,1,2},{-1,0,1},{0,1,2,3}}, DefaultFont->{"cmr10", 10},
   PlotPoints->100]
```

Figure 5.3: A Tilted Helix as a Periodic Curve.

5.1.5 Closed Cubic Splines

A closed cubic spline has an extra curve segment from $\mathbf{P}_n$ to $\mathbf{P}_1$ that closes the curve. In such a curve, every point is interior, so Equation (5.7) becomes a system of n equations in the same n unknowns. No user input is needed, which implies that the only way to control or modify such a curve is to move, add, or delete points. It is convenient to define the two additional points $\mathbf{P}_{n+1} \overset{\text{def}}{=} \mathbf{P}_1$ and $\mathbf{P}_{n+2} \overset{\text{def}}{=} \mathbf{P}_2$. Equation (5.7) then becomes

$$
n\left\{
\begin{bmatrix}
1 & 4 & 1 & \cdots & 0 & \cdots & 0 \\
0 & 1 & 4 & 1 & \cdots & \cdots & 0 \\
 & & & \ddots & \ddots & & \vdots \\
0 & \cdots & \cdots & \cdots & 1 & 4 & 1 \\
1 & \cdots & \cdots & \cdots & 0 & 1 & 4 \\
4 & 1 & 0 & \cdots & 0 & 0 & 1
\end{bmatrix}
}_{n}
\begin{bmatrix}
\mathbf{P}_1^t \\
\mathbf{P}_2^t \\
\vdots \\
\mathbf{P}_{n-1}^t \\
\mathbf{P}_n^t
\end{bmatrix}
=
\begin{bmatrix}
3(\mathbf{P}_3 - \mathbf{P}_1) \\
3(\mathbf{P}_4 - \mathbf{P}_2) \\
\vdots \\
3(\mathbf{P}_{n+1} - \mathbf{P}_{n-1}) \\
3(\mathbf{P}_{n+2} - \mathbf{P}_n)
\end{bmatrix}.
\tag{5.16}
$$

Example: Given the four points of Section 5.1.1, $\mathbf{P}_1 = (0,0)$, $\mathbf{P}_2 = (1,0)$, $\mathbf{P}_3 = (1,1)$, and $\mathbf{P}_4 = (0,1)$, we are looking for four Hermite segments $\mathbf{P}_1(t)$, $\mathbf{P}_2(t)$, $\mathbf{P}_3(t)$, and $\mathbf{P}_4(t)$ that would connect smoothly at the four points. Equation (5.16) becomes

$$
\begin{pmatrix}
1 & 4 & 1 & 0 \\
0 & 1 & 4 & 1 \\
1 & 0 & 1 & 4 \\
4 & 1 & 0 & 1
\end{pmatrix}
\begin{pmatrix}
\mathbf{P}_1^t \\
\mathbf{P}_2^t \\
\mathbf{P}_3^t \\
\mathbf{P}_4^t
\end{pmatrix}
=
\begin{bmatrix}
3(\mathbf{P}_3 - \mathbf{P}_1) \\
3(\mathbf{P}_4 - \mathbf{P}_2) \\
3(\mathbf{P}_1 - \mathbf{P}_3) \\
3(\mathbf{P}_2 - \mathbf{P}_4)
\end{bmatrix}.
\tag{5.17}
$$

Its solutions are $\mathbf{P}_1^t = (3/4, -3/4)$, $\mathbf{P}_2^t = (3/4, 3/4)$, $\mathbf{P}_3^t = (-3/4, 3/4)$, and $\mathbf{P}_4^t = (-3/4, -3/4)$, and the four spline segments are

$$
\begin{aligned}
\mathbf{P}_1(t) &= (-1/2, 0)t^3 + (3/4, 3/4)t^2 + (3/4, -3/4)t, \\
\mathbf{P}_2(t) &= (0, -1/2)t^3 + (-3/4, 3/4)t^2 + (3/4, 3/4)t + (1, 0), \\
\mathbf{P}_3(t) &= (1/2, 0)t^3 + (-3/4, -3/4)t^2 + (-3/4, 3/4)t + (1, 1), \\
\mathbf{P}_4(t) &= (0, 1/2)t^3 + (3/4, -3/4)t^2 + (-3/4, -3/4)t + (0, 1).
\end{aligned}
$$

5.1.6 Nonuniform Cubic Splines

All the different types of cubic splines discussed so far assume that the parameter t varies in the interval $[0, 1]$ in every segment. These types of cubic spline are therefore *uniform* or *normalized*. The *nonuniform cubic spline* is obtained by adding another parameter t_k to every spline segment and letting t vary in the interval $[0, t_k]$. Since there are $n - 1$ spline segments connecting the n data points, this adds $n - 1$ parameters to the curve, which makes it easier to fine-tune the shape of the curve. The nonuniform cubic splines are especially useful in cases where the data points are nonuniformly spaced. In regions where the points are closely spaced, the normalized cubic spline tends to develop loops and overshoots. In regions where the points are widely spaced, it tends to "cut corners," i.e., to be too tight. Careful selection of the t_k parameters can overcome these tendencies.

The calculation of the nonuniform cubic spline is based on that of the uniform version. We simply rewrite some of the basic equations, substituting t_k for 1 as the final value of t. We start with Equation (4.2) that becomes, for the first spline segment,

$$\mathbf{a}\cdot 0^3 + \mathbf{b}\cdot 0^2 + \mathbf{c}\cdot 0 + \mathbf{d} = \mathbf{P}_1,$$

$$\mathbf{a}(t_1)^3 + \mathbf{b}(t_1)^2 + \mathbf{c}(t_1) + \mathbf{d} = \mathbf{P}_2,$$

$$3\mathbf{a}\cdot 0^2 + 2\mathbf{b}\cdot 0 + \mathbf{c} = \mathbf{P}_1^t,$$

$$3\mathbf{a}(t_1)^2 + 2\mathbf{b}(t_1) + \mathbf{c} = \mathbf{P}_2^t,$$

with solutions

$$\begin{aligned}
\mathbf{a} &= \frac{2(\mathbf{P}_1 - \mathbf{P}_2)}{t_1^3} + \frac{\mathbf{P}_1^t}{t_1^2} + \frac{\mathbf{P}_2^t}{t_1^2}, \\
\mathbf{b} &= \frac{3(\mathbf{P}_2 - \mathbf{P}_1)}{t_1^2} - \frac{2\mathbf{P}_1^t}{t_1} - \frac{\mathbf{P}_2^t}{t_1}, \\
\mathbf{c} &= \mathbf{P}_1^t, \\
\mathbf{d} &= \mathbf{P}_1.
\end{aligned} \tag{5.18}$$

Equation (4.4) now becomes

$$\mathbf{P}(t) = \left[\frac{2(\mathbf{P}_1 - \mathbf{P}_2)}{t_1^3} + \frac{\mathbf{P}_1^t}{t_1^2} + \frac{\mathbf{P}_2^t}{t_1^2} \right] t^3 + \left[\frac{3(\mathbf{P}_2 - \mathbf{P}_1)}{t_1^2} - \frac{2\mathbf{P}_1^t}{t_1} - \frac{\mathbf{P}_2^t}{t_1} \right] t^2 + \mathbf{P}_1^t t + \mathbf{P}_1. \tag{5.19}$$

Equation (5.4) becomes

$$\mathbf{P}_k^{tt}(t_k) = \mathbf{P}_{k+1}^{tt}(0) \qquad \text{or} \qquad 6\mathbf{a}_k \times t_k + 2\mathbf{b}_k = 6\mathbf{a}_{k+1} \times 0 + 2\mathbf{b}_{k+1}, \tag{5.20}$$

and Equation (5.5) is now

$$\begin{aligned}
2 & \left[\frac{3(\mathbf{P}_{k+1} - \mathbf{P}_k)}{t_k^2} - \frac{2\mathbf{P}_k^t}{t_k} - \frac{\mathbf{P}_{k+1}^t}{t_k} \right] + 6t_k \left[\frac{2(\mathbf{P}_k - \mathbf{P}_{k+1})}{t_k^3} + \frac{\mathbf{P}_k^t}{t_k^2} + \frac{\mathbf{P}_{k+1}^t}{t_k^2} \right] \\
&= 2 \left[\frac{3(\mathbf{P}_{k+2} - \mathbf{P}_{k+1})}{t_{k+1}^2} - \frac{2\mathbf{P}_{k+1}^t}{t_{k+1}} - \frac{\mathbf{P}_{k+2}^t}{t_{k+1}} \right].
\end{aligned} \tag{5.21}$$

Equation (5.6) now becomes

$$\begin{aligned}
t_{k+1}\mathbf{P}_k^t &+ 2(t_k + t_{k+1})\mathbf{P}_{k+1}^t + t_k\mathbf{P}_{k+2}^t \\
&= \frac{3}{t_k t_{k+1}} \left[t_k^2(\mathbf{P}_{k+2} - \mathbf{P}_{k+1}) + t_{k+1}^2(\mathbf{P}_{k+1} - \mathbf{P}_k) \right].
\end{aligned} \tag{5.22}$$

This produces the new version of Equation (5.7)

$$
n-2\left\{
\begin{bmatrix}
t_2 & 2(t_1+t_2) & t_1 & 0 & 0 & \cdots & 0 \\
0 & t_3 & 2(t_2+t_3) & t_2 & 0 & \cdots & 0 \\
 & & & \ddots & & \ddots & \vdots \\
0 & 0 & \cdots & \cdots & t_{n-1} & 2(t_{n-1}+t_{n-2}) & t_{n-2}
\end{bmatrix}
\underbrace{}_{n}
\begin{bmatrix}
\mathbf{P}_1^t \\
\mathbf{P}_2^t \\
\vdots \\
\mathbf{P}_n^t
\end{bmatrix}
\right.
$$

$$
=
\begin{bmatrix}
\frac{3}{t_1 t_2}\left[t_1^2(\mathbf{P}_3 - \mathbf{P}_2) + t_2^2(\mathbf{P}_2 - \mathbf{P}_1)\right] \\
\frac{3}{t_2 t_3}\left[t_2^2(\mathbf{P}_4 - \mathbf{P}_3) + t_3^2(\mathbf{P}_3 - \mathbf{P}_2)\right] \\
\vdots \\
\frac{3}{t_{n-2}t_{n-1}}\left[t_{n-2}^2(\mathbf{P}_n - \mathbf{P}_{n-1}) + t_{n-1}^2(\mathbf{P}_{n-1} - \mathbf{P}_{n-2})\right]
\end{bmatrix}.
$$

$$(5.23)$$

This is again a system of $n-2$ equations in the n unknowns $\mathbf{P}_1^t, \mathbf{P}_2^t, \ldots, \mathbf{P}_n^t$. After the user inputs the guessed or estimated values for the two extreme tangent vectors $\mathbf{P}_1^t$ and $\mathbf{P}_n^t$, this system can be solved, yielding the values of the remaining $n-2$ tangent vectors. Each of the $n-1$ spline segments can now be calculated by means of Equation (5.18) that is written here for the first segment in compact form

$$
\begin{pmatrix} \mathbf{a} \\ \mathbf{b} \\ \mathbf{c} \\ \mathbf{d} \end{pmatrix}
=
\begin{pmatrix}
2/t_1^3 & -2/t_1^3 & 1/t_1^2 & 1/t_1^2 \\
-3/t_1^2 & 3/t_1^2 & -2/t_1 & -1/t_1 \\
0 & 0 & 1 & 0 \\
1 & 0 & 0 & 0
\end{pmatrix}
\begin{pmatrix} \mathbf{P}_1 \\ \mathbf{P}_2 \\ \mathbf{P}_1^t \\ \mathbf{P}_2^t \end{pmatrix}.
$$

$$(5.24)$$

Notice how each of Equations (5.18) through (5.24) reduces to the corresponding original equation when all the t_i are set to 1. The nonuniform cubic spline can now be calculated in the following steps:

1. The user inputs the values of the two extreme tangent vectors and the values of the $n-1$ parameters t_k. The software sets up and solves Equation (5.23) to calculate the remaining tangent vectors.

2. The software sets up and solves Equation (5.24) $n-1$ times, once for each of the spline segments.

3. Each segment $\mathbf{P}_k(t)$ is plotted by varying t from 0 to t_k.

Before looking at an example, it is useful to try to understand the advantage of having the extra parameters t_k. Equation (5.18) shows that a large value of t_k for spline segment $\mathbf{P}_k(t)$ means small $\mathbf{a}$ and $\mathbf{b}$ coefficients (since t_k appears in the denominators), and hence a small second derivative $\mathbf{P}_k^{tt}(t) = 6\mathbf{a}_k + 2\mathbf{b}_k$ for that segment. Since the second derivative can be interpreted as the acceleration of the curve, we can predict that a large t_k will result in small overall acceleration for segment k. Thus, most of the segment will be close to a straight line. This is also easy to see when we substitute small $\mathbf{a}$ and $\mathbf{b}$ in $\mathbf{P}_k(t) = \mathbf{a}t^3 + \mathbf{b}t^2 + \mathbf{c}t + \mathbf{d}$. The dominant part of the segment becomes $\mathbf{c}t + \mathbf{d}$, which brings it close to linear. If the start and end directions of the segment are very different, the entire segment cannot be a straight line, so, in order to minimize its overall second derivative, the segment will end up consisting of two or three parts, each close to a straight line, with short, highly-curved corners connecting them (Figure 5.4).

Such a geometry has a small overall second derivative. This knowledge is useful when designing curves, which is why the nonuniform cubic spline should not be dismissed as impractical. It may be the best method for certain curves.

Figure 5.4: Curves with Small Overall Second Derivative.

Example: The four points of Section 5.1.1 are used in this example. They are $\mathbf{P}_1 = (0,0)$, $\mathbf{P}_2 = (1,0)$, $\mathbf{P}_3 = (1,1)$, and $\mathbf{P}_4 = (0,1)$. We also select the same initial and final directions $\mathbf{P}_1^t = (1,-1)$ and $\mathbf{P}_4^t = (-1,-1)$. We decide to use $t_k = 2$ for each of the three spline segments to illustrate how large t_k values create a curve very different from the one of Section 5.1.1. Equation (5.23) becomes

$$\begin{bmatrix} t_2 & 2(t_1+t_2) & t_1 & 0 \\ 0 & t_3 & 2(t_2+t_3) & t_2 \end{bmatrix} \begin{bmatrix} (1,-1) \\ \mathbf{P}_2^t \\ \mathbf{P}_3^t \\ (-1,-1) \end{bmatrix} = \begin{bmatrix} \frac{3}{t_1 t_2}[t_1^2(\mathbf{P}_3 - \mathbf{P}_2) + t_2^2(\mathbf{P}_2 - \mathbf{P}_1)] \\ \frac{3}{t_2 t_3}[t_2^2(\mathbf{P}_4 - \mathbf{P}_3) + t_3^2(\mathbf{P}_3 - \mathbf{P}_2)] \end{bmatrix}.$$

For $t_1 = t_2 = t_3 = 2$, this yields $\mathbf{P}_2^t = (1/6, 1/2)$ and $\mathbf{P}_3^t = (-1/6, 1/2)$. Equation (5.24) is now written and solved three times:

Segment 1
$$\begin{pmatrix} \mathbf{a} \\ \mathbf{b} \\ \mathbf{c} \\ \mathbf{d} \end{pmatrix} = \begin{pmatrix} 2/t_1^3 & -2/t_1^3 & 1/t_1^2 & 1/t_1^2 \\ -3/t_1^2 & 3/t_1^2 & -2/t_1 & -1/t_1 \\ 0 & 0 & 1 & 0 \\ 1 & 0 & 0 & 0 \end{pmatrix} \begin{bmatrix} (0,0) \\ (1,0) \\ (1,-1) \\ (1/6,1/2) \end{bmatrix}.$$

Segment 2
$$\begin{pmatrix} \mathbf{a} \\ \mathbf{b} \\ \mathbf{c} \\ \mathbf{d} \end{pmatrix} = \begin{pmatrix} 2/t_2^3 & -2/t_2^3 & 1/t_2^2 & 1/t_2^2 \\ -3/t_2^2 & 3/t_2^2 & -2/t_2 & -1/t_2 \\ 0 & 0 & 1 & 0 \\ 1 & 0 & 0 & 0 \end{pmatrix} \begin{bmatrix} (1,0) \\ (1,1) \\ (1/6,1/2) \\ (-1/6,1/2) \end{bmatrix}.$$

Segment 3
$$\begin{pmatrix} \mathbf{a} \\ \mathbf{b} \\ \mathbf{c} \\ \mathbf{d} \end{pmatrix} = \begin{pmatrix} 2/t_3^3 & -2/t_3^3 & 1/t_3^2 & 1/t_3^2 \\ -3/t_3^2 & 3/t_3^2 & -2/t_3 & -1/t_3 \\ 0 & 0 & 1 & 0 \\ 1 & 0 & 0 & 0 \end{pmatrix} \begin{bmatrix} (1,1) \\ (0,1) \\ (-1/6,1/2) \\ (-1,-1) \end{bmatrix}.$$

This yields the coefficients for the three spline segments:

$$\mathbf{P}_1(t) = (1/24, -1/8)t^3 + (-1/3, 3/4)t^2 + (1,-1)t,$$
$$\mathbf{P}_2(t) = (0,0)t^3 + (-1/12, 0)t^2 + (1/6, 1/2)t + (1,0),$$
$$\mathbf{P}_3(t) = -(1/24, 1/8)t^3 + (-1/12, 0)t^2 + (-1/6, 1/2)t + (1,1).$$

The result is shown in Figure 5.5. It should be compared with the uniform curve of Figure 5.2 that's based on the same four points. (Recall that t varies from 0 to 2 in each of the segments above.)

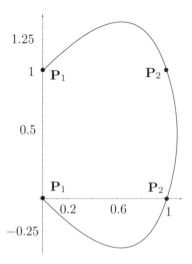

```
(* Nonuniform cubic spline example *)
C1:=ParametricPlot[{1/24,-1/8}t^3+{-1/3,3/4}t^2+{1,-1}t, {t,0,2},
PlotRange->All, Compiled->False, DisplayFunction->Identity];
C2:=ParametricPlot[{-1/12,0}t^2+{1/6,1/2}t+{1,0}, {t,0,2},
PlotRange->All, Compiled->False, DisplayFunction->Identity];
C3:=ParametricPlot[-{1/24,1/8}t^3+{-1/12,0}t^2+{-1/6,1/2}t+{1,1}, {t,0,2},
PlotRange->All, Compiled->False, DisplayFunction->Identity];
Show[C1, C2, C3, PlotRange->All, AspectRatio->Automatic,
DisplayFunction->$DisplayFunction, DefaultFont->{"cmr10", 10}];
```

Figure 5.5: A Nonuniform Cubic Spline Example.

5.2 The Quadratic Spline

The cubic spline curve is useful in certain practical applications, which raises the question of splines of different degrees based on the same concepts. It turns out that splines of degrees higher than 3 are useful only for special applications because they are more computationally intensive and tend to have many undesirable inflection points (i.e., they tend to wiggle excessively). Splines of degree 1 are, of course, just straight segments connected to form a polyline, but quadratic (degree-2) splines can be useful in some applications. Such a spline is easy to derive and to compute. Each spline segment is a quadratic polynomial, i.e., a parabolic arc, so it results in fewer oscillations in the curve. On the other hand, quadratic spline segments connect with at most C^1 continuity because their second derivative is a constant. Thus, a quadratic spline curve may not be as tight as a cubic spline that passes through the same points.

The quadratic spline curve is derived in this section based on the variant Hermite segment of Section 4.7. Each segment $\mathbf{P}_i(t)$ is therefore a quadratic polynomial defined by its two endpoints $\mathbf{P}_i$ and $\mathbf{P}_{i+1}$ and by its start tangent vector $\mathbf{P}_i^t$. Equation (4.33) shows that the end tangent of such a segment is $\mathbf{P}_i^t(1) = 2(\mathbf{P}_{i+1} - \mathbf{P}_i) - \mathbf{P}_i^t$. The first two spline segments are

$$\mathbf{P}_1(t) = (\mathbf{P}_2 - \mathbf{P}_1 - \mathbf{P}_1^t)t^2 + \mathbf{P}_1^t t + \mathbf{P}_1,$$
$$\mathbf{P}_2(t) = (\mathbf{P}_3 - \mathbf{P}_2 - \mathbf{P}_2^t)t^2 + \mathbf{P}_2^t t + \mathbf{P}_2.$$

At their joint point $\mathbf{P}_2$ they have the tangent vectors $\mathbf{P}_1^t(1) = 2(\mathbf{P}_2 - \mathbf{P}_1) - \mathbf{P}_1^t$ and $\mathbf{P}_2^t(0) = \mathbf{P}_2^t$. In order to achieve C^1 continuity we have to have the boundary condition $\mathbf{P}_1^t(1) = \mathbf{P}_2^t(0)$ or $2(\mathbf{P}_2 - \mathbf{P}_1) - \mathbf{P}_1^t = \mathbf{P}_2^t$. This equation can be written $\mathbf{P}_1^t + \mathbf{P}_2^t = 2(\mathbf{P}_2 - \mathbf{P}_1)$, and when duplicated $n-1$ times, for the points $\mathbf{P}_1$ through $\mathbf{P}_{n-1}$, the result is

$$n-1\left\{ \underbrace{\begin{bmatrix} 1 & 1 & 0 & 0 & \cdots & 0 & 0 \\ 0 & 1 & 1 & 0 & \cdots & 0 & 0 \\ & & \ddots & & \ddots & & \vdots \\ 0 & 0 & 0 & 0 & \cdots & 1 & 1 \end{bmatrix}}_{n} \begin{bmatrix} \mathbf{P}_1^t \\ \mathbf{P}_2^t \\ \vdots \\ \mathbf{P}_n^t \end{bmatrix} = 2 \begin{bmatrix} \mathbf{P}_2 - \mathbf{P}_1 \\ \mathbf{P}_3 - \mathbf{P}_2 \\ \vdots \\ \mathbf{P}_n - \mathbf{P}_{n-1} \end{bmatrix} \right. \tag{5.25}$$

As with the cubic spline, there are more unknowns than equations (n unknowns and $n-1$ equations), and the standard technique is to ask the user to provide a value for one of the unknown tangent vectors, normally $\mathbf{P}_1^t$.

Example: We select the four points of Section 5.1.1, namely $\mathbf{P}_1 = (0,0)$, $\mathbf{P}_2 = (1,0)$, $\mathbf{P}_3 = (1,1)$, and $\mathbf{P}_4 = (0,1)$. We also select the same start tangent $\mathbf{P}_1^t = (1,-1)$. Equation (5.25) becomes

$$\begin{pmatrix} 1 & 1 & 0 & 0 \\ 0 & 1 & 1 & 0 \\ 0 & 0 & 1 & 1 \end{pmatrix} \begin{pmatrix} \mathbf{P}_1^t \\ \mathbf{P}_2^t \\ \mathbf{P}_3^t \\ \mathbf{P}_4^t \end{pmatrix} = 2 \begin{pmatrix} \mathbf{P}_2 - \mathbf{P}_1 \\ \mathbf{P}_3 - \mathbf{P}_2 \\ \mathbf{P}_4 - \mathbf{P}_3 \end{pmatrix} = \begin{pmatrix} (2,0) \\ (0,2) \\ (-2,0) \end{pmatrix},$$

with solutions $\mathbf{P}_2^t = (1,1)$, $\mathbf{P}_3^t = (-1,1)$, and $\mathbf{P}_4^t = (-1,-1)$. The three spline segments become

$$\mathbf{P}_1(t) = (\mathbf{P}_2 - \mathbf{P}_1 - \mathbf{P}_1^t)t^2 + \mathbf{P}_1^t t + \mathbf{P}_1 = (t, t^2 - t),$$
$$\mathbf{P}_2(t) = (\mathbf{P}_3 - \mathbf{P}_2 - \mathbf{P}_2^t)t^2 + \mathbf{P}_2^t t + \mathbf{P}_2 = (-t^2 + t + 1, t),$$
$$\mathbf{P}_3(t) = (\mathbf{P}_4 - \mathbf{P}_3 - \mathbf{P}_3^t)t^2 + \mathbf{P}_3^t t + \mathbf{P}_3 = (-t + 1, -t^2 + t + 1).$$

Their tangent vectors are $\mathbf{P}_1^t(t) = (1, 2t-1)$, $\mathbf{P}_2^t(t) = (-2t+1, 1)$, and $\mathbf{P}_3^t(t) = (-1, -2t+1)$. It is easy to see that $\mathbf{P}_1^t(1) = \mathbf{P}_2^t(0) = (1,1)$ and $\mathbf{P}_2^t(1) = \mathbf{P}_3^t(0) = (-1,1)$. Also, the end tangent of the entire curve is $\mathbf{P}_3^t(1) = (-1,-1)$, the same as for the cubic case. The complete spline curve is shown in Figure 5.6.

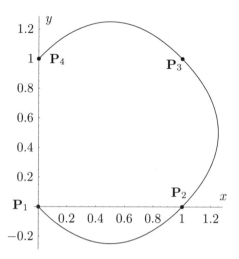

```
(* quadratic spline example *)
C1:=ParametricPlot[{t,t^2-t}, {t,0,1}, DisplayFunction->Identity];
C2:=ParametricPlot[{-t^2+t+1,t}, {t,0,1}, DisplayFunction->Identity];
C3:=ParametricPlot[{-t+1,-t^2+t+1}, {t,0,1}, DisplayFunction->Identity];
C4=Graphics[{AbsolutePointSize[3],
 Point[{0,0}], Point[{1,0}], Point[{1,1}], Point[{0,1}] }];
Show[C1, C2, C3, C4, DisplayFunction->$DisplayFunction,
DefaultFont->{"cmr10", 10}, AspectRatio->Automatic]
```

Figure 5.6: A Quadratic Spline Example.

5.3 The Quintic Spline

The derivation of the cubic spline is based on the requirement (boundary condition) that the second derivatives of the individual segments be equal at the interior points. This produces $n-2$ equations to compute the first derivatives, but makes it impossible to control the values of the second derivatives. In cases where the designer wants to specify the values of the second derivatives, higher-degree polynomials must be used. A degree-5 (quintic) polynomial is a natural choice. Section 4.3 discusses the similar case of the quintic Hermite segment.

The approach to the quintic spline is similar to that of the cubic spline. The spline is a set of $n-1$ segments, each a quintic polynomial, so we have to compute the coefficients of each segment from the boundary conditions. A general quintic spline segment from point $\mathbf{P}_k$ to point $\mathbf{P}_{k+1}$ is given by Equation (4.16), duplicated here

$$\mathbf{P}_k(t) = \mathbf{a}_k t^5 + \mathbf{b}_k t^4 + \mathbf{c}_k t^3 + \mathbf{d}_k t^2 + \mathbf{e}_k t + \mathbf{f}_k. \tag{4.16}$$

The six coefficients are computed from the following six boundary conditions

$$\mathbf{P}_k(0) = \mathbf{P}_k, \quad \mathbf{P}_k(1) = \mathbf{P}_{k+1}, \quad \mathbf{P}'_k(1) = \mathbf{P}'_{k+1}(0),$$

$$\mathbf{P}_k''(1) = \mathbf{P}_{k+1}''(0), \quad \mathbf{P}_k'''(1) = \mathbf{P}_{k+1}'''(0), \quad \mathbf{P}_k''''(1) = \mathbf{P}_{k+1}''''(0).$$

(Notice that these conditions involve the first four derivatives. Experience indicates that better-looking splines are obtained when the boundary conditions are based on an even number of derivatives, which is why the quintic, and not the quartic, polynomial is a natural choice.)

The boundary conditions can be written explicitly as follows:

$$
\begin{aligned}
\mathbf{f}_k &= \mathbf{P}_k, & a \\
\mathbf{a}_k + \mathbf{b}_k + \mathbf{c}_k + \mathbf{d}_k + \mathbf{e}_k + \mathbf{f}_k &= \mathbf{f}_{k+1} = \mathbf{P}_{k+1}, & b \\
5\mathbf{a}_k + 4\mathbf{b}_k + 3\mathbf{c}_k + 2\mathbf{d}_k + \mathbf{e}_k &= \mathbf{e}_{k+1}, & c \\
20\mathbf{a}_k + 12\mathbf{b}_k + 6\mathbf{c}_k + 2\mathbf{d}_k &= 2\mathbf{d}_{k+1}, & (5.26)d \\
60\mathbf{a}_k + 24\mathbf{b}_k + 6\mathbf{c}_k &= 6\mathbf{c}_{k+1}, & e \\
120\mathbf{a}_k + 24\mathbf{b}_k &= 24\mathbf{b}_{k+1}. & f
\end{aligned}
$$

These equations are now used to express the six coefficients of each of the $n - 1$ quintic polynomials in terms of the second and fourth derivatives.

Equation $(5.26)f$ results in $24\mathbf{b}_{k+1} = \mathbf{P}_{k+1}''''(0)$ or $\mathbf{b}_k = \frac{1}{24}\mathbf{P}_k''''(0)$. This also implies $\mathbf{a}_k = \frac{1}{120}[\mathbf{P}_{k+1}''''(0) - \mathbf{P}_k''''(0)]$. Equation $(5.26)d$ implies $2\mathbf{d}_{k+1} = \mathbf{P}_{k+1}''(0)$ or $\mathbf{d}_k = \frac{1}{2}\mathbf{P}_k''(0)$. Now that we have expressed $\mathbf{a}_k$, $\mathbf{b}_k$, and $\mathbf{d}_k$ in terms of the second and fourth derivatives, we substitute them in Equation $(5.26)d$ to get the following expression for $\mathbf{c}_k$

$$\mathbf{c}_k = \frac{1}{6}[\mathbf{P}_{k+1}''(0) - \mathbf{P}_k''(0)] - \frac{1}{36}[\mathbf{P}_{k+1}''''(0) + 2\mathbf{P}_k''''(0)].$$

The last coefficient to be expressed in terms of the (still unknown) second and fourth derivatives is $\mathbf{e}_k$. This is done from $\mathbf{P}_k(1) = \mathbf{P}_{k+1}$ and results in

$$\mathbf{e}_k = [\mathbf{P}_{k+1} - \mathbf{P}_k] - \frac{1}{6}[\mathbf{P}_{k+1}''(0) + 2\mathbf{P}_k''(0)] + \frac{1}{360}[7\mathbf{P}_{k+1}''''(0) + 8\mathbf{P}_k''''(0)].$$

When these expressions for the six coefficients are combined with $\mathbf{P}_{k-1}'(1) = \mathbf{P}_k'(0)$, all the terms with first and third derivatives are eliminated, and the result is a relation between the (unknown) second and fourth derivatives and the (known) data points

$$[\mathbf{P}_{k-1} - \mathbf{P}_k] + \frac{1}{6}[\mathbf{P}_{k-1}''(0) + 2\mathbf{P}_k''(0)] - \frac{1}{360}[7\mathbf{P}_{k-1}''''(0) + 8\mathbf{P}_k''''(0)] \qquad (5.27)$$
$$= [\mathbf{P}_{k+1} - \mathbf{P}_k] - \frac{1}{6}[\mathbf{P}_{k+1}''(0) + 2\mathbf{P}_k''(0)] + \frac{1}{360}[7\mathbf{P}_{k+1}''''(0) + 8\mathbf{P}_k''''(0)].$$

When these expressions for the six coefficients are similarly combined with $\mathbf{P}_{k-1}'''(1) = \mathbf{P}_k'''(0)$, the result is another relation between the second and fourth derivates

$$-\mathbf{P}_{k-1}''(0) + 2\mathbf{P}_k''(0) - \mathbf{P}_{k+1}''(0) + \frac{1}{6}\mathbf{P}_{k-1}''''(0) + \frac{2}{3}\mathbf{P}_k''''(0) + \frac{1}{6}\mathbf{P}_{k+1}''''(0) = 0. \qquad (5.28)$$

Each of Equations (5.27) and (5.28) is $n-1$ equations for $k = 1, 2, \ldots, n-1$, so we end up with $2(n-1)$ equations with the $2n$ second and fourth unknown derivatives. As in the case of the cubic spline, we complete this system of equations by guessing values for some extreme derivatives. The simplest end condition is to require

$$\mathbf{P}_1'''(0) = \mathbf{P}_{n-1}'''(1) = \mathbf{P}_1''''(0) = \mathbf{P}_{n-1}''''(1) = 0,$$

which implies $\mathbf{P}_1''(0) = \mathbf{P}_1''(1) - \frac{1}{6}\mathbf{P}_1''''(1)$ and $\mathbf{P}_n''(0) = \mathbf{P}_{n-1}''(1) - \frac{1}{6}\mathbf{P}_{n-1}''''(1)$ and makes it possible to eliminate $\mathbf{P}_1''(0)$ and $\mathbf{P}_n''(0)$ from Equations (5.27) and (5.28). [Späth 83] shows that the end result is the system of equations

$$\begin{bmatrix} \mathbf{A} & -\mathbf{B} \\ \mathbf{C} & \mathbf{A} \end{bmatrix} \begin{bmatrix} \mathbf{P}'' \\ \mathbf{P}'''' \end{bmatrix} = \begin{bmatrix} \mathbf{D} \\ \mathbf{0} \end{bmatrix}, \tag{5.29}$$

where

$$\mathbf{P}'' = \left(\mathbf{P}_1''(0), \ldots, \mathbf{P}_{n-1}''(0)\right)^T, \quad \mathbf{P}'''' = \left(\mathbf{P}_1''''(0), \ldots, \mathbf{P}_{n-1}''''(0)\right)^T,$$

$$\mathbf{D} = \left[6[(\mathbf{P}_2 - \mathbf{P}_1) - (\mathbf{P}_1 - \mathbf{P}_0)], \ldots, 6[(\mathbf{P}_n - \mathbf{P}_{n-1}) - (\mathbf{P}_{n-1} - \mathbf{P}_{n-2})]\right]^T,$$

and

$$\mathbf{A} = \begin{bmatrix} 5 & 1 & & & & \\ 1 & 4 & 1 & & & \\ & 1 & 4 & 1 & & \\ & & & \ddots & & \\ & & & 1 & 4 & 1 \\ & & & & 1 & 5 \end{bmatrix}, \; \mathbf{B} = \begin{bmatrix} 26 & 7 & & & & \\ 7 & 16 & 7 & & & \\ & 7 & 16 & 7 & & \\ & & & \ddots & & \\ & & & 7 & 16 & 7 \\ & & & & 7 & 26 \end{bmatrix}, \; \mathbf{C} = \begin{bmatrix} 6 & -6 & & & & \\ -6 & 12 & -6 & & & \\ & -6 & 12 & -6 & & \\ & & & \ddots & & \\ & & & -6 & 12 & -6 \\ & & & & 6 & 6 \end{bmatrix}.$$

Notice that matrices $\mathbf{A}$, $\mathbf{B}$, and $\mathbf{C}$ are tridiagonal and symmetric. In addition, $\mathbf{A}$ and $\mathbf{B}$ are diagonally dominant, while $\mathbf{C}$ is nonnegative definite. This guarantees that the block matrix of Equation (5.29) will have an inverse, which implies that the system of equations has a unique solution.

Solving the system of Equations (5.29) means expressing the second and fourth derivatives of the spline segments in terms of the data points (the known quantities). Once this is done, the six coefficients of each of the $n-1$ spline segments can be expressed in terms of the data points, and the segments can be constructed.

5.4 Cardinal Splines

The cardinal spline is another example of how Hermite interpolation is applied to construct a spline curve. The cardinal spline overcomes the main disadvantages of cubic splines, namely the lack of local control and the need to solve a system of linear equations that may be large (its size depends on the number of data points). Cardinal splines also offer a natural way to control the tension of the curve by modifying the magnitudes of the tangent vectors (Section 4.2.3). The price for all this is the loss of second-order continuity. Strictly speaking, this loss means that the cardinal spline isn't really a spline (see the definition of splines on page 141), but its form, its derivation, and its behavior are so similar to those of other splines that the name "cardinal spline" has stuck.

Figure 5.7a illustrates the principle of this method. The figure shows a curve that passes through seven points. The curve looks continuous but is constructed in segments, two of which are thicker than the others. The first thick segment, the one from $\mathbf{P}_2$ to $\mathbf{P}_3$, starts in the direction from $\mathbf{P}_1$ to $\mathbf{P}_3$ and ends going in the direction from $\mathbf{P}_2$ to $\mathbf{P}_4$. The second thick segment, from $\mathbf{P}_5$ to $\mathbf{P}_6$, features the same behavior. It starts in the direction from $\mathbf{P}_4$ to $\mathbf{P}_6$ and ends going in the direction from $\mathbf{P}_5$ to $\mathbf{P}_7$.

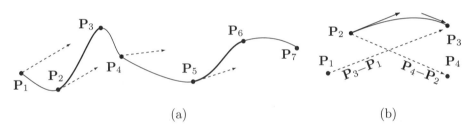

(a)   (b)

Figure 5.7: Tangent Vectors in a Cardinal Spline.

The cardinal spline for n given points is calculated and drawn in segments, each depending on four points only. Each point participates in at most four curve segments, so moving one point affects only those segments and not the entire curve. This is why the curve features *local control*. The individual segments connect smoothly; their first derivatives are equal at the connection points (the curve therefore has first-order continuity). However, the second derivatives of the segments are generally different at the connection points.

The first step in constructing the complete curve is to organize the points into $n-3$ highly-overlapping groups of four consecutive points each. The groups are

$$[\mathbf{P}_1, \mathbf{P}_2, \mathbf{P}_3, \mathbf{P}_4], \quad [\mathbf{P}_2, \mathbf{P}_3, \mathbf{P}_4, \mathbf{P}_5], \quad [\mathbf{P}_3, \mathbf{P}_4, \mathbf{P}_5, \mathbf{P}_6], \ldots, [\mathbf{P}_{n-3}, \mathbf{P}_{n-2}, \mathbf{P}_{n-1}, \mathbf{P}_n].$$

Hermite interpolation is then applied to construct a curve segment $\mathbf{P}(t)$ for each group. Denoting the four points of a group by $\mathbf{P}_1$, $\mathbf{P}_2$, $\mathbf{P}_3$, and $\mathbf{P}_4$, the two interior points $\mathbf{P}_2$ and $\mathbf{P}_3$ become the start and end points of the segment and the two tangent vectors become $s(\mathbf{P}_3 - \mathbf{P}_1)$ and $s(\mathbf{P}_4 - \mathbf{P}_2)$, where s is a real number. Thus, segment $\mathbf{P}(t)$ goes from $\mathbf{P}_2$ to $\mathbf{P}_3$ and its two extreme tangent vectors are proportional to the vectors

$\mathbf{P}_3 - \mathbf{P}_1$ and $\mathbf{P}_4 - \mathbf{P}_2$ (Figure 5.7b). The proportionality constant s is related to the tension parameter T. Note how there are no segments from $\mathbf{P}_1$ to $\mathbf{P}_2$ and from $\mathbf{P}_{n-1}$ to $\mathbf{P}_n$. These segments can be added to the curve by adding two new extreme points $\mathbf{P}_0$ and $\mathbf{P}_{n+1}$. These points can also be employed to edit the curve, because the first segment, from $\mathbf{P}_1$ to $\mathbf{P}_2$, starts going in the direction from $\mathbf{P}_0$ to $\mathbf{P}_2$, and similarly for the last segment.

The particular choice of the tangent vectors guarantees that the individual segments of the cardinal spline will connect smoothly. The end tangent $s(\mathbf{P}_4 - \mathbf{P}_2)$ of the segment for group $[\mathbf{P}_1, \mathbf{P}_2, \mathbf{P}_3, \mathbf{P}_4]$ is identical to the start tangent of the next group, $[\mathbf{P}_2, \mathbf{P}_3, \mathbf{P}_4, \mathbf{P}_5]$.

Segment $\mathbf{P}(t)$ is therefore defined by

$$
\begin{aligned}
\mathbf{P}(0) = \mathbf{P}_2, \quad &\mathbf{P}(1) = \mathbf{P}_3, \\
\mathbf{P}^t(0) = s(\mathbf{P}_3 - \mathbf{P}_1), \quad &\mathbf{P}^t(1) = s(\mathbf{P}_4 - \mathbf{P}_2)
\end{aligned}
\tag{5.30}
$$

and is easily calculated by applying Hermite interpolation [Equation (4.7)] to the four quantities of Equation (5.30)

$$
\begin{aligned}
\mathbf{P}(t) &= (t^3, t^2, t, 1)
\begin{pmatrix}
2 & -2 & 1 & 1 \\
-3 & 3 & -2 & -1 \\
0 & 0 & 1 & 0 \\
1 & 0 & 0 & 0
\end{pmatrix}
\begin{pmatrix}
\mathbf{P}_2 \\
\mathbf{P}_3 \\
s(\mathbf{P}_3 - \mathbf{P}_1) \\
s(\mathbf{P}_4 - \mathbf{P}_2)
\end{pmatrix} \\
&= (t^3, t^2, t, 1)
\begin{pmatrix}
-s & 2-s & s-2 & s \\
2s & s-3 & 3-2s & -s \\
-s & 0 & s & 0 \\
0 & 1 & 0 & 0
\end{pmatrix}
\begin{pmatrix}
\mathbf{P}_1 \\
\mathbf{P}_2 \\
\mathbf{P}_3 \\
\mathbf{P}_4
\end{pmatrix}.
\end{aligned}
\tag{5.31}
$$

Tension in the cardinal spline can now be controlled by changing the lengths of the tangent vectors by means of parameter s. A long tangent vector (obtained by a large s) causes the curve to continue longer in the direction of the tangent. A short tangent has the opposite effect; the curve moves a short distance in the direction of the tangent, then quickly changes direction and moves toward the end point. A zero-length tangent (corresponding to $s = 0$) produces a straight line between the endpoints (infinite tension). In principle, the parameter s can be varied from 0 to ∞. In practice, we use only values in the range $[0, 1]$. However, since $s = 0$ produces maximum tension, we cannot intuitively think of s as the tension parameter and we need to define another parameter, T inversely related to s.

The tension parameter T is defined as $s = (1-T)/2$, which implies $T = 1 - 2s$. The value $T = 0$ results in $s = 1/2$. The curve is defined as having tension zero in this case and is called the *Catmull–Rom spline* [Catmull and Rom 74]. (Reference [Salomon 99] has a detailed derivation of this type of spline as a blend of two parabolas.) Increasing T from 0 to 1 decreases s from $1/2$ to 0, thereby reducing the magnitude of the tangent vectors down to 0. This produces curves with more tension. Exercise 4.7 tells us that when the tangent vectors have magnitude zero, the Hermite curve segment is a straight line, so the entire cardinal spline curve becomes a set of straight segments, a polyline, the curve with maximum tension. Decreasing T from 0 to -1 increases s from $1/2$ to 1. The result is a curve with more slack at the data points.

To illustrate this behavior mathematically, we rewrite Equation (5.31) explicitly to show its dependence on s:

$$\mathbf{P}(t) = s(-t^3 + 2t^2 - t)\mathbf{P}_1 + s(-t^3 + t^2)\mathbf{P}_2 + (2t^3 - 3t^2 + 1)\mathbf{P}_2 \\ + s(t^3 - 2t^2 + t)\mathbf{P}_3 + (-2t^3 + 3t^2)\mathbf{P}_3 + s(t^3 - t^2)\mathbf{P}_4. \tag{5.32}$$

For $s = 0$, Equation (5.32) becomes $(2t^3 - 3t^2 + 1)\mathbf{P}_2 + (-2t^3 + 3t^2)\mathbf{P}_3$, which can be simplified to $(3t^2 - 2t^3)(\mathbf{P}_3 - \mathbf{P}_2) + \mathbf{P}_2$. Substituting $u = 3t^2 - 2t^3$ reduces this to $u(\mathbf{P}_3 - \mathbf{P}_2) + \mathbf{P}_2$, which is the straight line from $\mathbf{P}_2$ to $\mathbf{P}_3$.

For large s, we use Equation (5.32) to calculate the mid-curve value $\mathbf{P}(0.5)$:

$$\mathbf{P}(0.5) = \frac{s}{8}\left[(\mathbf{P}_3 - \mathbf{P}_1) + (\mathbf{P}_2 - \mathbf{P}_4)\right] + 0.5(\mathbf{P}_2 + \mathbf{P}_3)$$
$$= \frac{s}{8}\left[\mathbf{P}^t(0) - \mathbf{P}^t(1)\right] + 0.5(\mathbf{P}_2 + \mathbf{P}_3).$$

This is an extension of Equation (Ans.7). The first term is the difference of the two tangent vectors, multiplied by $s/8$. As s grows, this term grows without limit. The second term is the midpoint of $\mathbf{P}_2$ and $\mathbf{P}_3$. Adding the two terms (a vector and a point) produces a point that may be located far away (for large s) from the midpoint, showing that the curve moves a long distance away from the start point $\mathbf{P}_2$ before changing direction and starting toward the end point $\mathbf{P}_3$. Large values of s therefore feature a loose curve (low tension).

Thus, the tension of the curve can be increased by setting s close to 0 (or, equivalently, setting T close to 1); it can be decreased by increasing s (or, equivalently, decreasing T toward 0).

◇ **Exercise 5.7:** What happens when $T > 1$?

Setting $T = 0$ results in $s = 0.5$. Equation (5.31) reduces in this case to

$$\mathbf{P}(t) = (t^3, t^2, t, 1)\begin{pmatrix} -0.5 & 1.5 & -1.5 & 0.5 \\ 1 & -2.5 & 2 & -0.5 \\ -0.5 & 0 & 0.5 & 0 \\ 0 & 1 & 0 & 0 \end{pmatrix}\begin{pmatrix} \mathbf{P}_1 \\ \mathbf{P}_2 \\ \mathbf{P}_3 \\ \mathbf{P}_4 \end{pmatrix}, \tag{5.33}$$

a curve known as the Catmull–Rom spline. Its basis matrix is termed the parabolic blending matrix.

Example: Given the four points $(1,0)$, $(3,1)$, $(6,2)$, and $(2,3)$, we apply Equation (5.31) to calculate the cardinal spline segment from $(3,1)$ to $(6,2)$:

$$\mathbf{P}(t) = (t^3, t^2, t, 1)\begin{bmatrix} -s & 2-s & s-2 & s \\ 2s & s-3 & 3-2s & -s \\ -s & 0 & s & 0 \\ 0 & 1 & 0 & 0 \end{bmatrix}\begin{bmatrix} (1,0) \\ (3,1) \\ (6,2) \\ (2,3) \end{bmatrix}$$
$$= t^3(4s-6, 4s-2) + t^2(-9s+9, -6s+3) + t(5s, 2s) + (3,1).$$

For high tension (i.e., $T = 1$ or $s = 0$), this reduces to the straight line

$$\mathbf{P}(t) = (-6, -2)t^3 + (9, 3)t^2 + (3, 1) = (3, 1)(-2t^3 + 3t^2) + (3, 1) = (3, 1)u + (3, 1).$$

For $T = 0$ (or $s = 1/2$), this cardinal spline reduces to the Catmull–Rom curve

$$\mathbf{P}(t) = (-4, 0)t^3 + (4.5, 0)t^2 + (2.5, 1)t + (3, 1). \tag{5.34}$$

Figure 5.8 shows an example of a similar cardinal spline (the points are different) with four values 0, 1/6, 2/6, and 3/6 of the tension parameter.

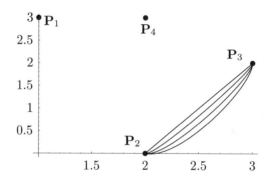

```
(* Cardinal spline example *)
T={t^3,t^2,t,1};
H[s_]:={{-s,2-s,s-2,s},{2s,s-3,3-2s,-s},{-s,0,s,0},{0,1,0,0}};
B={{1,3},{2,0},{3,2},{2,3}};
s=3/6; (* T=0 *)
g1=ParametricPlot[T.H[s].B,{t,0,1},PlotRange->All, Compiled->False,
DisplayFunction->Identity];
s=2/6; (* T=1/3 *)
g2=ParametricPlot[T.H[s].B,{t,0,1},PlotRange->All, Compiled->False,
DisplayFunction->Identity];
s=1/6; (* T=2/3 *)
g3=ParametricPlot[T.H[s].B,{t,0,1},PlotRange->All, Compiled->False,
DisplayFunction->Identity];
s=0; (* T=1 *)
g4=ParametricPlot[T.H[s].B,{t,0,1},PlotRange->All, Compiled->False,
DisplayFunction->Identity];
g5=Graphics[{AbsolutePointSize[4], Table[Point[B[[i]]],{i,1,4}] }];
Show[g1,g2,g3,g4,g5, DefaultFont->{"cmr10", 10},
DisplayFunction->$DisplayFunction]
```

Figure 5.8: A Cardinal Spline Example.

5.5 Catmull–Rom Surfaces

The cardinal spline or the Catmull–Rom curve can easily be extended to a surface that's fully defined by a rectangular grid of data points. In analogy to the Catmull–Rom curve segment—which involves four points but only passes through the two interior points—a single Catmull–Rom surface patch is specified by 16 points, the patch is anchored at the four middle points and spans the area delimited by them.

We start with a group of $m \times n$ data points roughly arranged in a rectangle. We look at all the overlapping groups that consist of 4×4 adjacent points, and we calculate a surface patch for each group. Some of the groups are shown in Figure 5.9.

$$
\begin{array}{llll}
\mathbf{P}_{40}\mathbf{P}_{41}\mathbf{P}_{42}\mathbf{P}_{43} & \mathbf{P}_{41}\mathbf{P}_{42}\mathbf{P}_{43}\mathbf{P}_{44} & \mathbf{P}_{42}\mathbf{P}_{43}\mathbf{P}_{44}\mathbf{P}_{45} & \cdots & \mathbf{P}_{4,n-3}\mathbf{P}_{4,n-2}\mathbf{P}_{4,n-1}\mathbf{P}_{4n} \\
\mathbf{P}_{30}\mathbf{P}_{31}\mathbf{P}_{32}\mathbf{P}_{33} & \mathbf{P}_{31}\mathbf{P}_{32}\mathbf{P}_{33}\mathbf{P}_{34} & \mathbf{P}_{32}\mathbf{P}_{33}\mathbf{P}_{34}\mathbf{P}_{35} & \cdots & \mathbf{P}_{3,n-3}\mathbf{P}_{3,n-2}\mathbf{P}_{3,n-1}\mathbf{P}_{3n} \\
\mathbf{P}_{20}\mathbf{P}_{21}\mathbf{P}_{22}\mathbf{P}_{23} & \mathbf{P}_{21}\mathbf{P}_{22}\mathbf{P}_{23}\mathbf{P}_{24} & \mathbf{P}_{22}\mathbf{P}_{23}\mathbf{P}_{24}\mathbf{P}_{25} & \cdots & \mathbf{P}_{2,n-3}\mathbf{P}_{2,n-2}\mathbf{P}_{2,n-1}\mathbf{P}_{2n} \\
\mathbf{P}_{10}\mathbf{P}_{11}\mathbf{P}_{12}\mathbf{P}_{13} & \mathbf{P}_{11}\mathbf{P}_{12}\mathbf{P}_{13}\mathbf{P}_{14} & \mathbf{P}_{12}\mathbf{P}_{13}\mathbf{P}_{14}\mathbf{P}_{15} & \cdots & \mathbf{P}_{1,n-3}\mathbf{P}_{1,n-2}\mathbf{P}_{1,n-1}\mathbf{P}_{1n} \\
\\
\mathbf{P}_{30}\mathbf{P}_{31}\mathbf{P}_{32}\mathbf{P}_{33} & \mathbf{P}_{31}\mathbf{P}_{32}\mathbf{P}_{33}\mathbf{P}_{34} & \mathbf{P}_{32}\mathbf{P}_{33}\mathbf{P}_{34}\mathbf{P}_{35} & \cdots & \mathbf{P}_{3,n-3}\mathbf{P}_{3,n-2}\mathbf{P}_{3,n-1}\mathbf{P}_{3n} \\
\mathbf{P}_{20}\mathbf{P}_{21}\mathbf{P}_{22}\mathbf{P}_{23} & \mathbf{P}_{21}\mathbf{P}_{22}\mathbf{P}_{23}\mathbf{P}_{24} & \mathbf{P}_{22}\mathbf{P}_{23}\mathbf{P}_{24}\mathbf{P}_{25} & \cdots & \mathbf{P}_{2,n-3}\mathbf{P}_{2,n-2}\mathbf{P}_{2,n-1}\mathbf{P}_{2n} \\
\mathbf{P}_{10}\mathbf{P}_{11}\mathbf{P}_{12}\mathbf{P}_{13} & \mathbf{P}_{11}\mathbf{P}_{12}\mathbf{P}_{13}\mathbf{P}_{14} & \mathbf{P}_{12}\mathbf{P}_{13}\mathbf{P}_{14}\mathbf{P}_{15} & \cdots & \mathbf{P}_{1,n-3}\mathbf{P}_{1,n-2}\mathbf{P}_{1,n-1}\mathbf{P}_{1n} \\
\mathbf{P}_{00}\mathbf{P}_{01}\mathbf{P}_{02}\mathbf{P}_{03} & \mathbf{P}_{01}\mathbf{P}_{02}\mathbf{P}_{03}\mathbf{P}_{04} & \mathbf{P}_{02}\mathbf{P}_{03}\mathbf{P}_{04}\mathbf{P}_{05} & \cdots & \mathbf{P}_{0,n-3}\mathbf{P}_{0,n-2}\mathbf{P}_{0,n-1}\mathbf{P}_{0n}
\end{array}
$$

Figure 5.9: Points for a Catmull–Rom Surface Patch.

The expression of the surface is obtained by applying the technique of Cartesian product (Section 1.9) to the Catmull–Rom curve. Equation (1.28) produces

$$
\mathbf{P}(u, w) = (u^3, u^2, u, 1)\mathbf{B}\mathbf{P}\mathbf{B}^T
\begin{pmatrix} w^3 \\ w^2 \\ w \\ 1 \end{pmatrix},
$$

where $\mathbf{B}$ is the parabolic blending matrix of Equation (5.33)

$$
\mathbf{B} = \begin{pmatrix}
-0.5 & 1.5 & -1.5 & 0.5 \\
1 & -2.5 & 2 & -0.5 \\
-0.5 & 0 & 0.5 & 0 \\
0 & 1 & 0 & 0
\end{pmatrix}
$$

and $\mathbf{P}$ is a matrix consisting of the 4×4 points participating in the patch

$$
\mathbf{P} = \begin{pmatrix}
\mathbf{P}_{i+3,j} & \mathbf{P}_{i+3,j+1} & \mathbf{P}_{i+3,j+2} & \mathbf{P}_{i+3,j+3} \\
\mathbf{P}_{i+2,j} & \mathbf{P}_{i+2,j+1} & \mathbf{P}_{i+2,j+2} & \mathbf{P}_{i+2,j+3} \\
\mathbf{P}_{i+1,j} & \mathbf{P}_{i+1,j+1} & \mathbf{P}_{i+1,j+2} & \mathbf{P}_{i+1,j+3} \\
\mathbf{P}_{i,j} & \mathbf{P}_{i,j+1} & \mathbf{P}_{i,j+2} & \mathbf{P}_{i,j+3}
\end{pmatrix}.
$$

Notice that the patch spans the area bounded by the four central points. In general, the entire surface spans the area bounded by the four points $\mathbf{P}_{11}$, $\mathbf{P}_{1,n-1}$, $\mathbf{P}_{m-1,1}$, and

$\mathbf{P}_{m-1,n-1}$. If we want the surface to span the area bounded by the four corner points $\mathbf{P}_{00}$, $\mathbf{P}_{0n}$, $\mathbf{P}_{m0}$, and $\mathbf{P}_{mn}$, we have to create two new extreme rows and two new extreme columns of points, by analogy with the Catmull–Rom curve.

Example: Given the following coordinates for 16 points in file `CRpoints`

0 0 0	1	0 0	2	0 0	3 0 0
0 1 0	.5	.5 1	2.5	.5 0	3 1 0
0 2 0	.5	2.5 0	2.5	2.5 1	3 2 0
0 3 0	1	3 0	2	3 0	3 3 0

the *Mathematica* code of Figure 5.10 reads the file and generates the Catmull–Rom patch. Note how the patch spans only the four center points and how the z coordinates of 0 and 1 create the particular shape of the patch.

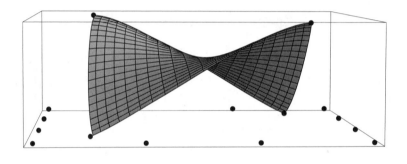

```
000 1 002  00 300
010 .5 .51 2.5 .50 310
020 .52.50 2.52.51 320
030 13 02 3 0 330
```

```
<<:Graphics:ParametricPlot3D.m;
Clear[Pt,Bm,CRpatch,g1,g2,g3];
Pt=ReadList["CRpoints",{Number,Number,Number},
 RecordLists->True];
Bm:={{-.5,1.5,-1.5,.5},{1,-2.5,2,-.5},
 {-.5,0,.5,0},{0,1,0,0}};
CRpatch[i_]:= (* 1st patch, rows 1-4 *)
{u^3,u^2,u,1}.Bm.Pt[[{1,2,3,4},{1,2,3,4},i]].
 Transpose[Bm].{w^3,w^2,w,1};
g1=Graphics3D[{AbsolutePointSize[4],
 Table[Point[Pt[[i,j]]],{i,1,4},{j,1,4}]}];
g2=ParametricPlot3D[{CRpatch[1],CRpatch[2],CRpatch[3]},
 {u,0,.98,.1},{w,0,1,.1}, DisplayFunction->Identity];
Show[g1,g2, ViewPoint->{-4.322, 0.242, 0.306},
 DisplayFunction->$DisplayFunction]
```

Figure 5.10: A Catmull–Rom Surface Patch.

Example: (extended) We now add four more points to file `CRpoints`, and use rows 2–5 to calculate and display another patch. Notice the five values of y compared to the four values of x. The code of Figure 5.11 reads the extended file and generates and

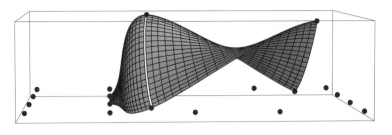

```
0 0 0  1 0 0 2  0 0 3 0 0
0 1 0 .5 .5 1 2.5 .5 0 3 1 0
0 2 0 .5 2.5 0 0 2.5 2.5 1 3 2 0
0 3 0  1 3 0 2  3 0 3 3 0
0 4 0  1 4 0 2  4 0 3 4 0
```

```
<<:Graphics:ParametricPlot3D.m;
Clear[Pt,Bm,CRpatch,CRpatchM,g1,g2,g3];
Pt=ReadList["CRpoints",{Number,Number,Number},
 RecordLists->True];
Bm:={{-.5,1.5,-1.5,.5},{1,-2.5,2,-.5},
 {-.5,0,.5,0},{0,1,0,0}};
CRpatch[i_]:= (* 1st patch, rows 1-4 *)
{u^3,u^2,u,1}.Bm.Pt[[{1,2,3,4},{1,2,3,4},i]].
 Transpose[Bm].{w^3,w^2,w,1};
CRpatchM[i_]:= (* 2nd patch, rows 2-5 *)
{u^3,u^2,u,1}.Bm.Pt[[{2,3,4,5},{1,2,3,4},i]].
 Transpose[Bm].{w^3,w^2,w,1};
g1=Graphics3D[{AbsolutePointSize[4],
Table[Point[Pt[[i,j]]],{i,1,5},{j,1,4}]}];
g2=ParametricPlot3D[{CRpatch[1],CRpatch[2],CRpatch[3]},
{u,0,.98,.1},{w,0,1,.1}, DisplayFunction->Identity];
g3=ParametricPlot3D[{CRpatchM[1],CRpatchM[2],CRpatchM[3]},
{u,0,1,.1},{w,0,1,.1}, DisplayFunction->Identity];
Show[g1,g2,g3, ViewPoint->{-4.322, 0.242, 0.306},
DisplayFunction->$DisplayFunction]
```

Figure 5.11: Two Catmull–Rom Surface Patches.

displays both patches. Each patch spans four points, but they share the two points $(0.5, 2.5, 0)$ and $(2.5, 2.5, 1)$. Note how they connect smoothly.

Tension can be added to a Catmull–Rom surface patch in the same way that it is added to a Catmull–Rom curve or to a cardinal spline. Figure 5.12 illustrates how smaller values of s create a surface closer to a flat plane.

5.6 Kochanek–Bartels Splines

The Kochanek–Bartels spline method [Kochanek and Bartels 84] is an extension of the cardinal spline. In addition to the tension parameter T, this method introduces two new parameters, c and b to control the *continuity* and *bias*, respectively, of individual curve segments. The curve is a spline computed from a set of n data points, and the three shape parameters can be specified separately for each point or can be global. Thus, the user/designer has to specify either 3 or $3n$ parameters.

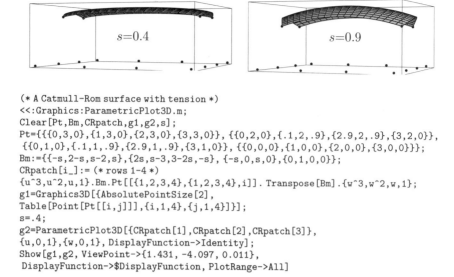

```
(* A Catmull-Rom surface with tension *)
<<:Graphics:ParametricPlot3D.m;
Clear[Pt,Bm,CRpatch,g1,g2,s];
Pt={{{0,3,0},{1,3,0},{2,3,0},{3,3,0}}, {{0,2,0},{.1,2,.9},{2.9,2,.9},{3,2,0}},
 {{0,1,0},{.1,1,.9},{2.9,1,.9},{3,1,0}}, {{0,0,0},{1,0,0},{2,0,0},{3,0,0}}};
Bm:={{-s,2-s,s-2,s},{2s,s-3,3-2s,-s}, {-s,0,s,0},{0,1,0,0}};
CRpatch[i_]:= (* rows 1-4 *)
{u^3,u^2,u,1}.Bm.Pt[[{1,2,3,4},{1,2,3,4},i]]. Transpose[Bm].{w^3,w^2,w,1};
g1=Graphics3D[{AbsolutePointSize[2],
Table[Point[Pt[[i,j]]],{i,1,4},{j,1,4}]}];
s=.4;
g2=ParametricPlot3D[{CRpatch[1],CRpatch[2],CRpatch[3]},
{u,0,1},{w,0,1}, DisplayFunction->Identity];
Show[g1,g2, ViewPoint->{1.431, -4.097, 0.011},
DisplayFunction->$DisplayFunction, PlotRange->All]
```

Figure 5.12: A Catmull–Rom Surface Patch With Tension.

Consider an interior point $\mathbf{P}_k$ where two spline segments meet. When the "arriving" segment arrives at the point it is moving in a certain direction that we call the arriving tangent vector. Similarly, the departing segment starts at the point while moving in a direction that we call the departing tangent vector. The three shape parameters control these two tangent vectors in various ways. The tension parameter varies the magnitudes of the arriving and departing vectors. The bias parameter rotates both tangents by the same amount from their "natural" direction, and the continuity parameter rotates each tangent separately, so they may no longer point in the same direction.

A complete Kochanek–Bartels spline passes through n given data points $\mathbf{P}_1$ through $\mathbf{P}_n$ and is computed and displayed in the following steps:

1. The designer (or user) adds two new points $\mathbf{P}_0$ and $\mathbf{P}_{n+1}$. Recall that each cardinal spline segment is determined by a group of four points but it goes from the second point to the third one. Adding point $\mathbf{P}_0$ makes it possible to have a segment from $\mathbf{P}_1$ to $\mathbf{P}_2$, and similarly for the new point $\mathbf{P}_{n+1}$. All the original n points are now interior.

2. Two tangent vectors, arriving and departing, are computed for each of the n interior points from Equations (5.35) and (5.36). The arriving tangent at $\mathbf{P}_1$ and the departing tangent at $\mathbf{P}_n$ are not used, so the total number of tangents to compute is $2n - 2$.

3. The $n+2$ points are divided into $n-1$ overlapping groups of four points each, and a Hermite curve segment is computed and displayed for each group. The computations are similar to those for the cardinal spline, the only difference being that the tangent vectors are computed in a special way.

Figure 5.13 shows two spline segments $\mathbf{P}_{k-1}(t)$ and $\mathbf{P}_k(t)$ that meet at interior point $\mathbf{P}_k$. This point is the last endpoint of segment $\mathbf{P}_{k-1}(t)$ and the first endpoint

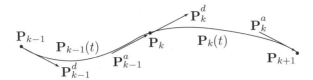

Figure 5.13: Two Kochanek–Bartels Spline Segments.

of segment $\mathbf{P}_k(t)$. We denote the two tangent vectors at $\mathbf{P}_k$ by $\mathbf{P}_{k-1}^a \overset{\text{def}}{=} \mathbf{P}_{k-1}^t(1)$ and $\mathbf{P}_k^d \overset{\text{def}}{=} \mathbf{P}_k^t(0)$. In a cardinal spline the two tangents $\mathbf{P}_{k-1}^a$ and $\mathbf{P}_k^d$ are identical and are proportional to the vector $\mathbf{P}_{k+1} - \mathbf{P}_{k-1}$ (the chord surrounding $\mathbf{P}_k$). This guarantees a smooth connection of the two segments. In a Kochanek–Bartels spline, the two tangents are computed as shown here, they have the same magnitude, but may point in different directions. Notice that the two endpoints of segment $\mathbf{P}_k(t)$ are $\mathbf{P}_k$ and $\mathbf{P}_{k+1}$ and its two extreme tangent vectors are $\mathbf{P}_k^d$ and $\mathbf{P}_k^a$. Here is how the tangent vectors are computed.

Tension. In a cardinal spline, tension is controlled by multiplying the tangent vectors by a parameter s. Small values of s produce high tension, so the tension parameter T is defined by $s = (1 - T)/2$. Thus, we can express the tangents as

$$\frac{1-T}{2}(\mathbf{P}_{k+1} - \mathbf{P}_{k-1}) = (1 - T)\frac{1}{2}\big((\mathbf{P}_{k+1} - \mathbf{P}_k) + (\mathbf{P}_k - \mathbf{P}_{k-1})\big).$$

This can be interpreted as $(1 - T)$ multiplied by the average of the "arriving" chord $(\mathbf{P}_k - \mathbf{P}_{k-1})$ and the "departing" chord $(\mathbf{P}_{k+1} - \mathbf{P}_k)$. In a Kochanek–Bartels spline, the tension parameter contributes the same quantity

$$(1 - T_k)\frac{1}{2}\big((\mathbf{P}_{k+1} - \mathbf{P}_k) + (\mathbf{P}_k - \mathbf{P}_{k-1})\big)$$

to the two tangents $\mathbf{P}_{k-1}^a$ and $\mathbf{P}_k^d$ at point $\mathbf{P}_k$. The value $T_k = 1$ results in tangent vectors of zero magnitude, which corresponds to maximum tension. The value $T_k = 0$ (zero tension) results in a contribution of $(\mathbf{P}_{k+1} - \mathbf{P}_{k-1})/2$ to both tangent vectors. The value $T_k = -1$ results in twice that contribution and therefore to long tangents and low tension.

Continuity. Curves are important in computer animation. An object being animated is often moved along a curve and the (virtual) camera may also move along a path. Sometimes, an animation path should not be completely smooth, but should feature jumps and jerks at certain points. This effect is achieved in a Kochanek–Bartels spline by separately rotating $\mathbf{P}_{k-1}^a$ and $\mathbf{P}_k^d$, so that they point in different directions. The contributions of the continuity parameter to these vectors are

$$\text{Contribution to}\quad \mathbf{P}_{k-1}^a \quad\text{is}\quad \frac{1-c_k}{2}(\mathbf{P}_k - \mathbf{P}_{k-1}) + \frac{1+c_k}{2}(\mathbf{P}_{k+1} - \mathbf{P}_k)\,,$$

$$\text{Contribution to}\quad \mathbf{P}_k^d \quad\text{is}\quad \frac{1+c_k}{2}(\mathbf{P}_k - \mathbf{P}_{k-1}) + \frac{1-c_k}{2}(\mathbf{P}_{k+1} - \mathbf{P}_k)\,,$$

where c_k is the continuity parameter at point $\mathbf{P}_k$. The value $c_k = 0$ results in $\mathbf{P}_{k-1}^a = \mathbf{P}_k^d$ and therefore in a smooth curve at $\mathbf{P}_k$. For $c_k \neq 0$, the two tangents are different and the curve has a sharp corner (a kink or a cusp) at point $\mathbf{P}_k$, a corner that becomes more pronounced for large values of c_k. The case $c_k = -1$ implies $\mathbf{P}_{k-1}^a = \mathbf{P}_k - \mathbf{P}_{k-1}$ (the arriving chord) and $\mathbf{P}_k^d = \mathbf{P}_{k+1} - \mathbf{P}_k$ (the departing chord). The case $c_k = 1$ produces tangent vectors in the opposite directions: $\mathbf{P}_{k-1}^a = \mathbf{P}_{k+1} - \mathbf{P}_k$ and $\mathbf{P}_k^d = \mathbf{P}_k - \mathbf{P}_{k-1}$. These three extreme cases are illustrated in Figure 5.14.

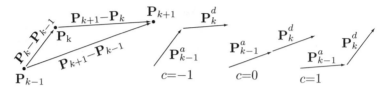

Figure 5.14: Effects of the Continuity Parameter.

Tension and continuity may have the same effect, yet they affect the dynamics of the curve in different ways as illustrated by Figure 5.15. Part (a) of the figure shows five points and a two-segment Kochanek–Bartels spline from $\mathbf{P}_1$ through $\mathbf{P}_2$ to $\mathbf{P}_3$. Both the tension and continuity parameters are set to zero at $\mathbf{P}_2$, so the direction of the curve at this point is the direction of the chord $\mathbf{P}_3 - \mathbf{P}_1$. Setting $T = 1$ at $\mathbf{P}_2$ increases the tension to maximum at that point, thereby changing the curve to two straight segments [part (b) of the figure]. However, if we leave T at zero and set $c = -1$ at $\mathbf{P}_2$, the resulting curve will have the same shape (the direction of the arriving tangent $\mathbf{P}_1^a$ is from $\mathbf{P}_1$ to $\mathbf{P}_2$ while the direction of the departing tangent $\mathbf{P}_2^d$ is from $\mathbf{P}_2$ to $\mathbf{P}_3$).

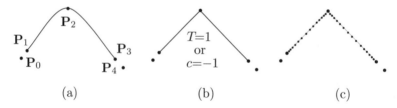

Figure 5.15: Different Dynamics of Tension and Continuity.

Thus, maximum tension and minimum continuity may result in identical geometries, but not in identical curves. These parameters have different effects on the *speed* of the curve as illustrated in Part (c) of the figure. Specifically, infinite tension results in nonuniform speed. If the first spline segment $\mathbf{P}_1(t)$ is plotted by incrementing t in equal steps, the resulting points are first bunched together, then feature larger gaps, and finally become dense again.

When the user specifies high (or maximum) tension at a point, the tangent vectors become short (or zero) at the point, but they get longer as the curve moves away from the point. The speed of the curve is determined by the size of its tangent vector, which is why high tension results in nonuniform speed. In contrast, low tension does not affect the magnitude of the tangent vectors, which is why it does not affect the speed. When low continuity results in a straight segment, the speed will be uniform. Curved segments, however, always have variable speed regardless of the continuity parameters at the endpoints of the segment.

⋄ **Exercise 5.8:** Compute the tangent vector of the cardinal spline for $s = 0$ and show that its length is zero for $t = 0$ and $t = 1$, but is nonzero elsewhere.

Bias. In a cardinal spline with zero tension, both tangent vectors at point $\mathbf{P}_k$ have the value

$$\frac{1}{2}(\mathbf{P}_{k+1} - \mathbf{P}_{k-1}) = \frac{1}{2}\big((\mathbf{P}_k - \mathbf{P}_{k-1}) + (\mathbf{P}_{k+1} - \mathbf{P}_k)\big),$$

implying that the direction of the curve at point $\mathbf{P}_k$ is the average of the two chords connecting at $\mathbf{P}_k$.

The Kochanek–Bartels spline introduces an additional (sometimes misunderstood) parameter b_k to control the direction of the curve at $\mathbf{P}_k$ by rotating $\mathbf{P}_{k-1}^a$ and $\mathbf{P}_k^d$ by the same amount. The contribution of the bias parameter to the arriving and departing tangents is set (somewhat arbitrarily) to

$$\frac{1+b_k}{2}(\mathbf{P}_k - \mathbf{P}_{k-1}) + \frac{1-b_k}{2}(\mathbf{P}_{k+1} - \mathbf{P}_k).$$

Setting $b_k = 1$ changes both tangents to $\mathbf{P}_k - \mathbf{P}_{k-1}$, the chord on the left of $\mathbf{P}_k$. The other extreme value, $b_k = -1$, changes them to the chord on the right of $\mathbf{P}_k$. Figure 5.16 illustrates the effects of the three extreme values of b_k.

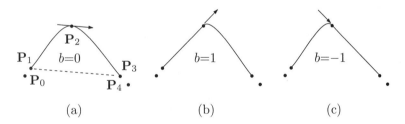

Figure 5.16: Effect of the Bias Parameter b.

Bias is used in computer animation to obtain the effect of overshooting a point ($b_k = 1$) or undershooting it ($b_k = -1$).

The three shape parameters are incorporated in the tangent vectors as follows: the tangent vector that departs point $\mathbf{P}_k$ is defined by

$$\mathbf{P}_k^d = \mathbf{P}_k^t(0) = \frac{1}{2}(1-T_k)(1+b_k)(1-c_k)(\mathbf{P}_k-\mathbf{P}_{k-1}) + \frac{1}{2}(1-T_k)(1-b_k)(1+c_k)(\mathbf{P}_{k+1}-\mathbf{P}_k). \tag{5.35}$$

Similarly, the tangent vector arriving at point $\mathbf{P}_{k+1}$ is defined by

$$\mathbf{P}_k^a = \mathbf{P}_k^t(1) = \frac{1}{2}(1 - T_{k+1})(1 + b_{k+1})(1 + c_{k+1})(\mathbf{P}_{k+1} - \mathbf{P}_k)$$

$$+ \frac{1}{2}(1 - T_{k+1})(1 - b_{k+1})(1 - c_{k+1})(\mathbf{P}_{k+2} - \mathbf{P}_{k+1}). \tag{5.36}$$

As a result, the Kochanek–Bartels curve segment $\mathbf{P}_k(t)$ from $\mathbf{P}_k$ to $\mathbf{P}_{k+1}$ is constructed by the familiar expression

$$\mathbf{P}_k(t) = (t^3, t^2, t, 1)\mathbf{H} \begin{pmatrix} \mathbf{P}_k \\ \mathbf{P}_{k+1} \\ \mathbf{P}_k^d \\ \mathbf{P}_k^a \end{pmatrix},$$

where $\mathbf{H}$ is the Hermite matrix, Equation (4.7). Notice that the segment depends on six shape parameters, three at $\mathbf{P}_k$ and three at $\mathbf{P}_{k+1}$. The segment also depends on four points $\mathbf{P}_{k-1}$, $\mathbf{P}_k$, $\mathbf{P}_{k+1}$, and $\mathbf{P}_{k+2}$.

Note also that the second derivatives of this curve are generally not continuous at the data points.

Example: The three points $\mathbf{P}_1 = (0,0)$, $\mathbf{P}_2 = (4,6)$, and $\mathbf{P}_3 = (10,-1)$ are given, together with the extra points $\mathbf{P}_0 = (-1,-1)$ and $\mathbf{P}_4 = (11,-2)$. Up to nine shape parameters can be specified (three parameters for each of the three interior points). Figure 5.17 shows the curve with all shape parameters set to zero, and the effects of setting T to 1 (maximum tension) and to -1 (a loose curve), setting c to 1, and setting b to 1 (overshoot) and -1 (undershoot), all in $\mathbf{P}_2$. The *Mathematica* code that computed the curves is also included.

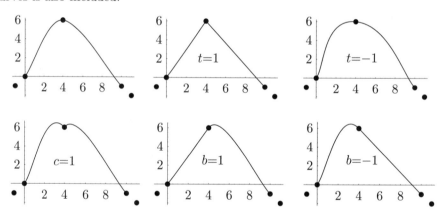

Figure 5.17: Effects of the Three Parameters in the Kochanek–Bartels Spline.

```
Clear[T,H,B,pts,Pa,Pd,te,bi,co]; (* Kochanek Bartels 3+2 points*)
T={t^3,t^2,t,1};
H={{2,-2,1,1},{-3,3,-2,-1},{0,0,1,0},{1,0,0,0}};
Pd[k_]:=(1-te[[k+1]])(1+bi[[k+1]])(1+co[[k+1]])(pts[[k+1]]-pts[[k]])/2+
 (1-te[[k+1]])(1-bi[[k+1]])(1-co[[k+1]])(pts[[k+2]]-pts[[k+1]])/2;
Pa[k_]:=(1-te[[k+2]])(1+bi[[k+2]])(1-co[[k+2]])(pts[[k+2]]-pts[[k+1]])/2+
 (1-te[[k+2]])(1-bi[[k+2]])(1+co[[k+2]])(pts[[k+3]]-pts[[k+2]])/2;
pts:={{-1,-1},{0,0},{4,6},{10,-1},{11,-2}};
te={0,0,0,0,0}; bi={0,0,0,0,0}; co={0,0,0,0,0};

B={pts[[2]],pts[[3]],Pd[1],Pa[1]};
Simplify[T.H.B]
Simplify[D[T.H.B,t]]
g1=ParametricPlot[T.H.B,{t,0,1},PlotRange->All];
```

```
B={pts[[3]],pts[[4]],Pd[2],Pa[2]};
Simplify[T.H.B]
Simplify[D[T.H.B,t]]
g2=ParametricPlot[T.H.B,{t,0,1},PlotRange->All];
g3=Graphics[{AbsolutePointSize[4], Table[Point[pts[[i]]],{i,1,5}] }];
Show[g1,g2,g3]
```

Code For Figure 5.17.

At the forward end of the crankshaft there is mounted
a master bevel gear on six splines; this bevel floats
on the splines against a ball thrust bearing, and, in
turn, the thrust is taken by the crank case cover.

—E. Charles Vivian, *A History of Aeronautics*

6
Bézier Approximation

Bézier methods for curves and surfaces are popular, are commonly used in practical work, and are described here in detail. Two approaches to the design of a Bézier curve are described, one using Bernstein polynomials and the other using the mediation operator. Both rectangular and triangular Bézier surface patches are discussed, with examples.

Historical Notes

Pierre Etienne Bézier (pronounced "Bez-yea" or "bez-ee-ay") was an applied mathematician with the French car manufacturer Renault. In the early 1960s, encouraged by his employer, he began searching for ways to automate the process of designing cars. His methods have been the basis of the modern field of Computer Aided Geometric Design (CAGD), a field with practical applications in many areas.

It is interesting to note that Paul de Faget de Casteljau, an applied mathematician with Citroën, was the first, in 1959, to develop the various Bézier methods but—because of the secretiveness of his employer—never published it (except for two internal technical memos that were discovered in 1975). This is why the entire field is named after the second person, Bézier, who developed it.

Bézier and de Casteljau did their work while working for car manufacturers. It is little known that Steven Anson Coons of MIT did most of his work on surfaces (around 1967) while a consultant for Ford. Another mathematician, William J. Gordon, has generalized the Coons surfaces, in 1969, as part of his work for General Motors research labs. In addition, airplane designer James Ferguson also came up with the same ideas for the construction of curves and surfaces. It seems that car and airplane manufacturers have been very innovative in the CAGD field. Detailed historical surveys of CAGD can be found in [Farin 04] and [Schumaker 81].

6.1 The Bézier Curve

The Bézier curve is a parametric curve $\mathbf{P}(t)$ that is a polynomial function of the parameter t. The degree of the polynomial depends on the number of points used to define the curve. The method employs *control points* and produces an approximating curve (note the title of this chapter). The curve does not pass through the interior points but is attracted by them (however, see Exercise 6.7 for an exception). It is as if the points exert a pull on the curve. Each point influences the direction of the curve by pulling it toward itself, and that influence is strongest when the curve gets nearest the point. Figure 6.1 shows some examples of cubic Bézier curves. Such a curve is defined by four points and is a cubic polynomial. Notice that one has a cusp and another one has a loop. The fact that the curve does not pass through the points implies that the points are not "set in stone" and can be moved. This makes it easy to edit, modify and reshape the curve, which is one reason for its popularity. The curve can also be edited by adding new points, or deleting points. These techniques are discussed in Sections 6.8 and 6.9, but they are cumbersome because the mathematical expression of the curve depends on the number of points, not just on the points themselves.

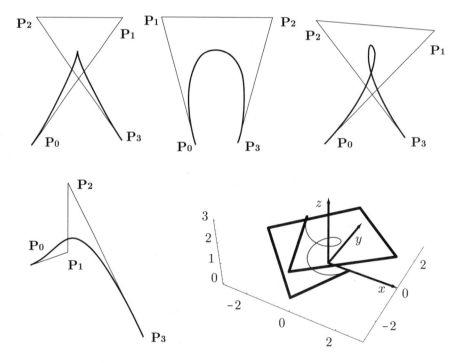

Figure 6.1: Four Plane Cubic and One Space Bézier Curves With Their Control Points and Polygons.

The *control polygon* of the Bézier curve is the polygon obtained when the control points are connected, in their natural order, with straight segments.

How does one go about deriving such a curve? We describe two approaches to the design—a weighted sum and a linear interpolation—and show that they are identical.

6.1.1 Pascal Triangle and the Binomial Theorem

The Pascal triangle and the binomial theorem are related because both employ the same numbers. The Pascal triangle is an infinite triangular matrix that's built from the edges inside

$$
\begin{array}{ccccccc}
 & & & 1 & & & \\
 & & 1 & & 1 & & \\
 & 1 & & 2 & & 1 & \\
 1 & & 3 & & 3 & & 1 \\
1 & & 4 & & 6 & & 4 & & 1 \\
1 & & 5 & & 10 & & 10 & & 5 & & 1 \\
\cdots & & & \cdots & & & \cdots
\end{array}
$$

We first fill the left and right edges with ones, then compute each interior element as the sum of the two elements directly above it. As can be expected, it is not hard to obtain an explicit expression for the general element of the Pascal triangle. We first number the rows from 0 starting at the top, and the columns from 0 starting on the left. A general element is denoted by $\binom{i}{j}$. We then observe that the top two rows (corresponding to $i = 0, 1$) consist of 1's and that every other row can be obtained as the sum of its predecessor and a shifted version of its predecessor. For example,

$$
\begin{array}{r}
1\ 3\ 3\ 1 \\
+ \quad 1\ 3\ 3\ 1 \\
\hline
1\ 4\ 6\ 4\ 1
\end{array}
$$

This shows that the elements of the triangle satisfy

$$
\binom{i}{0} = \binom{i}{i} = 1, \qquad i = 0, 1, \ldots,
$$

$$
\binom{i}{j} = \binom{i-1}{j-1} + \binom{i-1}{j}, \qquad i = 2, 3, \ldots, \quad j = 1, 2, \ldots, (i-1).
$$

From this it is easy to derive the explicit expression

$$
\begin{aligned}
\binom{i}{j} &= \binom{i-1}{j-1} + \binom{i-1}{j} \\
&= \frac{(i-1)!}{(j-1)!(i-j)!} + \frac{(i-1)!}{j!(i-1-j)!} \\
&= \frac{j(i-1)!}{j!(i-j)!} + \frac{(i-j)(i-1)!}{j!(i-j)!} \\
&= \frac{i!}{j!(i-j)!}.
\end{aligned}
$$

Thus, the general element of the Pascal triangle is the well-known *binomial coefficient*

$$
\binom{i}{j} = \frac{i!}{j!(i-j)!}.
$$

The binomial coefficient is one of Newton's many contributions to mathematics. His binomial theorem states that

$$(a + b)^n = \sum_{i=0}^{n} \binom{n}{i} a^i b^{n-i}. \tag{6.1}$$

This equation can be written in a symmetric way by denoting $j = n - i$. The result is

$$(a + b)^n = \sum_{\substack{i+j=n \\ i,j \geq 0}} \frac{(i + j)!}{i!j!} a^i b^j, \tag{6.2}$$

from which we can easily guess the *trinomial theorem* (which is used in Section 6.23)

$$(a + b + c)^n = \sum_{\substack{i+j+k=n \\ i,j,k \geq 0}} \frac{(i + j + k)!}{i!j!k!} a^i b^j c^k. \tag{6.3}$$

6.2 The Bernstein Form of the Bézier Curve

The first approach to the Bézier curve expresses it as a weighted sum of the points (with, of course, barycentric weights). Each control point is multiplied by a weight and the products are added. We denote the control points by $\mathbf{P}_0, \mathbf{P}_1, \ldots, \mathbf{P}_n$ (n is therefore defined as 1 less than the number of points) and the weights by B_i. The expression of weighted sum is

$$\mathbf{P}(t) = \sum_{i=0}^{n} \mathbf{P}_i B_i, \quad 0 \leq t \leq 1.$$

The result, $\mathbf{P}(t)$, depends on the parameter t. Since the points are given by the user, they are fixed, so it is the weights that must depend on t. We therefore denote them by $B_i(t)$. How should $B_i(t)$ behave as a function of t?

We first examine $B_0(t)$, the weight associated with the first point $\mathbf{P}_0$. We want that point to affect the curve mostly at the beginning, i.e., when t is close to 0. Thus, as t grows toward 1 (i.e., as the curve moves away from $\mathbf{P}_0$), $B_0(t)$ should drop down to 0. When $B_0(t) = 0$, the first point no longer influences the shape of the curve.

Next, we turn to $B_1(t)$. This weight function should start small, should have a maximum when the curve approaches the second point $\mathbf{P}_1$, and should then start dropping until it reaches zero. A natural question is: When (for what value of t) does the curve reach its closest approach to the second point? The answer is: It depends on the number of points. For three points (the case $n = 2$), the Bézier curve passes closest to the second point (the interior point) when $t = 0.5$. For four points, the curve is nearest the second point when $t = 1/3$. It is now clear that the weight functions must also depend on n and we denote them by $B_{n,i}(t)$. Hence, $B_{3,1}(t)$ should start at 0, have a maximum at $t = 1/3$, and go down to 0 from there. Figure 6.2 shows the desired behavior of $B_{n,i}(t)$

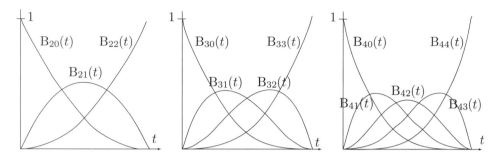

```
(* Just the base functions bern. Note how "pwr" handles 0^0 *)
Clear[pwr,bern];
pwr[x_,y_]:=If[x==0 && y==0, 1, x^y];
bern[n_,i_,t_]:=Binomial[n,i]pwr[t,i]pwr[1-t,n-i] (* t^i x (1-t)^(n-i) *)
Plot[Evaluate[Table[bern[5,i,t], {i,0,5}]], {t,0,1}, DefaultFont->{"cmr10", 10}];
```

Figure 6.2: The Bernstein Polynomials for $n = 2, 3, 4$.

for $n = 2$, 3, and 4. The five different weights $B_{4,i}(t)$ have their maxima at $t = 0$, 1/4, 1/2, 3/4, and 1.

The functions chosen by Bézier (and also by de Casteljau) were derived by the Russian mathematician Sergeĭ Natanovich Bernshteĭn in 1912, as part of his work on approximation theory (see Chapter 6 of [Davis 63]). They are known as the Bernstein polynomials and are defined by

$$B_{n,i}(t) = \binom{n}{i} t^i (1 - t)^{n-i}, \quad \text{where} \quad \binom{n}{i} = \frac{n!}{i!(n-i)!} \tag{6.4}$$

are the binomial coefficients. These polynomials feature the desired behavior and have a few more useful properties that are discussed here. (In calculating the curve, we assume that the quantity 0^0, which is normally undefined, equals 1.)

The Bézier curve is now defined as

$$\mathbf{P}(t) = \sum_{i=0}^{n} \mathbf{P}_i B_{n,i}(t), \quad \text{where } B_{n,i}(t) = \binom{n}{i} t^i (1 - t)^{n-i} \text{ and } 0 \le t \le 1. \tag{6.5}$$

Each control point (a pair or a triplet of coordinates) is multiplied by its weight, which is in the range $[0, 1]$. The weights act as *blending functions* that blend the contributions of the different points.

Here is *Mathematica* code to calculate and plot the Bernstein polynomials and the Bézier curve:

```
(* Just the base functions bern. Note how "pwr" handles 0^0 *)
Clear[pwr,bern,n,i,t]
pwr[x_,y_]:=If[x==0 && y==0, 1, x^y];
bern[n_,i_,t_]:=Binomial[n,i]pwr[t,i]pwr[1-t,n-i]
  (* t^i \[Times] (1-t)^(n-i) *)
Plot[Evaluate[Table[bern[5,i,t], {i,0,5}]], {t,0,1},
  DefaultFont->{"cmr10", 10}]
```

```
Clear[i,t,pnts,pwr,bern,bzCurve,g1,g2]; (* Cubic Bezier curve *)
(* either read points from file
pnts=ReadList["DataPoints",{Number,Number}]; *)
(* or enter them explicitly *)
pnts={{0,0},{.7,1},{.3,1},{1,0}}; (* 4 points for a cubic curve *)
pwr[x_,y_]:=If[x==0 && y==0, 1, x^y];
bern[n_,i_,t_]:=Binomial[n,i]pwr[t,i]pwr[1-t,n-i]
bzCurve[t_]:=Sum[pnts[[i+1]]bern[3,i,t], {i,0,3}]
g1=ListPlot[pnts, Prolog->AbsolutePointSize[4], PlotRange->All,
  AspectRatio->Automatic, DisplayFunction->Identity]
g2=ParametricPlot[bzCurve[t], {t,0,1}, DisplayFunction->Identity]
Show[g1,g2, DisplayFunction->$DisplayFunction]
```

Next is similar code for a three-dimensional Bézier curve. It was used to draw the space curve of Figure 6.1.

```
Clear[pnts,pwr,bern,bzCurve,g1,g2,g3]; (* General 3D Bezier curve *)
pnts={{1,0,0},{0,-3,0.5},{-3,0,0.75},{0,3,1},{3,0,1.5},{0,-3,1.75},{-1,0,2}};
n=Length[pnts]-1;
pwr[x_,y_]:=If[x==0 && y==0, 1, x^y];
bern[n_,i_,t_]:=Binomial[n,i]pwr[t,i]pwr[1-t,n-i] (* t^i x (1-t)^(n-i) *)
bzCurve[t_]:=Sum[pnts[[i+1]]bern[n,i,t], {i,0,n}];
g1=ParametricPlot3D[bzCurve[t], {t,0,1}, Compiled->False,
DisplayFunction->Identity];
g2=Graphics3D[{AbsolutePointSize[2], Map[Point,pnts]}];
g3=Graphics3D[{AbsoluteThickness[2], (* control polygon *)
Table[Line[{pnts[[j]],pnts[[j+1]]}], {j,1,n}]}];
g4=Graphics3D[{AbsoluteThickness[1.5], (* the coordinate axes *)
Line[{{0,0,3},{0,0,0},{3,0,0},{0,0,0},{0,3,0}}]}];
Show[g1,g2,g3,g4, AspectRatio->Automatic, PlotRange->All, DefaultFont->{"cmr10", 10},
Boxed->False, DisplayFunction->$DisplayFunction];
```

⋄ **Exercise 6.1:** Design a heart-shaped Bézier curve based on nine control points.

When Bézier started searching for such functions in the early 1960s, he set the following requirements [Bézier 86]:

1. The functions should be such that the curve passes through the first and last control points.

2. The tangent to the curve at the start point should be $P_1 - P_0$, i.e., the curve should start at point P_0 moving toward P_1. A similar property should hold at the last point.

3. The same requirement is generalized for higher derivatives of the curve at the two extreme endpoints. Hence, $P^{tt}(0)$ should depend only on the first point P_0 and its two neighbors P_1 and P_2. In general, $P^{(k)}(0)$ should only depend on P_0 and its k neighbors P_1 through P_k. This feature provides complete control over the continuity at the joints between separate Bézier curve segments (Section 6.5).

4. The weight functions should be symmetric with respect to t and $(1 - t)$. This means that a reversal of the sequence of control points would not affect the shape of the curve.

5. The weights should be barycentric, to guarantee that the shape of the curve is independent of the coordinate system.

6. The entire curve lies within the convex hull of the set of control points. (See property 8 of Section 6.4 for a discussion of this point.)

The definition shown in Equation (6.5), using Bernstein polynomials as the weights, satisfies all these requirements. In particular, requirement 5 is proved when Equation (6.1) is written in the form $[t + (1 - t)]^n = \cdots$ (see Equation (6.12) if you cannot figure this out). Following are the explicit expressions of these polynomials for $n = 2$, 3, and 4.

Example: For $n = 2$ (three control points), the weights are

$$
\begin{aligned}
B_{2,0}(t) &= \tbinom{2}{0} t^0 (1-t)^{2-0} = (1-t)^2, \\
B_{2,1}(t) &= \tbinom{2}{1} t^1 (1-t)^{2-1} = 2t(1-t), \\
B_{2,2}(t) &= \tbinom{2}{2} t^2 (1-t)^{2-2} = t^2,
\end{aligned}
$$

and the curve is

$$
\begin{aligned}
\mathbf{P}(t) &= (1-t)^2 \mathbf{P}_0 + 2t(1-t)\mathbf{P}_1 + t^2 \mathbf{P}_2 \\
&= \left((1-t)^2, 2t(1-t), t^2 \right) (\mathbf{P}_0, \mathbf{P}_1, \mathbf{P}_2)^T \\
&= (t^2, t, 1) \begin{pmatrix} 1 & -2 & 1 \\ -2 & 2 & 0 \\ 1 & 0 & 0 \end{pmatrix} \begin{pmatrix} \mathbf{P}_0 \\ \mathbf{P}_1 \\ \mathbf{P}_2 \end{pmatrix}.
\end{aligned}
\tag{6.6}
$$

This is the quadratic Bézier curve.

⋄ **Exercise 6.2:** Given three points $\mathbf{P}_1$, $\mathbf{P}_2$, and $\mathbf{P}_3$, calculate the parabola that goes from $\mathbf{P}_1$ to $\mathbf{P}_3$ and whose start and end tangent vectors point in directions $\mathbf{P}_2 - \mathbf{P}_1$ and $\mathbf{P}_3 - \mathbf{P}_2$, respectively.

In the special case $n = 3$, the four weight functions are

$$
\begin{aligned}
B_{3,0}(t) &= \tbinom{3}{0} t^0 (1-t)^{3-0} = (1-t)^3, \\
B_{3,1}(t) &= \tbinom{3}{1} t^1 (1-t)^{3-1} = 3t(1-t)^2, \\
B_{3,2}(t) &= \tbinom{3}{2} t^2 (1-t)^{3-2} = 3t^2(1-t), \\
B_{3,3}(t) &= \tbinom{3}{3} t^3 (1-t)^{3-3} = t^3,
\end{aligned}
$$

and the curve is

$$
\mathbf{P}(t) = (1-t)^3 \mathbf{P}_0 + 3t(1-t)^2 \mathbf{P}_1 + 3t^2(1-t)\mathbf{P}_2 + t^3 \mathbf{P}_3
\tag{6.7}
$$

$$
\begin{aligned}
&= \left[(1-t)^3, 3t(1-t)^2, 3t^2(1-t), t^3 \right] \left[\mathbf{P}_0, \mathbf{P}_1, \mathbf{P}_2, \mathbf{P}_3 \right]^T \\
&= \left[(1 - 3t + 3t^2 - t^3), (3t - 6t^2 + 3t^3), (3t^2 - 3t^3), t^3 \right] \left[\mathbf{P}_0, \mathbf{P}_1, \mathbf{P}_2, \mathbf{P}_3 \right]^{\mathbf{T}} \\
&= (t^3, t^2, t, 1) \begin{pmatrix} -1 & 3 & -3 & 1 \\ 3 & -6 & 3 & 0 \\ -3 & 3 & 0 & 0 \\ 1 & 0 & 0 & 0 \end{pmatrix} \begin{pmatrix} \mathbf{P}_0 \\ \mathbf{P}_1 \\ \mathbf{P}_2 \\ \mathbf{P}_3 \end{pmatrix}.
\end{aligned}
\tag{6.8}
$$

It is clear that $\mathbf{P}(t)$ is a cubic polynomial in t. It is the cubic Bézier curve. In general, the Bézier curve for points $\mathbf{P}_0$, $\mathbf{P}_1, \ldots, \mathbf{P}_n$ is a polynomial of degree n.

⋄ **Exercise 6.3:** Given the curve $\mathbf{P}(t) = (1 + t + t^2, t^3)$, find its control points.

⋄ **Exercise 6.4:** The cubic curve of Equation (6.8) is drawn when the parameter t varies in the interval $[0, 1]$. Show how to substitute t with a new parameter u such that the curve will be drawn when $-1 \leq u \leq +1$.

⋄ **Exercise 6.5:** Calculate the Bernstein polynomials for $n = 4$.

It can be proved by induction that the general, $(n + 1)$-point Bézier curve can be represented by

$$\mathbf{P}(t) = (t^n, t^{n-1}, \ldots, t, 1)\mathbf{N}\begin{pmatrix}\mathbf{P}_0 \\ \mathbf{P}_1 \\ \vdots \\ \mathbf{P}_{n-1} \\ \mathbf{P}_n\end{pmatrix} = \mathbf{T}(t) \cdot \mathbf{N} \cdot \mathbf{P}, \qquad (6.9)$$

where

$$\mathbf{N} = \begin{pmatrix} \binom{n}{0}\binom{n}{n}(-1)^n & \binom{n}{1}\binom{n-1}{n-1}(-1)^{n-1} & \cdots & \binom{n}{n}\binom{n-n}{n-n}(-1)^0 \\ \binom{n}{0}\binom{n}{n-1}(-1)^{n-1} & \binom{n}{1}\binom{n-1}{n-2}(-1)^{n-2} & \cdots & 0 \\ \vdots & \vdots & \cdots & 0 \\ \binom{n}{0}\binom{n}{1}(-1)^1 & \binom{n}{1}\binom{n-1}{0}(-1)^0 & \cdots & 0 \\ \binom{n}{0}\binom{n}{0}(-1)^0 & 0 & \cdots & 0 \end{pmatrix}. \qquad (6.10)$$

Matrix $\mathbf{N}$ is symmetric and its elements below the second diagonal are all zeros. Its determinant therefore equals (up to a sign) the product of the diagonal elements, which are all nonzero. A nonzero determinant implies a nonsingular matrix. Thus, matrix $\mathbf{N}$ always has an inverse. $\mathbf{N}$ can also be written as the product $\mathbf{AB}$, where

$$\mathbf{A} = \begin{pmatrix} \binom{n}{n}(-1)^n & \binom{n}{1}\binom{n-1}{n-1}(-1)^{n-1} & \cdots & \binom{n}{n}\binom{n-n}{n-n}(-1)^0 \\ \binom{n}{n-1}(-1)^{n-1} & \binom{n}{1}\binom{n-1}{n-2}(-1)^{n-2} & \cdots & 0 \\ \vdots & \vdots & \cdots & 0 \\ \binom{n}{1}(-1)^1 & \binom{n}{1}\binom{n-1}{0}(-1)^0 & \cdots & 0 \\ \binom{n}{0}(-1)^0 & 0 & \cdots & 0 \end{pmatrix}$$

and

$$\mathbf{B} = \begin{pmatrix} \binom{n}{0} & 0 & \cdots & 0 \\ 0 & \binom{n}{1} & \cdots & 0 \\ \vdots & & \ddots & \vdots \\ 0 & 0 & \cdots & \binom{n}{n} \end{pmatrix}.$$

Figure 6.3 shows the Bézier $\mathbf{N}$ matrices for $n = 1, 2, \ldots, 7$.

⋄ **Exercise 6.6:** Calculate the Bézier curve for the case $n = 1$ (two control points). What kind of a curve is it?

$$\mathbf{N}_1 = \begin{pmatrix} -1 & 1 \\ 1 & 0 \end{pmatrix},$$

$$\mathbf{N}_2 = \begin{pmatrix} 1 & -2 & 1 \\ -2 & 2 & 0 \\ 1 & 0 & 0 \end{pmatrix},$$

$$\mathbf{N}_3 = \begin{pmatrix} -1 & 3 & -3 & 1 \\ 3 & -6 & 3 & 0 \\ -3 & 3 & 0 & 0 \\ 1 & 0 & 0 & 0 \end{pmatrix},$$

$$\mathbf{N}_4 = \begin{pmatrix} 1 & -4 & 6 & -4 & 1 \\ -4 & 12 & -12 & 4 & 0 \\ 6 & -12 & 6 & 0 & 0 \\ -4 & 4 & 0 & 0 & 0 \\ 1 & 0 & 0 & 0 & 0 \end{pmatrix},$$

$$\mathbf{N}_5 = \begin{pmatrix} -1 & 5 & -10 & 10 & -5 & 1 \\ 5 & -20 & 30 & -20 & 5 & 0 \\ -10 & 30 & -30 & 10 & 0 & 0 \\ 10 & -20 & 10 & 0 & 0 & 0 \\ -5 & 5 & 0 & 0 & 0 & 0 \\ 1 & 0 & 0 & 0 & 0 & 0 \end{pmatrix},$$

$$\mathbf{N}_6 = \begin{pmatrix} 1 & -6 & 15 & -20 & 15 & -6 & 1 \\ -6 & 30 & -60 & 60 & -30 & 6 & 0 \\ 15 & -60 & 90 & -60 & 15 & 0 & 0 \\ -20 & 60 & -60 & 20 & 0 & 0 & 0 \\ 15 & -30 & 15 & 0 & 0 & 0 & 0 \\ -6 & 6 & 0 & 0 & 0 & 0 & 0 \\ 1 & 0 & 0 & 0 & 0 & 0 & 0 \end{pmatrix},$$

$$\mathbf{N}_7 = \begin{pmatrix} -1 & 7 & -21 & 35 & -35 & 21 & -7 & 1 \\ 7 & -42 & 105 & -140 & 105 & -42 & 7 & 0 \\ -21 & 105 & -210 & 210 & -105 & 21 & 0 & 0 \\ 35 & -140 & 210 & -140 & 35 & 0 & 0 & 0 \\ -35 & 105 & -105 & 35 & 0 & 0 & 0 & 0 \\ 21 & -42 & 21 & 0 & 0 & 0 & 0 & 0 \\ -7 & 7 & 0 & 0 & 0 & 0 & 0 & 0 \\ 1 & 0 & 0 & 0 & 0 & 0 & 0 & 0 \end{pmatrix}.$$

Figure 6.3: The First Seven Bézier Basis Matrices.

◇ **Exercise 6.7:** Generally, the Bézier curve passes through the first and last control points, but not through the intermediate points. Consider the case of three points $\mathbf{P}_0$, $\mathbf{P}_1$, and $\mathbf{P}_2$ on a straight line. Intuitively, it seems that the curve will be a straight line and would therefore pass through the interior point $\mathbf{P}_1$. Is that so?

The Bézier curve can also be represented in a very compact and elegant way as $\mathbf{P}(t) = (1 - t + tE)^n \mathbf{P}_0$, where E is the shift operator defined by $E\mathbf{P}_i = \mathbf{P}_{i+1}$ (i.e., applying E to point $\mathbf{P}_i$ produces point $\mathbf{P}_{i+1}$). The definition of E implies $E\mathbf{P}_0 = \mathbf{P}_1$, $E^2\mathbf{P}_0 = \mathbf{P}_2$, and $E^i\mathbf{P}_0 = \mathbf{P}_i$.

The Bézier curve can now be written

$$\mathbf{P}(t) = \sum_{i=0}^{n} \binom{n}{i} t^i (1-t)^{n-i} \mathbf{P}_i = \sum_{i=0}^{n} \binom{n}{i} t^i (1-t)^{n-i} E^i \mathbf{P}_0$$

$$= \sum_{i=0}^{n} \binom{n}{i} (tE)^i (1-t)^{n-i} \mathbf{P}_0 = \big(tE + (1-t)\big)^n \mathbf{P}_0,$$

where the last step is an application of the binomial theorem, Equation (6.1).

Example: For $n = 1$, this representation amounts to

$$\mathbf{P}(t) = (1 - t + tE)\mathbf{P}_0 = \mathbf{P}_0(1 - t) + \mathbf{P}_1 t.$$

For $n = 2$, we get

$$\begin{aligned}
\mathbf{P}(t) &= (1 - t + tE)^2 \mathbf{P}_0 \\
&= (1 - t + tE - t + t^2 - t^2 E + tE - t^2 E + t^2 E^2)\mathbf{P}_0 \\
&= \mathbf{P}_0(1 - 2t + t^2) + \mathbf{P}_1(2t - 2t^2) + \mathbf{P}_2 t^2 \\
&= \mathbf{P}_0(1 + t)^2 + \mathbf{P}_1 2t(1 - t) + \mathbf{P}_2 t^2.
\end{aligned}$$

Given $n + 1$ control points $\mathbf{P}_0$ through $\mathbf{P}_n$, we can represent the Bézier curve for the points by $\mathbf{P}_n^{(n)}(t)$, where the quantity $\mathbf{P}_i^{(j)}(t)$ is defined recursively by

$$\mathbf{P}_i^{(j)}(t) = \begin{cases} (1-t)\mathbf{P}_{i-1}^{(j-1)}(t) + t\mathbf{P}_i^{(j-1)}(t), & \text{for } j > 0, \\ \mathbf{P}_i, & \text{for } j = 0. \end{cases} \qquad (6.11)$$

The following examples show how the definition above is used to generate the quantities $\mathbf{P}_i^{(j)}(t)$ and why $\mathbf{P}_n^{(n)}(t)$ is the degree-n curve:

$$\begin{aligned}
\mathbf{P}_0^{(0)}(t) &= \mathbf{P}_0, \quad \mathbf{P}_1^{(0)}(t) = \mathbf{P}_1, \quad \mathbf{P}_2^{(0)}(t) = \mathbf{P}_2, \dots, \mathbf{P}_n^{(0)}(t) = \mathbf{P}_n, \\
\mathbf{P}_1^{(1)}(t) &= (1-t)\mathbf{P}_0^{(0)}(t) + t\mathbf{P}_1^{(0)}(t) = (1-t)\mathbf{P}_0 + t\mathbf{P}_1, \\
\mathbf{P}_2^{(2)}(t) &= (1-t)\mathbf{P}_1^{(1)}(t) + t\mathbf{P}_2^{(1)}(t) \\
&= (1-t)\big((1-t)\mathbf{P}_0 + t\mathbf{P}_1\big) + t\big((1-t)\mathbf{P}_1 + t\mathbf{P}_2\big) \\
&= (1-t)^2\mathbf{P}_0 + 2t(1-t)\mathbf{P}_1 + t^2\mathbf{P}_2, \\
\mathbf{P}_3^{(3)}(t) &= (1-t)\mathbf{P}_2^{(2)}(t) + t\mathbf{P}_3^{(2)}(t)
\end{aligned}$$

$$= (1-t)\big((1-t)\mathbf{P}_1^{(1)}(t) + t\mathbf{P}_2^{(1)}(t)\big) + t\big((1-t)\mathbf{P}_2^{(1)}(t) + t\mathbf{P}_3^{(1)}(t)\big)$$

$$= (1-t)^2\mathbf{P}_1^{(1)}(t) + 2t(1-t)\mathbf{P}_2^{(1)}(t) + t^2\mathbf{P}_3^{(1)}(t)$$

$$= (1-t)^2\big((1-t)\mathbf{P}_0 + t\mathbf{P}_1\big) + 2t(1-t)\big((1-t)\mathbf{P}_1 + t\mathbf{P}_2\big)$$

$$+ t^2\big((1-t)\mathbf{P}_2 + t\mathbf{P}_3\big)$$

$$= (1-t)^3\mathbf{P}_0 + 3t(1-t)^2\mathbf{P}_1 + 3t^2(1-t)\mathbf{P}_2 + t^3\mathbf{P}_3.$$

6.3 Fast Calculation of the Curve

Calculating the Bézier curve is straightforward but slow. However, with a little thinking, it can be speeded up considerably, a feature that makes this curve very useful in practice. This section discusses three methods.

Method 1: We notice the following:

■ The calculation involves the binomials $\binom{n}{i}$ for $i = 0, 1, \ldots, n$, which, in turn, require the factorials $0!, 1!, \ldots, n!$. The factorials can be precalculated once (each one from its predecessor) and stored in a table. They can then be used to calculate all the necessary binomials and those can also be stored in a table.

■ The calculation involves terms of the form t^i for $i = 0, 1, \ldots, n$ and for many t values in the interval $[0, 1]$. These can also be precalculated and stored in a two-dimensional table where they can be accessed later, using t and i as indexes. This has the advantage that the values of $(1-t)^{n-i}$ can be read from the same table (using $1-t$ and $n-i$ as row and column indexes).

The calculation now reduces to a sum where each term is a product of four quantities, one control point and three numbers from tables. Instead of computing

$$\sum_{i=0}^{n} \binom{n}{i} t^i (1-t)^{n-i}\mathbf{P}_i,$$

we need to compute the simple sum

$$\sum_{i=0}^{n} \text{Table}_1[i, n] \cdot \text{Table}_2[t, i] \cdot \text{Table}_2[1-t, n-i] \cdot \mathbf{P}_i.$$

The parameter t is a real number that varies from 0 to 1, so a practical implementation of this method should use an integer T related to t. For example, if we increment t in 100 steps, then T should be the integer $100t$.

Method 2: Once n is known, each of the $n + 1$ Bernstein polynomials $B_{n,i}(t)$, $i = 0, 1, \ldots, n$, can be precalculated for all the necessary values of t and stored in a table. The curve can now be calculated as the sum

$$\sum_{i=0}^{n} \text{Table}[t, i]\mathbf{P}_i,$$

indicating that each point on the computed curve requires $n + 1$ table lookups, $n + 1$ multiplications, and n additions. Again, an integer index T should be used instead of t.

Method 3: Use *forward differences* in combination with the Taylor series representation, to speed up the calculation significantly. The Bézier curve, which we denote by $\mathbf{B}(t)$, is drawn pixel by pixel in a loop where t is incremented from 0 to 1 in fixed, small steps of Δt. The principle of forward differences (Section 1.5.1) is to find a quantity $\mathbf{dB}$ such that $\mathbf{B}(t + \Delta t) = \mathbf{B}(t) + \mathbf{dB}$ for any value of t. If such a $\mathbf{dB}$ can be found, then it is enough to calculate $\mathbf{B}(0)$ (which, as we know, is simply $\mathbf{P}_0$) and use forward differences to calculate

$$\mathbf{B}(0 + \Delta t) = \mathbf{B}(0) + \mathbf{dB},$$
$$\mathbf{B}(2\Delta t) = \mathbf{B}(\Delta t) + \mathbf{dB} = \mathbf{B}(0) + 2\mathbf{dB},$$

and, in general,

$$\mathbf{B}(i\Delta t) = \mathbf{B}\big((i - 1)\Delta t\big) + \mathbf{dB} = \mathbf{B}(0) + i\,\mathbf{dB}.$$

The point is that $\mathbf{dB}$ should not depend on t. If $\mathbf{dB}$ turns out to depend on t, then as we advance t from 0 to 1, we would have to use different values of $\mathbf{dB}$, slowing down the calculations. The fastest way to calculate the curve is to precalculate $\mathbf{dB}$ before the loop starts and to repeatedly add this precalculated value to $\mathbf{B}(t)$ inside the loop.

We calculate $\mathbf{dB}$ by using the *Taylor series* representation of the Bézier curve. In general, the Taylor series representation of a function $f(t)$ at a point $f(t + \Delta t)$ is the infinite sum

$$f(t + \Delta t) = f(t) + f'(t)\Delta t + \frac{f''(t)\Delta^2 t}{2!} + \frac{f'''(t)\Delta^3 t}{3!} + \cdots.$$

In order to avoid dealing with an infinite sum, we limit our discussion to cubic Bézier curves. These are the most common Bézier curves and are used by many popular graphics applications. They are defined by four control points and are given by Equations (6.7) and (6.8):

$$\mathbf{B}(t) = (1 - t)^3\mathbf{P}_0 + 3t(1 - t)^2\mathbf{P}_1 + 3t^2(1 - t)\mathbf{P}_2 + t^3\mathbf{P}_3$$

$$= (t^3, t^2, t, 1)\begin{pmatrix} -1 & 3 & -3 & 1 \\ 3 & -6 & 3 & 0 \\ -3 & 3 & 0 & 0 \\ 1 & 0 & 0 & 0 \end{pmatrix}\begin{pmatrix} \mathbf{P}_0 \\ \mathbf{P}_1 \\ \mathbf{P}_2 \\ \mathbf{P}_3 \end{pmatrix}.$$

These curves are cubic polynomials in t, implying that only their first three derivatives are nonzero. In order to simplify the calculation of their derivatives, we need to express these curves in the form $\mathbf{B}(t) = \mathbf{a}t^3 + \mathbf{b}t^2 + \mathbf{c}t + \mathbf{d}$ [Equation (3.1)]. This is done by

$$\mathbf{B}(t) = (1 - t)^3\mathbf{P}_0 + 3t(1 - t)^2\mathbf{P}_1 + 3t^2(1 - t)\mathbf{P}_2 + t^3\mathbf{P}_3$$
$$= \big(3(\mathbf{P}_1 - \mathbf{P}_2) - \mathbf{P}_0 + \mathbf{P}_3\big)t^3 + \big(3(\mathbf{P}_0 + \mathbf{P}_2) - 6\mathbf{P}_1\big)t^2 + 3(\mathbf{P}_1 - \mathbf{P}_0)t + \mathbf{P}_0$$
$$= \mathbf{a}t^3 + \mathbf{b}t^2 + \mathbf{c}t + \mathbf{d},$$

so $\mathbf{a} = 3(\mathbf{P}_1 - \mathbf{P}_2) - \mathbf{P}_0 + \mathbf{P}_3$, $\mathbf{b} = 3(\mathbf{P}_0 + \mathbf{P}_2) - 6\mathbf{P}_1$, $\mathbf{c} = 3(\mathbf{P}_1 - \mathbf{P}_0)$, and $\mathbf{d} = \mathbf{P}_0$. These relations can also be expressed in matrix notation

$$
\begin{pmatrix} \mathbf{a} \\ \mathbf{b} \\ \mathbf{c} \\ \mathbf{d} \end{pmatrix} = \begin{pmatrix} -1 & 3 & -3 & 1 \\ 3 & -6 & 3 & 0 \\ -3 & 3 & 0 & 0 \\ 1 & 0 & 0 & 0 \end{pmatrix} \begin{pmatrix} \mathbf{P}_0 \\ \mathbf{P}_1 \\ \mathbf{P}_2 \\ \mathbf{P}_3 \end{pmatrix}.
$$

The curve is now easy to differentiate

$$
\mathbf{B}^t(t) = 3\mathbf{a}t^2 + 2\mathbf{b}t + \mathbf{c}, \quad \mathbf{B}^{tt}(t) = 6\mathbf{a}t + 2\mathbf{b}, \quad \mathbf{B}^{ttt}(t) = 6\mathbf{a};
$$

and the Taylor series representation yields

$$
\begin{aligned}
\mathbf{dB} &= \mathbf{B}(t + \Delta t) - \mathbf{B}(t) \\
&= \mathbf{B}^t(t)\Delta t + \frac{\mathbf{B}^{tt}(t)\Delta^2 t}{2} + \frac{\mathbf{B}^{ttt}(t)\Delta^3 t}{6} \\
&= 3\mathbf{a}\,t^2 \Delta t + 2\mathbf{b}\,t\Delta t + \mathbf{c}\Delta t + 3\mathbf{a}\,t\Delta^2 t + \mathbf{b}\Delta^2 t + \mathbf{a}\Delta^3 t.
\end{aligned}
$$

This seems like a failure since the value obtained for $\mathbf{dB}$ is a function of t (it should be denoted by $\mathbf{dB}(t)$ instead of just $\mathbf{dB}$) and is also slow to calculate. However, the original cubic curve $\mathbf{B}(t)$ is a degree-3 polynomial in t, whereas $\mathbf{dB}(t)$ is only a degree-2 polynomial. This suggests a way out of our dilemma. We can try to express $\mathbf{dB}(t)$ by means of the Taylor series, similar to what we did with the original curve $\mathbf{B}(t)$. This should result in a forward difference $\mathbf{ddB}(t)$ that's a polynomial of degree 1 in t. The quantity $\mathbf{ddB}(t)$ can, in turn, be represented by another Taylor series to produce a forward difference $\mathbf{dddB}$ that's a degree-0 polynomial, i.e., a constant. Once we do that, we will end up with an algorithm of the form

```
precalculate certain quantities;
B = P₀;
for t:=0 to 1 step Δt do
PlotPixel(B);
B:=B+dB; dB:=dB+ddB; ddB:=ddB+dddB;
endfor;
```

The quantity $\mathbf{ddB}(t)$ is obtained by

$$
\mathbf{dB}(t + \Delta t) = \mathbf{dB}(t) + \mathbf{ddB}(t) = \mathbf{dB}(t) + \mathbf{dB}^t(t)\Delta t + \frac{\mathbf{dB}(t)^{tt}\Delta^2 t}{2},
$$

yielding

$$
\begin{aligned}
\mathbf{ddB}(t) &= \mathbf{dB}^t(t)\Delta t + \frac{\mathbf{dB}(t)^{tt}\Delta^2 t}{2} \\
&= (6\mathbf{a}\,t\Delta t + 2\mathbf{b}\Delta t + 3\mathbf{a}\Delta^2 t)\Delta t + \frac{6\mathbf{a}\Delta t\Delta^2 t}{2} \\
&= 6\mathbf{a}\,t\Delta^2 t + 2\mathbf{b}\Delta^2 t + 6\mathbf{a}\Delta^3 t.
\end{aligned}
$$

Finally, the constant **dddB** is similarly obtained by

$$\mathbf{ddB}(t + \Delta t) = \mathbf{ddB}(t) + \mathbf{dddB} = \mathbf{ddB}(t) + \mathbf{ddB}^t(t)\Delta t,$$

yielding $\mathbf{dddB} = \mathbf{ddB}^t(t)\Delta t = 6\mathbf{a}\Delta^3 t$.

The four quantities involved in the calculation of the curve are therefore

$$\mathbf{B}(t) = \mathbf{a}t^3 + \mathbf{b}t^2 + \mathbf{c}t + \mathbf{d},$$
$$\mathbf{dB}(t) = 3\mathbf{a}\,t^2\Delta t + 2\mathbf{b}\,t\Delta t + \mathbf{c}\Delta t + 3\mathbf{a}\,t\Delta^2 t + \mathbf{b}\Delta^2 t + \mathbf{a}\Delta^3 t,$$
$$\mathbf{ddB}(t) = 6\mathbf{a}\,t\Delta^2 t + 2\mathbf{b}\Delta^2 t + 6\mathbf{a}\Delta^3 t,$$
$$\mathbf{dddB} = 6\mathbf{a}\Delta^3 t.$$

They all have to be calculated at $t = 0$, as functions of the four control points $\mathbf{P}_i$, before the loop starts:

$$\mathbf{B}(0) = \mathbf{d} = \mathbf{P}_0,$$
$$\begin{aligned}
\mathbf{dB}(0) &= \mathbf{c}\Delta t + \mathbf{b}\Delta^2 t + \mathbf{a}\Delta^3 t \\
&= 3\Delta t(\mathbf{P}_1 - \mathbf{P}_0) + \Delta^2 t\big(3(\mathbf{P}_0 + \mathbf{P}_2) - 6\mathbf{P}_1\big) \\
&\quad + \Delta^3 t\big(3(\mathbf{P}_1 - \mathbf{P}_2) - \mathbf{P}_0 + \mathbf{P}_3\big) \\
&= 3\Delta t(\mathbf{P}_1 - \mathbf{P}_0) + 3\Delta^2 t(\mathbf{P}_0 - 2\mathbf{P}_1 + \mathbf{P}_2) \\
&\quad + \Delta^3 t\big(3(\mathbf{P}_1 - \mathbf{P}_2) - \mathbf{P}_0 + \mathbf{P}_3\big),
\end{aligned}$$
$$\begin{aligned}
\mathbf{ddB}(0) &= 2\mathbf{b}\Delta^2 t + 6\mathbf{a}\Delta^3 t \\
&= 2\Delta^2 t\big(3(\mathbf{P}_0 + \mathbf{P}_2) - 6\mathbf{P}_1\big) + 6\Delta^3 t\big(3(\mathbf{P}_1 - \mathbf{P}_2) - \mathbf{P}_0 + \mathbf{P}_3\big) \\
&= 6\Delta^2 t(\mathbf{P}_0 - 2\mathbf{P}_1 + \mathbf{P}_2) + 6\Delta^3 t\big(3(\mathbf{P}_1 - \mathbf{P}_2) - \mathbf{P}_0 + \mathbf{P}_3\big),
\end{aligned}$$
$$\mathbf{dddB} = 6\mathbf{a}\Delta^3 t = 6\Delta^3 t\big(3(\mathbf{P}_1 - \mathbf{P}_2) - \mathbf{P}_0 + \mathbf{P}_3\big).$$

The above relations can be expressed in matrix notation as follows:

$$
\begin{pmatrix} \mathbf{dddB} \\ \mathbf{ddB}(0) \\ \mathbf{dB}(0) \\ \mathbf{B}(0) \end{pmatrix}
= \begin{pmatrix} 6 & 0 & 0 & 0 \\ 6 & 2 & 0 & 0 \\ 1 & 1 & 1 & 0 \\ 0 & 0 & 0 & 1 \end{pmatrix}
\begin{pmatrix} \Delta^3 t & 0 & 0 & 0 \\ 0 & \Delta^2 t & 0 & 0 \\ 0 & 0 & \Delta t & 0 \\ 0 & 0 & 0 & 1 \end{pmatrix}
\begin{pmatrix} \mathbf{a} \\ \mathbf{b} \\ \mathbf{c} \\ \mathbf{d} \end{pmatrix}
$$

$$
= \begin{pmatrix} 6 & 0 & 0 & 0 \\ 6 & 2 & 0 & 0 \\ 1 & 1 & 1 & 0 \\ 0 & 0 & 0 & 1 \end{pmatrix}
\begin{pmatrix} \Delta^3 t & 0 & 0 & 0 \\ 0 & \Delta^2 t & 0 & 0 \\ 0 & 0 & \Delta t & 0 \\ 0 & 0 & 0 & 1 \end{pmatrix}
\begin{pmatrix} -1 & 3 & -3 & 1 \\ 3 & -6 & 3 & 0 \\ -3 & 3 & 0 & 0 \\ 1 & 0 & 0 & 0 \end{pmatrix}
\begin{pmatrix} \mathbf{P}_0 \\ \mathbf{P}_1 \\ \mathbf{P}_2 \\ \mathbf{P}_3 \end{pmatrix}
$$

$$
= \begin{pmatrix}
-6\Delta^3 t & 18\Delta^3 t & -18\Delta^3 t & 6\Delta^3 t \\
6\Delta^2 t - 6\Delta^3 t & -12\Delta^2 t + 18\Delta^3 t & 6\Delta^2 t - 18\Delta^3 t & 6\Delta^3 t \\
3\Delta^2 t - \Delta^3 t - 3\Delta^t & -6\Delta^2 t + 3\Delta^3 t + 3\Delta^t & 3\Delta^2 t - 3\Delta^3 t & \Delta^3 t \\
1 & 0 & 0 & 0
\end{pmatrix}
\begin{pmatrix} \mathbf{P}_0 \\ \mathbf{P}_1 \\ \mathbf{P}_2 \\ \mathbf{P}_3 \end{pmatrix}
$$

$$= \mathbf{Q} \begin{pmatrix} \mathbf{P}_0 \\ \mathbf{P}_1 \\ \mathbf{P}_2 \\ \mathbf{P}_3 \end{pmatrix},$$

where $\mathbf{Q}$ is a 4×4 matrix that can be calculated once Δt is known.

A detailed examination of the above expressions shows that the following quantities have to be precalculated: $3\Delta t$, $3\Delta^2 t$, $\Delta^3 t$, $6\Delta^2 t$, $6\Delta^3 t$, $\mathbf{P}_0 - 2\mathbf{P}_1 + \mathbf{P}_2$, and $3(\mathbf{P}_1 - \mathbf{P}_2) - \mathbf{P}_0 + \mathbf{P}_3$. We therefore end up with the simple, fast algorithm shown in Figure 6.4. For those interested in a quick test, the corresponding *Mathematica* code is also included.

```
Q1:=3Δt;
Q2:=Q1×Δt;  // 3Δ²t
Q3:=Δ³t;
Q4:=2Q2;  // 6Δ²t
Q5:=6Q3;  // 6Δ³t
Q6:=P₀ − 2P₁ + P₂;
Q7:=3(P₁ − P₂) − P₀ + P₃;
B:=P₀;
dB:=(P₁ − P₀)Q1+Q6×Q2+Q7×Q3;
ddB:=Q6×Q4+Q7×Q5;
dddB:=Q7×Q5;
for t:=0 to 1 step Δt do
Pixel(B);
B:=B+dB; dB:=dB+ddB; ddB:=ddB+dddB;
endfor;
```

```
n=3; Clear[q1,q2,q3,q4,q5,Q6,Q7,B,dB,ddB,dddB,p0,p1,p2,p3,tabl];
p0={0,1}; p1={5,.5}; p2={0,.5}; p3={0,1}; (* Four points *)
dt=.01; q1=3dt; q2=3dt^2; q3=dt^3; q4=2q2; q5=6q3;
Q6=p0-2p1+p2; Q7=3(p1-p2)-p0+p3;
B=p0; dB=(p1-p0) q1+Q6 q2+Q7 q3; (* space indicates *)
ddB=Q6 q4+Q7 q5; dddB=Q7 q5;     (* multiplication *)
tabl={};
Do[{tabl=Append[tabl,B], B=B+dB, dB=dB+ddB, ddB=ddB+dddB},
                                        {t,0,1,dt}];

ListPlot[tabl];
```

Figure 6.4: A Fast Bézier Curve Algorithm.

Each point of the curve (i.e., each pixel in the loop) is calculated by three additions and three assignments only. There are no multiplications and no table lookups. This is a very fast algorithm indeed!

6.4 Properties of the Curve

The following useful properties are discussed in this section:

1. The weights add up to 1 (they are barycentric). This is easily shown from Newton's binomial theorem $(a+b)^n = \sum_{i=0}^{n} \binom{n}{i} a^i b^{n-i}$:

$$1 = \left(t + (1-t)\right)^n = \sum_{i=0}^{n} \binom{n}{i} t^i (1-t)^{n-i} = \sum_{i=0}^{n} B_{n,i}(t). \tag{6.12}$$

2. The curve passes through the two endpoints $\mathbf{P}_0$ and $\mathbf{P}_n$. We assume that $0^0 = 1$ and observe that

$$B_{n,0}(0) = \binom{n}{0} 0^0 (1-0)^{n-0} = 1 \cdot 1 \cdot 1^n = 1,$$

which implies

$$\mathbf{P}(0) = \sum_{i=0}^{n} \mathbf{P}_i B_{n,i}(0) = \mathbf{P}_0 B_{n,0}(0) = \mathbf{P}_0.$$

Also, the relation

$$B_{n,n}(1) = \binom{n}{n} 1^n (1-1)^{(n-n)} = 1 \cdot 1 \cdot 0^0 = 1,$$

implies

$$\mathbf{P}(1) = \sum_{i=0}^{n} \mathbf{P}_i B_{n,i}(1) = \mathbf{P}_n B_{n,n}(1) = \mathbf{P}_n.$$

3. Another interesting property of the Bézier curve is its symmetry with respect to the numbering of the control points. If we number the points $\mathbf{P}_n$, $\mathbf{P}_{n-1}, \ldots, \mathbf{P}_0$, we end up with the same curve, except that it proceeds from right (point $\mathbf{P}_0$) to left (point $\mathbf{P}_n$). The Bernstein polynomials satisfy the identity $B_{n,j}(t) = B_{n,n-j}(1-t)$, which can be proved directly and which can be used to prove the symmetry

$$\sum_{j=0}^{n} \mathbf{P}_j B_{n,j}(t) = \sum_{j=0}^{n} \mathbf{P}_{n-j} B_{n,j}(1-t).$$

4. The first derivative (the tangent vector) of the curve is straightforward to derive

$$\mathbf{P}^t(t) = \sum_{i=0}^{n} \mathbf{P}_i B'_{n,i}(t)$$

$$= \sum_{0}^{n} \mathbf{P}_i \binom{n}{i} \left[i\, t^{i-1}(1-t)^{n-i} + t^i (n-i)(1-t)^{n-i-1}(-1) \right]$$

$$= \sum_{0}^{n} \mathbf{P}_i \binom{n}{i} i\, t^{i-1}(1-t)^{n-i} - \sum_{0}^{n-1} \mathbf{P}_i \binom{n}{i} t^i (n-i)(1-t)^{n-1-i}$$

(using the identity $n\binom{n-1}{i-1} = i\binom{n}{i}$,   we get)

$$= n \sum_{1}^{n} \mathbf{P}_i \binom{n-1}{i-1} t^{i-1}(1-t)^{(n-1)-(i-1)} - n \sum_{0}^{n-1} \mathbf{P}_i \binom{n-1}{i} t^i (1-t)^{n-1-i}$$

(but $\binom{n-1}{i-1} t^{i-1}(1-t)^{(n-1)-(i-1)} = B_{n-1,i-1}(t)$, so)

$$= n \sum_{0}^{n-1} \mathbf{P}_{i+1} B_{n-1,i}(t) - n \sum_{0}^{n-1} \mathbf{P}_i B_{n-1,i}(t)$$

$$= n \sum_{0}^{n-1} [\mathbf{P}_{i+1} - \mathbf{P}_i] B_{n-1,i}(t)$$

$$= n \sum_{0}^{n-1} \Delta \mathbf{P}_i B_{n-1,i}(t), \quad \text{where} \quad \Delta \mathbf{P}_i = \mathbf{P}_{i+1} - \mathbf{P}_i. \tag{6.13}$$

Note that the tangent vector is a Bézier weighted sum (of n terms) where each Bernstein polynomial is the weight of a "control point" $\Delta \mathbf{P}_i$ ($\Delta \mathbf{P}_i$ is the difference of two points, hence it is a vector, but since it is represented by a pair or a triplet, we can conveniently consider it a point). As a result, the second derivative is obviously another Bézier sum based on the $n-1$ "control points" $\Delta^2 \mathbf{P}_i = \Delta \mathbf{P}_{i+1} - \Delta \mathbf{P}_i = \mathbf{P}_{i+2} - 2\mathbf{P}_{i+1} + \mathbf{P}_i$.

5. The weight functions $B_{n,i}(t)$ have a maximum at $t = i/n$. To see this, we first differentiate the weights

$$B'_{n,i}(t) = \binom{n}{i} \left[i\, t^{i-1}(1-t)^{n-i} + t^i (n-i)(1-t)^{n-i-1}(-1) \right]$$
$$= \binom{n}{i} i\, t^{i-1}(1-t)^{n-i} - \binom{n}{i} t^i (n-i)(1-t)^{n-1-i},$$

then equate the derivative to zero $\binom{n}{i} i\, t^{i-1}(1-t)^{n-i} - \binom{n}{i} t^i (n-i)(1-t)^{n-1-i} = 0$. Dividing by $t^{i-1}(1-t)^{n-i-1}$ yields $i(1-t) - t(n-i) = 0$ or $t = i/n$.

6. The two derivatives $\mathbf{P}^t(0)$ and $\mathbf{P}^t(1)$ are easy to derive from Equation (6.13) and are used to reshape the curve. They are $\mathbf{P}^t(0) = n(\mathbf{P}_1 - \mathbf{P}_0)$ and $\mathbf{P}^t(1) = n(\mathbf{P}_n - \mathbf{P}_{n-1})$. Since n is always positive, we conclude that $\mathbf{P}^t(0)$, the initial tangent of the curve, points in the direction from $\mathbf{P}_0$ to $\mathbf{P}_1$. This initial tangent can easily be controlled by moving point $\mathbf{P}_1$. The situation for the final tangent is similar.

7. The Bézier curve features global control. This means that moving one control point $\mathbf{P}_i$ modifies the entire curve. Most of the change, however, occurs at the vicinity of $\mathbf{P}_i$. This feature stems from the fact that the weight functions $B_{n,i}(t)$ are nonzero for all values of t except $t = 0$ and $t = 1$. Thus, any change in a control point $\mathbf{P}_i$ affects the contribution of the term $\mathbf{P}_i B_{n,i}(t)$ for all values of t. The behavior of the global control of the Bézier curve is easy to analyze. When a control point $\mathbf{P}_k$ is moved by a vector (α, β) to a new location $\mathbf{P}_k + (\alpha, \beta)$, the curve $\mathbf{P}(t)$ is changed from the original sum $\sum B_{ni}(t)\mathbf{P}_i$ to

$$\sum_{i=0}^{n} B_{ni}(t)\mathbf{P}_i + B_{nk}(t)(\alpha, \beta) = \mathbf{P}(t) + B_{nk}(t)(\alpha, \beta).$$

Thus, every point $\mathbf{P}(t_0)$ on the curve is moved by the vector $B_{nk}(t_0)(\alpha, \beta)$. The points are all moved in the same direction, but by different amounts, depending on t_0. This behavior is demonstrated by Figure 6.19b. (In principle, the figure is for a rational curve, but the particular choice of weights in the figure results in a standard curve.)

8. The concept of the *convex hull* of a set of points was introduced in Section 2.2.5. Here, we show a connection between the Bézier curve and the convex hull. Let $\mathbf{P}_1$, $\mathbf{P}_2, \ldots, \mathbf{P}_n$ be a given set of points and let a point $\mathbf{P}$ be constructed as a barycentric sum of these points with nonnegative weights, i.e.,

$$\mathbf{P} = \sum_{i=1}^{n} a_i \mathbf{P}_i, \quad \text{where } \sum a_i = 1 \text{ and } a_i \geq 0. \tag{6.14}$$

It can be shown that the set of all points $\mathbf{P}$ satisfying Equation (6.14) lies in the convex hull of $\mathbf{P}_1$, $\mathbf{P}_2$ through $\mathbf{P}_n$. The Bézier curve, Equation (6.5), satisfies Equation (6.14) for all values of t, so all its points lie in the convex hull of the set of control points. Thus, the curve is said to have the *convex hull property*. The significance of this property is that it makes the Bézier curve more predictable. A designer specifying a set of control points needs just a little experience to visualize the shape of the curve, since the convex hull property guarantees that the curve will not "stray" far from the control points.

9. The control polygon of a Bézier curve intersects the curve at the first and the last points and in general may intersect the curve at a certain number m, of points (Figure 6.1, where m is 2, 3, or 4, may help to visualize this). If we take a straight segment and maneuver it to intersect the curve as many times as possible, we find that the number of intersection points is always less than or equal m. This property of the Bézier curve may be termed variation diminution.

10. Imagine that each control point is moved 10 units to the left. Such a transformation will move every point on the curve to the left by the same amount. Similarly, if the control points are rotated, reflected, or are subject to any other *affine* transformation, the entire curve will be transformed in the same way. We say that the Bézier curve is invariant under affine transformations. However, the curve is not invariant under projections. If we compute a three-dimensional Bézier curve and project every point on the curve by a perspective projection, we end up with a two-dimensional curve $\mathbf{P}(t)$. If we then project the three-dimensional control points and compute a two-dimensional Bézier curve $\mathbf{Q}(t)$ from the projected, two-dimensional points, the two curves $\mathbf{P}(t)$ and $\mathbf{Q}(t)$ will be different. Invariance under projections can be achieved by switching from the standard Bézier curve to the rational Bézier curve (Section 6.15).

6.5 Connecting Bézier Curves

The Bézier curve is a polynomial of degree n, which makes it slow to compute for large values of n. It is therefore preferable to connect several Bézier segments, each defined by a few points, typically four to six, into one smooth curve. The condition for smooth connection of two such segments is easy to derive. We assume that the control points are divided into two sets $\mathbf{P}_0, \mathbf{P}_1, \ldots, \mathbf{P}_n$ and $\mathbf{Q}_0, \mathbf{Q}_1, \ldots, \mathbf{Q}_m$. In order for the two segments to connect, $\mathbf{P}_n$ must equal $\mathbf{Q}_0$. We already know that the extreme tangent vectors of the Bézier curve satisfy

$$\mathbf{Q}^t(0) = m(\mathbf{Q}_1 - \mathbf{Q}_0) \quad \text{and} \quad \mathbf{P}^t(1) = n(\mathbf{P}_n - \mathbf{P}_{n-1}).$$

The condition for a smooth connection is $\mathbf{Q}^t(0) = \mathbf{P}^t(1)$ or $m\mathbf{Q}_1 - m\mathbf{Q}_0 = n\mathbf{P}_n - n\mathbf{P}_{n-1}$. Substituting $\mathbf{Q}_0 = \mathbf{P}_n$ yields

$$\mathbf{P}_n = \frac{m}{m+n}\mathbf{Q}_1 + \frac{n}{m+n}\mathbf{P}_{n-1}. \tag{6.15}$$

The three points $\mathbf{P}_{n-1}$, $\mathbf{P}_n$, and $\mathbf{Q}_1$ must therefore be dependent. Hence, the condition for smooth linking is that the three points $\mathbf{P}_{n-1}$, $\mathbf{P}_n$, and $\mathbf{Q}_1$ be *collinear*. In the special case where $n = m$, Equation (6.15) reduces to $\mathbf{P}_n = 0.5\mathbf{Q}_1 + 0.5\mathbf{P}_{n-1}$, implying that $\mathbf{P}_n$ should be the midpoint between $\mathbf{Q}_1$ and $\mathbf{P}_{n-1}$.

Example: Given that $\mathbf{P}_4 = \mathbf{Q}_0 = (6, -1)$, $\mathbf{Q}_1 = (7, 0)$, and $m = 5$, we compute $\mathbf{P}_3$ by

$$(6, -1) = \frac{5}{4+5}(7, 0) + \frac{4}{4+5}\mathbf{P}_3,$$

which yields $\mathbf{P}_3 = (21/4, -9/4)$.

⋄ **Exercise 6.8:** A more general condition for a smooth connection of two curve segments is $\alpha\mathbf{Q}^t(0) = \mathbf{P}^t(1)$. The two tangents at the connection point are in the same direction, but have different magnitudes. Discuss this condition and what it means for the three control points $\mathbf{P}_{n-1}$, $\mathbf{P}_n = \mathbf{Q}_0$, and $\mathbf{Q}_1$.

Breaking large curves into short segments has the additional advantage of easy control. The Bézier curve offers only global control, but if it is constructed of separate segments, a change in the control points in one segment will not affect the other segments. Figure 6.5 is an example of two Bézier segments connected smoothly.

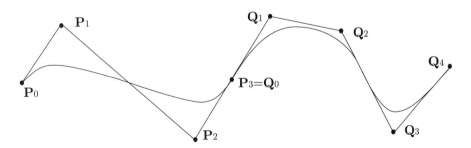

Figure 6.5: Connecting Bézier Segments.

6.6 The Bézier Curve as a Linear Interpolation

The original form of the Bézier curve, as developed by de Casteljau in 1959, is based on an approach entirely different from that of Bézier. Specifically, it employs linear interpolation and the *mediation operator*. Before we start, Figure 6.6 captures the essence of the concepts discussed here. The figure shows how a set of straight segments (or, equivalently, a single segment that slides along the base lines) creates the illusion (some would say, the magic) of a curve. Such a curve is called the envelope of the set, and the linear interpolation method of this section shows how to extend this simple construction to more than three points and two segments.

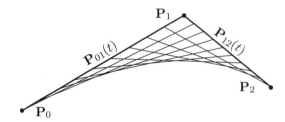

Figure 6.6: A Curve as an Envelope of Straight Segments.

Figure 6.6 involves only three points, which makes it easy to derive the expression of the envelope. The equation of the straight segment from $\mathbf{P}_0$ to $\mathbf{P}_1$ is $\mathbf{P}_{01}(t) = (1-t)\mathbf{P}_0 + t\mathbf{P}_1$ and the equation of the segment between $\mathbf{P}_1$ and $\mathbf{P}_2$ is similarly $\mathbf{P}_{12}(t) = (1-t)\mathbf{P}_1 + t\mathbf{P}_2$. If we fix t at a certain value, then $\mathbf{P}_{01}(t)$ and $\mathbf{P}_{12}(t)$ become points on the two segments. The straight segment connecting these points has the familiar form

$$\mathbf{P}(t) = (1-t)\mathbf{P}_{01}(t) + t\,\mathbf{P}_{12}(t) = (1-t)^2\mathbf{P}_0 + 2t(1-t)\mathbf{P}_1 + t^2\mathbf{P}_2.$$

For a fixed t, this is a point on the Bézier curve defined by $\mathbf{P}_0$, $\mathbf{P}_1$, and $\mathbf{P}_2$. When t is varied, the entire curve segment is obtained. Thus, the magical envelope has become a familiar curve. We can call this envelope a multilinear curve. *Linear*, because it is constructed from straight segments, and *multi*, because several such segments are required.

In order to extend this method to more than three points, we need appropriate notation. We start with a simple definition. The mediation operator $t[\![\mathbf{P}_0, \mathbf{P}_1]\!]$ between two points $\mathbf{P}_0$ and $\mathbf{P}_1$ is defined as the familiar linear interpolation*

$$t[\![\mathbf{P}_0, \mathbf{P}_1]\!] = (1-t)\mathbf{P}_0 + t\mathbf{P}_1 = t(\mathbf{P}_1 - \mathbf{P}_0) + \mathbf{P}_0, \quad \text{where} \quad 0 \le t \le 1.$$

The general definition, for any number of points, is recursive. The mediation operator

---

* The term "mediation" seems to have originated in [Knuth 86].

can be applied to any number of points according to

$$t[\![\mathbf{P}_0,\ldots,\mathbf{P}_n]\!] = t[\![\,t[\![\mathbf{P}_0,\ldots,\mathbf{P}_{n-1}]\!], t[\![\mathbf{P}_1,\ldots,\mathbf{P}_n]\!]\,]\!],$$

$$\vdots$$

$$t[\![\mathbf{P}_0,\mathbf{P}_1,\mathbf{P}_2,\mathbf{P}_3]\!] = t[\![\,t[\![\mathbf{P}_0,\mathbf{P}_1,\mathbf{P}_2]\!], t[\![\mathbf{P}_1,\mathbf{P}_2,\mathbf{P}_3]\!]\,]\!],$$

$$t[\![\mathbf{P}_0,\mathbf{P}_1,\mathbf{P}_2]\!] = t[\![\,t[\![\mathbf{P}_0,\mathbf{P}_1]\!], t[\![\mathbf{P}_1,\mathbf{P}_2]\!]\,]\!],$$

$$t[\![\mathbf{P}_0,\mathbf{P}_1]\!] = (1-t)\mathbf{P}_0 + t\mathbf{P}_1 = t(\mathbf{P}_1 - \mathbf{P}_0) + \mathbf{P}_0, \quad \text{where} \quad 0 \le t \le 1.$$

This operator creates curves that interpolate between the points. It has the advantages of being a simple mathematical function (and therefore fast to calculate) and of producing interpolation curves whose shape can easily be predicted. We examine cases involving more and more points.

Case 1. Two points. Given the two points $\mathbf{P}_0$ and $\mathbf{P}_1$, we denote the straight segment connecting them by $\mathbf{L}_{01}$. It is easy to see that $\mathbf{L}_{01} = t[\![\mathbf{P}_0,\mathbf{P}_1]\!]$, because the mediation operator is a linear function of t and because $0[\![\mathbf{P}_0,\mathbf{P}_1]\!] = \mathbf{P}_0$ and $1[\![\mathbf{P}_0,\mathbf{P}_1]\!] = \mathbf{P}_1$. Notice that values of t below 0 or above 1 correspond to those parts of the line that don't lie between the two points. Such values may be of interest in certain cases but not in the present context. The interpolation curve between the two points is denoted by $\mathbf{P}_1(t)$ and is simply selected as the line $\mathbf{L}_{01}$ connecting the points. Hence, $\mathbf{P}_1(t) = \mathbf{L}_{01} = t[\![\mathbf{P}_0,\mathbf{P}_1]\!]$. Notice that a straight line is also a polynomial of degree 1.

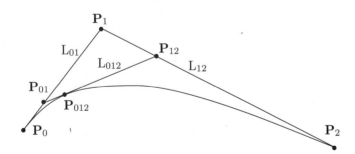

Figure 6.7: Repeated Linear Interpolation.

Case 2. Three points. Given the three points $\mathbf{P}_0$, $\mathbf{P}_1$, and $\mathbf{P}_2$ (Figure 6.7), the mediation operator can be used to construct an interpolation curve between them in the following steps:

1. Construct the two lines $\mathbf{L}_{01} = t[\![\mathbf{P}_0,\mathbf{P}_1]\!]$ and $\mathbf{L}_{12} = t[\![\mathbf{P}_1,\mathbf{P}_2]\!]$.

2. For some $0 \le t_0 \le 1$, consider the two points $\mathbf{P}_{01} = t_0[\![\mathbf{P}_0,\mathbf{P}_1]\!]$ and $\mathbf{P}_{12} = t_0[\![\mathbf{P}_1,\mathbf{P}_2]\!]$. Connect the points with a line $\mathbf{L}_{012}$. The equation of this line is, of course, $t[\![\mathbf{P}_{01},\mathbf{P}_{12}]\!]$ and it equals

$$\mathbf{L}_{012} = t[\![\mathbf{P}_{01},\mathbf{P}_{12}]\!] = t[\![\,t[\![\mathbf{P}_0,\mathbf{P}_1]\!], t[\![\mathbf{P}_1,\mathbf{P}_2]\!]\,]\!] = t[\![\mathbf{P}_0,\mathbf{P}_1,\mathbf{P}_2]\!].$$

3. For the same t_0, select point $\mathbf{P}_{012} = t_0[\![\mathbf{P}_0, \mathbf{P}_1, \mathbf{P}_2]\!]$ on $\mathbf{L}_{\mathbf{012}}$. The point can be expressed as

$$\mathbf{P}_{012} = t_0[\![\mathbf{P}_0, \mathbf{P}_1, \mathbf{P}_2]\!] = t_0[\![\mathbf{P}_{01}, \mathbf{P}_{12}]\!] = t_0[\![t_0[\![\mathbf{P}_0, \mathbf{P}_1]\!], t_0[\![\mathbf{P}_1, \mathbf{P}_2]\!]]\!].$$

Now, release t_0 and let it vary from 0 to 1. Point $\mathbf{P}_{012}$ slides along the line $\mathbf{L}_{012}$, whose endpoints will, in turn, slide along $\mathbf{L}_{01}$ and $\mathbf{L}_{12}$. The curve described by point $\mathbf{P}_{012}$ as it is sliding is the interpolation curve for $\mathbf{P}_0$, $\mathbf{P}_1$, and $\mathbf{P}_2$ that we are seeking. It is the equivalent of the envelope curve of Figure 6.6. We denote it by $\mathbf{P}_2(t)$ and its expression is easy to calculate, using the definition of $t[\![\mathbf{P}_i, \mathbf{P}_j]\!]$:

$$
\begin{aligned}
\mathbf{P}_2(t) &= t[\![\mathbf{P}_0, \mathbf{P}_1, \mathbf{P}_2]\!] \\
&= t[\![t[\![\mathbf{P}_0, \mathbf{P}_1]\!], t[\![\mathbf{P}_1, \mathbf{P}_2]\!]]\!] \\
&= t[\![t\mathbf{P}_1 + (1 - t)\mathbf{P}_0, t\mathbf{P}_2 + (1 - t)\mathbf{P}_1]\!] \\
&= t[t\mathbf{P}_2 + (1 - t)\mathbf{P}_1] + (1 - t)[t\mathbf{P}_1 + (1 - t)\mathbf{P}_0] \\
&= \mathbf{P}_0(1 - t)^2 + 2\mathbf{P}_1 t(1 - t) + \mathbf{P}_2 t^2.
\end{aligned}
$$

$\mathbf{P}_2(t)$ is therefore the Bézier curve for three points.

Case 3. Four points. Given the four points $\mathbf{P}_0$, $\mathbf{P}_1$, $\mathbf{P}_2$, and $\mathbf{P}_3$, we follow similar steps:

1. Construct the three lines $\mathbf{L}_{01} = t[\![\mathbf{P}_0, \mathbf{P}_1]\!]$, $\mathbf{L}_{12} = t[\![\mathbf{P}_1, \mathbf{P}_2]\!]$, and $\mathbf{L}_{23} = t[\![\mathbf{P}_2, \mathbf{P}_3]\!]$.

2. Select three points, $\mathbf{P}_{01} = t_0[\![\mathbf{P}_0, \mathbf{P}_1]\!]$, $\mathbf{P}_{12} = t_0[\![\mathbf{P}_1, \mathbf{P}_2]\!]$, and $\mathbf{P}_{23} = t_0[\![\mathbf{P}_2, \mathbf{P}_3]\!]$, and construct lines $\mathbf{L}_{012} = t[\![\mathbf{P}_0, \mathbf{P}_1, \mathbf{P}_2]\!] = t[\![\mathbf{P}_{01}, \mathbf{P}_{12}]\!]$ and $\mathbf{L}_{123} = t[\![\mathbf{P}_1, \mathbf{P}_2, \mathbf{P}_3]\!] = t[\![\mathbf{P}_{12}, \mathbf{P}_{23}]\!]$.

3. Select two points, $\mathbf{P}_{012} = t_0[\![\mathbf{P}_{01}, \mathbf{P}_{12}]\!]$ on segment $\mathbf{L}_{012}$ and $\mathbf{P}_{123} = t_0[\![\mathbf{P}_{12}, \mathbf{P}_{23}]\!]$ on segment $\mathbf{L}_{123}$. Construct a new segment $\mathbf{L}_{0123}$ as the mediation $t[\![\mathbf{P}_0, \mathbf{P}_1, \mathbf{P}_2, \mathbf{P}_3]\!] = t[\![\mathbf{P}_{012}, \mathbf{P}_{123}]\!]$.

4. Select point $\mathbf{P}_{0123} = t_0[\![\mathbf{P}_{012}, \mathbf{P}_{123}]\!]$ on $\mathbf{L}_{0123}$.

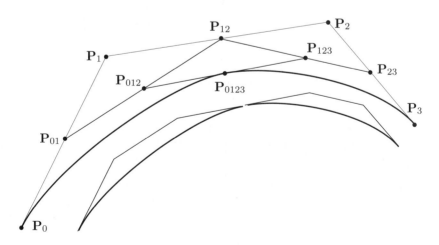

Figure 6.8: Scaffolding for $k = 3$.

When t_0 varies from 0 to 1, point $\mathbf{P}_{0123}$ slides along $\mathbf{L}_{0123}$, whose endpoints, in turn, slide along $\mathbf{L}_{012}$ and $\mathbf{L}_{123}$, which also slide. The entire structure, which resembles a *scaffolding* (Figure 6.8), slides along the original three lines. The interpolation curve for the four original points is denoted by $\mathbf{P}_3(t)$ and its expression is not hard to calculate, using the expression for $\mathbf{P}_2(t) = t[\![\mathbf{P}_0, \mathbf{P}_1, \mathbf{P}_2]\!]$:

$$
\begin{aligned}
\mathbf{P}_3(t) = t[\![\mathbf{P}_0, \mathbf{P}_1, \mathbf{P}_2, \mathbf{P}_3]\!] &= t[\![\, t[\![\mathbf{P}_0, \mathbf{P}_1, \mathbf{P}_2]\!], t[\![\mathbf{P}_1, \mathbf{P}_2, \mathbf{P}_3]\!]\,]\!] \\
&= t[t^2\mathbf{P}_3 + 2t(1-t)\mathbf{P}_2 + (1-t)^2\mathbf{P}_1] \\
&\quad + (1-t)[t^2\mathbf{P}_2 + 2t(1-t)\mathbf{P}_1 + (1-t)^2\mathbf{P}_0] \\
&= t^3\mathbf{P}_3 + 3t^2(1-t)\mathbf{P}_2 + 3t(1-t)^2\mathbf{P}_1 + (1-t)^3\mathbf{P}_0.
\end{aligned}
$$

$\mathbf{P}_3(t)$ is therefore the Bézier curve for four points.

Case 4. In the general case, $n + 1$ points $\mathbf{P}_0, \mathbf{P}_1, \ldots, \mathbf{P}_n$ are given. The interpolation curve is, similarly, $t[\![\mathbf{P}_0, \mathbf{P}_1, \ldots, \mathbf{P}_n]\!] = t[\![\mathbf{P}_{01\ldots n-1}, \mathbf{P}_{12\ldots n}]\!]$. It can be proved by induction that its value is the degree-n polynomial

$$
\mathbf{P}_n(t) = \sum_{i=0}^{n} \mathbf{P}_i B_{n,i}(t), \quad \text{where} \quad B_{n,i}(t) = \binom{n}{i} t^i (1-t)^{n-i},
$$

that is the Bézier curve for $n + 1$ points. The two approaches to curve construction, using Bernstein polynomials and using scaffolding, are therefore equivalent.

◇ **Exercise 6.9:** The scaffolding algorithm illustrated in Figure 6.8 is easy to understand because of the special placement of the four control points. The resulting curve is similar to a circular arc and doesn't have an inflection point (Section 1.6.8). Prove your grasp of this algorithm by performing it on the curve of Figure 6.9. Try to select the intermediate points so as to end up with the inflection point.

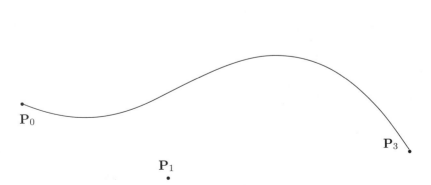

Figure 6.9: Scaffolding With an Inflection Point.

Figure 6.10 summarizes the process of scaffolding in the general case. The process takes n steps. In the first step, n new points are constructed between the original $n+1$ control points. In the second step, $n-1$ new points are constructed, between the n points of step 1 and so on, up to step n, where one point is constructed. The total number of points constructed during the entire process is therefore

$$n + (n-1) + (n-2) + \cdots + 2 + 1 = n(n+1)/2.$$

Step	Points constructed	# of points
1	$\mathbf{P}_{01}\,\mathbf{P}_{12}\,\mathbf{P}_{23}\ldots\mathbf{P}_{n-1,n}$	n
2	$\mathbf{P}_{012}\,\mathbf{P}_{123}\,\mathbf{P}_{234}\ldots\mathbf{P}_{n-2,n-1,n}$	$n-1$
3	$\mathbf{P}_{0123}\,\mathbf{P}_{1234}\,\mathbf{P}_{2345}\ldots\mathbf{P}_{n-3,n-2,n-1,n}$	$n-2$
$\vdots$	$\vdots$	$\vdots$
n	$\mathbf{P}_{0123\ldots n}$	1

Figure 6.10: The n Steps of Scaffolding.

6.7 Blossoming

The curves derived and discussed in the preceding chapters are based on polynomials. A typical curve is a pair or a triplet of polynomials of a certain degree n in t. Mathematicians know that a degree-n polynomial $P_n(t)$ of a single variable can be associated with a function $f(u_1, u_2, \ldots, u_n)$ in n variables that's linear (i.e., degree-1) in each variable and is symmetric with respect to the order of its variables. Such functions were named *blossom* by Lyle Ramshaw in [Ramshaw 87] to denote arrival at a promising stage. (The term *pole* was originally used by de Casteljau for those functions.) [Gallier 00] is a general, detailed reference for this topic.

Given a Bézier curve, this section shows how to derive its blossom and how to use the blossom to label the intermediate points obtained in the scaffolding construction. Other sections show how to apply blossoms to curve algorithms, such as curve subdivision (Section 6.8) and degree elevation (Section 6.9).

Dictionary definitions

Blossom:

Noun: The period of greatest prosperity or productivity.

Verb: To develop or come to a promising stage (Youth blossomed into maturity).

Blossoming: The process of budding and unfolding of blossoms.

We start by developing a special notation for use with blossoms. The equation of the straight segment from point $\mathbf{P}_0$ to point $\mathbf{P}_1$ is the familiar linear interpolation $\mathbf{P}(u) = (1-u)\mathbf{P}_0 + u\mathbf{P}_1$. Its start point is $\mathbf{P}(0)$, its end point is $\mathbf{P}(1)$, and a general point on this segment is $\mathbf{P}(u)$ for $0 \le u \le 1$. Because a straight segment has zero curvature, parameter values indicate arc lengths. Thus, the distance between $\mathbf{P}(0)$ and $\mathbf{P}(u)$ is proportional to u and the distance between $\mathbf{P}(u)$ and $\mathbf{P}(1)$ is proportional to $1 - u$. We can therefore consider parameter values u in the interval $[0, 1]$ a measure of distance (called affine distance) from the start of the segment. We introduce the symbol $\langle \mathbf{u} \rangle$ to denote point $\mathbf{P}(u)$. Similarly, points $\mathbf{P}(0)$ and $\mathbf{P}(1)$ are denoted by $\langle \mathbf{0} \rangle$ and $\langle \mathbf{1} \rangle$, respectively (Figure 6.11a).

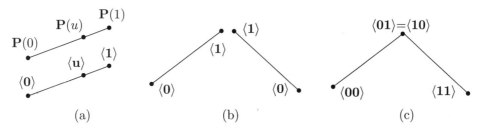

Figure 6.11: Blossom Notation For Points (Two Segments).

A spline consists of segments connected at the interior points, so we consider two straight segments connected at a common point. The endpoints of each segment are denoted by $\langle \mathbf{0} \rangle$ and $\langle \mathbf{1} \rangle$, but this creates an ambiguity. There are now two points labeled $\langle \mathbf{0} \rangle$ (Figure 6.11b). We distinguish between them by appending a bit to the symbol of each point. The two endpoints of one segment are now denoted by $\langle \mathbf{00} \rangle$ and $\langle \mathbf{01} \rangle$, while the two endpoints of the other segment are denoted by $\langle \mathbf{10} \rangle$ and $\langle \mathbf{11} \rangle$ (Figure 6.11c). The common point can be denoted by either $\langle \mathbf{01} \rangle$ or $\langle \mathbf{10} \rangle$. So far, it seems that the order of the individual indexes, 01 or 10, is immaterial. The new notation is symmetric with respect to the order of point indexes.

We now select a point with a parameter value u on each segment. The two new points are denoted by $\langle \mathbf{0u} \rangle$ and $\langle \mathbf{1u} \rangle$ (Figure 6.12a), but they can also be denoted by $\langle \mathbf{u0} \rangle$ and $\langle \mathbf{u1} \rangle$, respectively. The two points are now connected by a segment and a new point selected at affine distance u on that segment (Figure 6.12b). The new point deserves the label $\langle \mathbf{uu} \rangle$ because the endpoints of its segment have the common index $\mathbf{u}$.

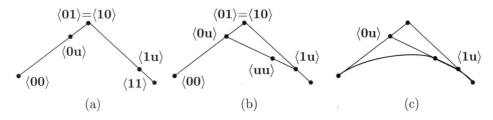

Figure 6.12: Blossom Notation For Points (Two Segments).

At this point it is clear that the simple scaffolding construction of Figure 6.12b is identical to the de Casteljau algorithm of Section 6.6, which implies that point $\langle \mathbf{uu} \rangle$ is located on the Bézier curve defined by the three points $\langle \mathbf{00} \rangle$, $\langle \mathbf{01} \rangle$, and $\langle \mathbf{11} \rangle$ (Figure 6.12c).

To illustrate this process for more points, it is applied to three line segments in Figure 6.13. Two bits are appended to each point in order to distinguish between the segments. Thus, a point is denoted by a triplet of the form $\langle \mathbf{00x} \rangle$, $\langle \mathbf{01x} \rangle$, or $\langle \mathbf{11x} \rangle$. Notice that our indexes are symmetric, so $\langle \mathbf{01x} \rangle = \langle \mathbf{10x} \rangle$, which is why we use $\langle \mathbf{11x} \rangle$ instead of $\langle \mathbf{10x} \rangle$ to identify the third segment.

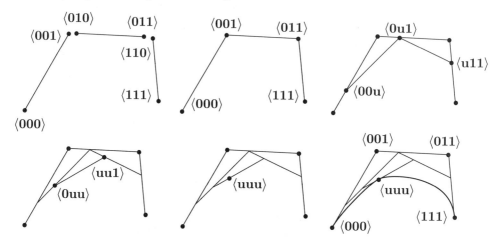

Figure 6.13: Blossom Notation For Points (Three Segments).

Again, our familiarity with the Bézier curve and the de Casteljau algorithm indicates intuitively that point $\langle \mathbf{uuu} \rangle$ is located on the Bézier curve defined by the four control points $\langle \mathbf{000} \rangle$, $\langle \mathbf{001} \rangle$, $\langle \mathbf{011} \rangle$, and $\langle \mathbf{111} \rangle$.

> Let us be grateful to people who make us happy, they are the charming gardeners who make our souls blossom.
>
> —Marcel Proust.

An actual construction of the scaffolding for this case verifies our intuitive feeling. Given points $\langle \mathbf{0uu} \rangle$ and $\langle \mathbf{uu1} \rangle$, we can write them as $\langle \mathbf{0uu} \rangle$ and $\langle \mathbf{1uu} \rangle$, which immediately produces point $\langle \mathbf{uuu} \rangle$ (it's located an affine distance u from $\langle \mathbf{0uu} \rangle$). Similarly, given points $\langle \mathbf{00u} \rangle$ and $\langle \mathbf{0u1} \rangle$, we can write them as $\langle \mathbf{00u} \rangle$ and $\langle \mathbf{01u} \rangle$, which immediately produces point $\langle \mathbf{0uu} \rangle$. A similar step produces $\langle \mathbf{00u} \rangle$ if points $\langle \mathbf{000} \rangle$ and $\langle \mathbf{001} \rangle$ are given. Thus, we conclude that knowledge of the four control points can produce all the intermediate points in the scaffolding construction and lead to one point $\langle \mathbf{uuu} \rangle$ that's located on the Bézier curve defined by the control points. This is an informal statement of the *blossoming principle*.

This principle can be illustrated in a different way. We know that point $\langle \mathbf{0u1} \rangle$ is obtained from points $\langle \mathbf{001} \rangle$ and $\langle \mathbf{011} \rangle$ as the linear interpolation $\langle \mathbf{0u1} \rangle = (1-u)\langle \mathbf{001} \rangle + u\langle \mathbf{011} \rangle$. We can therefore start from point $\langle \mathbf{uuu} \rangle$ and figure out its dependence on the four original points $\langle\mathbf{000}\rangle$, $\langle\mathbf{001}\rangle$, $\langle\mathbf{011}\rangle$, and $\langle\mathbf{111}\rangle$ as follows:

$$
\begin{aligned}
\langle\mathbf{uuu}\rangle &= (1-u)\langle\mathbf{0uu}\rangle + u\langle\mathbf{1uu}\rangle \\
&= (1-u)\big[(1-u)\langle\mathbf{00u}\rangle + u\langle\mathbf{01u}\rangle\big] + u\big[(1-u)\langle\mathbf{10u}\rangle + u\langle\mathbf{11u}\rangle\big] \\
&= (1-u)^2\langle\mathbf{00u}\rangle + 2u(1-u)\langle\mathbf{01u}\rangle + u^2\langle\mathbf{11u}\rangle \\
&= (1-u)^2\big[(1-u)\langle\mathbf{000}\rangle + u\langle\mathbf{001}\rangle\big] + 2u(1-u)\big[(1-u)\langle\mathbf{010}\rangle + u\langle\mathbf{011}\rangle\big] \\
&\quad + u^2\big[(1-u)\langle\mathbf{110}\rangle + u\langle\mathbf{111}\rangle\big] \\
&= (1-u)^3\langle\mathbf{000}\rangle + 3u(1-u)^2\langle\mathbf{001}\rangle + 3u^2(1-u)\langle\mathbf{011}\rangle + u^3\langle\mathbf{111}\rangle \\
&= B_{3,0}(u)\langle\mathbf{000}\rangle + B_{3,1}(u)\langle\mathbf{001}\rangle + B_{3,2}(u)\langle\mathbf{011}\rangle + B_{3,3}(u)\langle\mathbf{111}\rangle,
\end{aligned}
$$

where $B_{3,i}$ are the Bernstein polynomials for $n = 3$. This again shows that point $\langle\mathbf{uuu}\rangle$ lies on the Bézier curve whose control points are $\langle\mathbf{000}\rangle$, $\langle\mathbf{001}\rangle$, $\langle\mathbf{011}\rangle$, and $\langle\mathbf{111}\rangle$.

So far, blossoming has been used to assign labels to the control points and to the intermediate points. Even this simple application illustrates some of the power and elegance of the blossoming approach. Section 6.6 employs the notation $\mathbf{P}_{234}$, while various authors denote intermediate point i of scaffolding step j by d_i^j. The blossom labels $\langle\mathbf{u_1 u_2 \ldots u_n}\rangle$ are much more natural and useful.

We are now ready to see the actual blossom associated with the degree-n polynomial $P_n(t)$ as given by [Ramshaw 87]. The blossom of $P_n(t)$ is a function $f(u_1, u_2, \ldots, u_n)$ that satisfies the following:

1. f is linear in each variable u_i.
2. f is symmetric; the order of variables is irrelevant. Thus, $f(u_1, u_2, \ldots, u_n) = f(u_2, u_1, \ldots, u_n)$ or any other permutation of the n variables.
3. The diagonal $f(u, u, \ldots, u)$ of f equals $P_n(u)$.

Requirement 1 suggests the name "multilinear function" but [Ramshaw 87] explains why the term "multiaffine" is more appropriate.

Given $P_n(t)$, such a multiaffine function is easy to derive and is also unique. Here is an example for $n = 3$. Given the cubic polynomial $P(t) = -3t^3 + 6t^2 + 3t$, we are looking for a function $f(u, v, w)$ that's linear in each of its three parameters and is symmetric with respect to their order. The general form of such a function is

$$
f(u, v, w) = a_1 uvw + a_2 uv + a_3 uw + a_4 vw + a_5 u + a_6 v + a_7 w + a_8.
$$

If we also require that $f(u, v, w)$ satisfies $f(t, t, t) = P(t)$ for any t, it becomes obvious that a_1 must equal the coefficient of t^3. Because of the required symmetry, the sum $a_2 + a_3 + a_4$ must equal the coefficient of t^2 and the sum $a_5 + a_6 + a_7$ must equal the coefficient of t. Finally, a_8 must equal the free term of $P(t)$. Thus, we end up with the blossom $f(u, v, w) = -3uvw + 2(uv + uw + vw) + (u + v + w) + 0$. This blossom is unique.

In general, given an n-degree polynomial, the corresponding multiaffine blossom function is easy to construct in this way. Here are some examples.

Degree-0. $P(t) = a \rightarrow f(u, v, w) = a$,

Degree-1. $P(t) = at \rightarrow f(u, v, w) = \dfrac{a}{3}(u + v + w)$,

Degree-2. $P(t) = at^2 \rightarrow f(u,v,w) = \dfrac{a}{3}(uv + uw + vw),$ (6.16)

Degree-3. $P(t) = a_3 t^3 + a_2 t^2 + a_1 + a_0$

$$\rightarrow f(u,v,w) = a_3 uvw + \dfrac{a_2}{3}(uv + uw + vw) + \dfrac{a_1}{3}(u + v + w) + a_0.$$

The discussion above shows that the kth control point of the degree-n polynomial is associated with blossom value $f(\underbrace{00\ldots0}_{n-k}\underbrace{11\ldots1}_{k})$. Notice that there are $n+1$ such values, corresponding to the $n+1$ control points, and that blossom symmetry implies $f(011) = f(101) = f(110)$. If t varies in the general interval $[a,b]$ instead of in $[0,1]$, then the kth control point is associated with the blossom value $f(\underbrace{aa\ldots a}_{n-k}\underbrace{bb\ldots b}_{k})$.

⋄ **Exercise 6.10:** Given the four points $\mathbf{P}_0 = (0,1,1)$, $\mathbf{P}_1 = (1,1,0)$, $\mathbf{P}_2 = (4,2,0)$, and $\mathbf{P}_3 = (6,1,1)$, compute the Bézier curve defined by them, construct the three blossoms associated with this curve, and show that the four blossom values $f(0,0,0)$, $f(0,0,1)$, $f(0,1,1)$, and $f(1,1,1)$ yield the control points.

6.8 Subdividing the Bézier Curve

Bézier methods are interactive. It is possible to control the shape of the curve by moving the control points and by smoothly connecting individual segments. Imagine a situation where the points are moved and maneuvered for a while, but the curve "refuses" to get the right shape. This indicates that there are not enough points. There are two ways to increase the number of points. One is to add a point to a segment while increasing its degree. This is called *degree elevation* and is discussed in Section 6.9.

An alternative is to subdivide a Bézier curve segment into two segments such that there is no change in the shape of the curve. If the original segment is of degree n (i.e., based on $n+1$ control points), this is done by adding $2n-1$ new control points and deleting $n-1$ of the original points, bringing the number of points to $(n+1) + (2n-1) - (n-1) = 2n+1$. Each new segment is based on $n+1$ points and they share one of the new points. With more points, it is now possible to manipulate the control points of the two segments in order to fine-tune the shape of the segments. The advantage of this approach is that both the original and the new curves are based on $n+1$ points, so only one set of Bernstein polynomials is needed.

The new points being added consist of some of the ones constructed in the last k steps of the scaffolding process. For the case $k = 2$ (quadratic curve segments), the three points $\mathbf{P}_{01}$, $\mathbf{P}_{12}$, and $\mathbf{P}_{012}$ are added and the single point $\mathbf{P}_1$ is deleted (Figure 6.7). The two new segments consist of points $\mathbf{P}_0$, $\mathbf{P}_{01}$, and $\mathbf{P}_{012}$, and $\mathbf{P}_{012}$, $\mathbf{P}_{12}$, and $\mathbf{P}_2$. For the case $k = 3$ (cubic segments), the five points $\mathbf{P}_{01}$, $\mathbf{P}_{23}$, $\mathbf{P}_{012}$, $\mathbf{P}_{123}$, and $\mathbf{P}_{0123}$ are added and the two points $\mathbf{P}_1$ and $\mathbf{P}_2$ are deleted (Figure 6.8, duplicated here, where the inset shows the two segments with their control polygons). The two new segments consist of points $\mathbf{P}_0$, $\mathbf{P}_{01}$, $\mathbf{P}_{012}$, and $\mathbf{P}_{0123}$ and $\mathbf{P}_{0123}$, $\mathbf{P}_{123}$, $\mathbf{P}_{23}$, and $\mathbf{P}_3$.

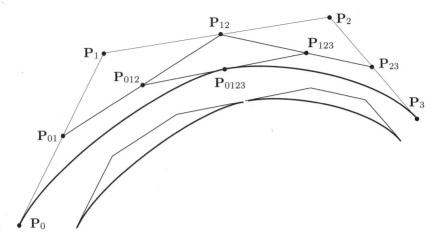

Figure 6.8: Scaffolding and Subdivision for $k = 3$ (Duplicate).

Using the mediation operator to express the new points in the scaffolding in terms of the original control points produces, for the quadratic case

$$\mathbf{P}_{01} = \alpha\mathbf{P}_0 + (1-\alpha)\mathbf{P}_1, \ \mathbf{P}_{12} = \alpha\mathbf{P}_1 + (1-\alpha)\mathbf{P}_2, \ \mathbf{P}_{012} = \alpha^2\mathbf{P}_0 + 2\alpha(1-\alpha)\mathbf{P}_1 + (1-\alpha)^2\mathbf{P}_2,$$

where α is any value in the range $[0, 1]$. We can therefore write

$$\begin{pmatrix} \mathbf{P}_0 \\ \mathbf{P}_{01} \\ \mathbf{P}_{012} \end{pmatrix} = \begin{pmatrix} 1 & 0 & 0 \\ \alpha & 1-\alpha & 0 \\ \alpha^2 & 2\alpha(1-\alpha) & (1-\alpha)^2 \end{pmatrix} \begin{pmatrix} \mathbf{P}_0 \\ \mathbf{P}_1 \\ \mathbf{P}_2 \end{pmatrix},$$

$$\begin{pmatrix} \mathbf{P}_{012} \\ \mathbf{P}_{12} \\ \mathbf{P}_2 \end{pmatrix} = \begin{pmatrix} \alpha^2 & 2\alpha(1-\alpha) & (1-\alpha)^2 \\ 0 & \alpha & 1-\alpha \\ 0 & 0 & 1 \end{pmatrix} \begin{pmatrix} \mathbf{P}_0 \\ \mathbf{P}_1 \\ \mathbf{P}_2 \end{pmatrix},$$

for the left and right segments, respectively.

⬦ **Exercise 6.11:** Use the mediation operator to calculate the scaffolding for the cubic case (four control points). Use $\alpha = 1/2$ and write the results in terms of matrices, as above.

In the general case where an $(n + 1)$-point Bézier curve is subdivided, the $n - 1$ points being deleted are $\mathbf{P}_1, \mathbf{P}_2, \ldots, \mathbf{P}_{n-1}$ (the original $n - 1$ interior control points). The $2n - 1$ points added are the first and last points constructed in each scaffolding step (except the last step, where only one point is constructed). Figure 6.10 shows that these are points $\mathbf{P}_{01}, \mathbf{P}_{n-1,n}$ (from step 1), $\mathbf{P}_{012}, \mathbf{P}_{n-2,n-1,n}$ (from step 2), $\mathbf{P}_{0123}$, $\mathbf{P}_{n-3,n-2,n-1,n}$ (from step 3), up to $\mathbf{P}_{0123\ldots n}$ from step n.

The $2n - 1$ points being added are therefore

$$\mathbf{P}_{01}, \mathbf{P}_{012}, \mathbf{P}_{0123}, \ldots, \mathbf{P}_{0123\ldots n}, \mathbf{P}_{123\ldots n}, \mathbf{P}_{23\ldots n}, \ldots, \mathbf{P}_{n-1,n}.$$

These points can be computed in two ways as follows:

1. Perform the entire scaffolding procedure and save all the points, then use only the appropriate $2n - 1$ points.

2. Compute just the required points. This is done by means of the two relations

$$(a) \ \mathbf{P}_{0123...k} = \sum_{j=0}^{k} B_{k,j}(t)\mathbf{P}_j, \quad \text{and} \quad (b) \ \mathbf{P}_{n-k,n-k+1,...,n} = \sum_{j=0}^{k} B_{k,j}(t)\mathbf{P}_{n-k+j}. \ (6.17)$$

(These expressions can be proved by induction.)

The first decision that has to be made when subdividing a curve, is at what point (what value of t) to break the original curve into two segments. Breaking a curve $\mathbf{P}(t)$ into two segments at $t = 0.1$ will result in a short segment followed by a long segment, each defined by $n + 1$ control points. Obviously, the first segment will be easier to edit. Once the value of t has been determined, the software computes the $2n - 1$ new points. The original $n - 1$ interior control points are easy to delete, and the set of $2n + 1$ points is partitioned into two sets. The procedure that computed the original curve is now invoked twice, to compute and display the two segments.

⋄ **Exercise 6.12:** Given the four points $\mathbf{P}_0 = (0, 1, 1)$, $\mathbf{P}_1 = (1, 1, 0)$, $\mathbf{P}_2 = (4, 2, 0)$, and $\mathbf{P}_3 = (6, 1, 1)$, apply Equation (6.17)a,b to subdivide the Bézier curve $\sum B_{3,i}(t)\mathbf{P}_i$ at $t = 1/3$.

Figure 6.14 illustrates how blossoms are applied to the problem of curve subdivision. The points on the left edge of the triangle become the control points of the first segment. In blossom notation these are points $\langle \underbrace{\mathbf{00} \ldots \mathbf{0}}_{n-k} \underbrace{\mathbf{tt} \ldots \mathbf{t}}_{k} \rangle$. Similarly, the points on the right edge of the triangle become the control points of the second segment. In blossom notation these are points $\langle \underbrace{\mathbf{11} \ldots \mathbf{1}}_{n-k} \underbrace{\mathbf{tt} \ldots \mathbf{t}}_{k} \rangle$. There are $n + 1$ points on each edge, but the total is $2n - 1$ because the top of the triangle has just one point, namely $\langle \mathbf{ttt} \rangle$.

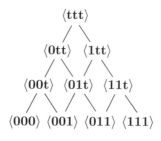

Figure 6.14: Blossoming for Subdivision.

6.9 Degree Elevation

Degree elevation of the Bézier curve is a process that starts with a Bézier curve $\mathbf{P}_n(t)$ of degree n (i.e., defined by $n+1$ control points) and adds a control point, thereby ending up with a curve $\mathbf{P}_{n+1}(t)$.

The advantage of degree elevation is that the new curve is based on more control points and is therefore easier to edit by maneuvering the points. Its shape can be better fine-tuned than that of the original curve.

Just adding a control point is not very useful because the new point will change the shape of the curve globally. Degree elevation is useful only if it is done without modifying the shape of the curve. The principle of degree elevation is therefore to compute a new set of $n+2$ control points $\mathbf{Q}_i$ from the original set of $n+1$ points $\mathbf{P}_i$, such that the Bézier curve $\mathbf{P}_{n+1}(t)$ defined by the new points will have the same shape as the original curve $\mathbf{P}_n(t)$.

We start with the innocuous identity that's true for any Bézier curve $\mathbf{P}(t)$

$$\mathbf{P}(t) = \big(t + (1-t)\big)\mathbf{P}(t) = t\mathbf{P}(t) + (1-t)\mathbf{P}(t).$$

The two Bézier curves on the right-hand side are polynomials of degree n, but because each is multiplied by t, the polynomial on the left-hand side is of degree $n+1$. Thus, we can represent a degree-$(n+1)$ curve as the weighted sum of two degree-n curves and write the identity in the form $\mathbf{P}_{n+1}(t) = (1-t)\mathbf{P}_n(t) + t\mathbf{P}_n(t)$. We use the notation

$$\mathbf{P}_n(t) = \sum_{i=0}^{n} \binom{n}{i} t^i (1-t)^{n-i} \mathbf{P}_i \stackrel{\text{def}}{=} \langle\langle \mathbf{P}_0, \mathbf{P}_1, \ldots, \mathbf{P}_n \rangle\rangle.$$

(Recall that the angle bracket notation indicates blossoms. The double-angle bracket notation used here implies that each point should be multiplied by the corresponding Bernstein polynomial and the products summed.)

The first step is to express $t\mathbf{P}_n(t)$ in the new notation

$$t\mathbf{P}_n(t) = \sum_{i=0}^{n} \binom{n}{i} t^{i+1}(1-t)^{n-i}\mathbf{P}_i = \sum_{k=1}^{m} \binom{m-1}{k-1} t^k (1-t)^{m-k}\mathbf{P}_{k-1}$$

$$= \sum_{k=0}^{m} \binom{m}{k} t^k (1-t)^{m-k} \frac{k}{m}\mathbf{P}_{k-1} = \left\langle\left\langle 0, \frac{\mathbf{P}_0}{n+1}, \frac{2\mathbf{P}_1}{n+1}, \cdots, \frac{n\mathbf{P}_{n-1}}{n+1}, \mathbf{P}_n \right\rangle\right\rangle.$$

Here, we first use the substitutions $k = i+1$ and $m = n+1$, and then the identity

$$\binom{m-1}{k-1} = \frac{k}{m}\binom{m}{k}.$$

The next step is to similarly express $(1-t)\mathbf{P}_n(t)$ in the new notation:

$$(1-t)\mathbf{P}_n(t) = \left\langle\left\langle \mathbf{P}_0, \frac{n\mathbf{P}_1}{n+1}, \frac{(n-1)\mathbf{P}_2}{n+1}, \cdots, \frac{\mathbf{P}_n}{n+1}, 0 \right\rangle\right\rangle.$$

Adding the two expressions produces

$$\mathbf{P}_{n+1}(t) = (1-t)\mathbf{P}_n(t) + t\mathbf{P}_n(t)$$

$$= \left\langle\!\!\left\langle 0, \frac{\mathbf{P}_0}{n+1}, \frac{2\mathbf{P}_1}{n+1}, \cdots, \frac{n\mathbf{P}_{n-1}}{n+1}, \mathbf{P}_n \right\rangle\!\!\right\rangle$$

$$+ \left\langle\!\!\left\langle \mathbf{P}_0, \frac{n\mathbf{P}_1}{n+1}, \frac{(n-1)\mathbf{P}_2}{n+1}, \cdots, \frac{\mathbf{P}_n}{n+1}, 0 \right\rangle\!\!\right\rangle$$

$$= \left\langle\!\!\left\langle \mathbf{P}_0, \frac{\mathbf{P}_0 + n\mathbf{P}_1}{n+1}, \frac{2\mathbf{P}_1 + (n-1)\mathbf{P}_2}{n+1}, \cdots, \frac{n\mathbf{P}_{n-1} + \mathbf{P}_n}{n+1}, \mathbf{P}_n \right\rangle\!\!\right\rangle, \quad (6.18)$$

which shows the $n+2$ control points that define the new, degree-elevated Bézier curve.

If the new control points are denoted by $\mathbf{Q}_i$, then the expression above can be summarized by the following notation:

$$\mathbf{Q}_0 = \mathbf{P}_0,$$

$$\mathbf{Q}_i = a_i\mathbf{P}_{i-1} + (1-a_i)\mathbf{P}_i, \quad \text{where} \quad a_i = \frac{i}{n+1}, \quad i = 1, 2, \ldots, n, \quad (6.19)$$

$$\mathbf{Q}_{n+1} = \mathbf{P}_n.$$

⋄ **Exercise 6.13:** Given the quadratic Bézier curve defined by the three control points $\mathbf{P}_0$, $\mathbf{P}_1$, and $\mathbf{P}_2$, elevate its degree twice and list the five new control points.

It is possible to elevate the degree of a curve many times. Each time the degree is elevated, the new set of control points grows by one point and also approaches the curve. At the limit, the set consists of infinitely many points that are located on the curve.

⋄ **Exercise 6.14:** Given the four control points $\mathbf{P}_0 = (0,0)$, $\mathbf{P}_1 = (1,2)$, $\mathbf{P}_2 = (3,2)$, and $\mathbf{P}_3 = (2,0)$, elevate the degree of the Bézier curve defined by them.

The degree elevation algorithm summarized by Equation (6.19) can also be derived as an application of blossoms. We define a three-parameter function $f_?(u_1, u_2, u_3)$ as a sum of blossoms of two parameters

$$f_?(u_1, u_2, u_3) = \frac{1}{3}\big[f_2(u_1, u_2) + f_2(u_1, u_3) + f_2(u_2, u_3) \big]$$

$$= \frac{1}{3}\big[[a_2 u_1 u_2 + \frac{a_1}{2}(u_1 + u_2) + a_0] + [a_2 u_1 u_3 + \frac{a_1}{2}(u_1 + u_3) + a_0]$$

$$+ [a_2 u_2 u_3 + \frac{a_1}{2}(u_2 + u_3) + a_0]\big]$$

$$= \frac{a_2}{3}(u_1 u_2 + u_1 u_3 + u_2 u_3) + a_1(u_1 + u_2 + u_3) + a_0. \quad (6.20)$$

We notice that $f_?(u_1, u_2, u_3)$ satisfies the following three conditions

1. It is linear in each of its three parameters.
2. It is symmetric with respect to the order of the parameters.
3. Its diagonal, $f_?(u, u, u)$, yields the polynomial $P_2(t) = a_2 t^2 + a_1 t + a_0$.

We therefore conclude that $f_?(u_1, u_2, u_3)$ is the $(n+1)$-blossom of $P_2(t)$. It should be denoted by $f_3(u_1, u_2, u_3)$. It can be shown that the extension of Equation (6.20) to any $f_{n+1}(u_1, u_2, \ldots, u_{n+1})$ is

$$f_{n+1}(u_1, \ldots, u_{n+1}) = \frac{1}{n+1} \sum_{i=1}^{n+1} f_n(u_1, \ldots, \underline{u_i}, \ldots, u_{n+1}). \tag{6.21}$$

(where the underline indicates a missing parameter).

Section 6.7 shows that control point $\mathbf{P}_k$ of a Bézier curve $\mathbf{P}_n(t)$ is given by the blossom $f(\underbrace{0\ldots0}_{n-k}\underbrace{1\ldots1}_{k})$. Equation (6.21) implies that the same control point $\mathbf{Q}_k$ of a Bézier curve $\mathbf{P}_{n+1}(t)$ is given as the sum

$$\mathbf{Q}_k = \frac{n+1-k}{n+1}\mathbf{P}_k + \frac{k}{n+1}\mathbf{P}_{k-1},$$

which is identical to Equation (6.19).

6.10 Reparametrizing the Curve

The parameter t varies normally in the range $[0, 1]$. It is, however, easy to reparametrize the Bézier curve such that its parameter varies in an arbitrary range $[a, b]$, where a and b are real and $a \le b$. The new curve is denoted by $\mathbf{P}_{ab}(t)$ and is simply the original curve with a different parameter:

$$\mathbf{P}_{ab}(t) = \mathbf{P}\left(\frac{t-a}{b-a}\right).$$

The two functions $\mathbf{P}_{ab}(t)$ and $\mathbf{P}(t)$ produce the same curve when t varies from a to b in the former and from 0 to 1 in the latter. Notice that the new curve has tangent vector

$$\mathbf{P}_{ab}^t(t) = \frac{1}{b-a}\mathbf{P}^t\left(\frac{t-a}{b-a}\right).$$

Reparametrization can also be used to answer the question: Given a Bézier curve $\mathbf{P}(t)$ where $0 \le t \le 1$, how can we calculate a curve $\mathbf{Q}(t)$ that's defined on an arbitrary part of $\mathbf{P}(t)$? More specifically, if $\mathbf{P}(t)$ is defined by control points $\mathbf{P}_i$ and if we select an interval $[a, b]$, how can we calculate control points $\mathbf{Q}_i$ such that the curve $\mathbf{Q}(t)$ based on them will go from $\mathbf{P}(a)$ to $\mathbf{P}(b)$ [i.e., $\mathbf{Q}(0) = \mathbf{P}(a)$ and $\mathbf{Q}(1) = \mathbf{P}(b)$] and will be identical in shape to $\mathbf{P}(t)$ in that interval? As an example, if $[a, b] = [0, 0.5]$, then $\mathbf{Q}(t)$ will be identical to the first half of $\mathbf{P}(t)$. The point is that the interval $[a, b]$ does not have to be inside $[0, 1]$. We may select, for example, $[a, b] = [0.9, 1.5]$ and end up with a curve $\mathbf{Q}(t)$ that will go from $\mathbf{P}(0.9)$ to $\mathbf{P}(1.5)$ as t varies from 0 to 1. Even though the Bézier curve was originally designed with $0 \le t \le 1$ in mind, it can still be calculated for t values outside this range. If we like its shape in the range $[0.2, 1.1]$, we may want to calculate new control points $\mathbf{Q}_i$ and obtain a new curve $\mathbf{Q}(t)$ that has this shape when its parameter varies in the standard range $[0, 1]$.

Our approach is to define the new curve $\mathbf{Q}(t)$ as $\mathbf{P}([b - a]t + a)$ and express the control points $\mathbf{Q}_i$ of $\mathbf{Q}(t)$ in terms of the control points $\mathbf{P}_i$ and a and b. We illustrate this technique with the cubic Bézier curve. This curve is given by Equation (6.8) and we can therefore write

$$\mathbf{Q}(t) = \mathbf{P}([b - a]t + a)$$

$$= \left(([b-a]t + a)^3, ([b-a]t + a)^2, ([b-a]t + a), 1\right) \begin{pmatrix} -1 & 3 & -3 & 1 \\ 3 & -6 & 3 & 0 \\ -3 & 3 & 0 & 0 \\ 1 & 0 & 0 & 0 \end{pmatrix} \begin{pmatrix} \mathbf{P}_0 \\ \mathbf{P}_1 \\ \mathbf{P}_2 \\ \mathbf{P}_3 \end{pmatrix}$$

$$= (t^3, t^2, t, 1) \begin{pmatrix} (b-a)^3 & 0 & 0 & 0 \\ 3a(b-a)^2 & (b-a)^2 & 0 & 0 \\ 3a^2(b-a) & 2a(b-a) & b-a & 0 \\ a^3 & a^2 & a & 1 \end{pmatrix} \begin{pmatrix} -1 & 3 & -3 & 1 \\ 3 & -6 & 3 & 0 \\ -3 & 3 & 0 & 0 \\ 1 & 0 & 0 & 0 \end{pmatrix} \begin{pmatrix} \mathbf{P}_0 \\ \mathbf{P}_1 \\ \mathbf{P}_2 \\ \mathbf{P}_3 \end{pmatrix}$$

$$= \mathbf{T}(t) \cdot \mathbf{A} \cdot \mathbf{M} \cdot \mathbf{P}$$

$$= \mathbf{T}(t) \cdot \mathbf{M} \cdot \mathbf{M}^{-1} \cdot \mathbf{A} \cdot \mathbf{M} \cdot \mathbf{P}$$

$$= \mathbf{T}(t) \cdot \mathbf{M} \cdot (\mathbf{M}^{-1} \cdot \mathbf{A} \cdot \mathbf{M}) \cdot \mathbf{P}$$

$$= \mathbf{T}(t) \cdot \mathbf{M} \cdot \mathbf{B} \cdot \mathbf{P}$$

$$= \mathbf{T}(t) \cdot \mathbf{M} \cdot \mathbf{Q},$$

where

$$\mathbf{B} = \mathbf{M}^{-1} \cdot \mathbf{A} \cdot \mathbf{M}$$

$$= \begin{pmatrix} (1-a)^3 & 3(a-1)^2 a & 3(1-a)a^2 & a^3 \\ (a-1)^2(1-b) & (a-1)(-2a-b+3ab) & a(a+2b-3ab) & a^2b \\ (1-a)(-1+b)^2 & (b-1)(-a-2b+3ab) & b(2a+b-3ab) & ab^2 \\ (1-b)^3 & 3(b-1)^2 b & 3(1-b)b^2 & b^3 \end{pmatrix}. \tag{6.22}$$

The four new control points $\mathbf{Q}_i$, $i = 0, 1, 2, 3$ are therefore obtained by selecting specific values for a and b, calculating matrix $\mathbf{B}$, and multiplying it by the column $\mathbf{P} = (\mathbf{P}_0, \mathbf{P}_1, \mathbf{P}_2, \mathbf{P}_3)^T$.

⋄ **Exercise 6.15:** Show that the new curve $\mathbf{Q}(t)$ is independent of the particular coordinate system used.

Example: We select values $b = 2$ and $a = 1$. The new curve $\mathbf{Q}(t)$ will be identical to the part of $\mathbf{P}(t)$ from $\mathbf{P}(1)$ to $\mathbf{P}(2)$ (normally, of course, we don't calculate this part, but this example assumes that we are interested in it). Matrix $\mathbf{B}$ becomes, in this case

$$\mathbf{B} = \begin{pmatrix} 0 & 0 & 0 & 1 \\ 0 & 0 & -1 & 2 \\ 0 & 1 & -4 & 4 \\ -1 & 6 & -12 & 8 \end{pmatrix}$$

(it is easy to verify that each row sums up to 1) and the new control points are

$$
\begin{pmatrix} \mathbf{Q}_0 \\ \mathbf{Q}_1 \\ \mathbf{Q}_2 \\ \mathbf{Q}_3 \end{pmatrix} = \mathbf{B} \begin{pmatrix} \mathbf{P}_0 \\ \mathbf{P}_1 \\ \mathbf{P}_2 \\ \mathbf{P}_3 \end{pmatrix} = \begin{pmatrix} \mathbf{P}_3 \\ -\mathbf{P}_2 + 2\mathbf{P}_3 \\ \mathbf{P}_1 - 4\mathbf{P}_2 + 4\mathbf{P}_3 \\ -\mathbf{P}_0 + 6\mathbf{P}_1 - 12\mathbf{P}_2 + 8\mathbf{P}_3 \end{pmatrix}.
$$

To understand the geometrical meaning of these points, we define three auxiliary points $\mathbf{R}_i$ as follows:

$$
\begin{aligned}
\mathbf{R}_1 &= \mathbf{P}_1 + (\mathbf{P}_1 - \mathbf{P}_0), \\
\mathbf{R}_2 &= \mathbf{P}_2 + (\mathbf{P}_2 - \mathbf{P}_1), \\
\mathbf{R}_3 &= \mathbf{R}_2 + (\mathbf{R}_2 - \mathbf{R}_1) = \mathbf{P}_0 - 4\mathbf{P}_1 + 4\mathbf{P}_2,
\end{aligned}
$$

and write the $\mathbf{Q}_i$'s in the form

$$
\begin{aligned}
\mathbf{Q}_0 &= \mathbf{P}_3, \\
\mathbf{Q}_1 &= \mathbf{P}_3 + (\mathbf{P}_3 - \mathbf{P}_2), \\
\mathbf{Q}_2 &= \mathbf{Q}_1 + (\mathbf{Q}_1 - \mathbf{R}_2) = \mathbf{P}_1 - 4\mathbf{P}_2 + 4\mathbf{P}_3, \\
\mathbf{Q}_3 &= \mathbf{Q}_2 + (\mathbf{Q}_2 - \mathbf{R}_3) = -\mathbf{P}_0 + 6\mathbf{P}_1 - 12\mathbf{P}_2 + 8\mathbf{P}_3.
\end{aligned}
$$

Figure 6.15 illustrates how the four new points $\mathbf{Q}_i$ are obtained from the four original points $\mathbf{P}_i$.

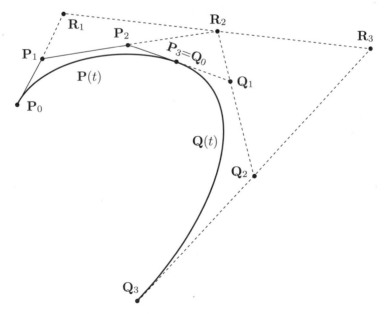

Figure 6.15: Control Points for the Case $[a, b] = [1, 2]$.

Example: We select $b = 2$ and $a = 0$. The new curve $\mathbf{Q}(t)$ will be identical to $\mathbf{P}(t)$ from $\mathbf{P}(0)$ to $\mathbf{P}(2)$. Matrix $\mathbf{B}$ becomes

$$\mathbf{B} = \begin{pmatrix} 1 & 0 & 0 & 0 \\ -1 & 2 & 0 & 0 \\ 1 & -4 & 4 & 0 \\ -1 & 6 & -12 & 8 \end{pmatrix},$$

and the new control points $\mathbf{V}_i$ are

$$\begin{pmatrix} \mathbf{V}_0 \\ \mathbf{V}_1 \\ \mathbf{V}_2 \\ \mathbf{V}_3 \end{pmatrix} = \mathbf{B} \begin{pmatrix} \mathbf{P}_0 \\ \mathbf{P}_1 \\ \mathbf{P}_2 \\ \mathbf{P}_3 \end{pmatrix} = \begin{pmatrix} \mathbf{P}_0 \\ -\mathbf{P}_0 + 2\mathbf{P}_1 \\ \mathbf{P}_0 - 4\mathbf{P}_1 + 4\mathbf{P}_2 \\ -\mathbf{P}_0 + 6\mathbf{P}_1 - 12\mathbf{P}_2 + 8\mathbf{P}_3 \end{pmatrix},$$

and it is easy to see that they satisfy $\mathbf{V}_0 = \mathbf{P}_0$, $\mathbf{V}_1 = \mathbf{R}_1$, $\mathbf{V}_2 = \mathbf{R}_3$, and $\mathbf{V}_3 = \mathbf{Q}_3$.

⋄ **Exercise 6.16:** (1) Calculate matrix $\mathbf{B}$ for $a = 1$ and $b = a + x$ (where x is positive); (2) calculate the four new control points $\mathbf{Q}_i$ as functions of the $\mathbf{P}_i$'s and of b; and (3) recalculate them for $x = 0.75$.

⋄ **Exercise 6.17:** Calculate matrix $\mathbf{B}$ and the four new control points $\mathbf{Q}_i$ for $a = 0$ and $b = 0.5$ (the first half of the curve).

6.11 Cubic Bézier Segments with Tension

Adding a tension parameter to a cubic Bézier segment is done by manipulating tangent vectors similar to how tension is added to the Cardinal spline (Section 5.4). We use Hermite interpolation [Equation (4.7)] to calculate a PC segment that starts at point $\mathbf{P}_0$ and ends at point $\mathbf{P}_3$ and whose extreme tangent vectors are $s(\mathbf{P}_1 - \mathbf{P}_0)$ and $s(\mathbf{P}_3 - \mathbf{P}_2)$ [see Equation (6.23).]

⋄ **Exercise 6.18:** Any set of four given control points $\mathbf{P}_0$, $\mathbf{P}_1$, $\mathbf{P}_2$, and $\mathbf{P}_3$ determines a unique (cubic) Bézier curve. Show that there is a Hermite curve that has an identical shape and is determined by the 4-tuple

$$(\mathbf{P}_0, \mathbf{P}_3, 3(\mathbf{P}_1 - \mathbf{P}_0), 3(\mathbf{P}_3 - \mathbf{P}_2)). \tag{6.23}$$

Substituting these values in Equation (4.7), we manipulate it so that it ends up looking like a cubic Bézier segment, Equation (6.8)

$$\mathbf{P}(t) = (t^3, t^2, t, 1) \begin{pmatrix} 2 & -2 & 1 & 1 \\ -3 & 3 & -2 & -1 \\ 0 & 0 & 1 & 0 \\ 1 & 0 & 0 & 0 \end{pmatrix} \begin{pmatrix} \mathbf{P}_0 \\ \mathbf{P}_3 \\ s(\mathbf{P}_1 - \mathbf{P}_0) \\ s(\mathbf{P}_3 - \mathbf{P}_2) \end{pmatrix}$$

$$= (t^3, t^2, t, 1) \begin{pmatrix} 2 - s & s & -s & s - 2 \\ 2s - 3 & -2s & s & 3 - s \\ -s & s & 0 & 0 \\ 1 & 0 & 0 & 0 \end{pmatrix} \begin{pmatrix} \mathbf{P}_0 \\ \mathbf{P}_1 \\ \mathbf{P}_2 \\ \mathbf{P}_3 \end{pmatrix}. \tag{6.24}$$

A quick check verifies that Equation (6.24) reduces to the cubic Bézier segment, Equation (6.8), for $s = 3$. This value is therefore considered the "neutral" or "standard" value of the tension parameter s. Since s controls the length of the tangent vectors, small values of s should produce the effects of higher tension and, in the extreme, the value $s = 0$ should result in indefinite tangent vectors and in the curve segment becoming a straight line. To show this, we rewrite Equation (6.24) for $s = 0$:

$$\mathbf{P}(t) = (t^3, t^2, t, 1) \begin{pmatrix} 2 & 0 & 0 & -2 \\ -3 & 0 & 0 & 3 \\ 0 & 0 & 0 & 0 \\ 1 & 0 & 0 & 0 \end{pmatrix} \begin{pmatrix} \mathbf{P}_0 \\ \mathbf{P}_1 \\ \mathbf{P}_2 \\ \mathbf{P}_3 \end{pmatrix}$$

$$= (2t^3 - 3t^2 + 1)\mathbf{P}_0 + (-2t^3 + 3t^2)\mathbf{P}_3.$$

Substituting $T = 3t^2 - 2t^3$ for t changes the expression above to the form $\mathbf{P}(T) = (\mathbf{P}_3 - \mathbf{P}_0)T + \mathbf{P}_0$, i.e., a straight line from $\mathbf{P}(0) = \mathbf{P}_0$ to $\mathbf{P}(1) = \mathbf{P}_3$.

The tangent vector of Equation (6.24) is

$$\mathbf{P}^t(t) = (3t^2, 2t, 1, 0) \begin{pmatrix} 2-s & s & -s & s-2 \\ 2s-3 & -2s & s & 3-s \\ -s & s & 0 & 0 \\ 1 & 0 & 0 & 0 \end{pmatrix} \begin{pmatrix} \mathbf{P}_0 \\ \mathbf{P}_1 \\ \mathbf{P}_2 \\ \mathbf{P}_3 \end{pmatrix} \qquad (6.25)$$

$$= \left(3t^2(2-s) + 2t(2s-3) - s\right)\mathbf{P}_0 + \left(3st^2 - 4st + s\right)\mathbf{P}_1$$
$$+ \left(-3st^2 + 2st\right)\mathbf{P}_2 + \left(3t^2(s-2) + 2t(3-s)\right)\mathbf{P}_3.$$

The extreme tangents are $\mathbf{P}^t(0) = s(\mathbf{P}_1 - \mathbf{P}_0)$ and $\mathbf{P}^t(1) = s(\mathbf{P}_3 - \mathbf{P}_2)$. Substituting $s = 0$ in Equation (6.25) yields the tangent vector for the case of infinite tension (compare with Exercise 5.8)

$$\mathbf{P}^t(t) = 6(t^2 - t)\mathbf{P}_0 - 6(t^2 - t)\mathbf{P}_3 = 6(t - t^2)(\mathbf{P}_3 - \mathbf{P}_0). \qquad (6.26)$$

⋄ **Exercise 6.19:** Since the spline segment is a straight line in this case, its tangent vector should always point in the same direction. Use Equation (6.26) to show that this is so.

See also Section 7.4 for a discussion of cubic B-spline with tension.

> We interrupt this program to increase dramatic tension.
>
> —Joe Leahy (as the Announcer) in *Freakazoid!* (1995).

6.12 An Interpolating Bézier Curve: I

Any set of four control points $\mathbf{P}_1$, $\mathbf{P}_2$, $\mathbf{P}_3$, and $\mathbf{P}_4$ determines a unique Catmull–Rom segment that's a cubic polynomial going from point $\mathbf{P}_2$ to point $\mathbf{P}_3$. It turns out that such a segment can also be written as a four-point Bézier curve from $\mathbf{P}_2$ to $\mathbf{P}_3$. All that we have to do is find two points, $\mathbf{X}$ and $\mathbf{Y}$, located between $\mathbf{P}_2$ and $\mathbf{P}_3$, such that the Bézier curve based on $\mathbf{P}_2$, $\mathbf{X}$, $\mathbf{Y}$, and $\mathbf{P}_3$ will be identical to the Catmull–Rom segment. This turns out to be an easy task. We start with the expressions for a Catmull–Rom segment defined by $\mathbf{P}_1$, $\mathbf{P}_2$, $\mathbf{P}_3$, and $\mathbf{P}_4$, and for a four-point Bézier curve defined by $\mathbf{P}_2$, $\mathbf{X}$, $\mathbf{Y}$, and $\mathbf{P}_3$ [Equations (5.33) and (6.8)]:

$$
(t^3, t^2, t, 1)
\begin{pmatrix}
-0.5 & 1.5 & -1.5 & 0.5 \\
1 & -2.5 & 2 & -0.5 \\
-0.5 & 0 & 0.5 & 0 \\
0 & 1 & 0 & 0
\end{pmatrix}
\begin{pmatrix}
\mathbf{P}_1 \\
\mathbf{P}_2 \\
\mathbf{P}_3 \\
\mathbf{P}_4
\end{pmatrix},
$$

$$
(t^3, t^2, t, 1)
\begin{pmatrix}
-1 & 3 & -3 & 1 \\
3 & -6 & 3 & 0 \\
-3 & 3 & 0 & 0 \\
1 & 0 & 0 & 0
\end{pmatrix}
\begin{pmatrix}
\mathbf{P}_2 \\
\mathbf{X} \\
\mathbf{Y} \\
\mathbf{P}_3
\end{pmatrix}.
$$

These have to be equal for each power of t, which yields the four equations

$$
\begin{aligned}
-0.5\mathbf{P}_1 + 1.5\mathbf{P}_2 - 1.5\mathbf{P}_3 + 0.5\mathbf{P}_4 &= -\mathbf{P}_2 + 3\mathbf{X} - 3\mathbf{Y} + \mathbf{P}_3, \\
\mathbf{P}_1 - 2.5\mathbf{P}_2 + 2.0\mathbf{P}_3 - 0.5\mathbf{P}_4 &= 3\mathbf{P}_2 - 6\mathbf{X} + 3\mathbf{Y}, \\
-0.5\mathbf{P}_1 \qquad\quad +0.5\mathbf{P}_3 \qquad &= -3\mathbf{P}_2 + 3\mathbf{X}, \\
\mathbf{P}_2 \qquad\qquad &= \mathbf{P}_2.
\end{aligned}
$$

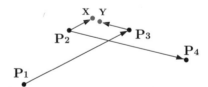

Figure 6.16: Calculating Points $\mathbf{X}$ and $\mathbf{Y}$.

These are easily solved to produce

$$
\mathbf{X} = \mathbf{P}_2 + \frac{1}{6}(\mathbf{P}_3 - \mathbf{P}_1) \quad \text{and} \quad \mathbf{Y} = \mathbf{P}_3 - \frac{1}{6}(\mathbf{P}_4 - \mathbf{P}_2). \tag{6.27}
$$

The difference $(\mathbf{P}_3 - \mathbf{P}_1)$ is the vector from $\mathbf{P}_1$ to $\mathbf{P}_3$. Thus, point $\mathbf{X}$ is obtained by adding $1/6$ of this vector to point $\mathbf{P}_2$ (Figure 6.16). Similarly, $\mathbf{Y}$ is obtained by subtracting $1/6$ of the difference $(\mathbf{P}_4 - \mathbf{P}_2)$ from point $\mathbf{P}_3$.

This simple result suggests a novel approach to the problem of interactive curve design, an approach that combines the useful features of both cubic splines and Bézier curves. A cubic spline passes through the (data) points but is not highly interactive. It can be edited only by modifying the two extreme tangent vectors. A Bézier curve does not pass through the (control) points, but it is easy to manipulate and edit by moving the points. The new approach constructs an *interpolating* Bézier curve in the following steps:

1. The user is asked to input n points, through which the final curve will pass.

2. The program divides the points into overlapping groups of four points each and applies Equation (6.27) to compute two auxiliary points $\mathbf{X}$ and $\mathbf{Y}$ for each group.

3. A Bézier segment is then drawn from the second to the third point of each group, using points $\mathbf{X}$ and $\mathbf{Y}$ as its other two control points. Note that points $\mathbf{Y}$ and $\mathbf{P}_3$ of a group are on a straight line with point $\mathbf{X}$ of the next group. This guarantees that the individual segments will connect smoothly.

4. It is also possible to draw a Bézier segment from $\mathbf{P}_1$ to $\mathbf{P}_2$ (and, similarly, from $\mathbf{P}_{n-1}$ to $\mathbf{P}_n$). This segment uses the two auxiliary control points $\mathbf{X} = \mathbf{P}_1 + \frac{1}{6}(\mathbf{P}_2 - \mathbf{P}_1)$ and $\mathbf{Y} = \mathbf{P}_2 - \frac{1}{6}(\mathbf{P}_3 - \mathbf{P}_1)$.

Users find it natural to specify such a curve, because they don't have to worry about the positions of the control points. The curve consists of $n - 1$ segments and the two auxiliary control points of each segment are calculated automatically.

Such a curve is usually pleasing to the eye and rarely needs to be edited. However, if it is not satisfactory, it can be modified by moving the auxiliary control points. There are $2(n - 1)$ of them, which allows for flexible control. A good program should display the auxiliary points and should make it easy for the user to grab and move any of them.

The well-known drawing program *Adobe Illustrator* [Adobe 04] uses a similar approach. The user specifies points with the mouse. At each point $\mathbf{P}_i$, the user presses the mouse button to fix $\mathbf{P}_i$, then drags the mouse before releasing the button, which defines two symmetrical points, $\mathbf{X}$ (following $\mathbf{P}_i$) and $\mathbf{Y}$ (preceding it). Releasing the button is a signal to the program to draw the segment from $\mathbf{P}_{i-1}$ to $\mathbf{P}_i$ (Figure 6.17).

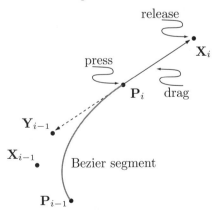

Figure 6.17: Construction of $\mathbf{X}_i$ and $\mathbf{Y}_i$ by Click and Drag.

Example: We apply this method to the six points $\mathbf{P}_0 = (1/2, 0)$, $\mathbf{P}_1 = (1/2, 1/2)$, $\mathbf{P}_2 = (0, 1)$, $\mathbf{P}_3 = (1, 3/2)$, $\mathbf{P}_4 = (3/2, 1)$, and $\mathbf{P}_5 = (1, 1/2)$. The six points yield three curve segments and the main step is to calculate the two intermediate points for each of the three segments. This is trivial and it results in:

$$\mathbf{X}_1 = \mathbf{P}_1 + (\mathbf{P}_2 - \mathbf{P}_0)/6 = (5/12, 2/3), \quad \mathbf{Y}1 = \mathbf{P}_2 - (\mathbf{P}_3 - \mathbf{P}_1)/6 = (-1/12, 5/6),$$
$$\mathbf{X}_2 = \mathbf{P}_2 + (\mathbf{P}_3 - \mathbf{P}_1)/6 = (1/12, 7/6), \quad \mathbf{Y}2 = \mathbf{P}_3 - (\mathbf{P}_4 - \mathbf{P}_2)/6 = (3/4, 3/2),$$
$$\mathbf{X}_3 = \mathbf{P}_3 + (\mathbf{P}_4 - \mathbf{P}_2)/6 = (5/4, 3/2), \quad \mathbf{Y}3 = \mathbf{P}_4 - (\mathbf{P}_5 - \mathbf{P}_3)/6 = (3/2, 7/6).$$

Once the points are available, the three segments can easily be calculated. Each is a cubic Bézier segment based on a group of four points. The groups are

$$[\mathbf{P}_1, \mathbf{X}_1, \mathbf{Y}_1, \mathbf{P}_2], \quad [\mathbf{P}_2, \mathbf{X}_2, \mathbf{Y}_2, \mathbf{P}_3], \quad [\mathbf{P}_3, \mathbf{X}_3, \mathbf{Y}_3, \mathbf{P}_4],$$

and the three curve segments are

$$\mathbf{P}_1(t) = (1 - t)^3 \mathbf{P}_1 + 3t(1 - t)^2 \mathbf{X}_1 + 3t^2(1 - t)\mathbf{Y}_1 + t^3 \mathbf{P}_2$$
$$= \left((2 - t - 5t^2 + 4t^3)/4, (1 + t)/2\right),$$
$$\mathbf{P}_2(t) = (1 - t)^3 \mathbf{P}_2 + 3t(1 - t)^2 \mathbf{X}_2 + 3t^2(1 - t)\mathbf{Y}_2 + t^3 \mathbf{P}_3$$
$$= \left((t + 7t^2 - 4t^3)/4, (2 + t + t^2 - t^3)/2\right),$$
$$\mathbf{P}_3(t) = (1 - t)^3 \mathbf{P}_3 + 3t(1 - t)^2 \mathbf{X}_3 + 3t^2(1 - t)\mathbf{Y}_3 + t^3 \mathbf{P}_4$$
$$= \left((4 + 3t - t^3)/4, (3 - 2t^2 + t^3)/2\right).$$

The 12 points and the three segments are shown in Figure 6.18 (where the segments have been separated intentionally), as well as the code for the entire example.

6.13 An Interpolating Bézier Curve: II

The approach outlined in this section calculates an interpolating Bézier curve by solving equations. Given a set of $n+1$ data points $\mathbf{Q}_0, \mathbf{Q}_1, \ldots, \mathbf{Q}_n$, we select $n+1$ values t_i such that $\mathbf{P}(t_i) = \mathbf{Q}_i$. We require that whenever t reaches one of the values t_i, the curve will pass through a point $\mathbf{Q}_i$. The values t_i don't have to be equally spaced, which provides control over the "speed" of the curve. All that's needed to calculate the curve is to compute the right set of $n+1$ control points $\mathbf{P}_i$. This is done by setting and solving the set of $n+1$ linear equations $\mathbf{P}(t_0) = \mathbf{Q}_0, \mathbf{P}(t_1) = \mathbf{Q}_1, \ldots, \mathbf{P}(t_n) = \mathbf{Q}_n$ that's expressed in matrix notation as follows:

$$\begin{pmatrix} B_{n,0}(t_0) & B_{n,1}(t_0) & \cdots & B_{n,n}(t_0) \\ B_{n,0}(t_1) & B_{n,1}(t_1) & \cdots & B_{n,n}(t_1) \\ \vdots & \vdots & \ddots & \vdots \\ B_{n,0}(t_n) & B_{n,1}(t_n) & \cdots & B_{n,n}(t_n) \end{pmatrix} \begin{pmatrix} \mathbf{P}_0 \\ \mathbf{P}_1 \\ \vdots \\ \mathbf{P}_n \end{pmatrix} = \begin{pmatrix} \mathbf{Q}_0 \\ \mathbf{Q}_1 \\ \vdots \\ \mathbf{Q}_n \end{pmatrix}. \quad (6.28)$$

This set of equations can be expressed as $\mathbf{MP} = \mathbf{Q}$ and it is easily solved by inverting $\mathbf{M}$ numerically. The solution is $\mathbf{P} = \mathbf{M}^{-1}\mathbf{Q}$. If we select $t_0 = 0$, the top row of Equation (6.28) yields $\mathbf{P}_0 = \mathbf{Q}_0$. Similarly, if we select $t_n = 1$, the bottom row of

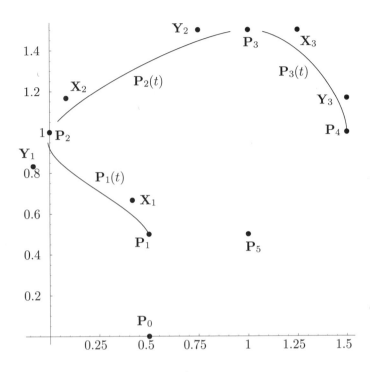

```
(* Interpolating Bezier Curve: I *)
Clear[p0,p1,p2,p3,p4,p5,x1,x2,x3,y1,y2,y3,c1,c2,c3,g1,g2,g3,g4];
p0={1/2,0}; p1={1/2,1/2}; p2={0,1};
 p3={1,3/2}; p4={3/2,1}; p5={1,1/2};
x1=p1+(p2-p0)/6;
x2=p2+(p3-p1)/6;
x3=p3+(p4-p2)/6;
y1=p2-(p3-p1)/6;
y2=p3-(p4-p2)/6;
y3=p4-(p5-p3)/6;
c1[t_]:=Simplify[(1-t)^3 p1+3t(1-t)^2 x1+3t^2(1-t) y1+t^3 p2]
c2[t_]:=Simplify[(1-t)^3 p2+3t(1-t)^2 x2+3t^2(1-t) y2+t^3 p3]
c3[t_]:=Simplify[(1-t)^3 p3+3t(1-t)^2 x3+3t^2(1-t) y3+t^3 p4]
g1=ListPlot[{p0,p1,p2,p3,p4,p5,x1,x2,x3,y1,y2,y3},
 Prolog->AbsolutePointSize[4], PlotRange->All,
 AspectRatio->Automatic, DisplayFunction->Identity]
g2=ParametricPlot[c1[t], {t,0,.9}, DisplayFunction->Identity]
g3=ParametricPlot[c2[t], {t,0.1,.9}, DisplayFunction->Identity]
g4=ParametricPlot[c3[t], {t,0.1,1}, DisplayFunction->Identity]
Show[g1,g2,g3,g4, DisplayFunction->$DisplayFunction]
```

Figure 6.18: An Interpolating Bézier Curve.

Equation (6.28) yields $\mathbf{P}_n = \mathbf{Q}_n$. This decreases the number of equations from $n+1$ to $n-1$.

The disadvantage of this approach is that any changes in the t_i's require a recalculation of $\mathbf{M}$ and, consequently, of $\mathbf{M}^{-1}$.

If controlling the speed of the curve is not important, we can select the $n+1$ equally-spaced values $t_i = i/n$. Equation (6.28) can now be written

$$
\begin{pmatrix}
B_{n,0}(0/n) & B_{n,1}(0/n) & \cdots & B_{n,n}(0/n) \\
B_{n,0}(1/n) & B_{n,1}(1/n) & \cdots & B_{n,n}(1/n) \\
\vdots & \vdots & \ddots & \vdots \\
B_{n,0}(n/n) & B_{n,1}(n/n) & \cdots & B_{n,n}(n/n)
\end{pmatrix}
\begin{pmatrix}
\mathbf{P}_0 \\ \mathbf{P}_1 \\ \vdots \\ \mathbf{P}_n
\end{pmatrix}
=
\begin{pmatrix}
\mathbf{Q}_0 \\ \mathbf{Q}_1 \\ \vdots \\ \mathbf{Q}_n
\end{pmatrix}.
\tag{6.29}
$$

Now, if the data points $\mathbf{Q}_i$ are moved, matrix $\mathbf{M}$ (or, rather, $\mathbf{M}^{-1}$) doesn't have to be recalculated. If we number the rows and columns of $\mathbf{M}$ 0 through n, then a general element of $\mathbf{M}$ is given by

$$
M_{ij} = B_{n,j}(i/n) = \binom{n}{j}(i/n)^j(1-i/n)^{n-j} = \frac{n!(n-i)^{n-j}i^j}{j!(n-j)!n^n}.
$$

Such elements can be calculated, if desired, as exact rational integers, instead of (approximate) floating-point numbers.

Example: We use Equation (6.29) to compute the interpolating Bézier curve that passes through the four points $\mathbf{Q}_0 = (0,0)$, $\mathbf{Q}_1 = (1,1)$, $\mathbf{Q}_2 = (2,1)$, and $\mathbf{Q}_3 = (3,0)$. Since the curve has to pass through the first and last point, we get $\mathbf{P}_0 = \mathbf{Q}_0 = (0,0)$ and $\mathbf{P}_3 = \mathbf{Q}_3 = (3,0)$. Since the four given points are equally spaced, it makes sense to assume that $\mathbf{P}(1/3) = \mathbf{Q}_1$ and $\mathbf{P}(2/3) = \mathbf{Q}_2$. We therefore end up with the two equations

$$
3(1/3)(1-1/3)^2\mathbf{P}_1 + 3(1/3)^2(1-1/3)\mathbf{P}_2 + (1/3)^3(3,0) = (1,1),
$$
$$
3(2/3)(1-2/3)^2\mathbf{P}_1 + 3(2/3)^2(1-2/3)\mathbf{P}_2 + (2/3)^3(3,0) = (2,1),
$$

that are solved to yield $\mathbf{P}_1 = (1,3/2)$ and $\mathbf{P}_2 = (2,3/2)$. The curve is

$$
\mathbf{P}(t) = (1-t)^3(0,0) + 3t(1-t)^2(1,3/2) + 3t^2(1-t)(2,3/2) + t^3(3,0).
$$

⋄ **Exercise 6.20:** Plot the curve and the eight points.

6.14 Nonparametric Bézier Curves

The explicit representation of a curve (Section 1.3) has the familiar form $y = f(x)$. The Bézier curve is, of course, parametric, but it can be represented in a nonparametric form, similar to explicit curves. Given $n + 1$ real values (not points) P_i, we start with the polynomial $c(t) = \sum P_i B_{ni}(t)$ and employ the identity

$$\sum_{i=0}^{n} (i/n) B_{ni}(t) = t \tag{6.30}$$

to create the curve

$$\mathbf{P}(t) = \big(t, c(t)\big) = \sum_{i=0}^{n} (i/n, P_i) B_{ni}(t).$$

(This identity is satisfied by the Bernstein polynomials and can be proved by induction.) It is clear that this version of the curve is defined by the control points $(i/n, P_i)$ which are equally-spaced on the x axis.

This version of the Bézier curve exists only for two-dimensional curves. In the general case, where t varies in the interval $[a, b]$, the control points are $\big((a + i(b - a))/n, P_i\big)$.

6.15 Rational Bézier Curves

The rational Bézier curve is an extension of the original Bézier curve [Equation (6.5)] to

$$\mathbf{P}(t) = \frac{\sum_{i=0}^{n} w_i \mathbf{P}_i B_{n,i}(t)}{\sum_{j=0}^{n} w_j B_{n,j}(t)} = \sum_{i=0}^{n} \mathbf{P}_i \left[\frac{w_i B_{n,i}(t)}{\sum_{j=0}^{n} w_j B_{n,j}(t)} \right] = \sum_{i=0}^{n} \mathbf{P}_i R_{n,i}(t), \qquad 0 \le t \le 1.$$

The new weight functions $R_{n,i}(t)$ are ratios of polynomials (which is the reason for the term *rational*) and they also depend on weights w_i that act as additional parameters that control the shape of the curve. Note that negative weights might lead to a zero denominator, which is why nonnegative weights are normally used. A rational curve seems unnecessarily complicated (and for many applications, it is), but it has the following advantages:

1. It is invariant under projections. Section 6.4 mentions that the Bézier curve is invariant under affine transformations. If we want to rotate, reflect, scale, or shear such a curve, we can apply the affine transformation to the control points, then use the new points to compute the transformed curve. The Bézier curve, however, is not invariant under projections. If we compute a three-dimensional Bézier curve and project every point of the curve by a perspective projection, we end up with a plane curve $\mathbf{P}(t)$. If we then project the three-dimensional control points and compute a plane Bézier curve $\mathbf{Q}(t)$ from the projected, two-dimensional points, the two curves $\mathbf{P}(t)$ and $\mathbf{Q}(t)$ will generally be different. One advantage of the rational Bézier curve is its invariance under projections.

2. The rational Bézier curve provides for accurate control of curve shape, such as precise representation of conic sections (Appendix A).

Section 7.5 shows that the Bézier curve is a special case of the B-spline curve. As a result, many current software systems use the rational B-spline (Section 7.14) when rational curves are required. Such a system can produce the rational Bézier curve as a special case.

Here is a quick example showing how the rational Bézier curve can be useful. Given the three points $\mathbf{P}_0 = (1,0)$, $\mathbf{P}_1 = (1,1)$, and $\mathbf{P}_2 = (0,1)$, The Bézier curve defined by the points is quadratic and is therefore a parabola $\mathbf{P}(t) = (1-t)^2\mathbf{P}_0 + 2t(1-t)\mathbf{P}_1 + t^2\mathbf{P}_2 = (1 - t^2, 2t(1 - t))$, but the rational Bézier curve with weights $w_0 = w_1 = 1$ and $w_2 = 2$ results in the more complex expression

$$\mathbf{P}(t) = \frac{(1-t)^2\mathbf{P}_0 + 2t(1-t)\mathbf{P}_1 + 2t^2\mathbf{P}_2}{(1-t)^2 + 2t(1-t) + 2t^2} = \left(\frac{1 - t^2}{1 + t^2}, \frac{2t}{1 + t^2}\right)$$

which is a circle, as illustrated by Figure 1.6a.

In general, a quadratic rational Bézier curve with weights $w_0 = w_2 = 1$ is a parabola when $w_1 = 1$, an ellipse for $w_1 < 1$, and a hyperbola for $w_1 > 1$. A quarter circle is obtained when $w_1 = \cos(\alpha/2)$ where α is the angle formed by the three control points $\mathbf{P}_0$, $\mathbf{P}_1$, and $\mathbf{P}_2$ (the control points must also be placed as the three corners of an isosceles triangle). Page 261 of [Beach 91] proves this construction for the special case $\alpha = 90°$.

Appendix A shows, among other features, that the canonical ellipse is represented as the rational expression

$$\left(a\frac{1 - t^2}{1 + t^2}, b\frac{2t}{1 + t^2}\right), \qquad -\infty < t < \infty, \qquad (A.7)$$

and the canonical hyperbola is represented as the rational

$$\left(a\frac{1 + t^2}{1 - t^2}, b\frac{2t}{1 - t^2}\right), \qquad -\infty < t < \infty. \qquad (A.8)$$

Accurate control of the shape of the curve is provided by either moving the control points or varying the weights, and Figure 6.19 illustrates the different responses of the curve to these changes. Part (a) of the figure shows four curves where weight w_1 is increased from 1 to 4. The curve is pulled toward $\mathbf{P}_1$ in such a way that individual points on the curve *converge* at $\mathbf{P}_1$. In contrast, part (b) of the figure illustrates how the curve behaves when $\mathbf{P}_1$ itself is moved (while all the weights remain set to 1). The curve is again pulled toward $\mathbf{P}_1$, but in such a way that every point on the curve moves in the same direction as $\mathbf{P}_1$ itself.

⬦ **Exercise 6.21:** Use mathematical software to compute Figure 6.19 or a similar illustration.

Section 6.22 extends the techniques presented here to rectangular Bézier surface patches.

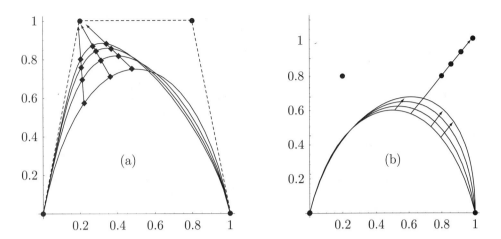

Figure 6.19: (a) Varying Weights and (b) Moving Points in a Rational Bézier Curve.

6.16 Rectangular Bézier Surfaces

The Bézier surface patch, like its relative the Bézier curve, is popular and is commonly used in practice. We discuss the rectangular and the triangular Bézier surface methods, and this section covers the former.

We start with an $(m + 1) \times (n + 1)$ grid of control points arranged in a roughly rectangular grid

$$
\begin{array}{cccc}
\mathbf{P}_{m,0} & \mathbf{P}_{m,1} & \dots & \mathbf{P}_{m,n} \\
\vdots & \vdots & & \vdots \\
\mathbf{P}_{1,0} & \mathbf{P}_{1,1} & \dots & \mathbf{P}_{1,n} \\
\mathbf{P}_{0,0} & \mathbf{P}_{0,1} & \dots & \mathbf{P}_{0,n}
\end{array}
$$

and construct the rectangular Bézier surface patch for the points by applying the technique of Cartesian product (Section 1.9) to the Bézier curve. Equation (1.28) produces

$$
\begin{aligned}
\mathbf{P}(u, w) &= \sum_{i=0}^{m} \sum_{j=0}^{n} B_{m,i}(u) \mathbf{P}_{i,j} B_{n,j}(w) \\
&= (B_{m,0}(u), B_{m,1}(u), \dots, B_{m,m}(u)) \, \mathbf{P} \begin{pmatrix} B_{n,0}(w) \\ B_{n,1}(w) \\ \vdots \\ B_{n,n}(w) \end{pmatrix} \\
&= \mathbf{B}_m(u) \, \mathbf{P} \, \mathbf{B}_n(w), \quad\quad\quad\quad\quad\quad (6.31)
\end{aligned}
$$

where
$$
\mathbf{P} = \begin{pmatrix} \mathbf{P}_{0,0} & \mathbf{P}_{0,1} & \dots & \mathbf{P}_{0,n} \\ \mathbf{P}_{1,0} & \mathbf{P}_{1,1} & \dots & \mathbf{P}_{1,n} \\ \vdots & \vdots & \ddots & \vdots \\ \mathbf{P}_{m,0} & \mathbf{P}_{m,1} & \dots & \mathbf{P}_{m,n} \end{pmatrix}.
$$

The surface can also be expressed, by analogy with Equation (6.9), as

$$\mathbf{P}(u, w) = \mathbf{UNPN}^T\mathbf{W}^T, \qquad (6.32)$$

where $\mathbf{U} = (u^m, u^{m-1}, \ldots, u, 1)$, $\mathbf{W} = (w^n, w^{n-1}, \ldots, w, 1)$, and $\mathbf{N}$ is defined by Equation (6.10).

Notice that both $\mathbf{P}(u_0, w)$ and $\mathbf{P}(u, w_0)$ (for constants u_0 and w_0) are Bézier curves on the surface. A Bézier curve is defined by $n+1$ control points, it passes through the two extreme points, and employs the interior points to determine its shape. Similarly, a rectangular Bézier surface patch is defined by a rectangular grid of $(m+1) \times (n+1)$ control points, it is anchored at the four corner points and employs the other grid points to determine its shape.

Figure 6.20 is an example of a biquadratic Bézier surface patch with the *Mathematica* code that generated it. Notice how the surface is anchored at the four corner points and how the other control points pull the surface toward them.

Example: Given the six three-dimensional points

$$\begin{array}{ccc} \mathbf{P}_{10} & \mathbf{P}_{11} & \mathbf{P}_{12} \\ \mathbf{P}_{00} & \mathbf{P}_{01} & \mathbf{P}_{02} \end{array}$$

the corresponding Bézier surface is generated in the following three steps:

1. Find the orders m and n of the surface. Since the points are numbered starting from 0, the two orders of the surface are $m = 1$ and $n = 2$.

2. Calculate the weight functions $B_{1i}(w)$ and $B_{2j}(u)$. For $m = 1$, we get

$$B_{1i}(w) = \binom{1}{i} w^i (1 - w)^{1-i},$$

which yields the two functions

$$B_{10}(w) = \binom{1}{0} w^0 (1 - w)^{1-0} = 1 - w, \quad B_{11}(w) = \binom{1}{1} w^1 (1 - w)^{1-1} = w.$$

For $n = 2$, we get

$$B_{2j}(u) = \binom{2}{j} u^j (1 - u)^{2-j},$$

which yields the three functions

$$B_{20}(u) = \binom{2}{0} u^0 (1 - u)^{2-0} = (1 - u)^2,$$

$$B_{21}(u) = \binom{2}{1} u^1 (1 - u)^{2-1} = 2u(1 - u),$$

$$B_{22}(u) = \binom{2}{2} u^2 (1 - u)^{2-2} = u^2.$$

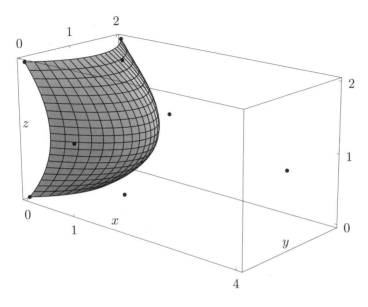

```
(* biquadratic bezier surface patch *)
Clear[pwr,bern,spnts,n,bzSurf,g1,g2];
n=2;
<<:Graphics:ParametricPlot3D.m
spnts={{{0,0,0},{1,0,1},{0,0,2}},
 {{1,1,0},{4,1,1},{1,1,2}}, {{0,2,0},{1,2,1},{0,2,2}}};
(* Handle Indeterminate condition *)
pwr[x_,y_]:=If[x==0 && y==0, 1, x^y];
bern[n_,i_,u_]:=Binomial[n,i]pwr[u,i]pwr[1-u,n-i]
bzSurf[u_,w_]:=Sum[bern[n,i,u] spnts[[i+1,j+1]] bern[n,j,w],
 {i,0,n}, {j,0,n}]
g1=ParametricPlot3D[bzSurf[u,w],{u,0,1}, {w,0,1},
 Ticks->{{0,1,4},{0,1,2},{0,1,2}},
 Compiled->False, DisplayFunction->Identity];
g2=Graphics3D[{AbsolutePointSize[3],
 Table[Point[spnts[[i,j]]],{i,1,n+1},{j,1,n+1}]}];
Show[g1,g2, ViewPoint->{2.783, -3.090, 1.243}, PlotRange->All,
 DefaultFont->{"cmr10", 10}, DisplayFunction->$DisplayFunction];
```

Figure 6.20: A Biquadratic Bézier Surface Patch.

3. Substitute the weight functions in the general expression for the surface [Equation (6.31)]:

$$\mathbf{P}(u,w) = \sum_{i=0}^{1} \sum_{j=0}^{2} B_{1i}(w) \mathbf{P}_{ij} B_{2j}(u)$$

$$= B_{10}(w) \sum_{j=0}^{2} \mathbf{P}_{0j} B_{2j}(u) + B_{11}(w) \sum_{j=0}^{2} \mathbf{P}_{1j} B_{2j}(u)$$

$$= (1-w) \left[\mathbf{P}_{00} B_{20}(u) + \mathbf{P}_{01} B_{21}(u) + \mathbf{P}_{02} B_{22}(u) \right]$$

$$+ w \left[\mathbf{P}_{10} B_{20}(u) + \mathbf{P}_{11} B_{21}(u) + \mathbf{P}_{12} B_{22}(u) \right]$$

$$= (1 - w) \left[\mathbf{P}_{00}(1 - u)^2 + \mathbf{P}_{01} 2u(1 - u) + \mathbf{P}_{02} u^2 \right]$$
$$+ w \left[\mathbf{P}_{10}(1 - u)^2 + \mathbf{P}_{11} 2u(1 - u) + \mathbf{P}_{12} u^2 \right]$$
$$= \mathbf{P}_{00}(1 - w)(1 - u)^2 + \mathbf{P}_{01}(1 - w)2u(1 - u) + \mathbf{P}_{02}(1 - w)u^2$$
$$+ \mathbf{P}_{10}w(1 - u)^2 + \mathbf{P}_{11}w2u(1 - u) + \mathbf{P}_{12}wu^2. \tag{6.33}$$

The final expression is linear in w since the surface is defined by just two points in the w direction. Surface lines in this direction are straight. In the u direction, where the surface is defined by three points, each line is a polynomial of degree 2 in u. This expression can also be written in the form

$$(1 - w) \sum B_{2,i}(u)\mathbf{P}_{0i} + w \sum B_{2,i}(u)\mathbf{P}_{1i} = (1 - w)\mathbf{P}(u, 0) + w\mathbf{P}(u, 1),$$

which is a lofted surface [Equation (2.14)].

A good technique to check the final expression is to calculate it for the four values $(u, w) = (0, 0)$, $(0, 1)$, $(1, 0)$, and $(1, 1)$. This should yield the coordinates of the four original corner points.

The entire surface can now be easily displayed, as a wire frame, by performing two loops. One draws curves in the u direction and the other draws the curves in the w direction. Notice that the expression of the patch is the same regardless of the particular points used. The user may change the points to modify the surface, and the new surface can be displayed (Figure 6.21) by calculating Equation (6.33).

⋄ **Exercise 6.22:** Given the 3×4 array of control points

$$\mathbf{P}_{20} = (0, 2, 0) \quad \mathbf{P}_{21} = (1, 2, 1) \quad \mathbf{P}_{22} = (2, 2, 1) \quad \mathbf{P}_{23} = (3, 2, 0)$$
$$\mathbf{P}_{10} = (0, 1, 0) \quad \mathbf{P}_{11} = (1, 1, 1) \quad \mathbf{P}_{12} = (2, 1, 1) \quad \mathbf{P}_{13} = (3, 1, 0)$$
$$\mathbf{P}_{00} = (0, 0, 0) \quad \mathbf{P}_{01} = (1, 0, 1) \quad \mathbf{P}_{02} = (2, 0, 1) \quad \mathbf{P}_{03} = (3, 0, 0),$$

calculate the order-2×3 Bézier surface patch defined by them.

Notice that the order-2×2 Bézier surface patch defined by only four control points is a bilinear patch. Its form is given by Equation (2.8).

6.16.1 Scaffolding Construction

The scaffolding construction (or de Casteljau algorithm) of Section 6.6 can be directly extended to the rectangular Bézier patch. Figure 6.22 illustrates the principle. Part (a) of the figure shows a rectangular Bézier patch defined by 3×4 control points (the circles). The de Casteljau algorithm for curves is applied to each row of three points to compute two intermediate points (the squares), followed by a final point (the triangle). The final point is located on the Bézier curve defined by the row of three points. The result of applying the de Casteljau algorithm to the four rows is four points (the triangles). The algorithm is now applied to those four points (Figure 6.22b) to compute one point (the heavy circle) that's located both on the curve defined by the four (black triangle) points and on the Bézier surface patch defined by the 3×4 control points. (This is one of the many curve algorithms that can be directly extended to surfaces.)

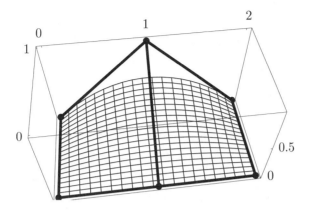

```
(* A Bezier surface example. Given the six two-dimensional... *)
Clear[pnts,b1,b2,g1,g2,vlines,hlines];
pnts={{{0,1,0},{1,1,1},{2,1,0}},{{0,0,0},{1,0,0},{2,0,0}}};
b1[w_]:={1-w,w}; b2[u_]:={(1-u)^2,2u(1-u),u^2};
comb[i_]:=(b1[w].pnts)[[i]] b2[u][[i]];
g1=ParametricPlot3D[comb[1]+comb[2]+comb[3], {u,0,1},{w,0,1}, Compiled->False,
DefaultFont->{"cmr10", 10}, DisplayFunction->Identity,
 AspectRatio->Automatic, Ticks->{{0,1,2},{0,1},{0,.5}}];
g2=Graphics3D[{AbsolutePointSize[5],
 Table[Point[pnts[[i,j]]],{i,1,2},{j,1,3}]}];
vlines=Graphics3D[{AbsoluteThickness[2],
 Table[Line[{pnts[[1,j]],pnts[[2,j]]}], {j,1,3}]}];
hlines=Graphics3D[{AbsoluteThickness[2],
 Table[Line[{pnts[[i,j]],pnts[[i,j+1]]}], {i,1,2}, {j,1,2}]}];
Show[g1,g2,vlines,hlines, ViewPoint->{-0.139, -1.179, 1.475},
DisplayFunction->$DisplayFunction, PlotRange->All, Shading->False,
DefaultFont->{"cmr10", 10}];
```

Figure 6.21: A Lofted Bézier Surface Patch.

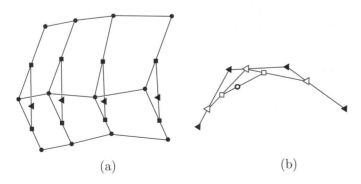

(a)   (b)

Figure 6.22: Scaffolding in a Rectangular Bézier Patch.

Referring to Equation (6.31), we can summarize this process as follows:

1. Construct the $n + 1$ curves

$$\mathbf{P}_j(u) = \sum_{i=0}^{m} B_{mi}(u)\mathbf{P}_{ij}, \quad j = 0, 1, \ldots, n.$$

2. Apply the de Casteljau algorithm to each curve to end up with $n + 1$ points, one on each curve.

3. Apply the same algorithm to the $n + 1$ points to end up with one point.

Alternatively, we can first construct the $m + 1$ curves

$$\mathbf{P}_i(w) = \sum_{j=0}^{n} \mathbf{P}_{ij} B_{nj}(w), \quad i = 0, 1, \ldots, m,$$

then apply the de Casteljau algorithm to each curve to end up with $m + 1$ points, and finally apply the same algorithm to the $m + 1$ points, and end up with one point.

6.17 Subdividing Rectangular Patches

A rectangular Bézier patch is computed from a given rectangular array of $m \times n$ control points. If there are not enough points, the patch may not have the right shape. Just adding points is not a good solution because this changes the shape of the surface, forcing the designer to start reshaping it from the beginning. A better solution is to subdivide the patch into four connected surface patches, each based on $m \times n$ control points. The technique described here is similar to that presented in Section 6.8 for subdividing the Bézier curve. It employs the scaffolding construction of Section 6.6.

Figure 6.22a shows a grid of 4×3 control points. The first step in subdividing the surface patch defined by this grid is for the user to select values for u and w. This determines a point on the surface, a point that will be common to the four new patches. The de Casteljau algorithm is then applied to each of the three columns of control points (the black circles of Figure 6.23a) separately. Each column of four control points $\mathbf{P}_0$, $\mathbf{P}_1$, $\mathbf{P}_2$, and $\mathbf{P}_3$ results in several points, of which the following seven are used for the subdivision (refer to Figure 6.8) $\mathbf{P}_0$, $\mathbf{P}_{01}$, $\mathbf{P}_{012}$, $\mathbf{P}_{0123}$, $\mathbf{P}_{123}$, $\mathbf{P}_{23}$, and $\mathbf{P}_3$. The result of this step is three columns of seven points each (Figure 6.23b where the black circles indicate original control points).

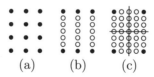

Figure 6.23: Subdividing a Rectangular 3×4 Bézier Patch.

The next step is to apply the de Casteljau algorithm to each of the seven rows of three points, to obtain five points (refer to Figure 6.7). The resulting grid of 7×5 is shown in Figure 6.23c. This grid is divided into four overlapping subgrids of 4×3 control points each, and each subgrid serves to compute a new rectangular Bézier patch.

6.18 Degree Elevation

Degree elevation of the rectangular Bézier surface is similar to elevating the degree of the Bézier curve (Section 6.9). Specifically, Equation (6.19) is extended in the following way. Given a rectangular Bézier patch of degree $m \times n$ (i.e., defined by $(m+1) \times (n+1)$ control points), expressed as a double-polynomial by Equation (6.31)

$$\mathbf{P}_{mn}(u, w) = \sum_{i=0}^{m} \sum_{j=0}^{n} B_{m,i}(u)\mathbf{P}_{i,j}B_{n,j}(w), \qquad (6.31)$$

we first write the patch as a double polynomial of degree $(m+1) \times n$ defined by intermediate control points $\mathbf{R}_{ij}$

$$\sum_{j=0}^{n} \left[\sum_{i=0}^{m+1} B_{m+1,i}(u)\mathbf{R}_{i,j} \right] B_{n,j}(w).$$

Based on the result of Section 6.9 the intermediate points are given by

$$\mathbf{R}_{ij} = \frac{i}{m+1}\mathbf{P}_{i-1,j} + (1 - \frac{i}{m+1})\mathbf{P}_{i,j}. \qquad (6.34)$$

We then repeat this process to increase the degree to $(m+1) \times (n+1)$ and write

$$\mathbf{P}_{m+1,n+1}(u, w) = \sum_{i=0}^{m+1} \sum_{j=0}^{n+1} B_{m+1,i}(u)\mathbf{Q}_{i,j}B_{n+1,j}(w),$$

where the new $(m+2) \times (n+2)$ control points $\mathbf{Q}_{ij}$ can be obtained either from the intermediate points $\mathbf{R}_{ij}$ by an expression similar to Equation (6.34) or directly from the original control points $\mathbf{P}_{ij}$ by a bilinear interpolation

$$\mathbf{Q}_{ij} = \left(\frac{i}{m+1}, 1 - \frac{i}{m+1} \right) \begin{bmatrix} \mathbf{P}_{i-1,j-1} & \mathbf{P}_{i-1,j} \\ \mathbf{P}_{i,j-1} & \mathbf{P}_{i,j} \end{bmatrix} \begin{bmatrix} \frac{j}{n+1} \\ 1 - \frac{j}{n+1} \end{bmatrix}, \qquad (6.35)$$

$$\text{for} \quad i = 0, 1, \dots, m+1, \quad \text{and} \quad j = 0, 1, \dots, n+1.$$

If $i = 0$ or $j = 0$, indexes of the form $i-1$ or $j-1$ are negative, but (the nonexistent) points with such indexes are multiplied by zero, which is why this bilinear interpolation works well in this case. Similarly, when $i = m+1$, point $\mathbf{P}_{i,j}$ does not exist, but the factor $1 - i/(m+1)$ that multiplies it is zero and when $j = n+1$, point $\mathbf{P}_{i,j}$ does not exist, but the factor $1 - j/(n+1)$ that multiplies it is also zero. Thus, Equation (6.35) always works.

Example: Starting with the 2×3 control points

$$
\begin{array}{ccc}
\mathbf{P}_{10} & \mathbf{P}_{11} & \mathbf{P}_{12} \\
\mathbf{P}_{00} & \mathbf{P}_{01} & \mathbf{P}_{02}
\end{array},
$$

(this implies that $m = 1$ and $n = 2$), we perform two steps to elevate the degree of the rectangular patch defined by them from 1×2 to 2×3. The first step is to elevate the degree of each of the three columns from 1 (two control points $\mathbf{P}_{0i}$ and $\mathbf{P}_{1i}$) to 2 (three intermediate points $\mathbf{R}_{0i}$, $\mathbf{R}_{1i}$, and $\mathbf{R}_{2i}$). This step produces the nine intermediate points

$$
\begin{array}{ccc}
\mathbf{R}_{20} & \mathbf{R}_{21} & \mathbf{R}_{22} \\
\mathbf{R}_{10} & \mathbf{R}_{11} & \mathbf{R}_{12} \\
\mathbf{R}_{00} & \mathbf{R}_{01} & \mathbf{R}_{02}
\end{array}.
$$

For the leftmost column, the two extreme points $\mathbf{R}_{00}$ and $\mathbf{R}_{20}$ equal the two original control points $\mathbf{P}_{00}$ and $\mathbf{P}_{10}$, respectively. The middle point $\mathbf{R}_{10}$ is computed from Equation (6.34) as

$$
\mathbf{R}_{10} = \tfrac{1}{2}\mathbf{P}_{00} + (1 - \tfrac{1}{2})\mathbf{P}_{10}.
$$

Similarly, the middle column yields

$$
\mathbf{R}_{01} = \mathbf{P}_{01}, \quad \mathbf{R}_{21} = \mathbf{P}_{11}, \quad \mathbf{R}_{11} = \tfrac{1}{2}\mathbf{P}_{01} + (1 - \tfrac{1}{2})\mathbf{P}_{11}
$$

and the rightmost column results in

$$
\mathbf{R}_{02} = \mathbf{P}_{02}, \quad \mathbf{R}_{22} = \mathbf{P}_{12}, \quad \mathbf{R}_{12} = \tfrac{1}{2}\mathbf{P}_{02} + (1 - \tfrac{1}{2})\mathbf{P}_{12}.
$$

The second step is to elevate the degree of each of the three rows from 2 (three points $\mathbf{R}_{i0}$, $\mathbf{R}_{i1}$, and $\mathbf{R}_{i2}$) to 3 (four new points $\mathbf{Q}_{i0}$, $\mathbf{Q}_{i1}$, $\mathbf{Q}_{i2}$, and $\mathbf{Q}_{i3}$). This step produces the 12 new control points

$$
\begin{array}{cccc}
\mathbf{Q}_{20} & \mathbf{Q}_{21} & \mathbf{Q}_{22} & \mathbf{Q}_{23} \\
\mathbf{Q}_{10} & \mathbf{Q}_{11} & \mathbf{Q}_{12} & \mathbf{Q}_{13} \\
\mathbf{Q}_{00} & \mathbf{Q}_{01} & \mathbf{Q}_{02} & \mathbf{Q}_{03}
\end{array}.
$$

For the bottom row, the two extreme points $\mathbf{Q}_{00}$ and $\mathbf{Q}_{03}$ equal the two intermediate control points $\mathbf{R}_{00}$ and $\mathbf{R}_{02}$, respectively. These, together with the two interior points $\mathbf{Q}_{01}$ and $\mathbf{Q}_{02}$ are computed from Equations (6.34) and (6.35) as

$$
\mathbf{Q}_{00} = \mathbf{R}_{00} = \mathbf{P}_{00} = (0, 1 - 0) \begin{pmatrix} \mathbf{P}_{-1,-1} & \mathbf{P}_{-1,0} \\ \mathbf{P}_{0,-1} & \mathbf{P}_{00} \end{pmatrix} \begin{pmatrix} 0 \\ 1 \end{pmatrix},
$$

$$
\mathbf{Q}_{01} = \tfrac{1}{3}\mathbf{R}_{00} + \tfrac{2}{3}\mathbf{R}_{01} = \tfrac{1}{3}\mathbf{P}_{00} + \tfrac{2}{3}\mathbf{P}_{01} = (0, 1) \begin{pmatrix} \mathbf{P}_{-1,0} & \mathbf{P}_{-1,1} \\ \mathbf{P}_{0,-1} & \mathbf{P}_{01} \end{pmatrix} \begin{pmatrix} 1/3 \\ 1 - 1/3 \end{pmatrix},
$$

$$
\mathbf{Q}_{02} = \tfrac{2}{3}\mathbf{R}_{01} + \tfrac{1}{3}\mathbf{R}_{02} = \tfrac{2}{3}\mathbf{P}_{01} + \tfrac{1}{3}\mathbf{P}_{02} = (0, 1) \begin{pmatrix} \mathbf{P}_{-1,1} & \mathbf{P}_{-1,2} \\ \mathbf{P}_{01} & \mathbf{P}_{02} \end{pmatrix} \begin{pmatrix} 2/3 \\ 1 - 2/3 \end{pmatrix},
$$

$$
\mathbf{Q}_{03} = \mathbf{R}_{02} = \mathbf{P}_{02} = (0, 1 - 0) \begin{pmatrix} \mathbf{P}_{-1,2} & \mathbf{P}_{-1,3} \\ \mathbf{P}_{0,2} & \mathbf{P}_{03} \end{pmatrix} \begin{pmatrix} 1 \\ 0 \end{pmatrix}.
$$

The middle row yields

$$\mathbf{Q}_{10} = \mathbf{R}_{10} = \tfrac{1}{2}\mathbf{P}_{00} + (1 - \tfrac{1}{2})\mathbf{P}_{10} = (\tfrac{1}{2}, 1 - \tfrac{1}{2}) \begin{pmatrix} \mathbf{P}_{0,-1} & \mathbf{P}_{00} \\ \mathbf{P}_{1,-1} & \mathbf{P}_{10} \end{pmatrix} \begin{pmatrix} 0 \\ 1 \end{pmatrix},$$

$$\mathbf{Q}_{11} = \tfrac{1}{3}\mathbf{R}_{10} + \tfrac{2}{3}\mathbf{R}_{11} = \tfrac{1}{3}(\tfrac{1}{2}\mathbf{P}_{00} + \tfrac{1}{2}\mathbf{P}_{10}) + \tfrac{2}{3}(\tfrac{1}{2}\mathbf{P}_{01} + \tfrac{1}{2}\mathbf{P}_{11})$$
$$= (\tfrac{1}{2}, 1 - \tfrac{1}{2}) \begin{pmatrix} \mathbf{P}_{00} & \mathbf{P}_{01} \\ \mathbf{P}_{10} & \mathbf{P}_{11} \end{pmatrix} \begin{pmatrix} 1/3 \\ 1 - 1/3 \end{pmatrix},$$

$$\mathbf{Q}_{12} = \tfrac{2}{3}\mathbf{R}_{11} + \tfrac{1}{3}\mathbf{R}_{12} = \tfrac{2}{3}(\tfrac{1}{2}\mathbf{P}_{01} + \tfrac{1}{2}\mathbf{P}_{11}) + \tfrac{1}{3}(\tfrac{1}{2}\mathbf{P}_{02} + \tfrac{1}{2}\mathbf{P}_{12})$$
$$= (\tfrac{1}{2}, 1 - \tfrac{1}{2}) \begin{pmatrix} \mathbf{P}_{01} & \mathbf{P}_{02} \\ \mathbf{P}_{11} & \mathbf{P}_{12} \end{pmatrix} \begin{pmatrix} 2/3 \\ 1 - 2/3 \end{pmatrix},$$

$$\mathbf{Q}_{13} = \mathbf{R}_{12} = \tfrac{1}{2}\mathbf{P}_{02} + (1 - \tfrac{1}{2})\mathbf{P}_{12} = (\tfrac{1}{2}, 1 - \tfrac{1}{2}) \begin{pmatrix} \mathbf{P}_{02} & \mathbf{P}_{03} \\ \mathbf{P}_{12} & \mathbf{P}_{13} \end{pmatrix} \begin{pmatrix} 1 \\ 0 \end{pmatrix}.$$

Finally, the third row of intermediate points produces the four new control points

$$\mathbf{Q}_{20} = \mathbf{R}_{20} = \mathbf{P}_{10} = (1, 0) \begin{pmatrix} \mathbf{P}_{1,-1} & \mathbf{P}_{10} \\ \mathbf{P}_{2,-1} & \mathbf{P}_{20} \end{pmatrix} \begin{pmatrix} 0 \\ 1 \end{pmatrix},$$

$$\mathbf{Q}_{21} = \tfrac{1}{3}\mathbf{R}_{20} + \tfrac{2}{3}\mathbf{R}_{21} = \tfrac{1}{3}\mathbf{P}_{10} + \tfrac{2}{3}\mathbf{P}_{11} = (1, 0) \begin{pmatrix} \mathbf{P}_{10} & \mathbf{P}_{11} \\ \mathbf{P}_{20} & \mathbf{P}_{21} \end{pmatrix} \begin{pmatrix} 1/3 \\ 1 - 1/3 \end{pmatrix},$$

$$\mathbf{Q}_{22} = \tfrac{2}{3}\mathbf{R}_{21} + \tfrac{1}{3}\mathbf{R}_{22} = \tfrac{2}{3}\mathbf{P}_{11} + \tfrac{1}{3}\mathbf{P}_{12} = (1, 0) \begin{pmatrix} \mathbf{P}_{11} & \mathbf{P}_{12} \\ \mathbf{P}_{21} & \mathbf{P}_{22} \end{pmatrix} \begin{pmatrix} 2/3 \\ 1/3 \end{pmatrix},$$

$$\mathbf{Q}_{23} = \mathbf{R}_{22} = \mathbf{P}_{12} = (1, 0) \begin{pmatrix} \mathbf{P}_{12} & \mathbf{P}_{13} \\ \mathbf{P}_{22} & \mathbf{P}_{23} \end{pmatrix} \begin{pmatrix} 1 \\ 0 \end{pmatrix}.$$

Figure 6.24 lists code for elevating the degree of a rectangular Bézier patch based on 2×3 control points. In part (a) of the figure each point is a symbol, such as p00, and in part (b) each point is a triplet of coordinates. The points are stored in a 2×3 array p and are transferred to a 4×5 array r, parts of which remain undefined.

6.19 Nonparametric Rectangular Patches

The explicit representation of a surface (Section 1.8) is $z = f(x, y)$. The rectangular Bézier surface is, of course, parametric, but it can be represented in a nonparametric form, similar to explicit surfaces. The derivation in this section is similar to that of Section 6.14. Given $(n + 1) \times (m + 1)$ real values (not points) P_{ij}, we start with the double polynomial

$$s(u, w) = \sum_{i=0}^{n} \sum_{j=0}^{m} B_{ni}(u) P_{ij} B_{mj}(w)$$

and employ the identity of Equation (6.30) twice, for u and for w, to create the surface patch

$$\mathbf{P}(u, w) = \big(u, w, s(u, w)\big) = \sum_{i=0}^{n} \sum_{j=0}^{m} B_{ni}(u)(i/m, j/n, P_{ij}) B_{mj}(w).$$

```
(* Degree elevation of a rect Bezier surface from 2x3 to 4x5 *)
Clear[p,q,r];
m=1; n=2;
p={{p00,p01,p02},{p10,p11,p12}}; (* array of points *)
r=Array[a, {m+3,n+3}]; (* extended array, still undefined *)
Part[r,1]=Table[a, {i,-1,m+2}];
Part[r,2]=Append[Prepend[Part[p,1],a],a];
Part[r,3]=Append[Prepend[Part[p,2],a],a];
Part[r,n+2]=Table[a, {i,-1,m+2}];
MatrixForm[r] (* display extended array *)
q[i_,j_]:=({i/(m+1),1-i/(m+1)}. (* dot product *)
 {{r[[i+1,j+1]],r[[i+1,j+2]]},{r[[i+2,j+1]],r[[i+2,j+2]]}}).
 {j/(n+1),1-j/(n+1)}
q[2,3] (* test *)
```

(a)

```
(* Degree elevation of a rect Bezier surface from 2x3 to 4x5 *)
Clear[p,r,comb];
m=1; n=2; (* set p to an array of 3D points *)
p={{{0,0,0},{1,0,1},{2,0,0}},{{0,1,0},{1,1,.5},{2,1,0}}};
r=Array[a, {m+3,n+3}]; (* extended array, still undefined *)
Part[r,1]=Table[{a,a,a}, {i,-1,m+2}];
Part[r,2]=Append[Prepend[Part[p,1],{a,a,a}],{a,a,a}];
Part[r,3]=Append[Prepend[Part[p,2],{a,a,a}],{a,a,a}];
Part[r,n+2]=Table[{a,a,a}, {i,-1,m+2}];
MatrixForm[r] (* display extended array *)
comb[i_,j_]:=({i/(m+1),1-i/(m+1)}.
{{r[[i+1,j+1]],r[[i+1,j+2]]},{r[[i+2,j+1]],r[[i+2,j+2]]}})[[1]]{j/(n+1),1-j/(n+1)}[[1]]+
({i/(m+1),1-i/(m+1)}.
{{r[[i+1,j+1]],r[[i+1,j+2]]},{r[[i+2,j+1]],r[[i+2,j+2]]}})[[2]]{j/(n+1),1-j/(n+1)}[[2]];
MatrixForm[Table[comb[i,j], {i,0,2},{j,0,3}]]
```

(b)

Figure 6.24: Code for Degree Elevation of a Rectangular Bézier Surface.

This version of the Bézier surface is defined by the control points $(i/m, j/n, P_{ij})$ which form a regular grid on the xy plane.

6.20 Joining Rectangular Bézier Patches

It is easy, although tedious, to explore the conditions for the smooth joining of two Bézier surface patches. Figure 6.25 shows a typical example of this problem. It shows parts of two patches **P** and **Q**. It is not difficult to see that the former is based on 4×5 control points and the latter on $4 \times n$ points, where $n \geq 2$. It is also easy to see that they are joined such that the eight control points along the joint satisfy $\mathbf{P}_{i4} = \mathbf{Q}_{i0}$ for $i = 0, 1, 2, 3$.

The condition for smooth joining of the two surface patches is that the two tangent vectors at the common boundary are in the same direction, although they may have different magnitudes. This condition is expressed as

$$\frac{\partial \mathbf{P}(u,w)}{\partial w}\bigg|_{w=1} = \alpha \frac{\partial \mathbf{Q}(u,w)}{\partial w}\bigg|_{w=0}.$$

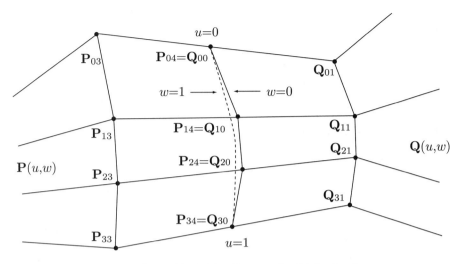

Figure 6.25: Smoothly Joining Rectangular Bézier Patches.

The two tangents are calculated from Equation (6.32) (and the $\mathbf{B}_3$ and $\mathbf{B}_4$ matrices given by Figure 6.3). For the first patch, we have

$$
\left.\frac{\partial \mathbf{P}(u,w)}{\partial w}\right|_{w=1} = (u^3, u^2, u, 1)\mathbf{B}_3 \begin{pmatrix} \mathbf{P}_{00} & \mathbf{P}_{01} & \mathbf{P}_{02} & \mathbf{P}_{03} & \mathbf{P}_{04} \\ \mathbf{P}_{10} & \mathbf{P}_{11} & \mathbf{P}_{12} & \mathbf{P}_{13} & \mathbf{P}_{14} \\ \mathbf{P}_{20} & \mathbf{P}_{21} & \mathbf{P}_{22} & \mathbf{P}_{23} & \mathbf{P}_{24} \\ \mathbf{P}_{30} & \mathbf{P}_{31} & \mathbf{P}_{32} & \mathbf{P}_{33} & \mathbf{P}_{34} \end{pmatrix} \mathbf{B}_4^T \left. \begin{pmatrix} 4w^3 \\ 3w^2 \\ 2w \\ 1 \\ 0 \end{pmatrix} \right|_{w=1}
$$

$$
= 4(u^3, u^2, u, 1)\mathbf{B}_3 \begin{pmatrix} \mathbf{P}_{04} - \mathbf{P}_{03} \\ \mathbf{P}_{14} - \mathbf{P}_{13} \\ \mathbf{P}_{24} - \mathbf{P}_{23} \\ \mathbf{P}_{34} - \mathbf{P}_{33} \end{pmatrix}.
$$

Similarly, for the second patch,

$$
\left.\frac{\partial \mathbf{Q}(u,w)}{\partial w}\right|_{w=0} = 4(u^3, u^2, u, 1)\mathbf{B}_3 \begin{pmatrix} \mathbf{Q}_{01} - \mathbf{Q}_{00} \\ \mathbf{Q}_{11} - \mathbf{Q}_{10} \\ \mathbf{Q}_{21} - \mathbf{Q}_{20} \\ \mathbf{Q}_{31} - \mathbf{Q}_{30} \end{pmatrix}.
$$

The conditions for a smooth join are therefore

$$
\begin{pmatrix} \mathbf{P}_{04} - \mathbf{P}_{03} \\ \mathbf{P}_{14} - \mathbf{P}_{13} \\ \mathbf{P}_{24} - \mathbf{P}_{23} \\ \mathbf{P}_{34} - \mathbf{P}_{33} \end{pmatrix} = \alpha \begin{pmatrix} \mathbf{Q}_{01} - \mathbf{Q}_{00} \\ \mathbf{Q}_{11} - \mathbf{Q}_{10} \\ \mathbf{Q}_{21} - \mathbf{Q}_{20} \\ \mathbf{Q}_{31} - \mathbf{Q}_{30} \end{pmatrix},
$$

or $\mathbf{P}_{i4} - \mathbf{P}_{i3} = \alpha(\mathbf{Q}_{i1} - \mathbf{Q}_{i0})$ for $i = 0, 1, 2$, and 3. This can also be expressed by saying that the three points $\mathbf{P}_{i3}$, $\mathbf{P}_{i4} = \mathbf{Q}_{i0}$, and $\mathbf{Q}_{i1}$ should be on a straight line, although not necessarily equally spaced.

Example: Each of the two patches in Figure 6.26 is based on 3×3 points ($n = 2$). The patches are smoothly connected along the curve defined by the common points $(0, 2, 0)$, $(0, 0, 0)$, and $(0, -2, 0)$. Note that in the diagram they are slightly separated, but this was done intentionally. The smooth connection is obtained by making sure that the points $(-2, 2, 0)$, $(0, 2, 0)$, and $(2, 2, 0)$ are collinear (find the other two collinear triplets). The coordinates of the points are

$$
\begin{array}{ccc ccc}
-2,2,2 & -2,2,0 & 0,2,0 & 0,2,0 & 2,2,0 & 2,2,-2 \\
-4,0,2 & -4,0,0 & 0,0,0 & 0,0,0 & 4,0,0 & 4,0,-2 \\
-2,-2,2 & -2,-2,0 & 0,-2,0 & 0,-2,0 & 2,-2,0 & 2,-2,-2
\end{array}
$$

The famous Utah teapot was designed in the 1960s at the University of Utah by digitizing a real teapot (now at the computer museum in Boston) and creating 32 smoothly-connected Bézier patches defined by a total of 306 control points. [Crow 87] has a detailed description. The coordinates of the points are publicly available, as is a program to display the entire surface. The program is part of a public-domain general three-dimensional graphics package called SIPP (SImple Polygon Processor). SIPP was originally written in Sweden and is distributed by the Free Software Foundation [Free 04]. It can be downloaded anonymously from several sources and for different platforms. A more recent source for this important surface is a *Mathematica* notebook by Jan Mangaldan, available at [MathSource 05].

She finished pouring the tea and put down the pot.

"That's an old teapot," remarked Harold.

"Sterling silver," said Maude wistfully. "It was my dear mother-in-law's, part of a dinner set of fifty pieces. It was sent to me, one of the few things that survived." Her voice trailed off and she absently sipped her tea.

— Colin Higgins, *Harold and Maude* (1971).

6.21 An Interpolating Bézier Surface Patch

An interpolating rectangular Bézier surface patch solves the following problem. Given a set of $(m + 1) \times (n + 1)$ data points $\mathbf{Q}_{kl}$, compute a set of $(m + 1) \times (n + 1)$ control points $\mathbf{P}_{ij}$, such that the rectangular Bézier surface patch $\mathbf{P}(u, w)$ defined by the $\mathbf{P}_{ij}$'s will pass through all the data points $\mathbf{Q}_{kl}$.

Section 6.13 discusses the same problem for the Bézier curve, and here we apply the same approach to the rectangular Bézier surface. We select $m + 1$ values u_k and $n + 1$ values w_l and require that the $(m + 1) \times (n + 1)$ surface points $\mathbf{P}(u_k, w_l)$ equal the data points $\mathbf{Q}_{kl}$ for $k = 0, 1, \ldots, m$ and $l = 0, 1, \ldots, n$. This results in a set of $(m+1) \times (n+1)$ equations with the control points $\mathbf{P}_{ij}$ as the unknowns. Such a set of equations may be

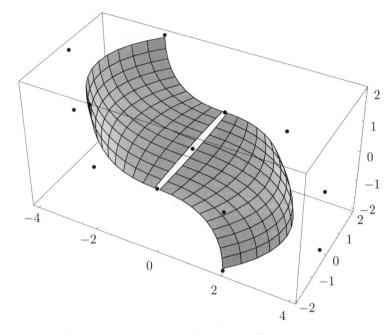

```
n=2; Clear[n,bern,p1,p2,g3,bzSurf,patch];
<<:Graphics:ParametricPlot3D.m
p1={{{-2,2,2},{-2,2,0},{0,2,0}},
 {{-4,0,2},{-4,0,0},{0,0,0}},
 {{-2,-2,2},{-2,-2,0},{0,-2,0}}};
p2={{{0,2,0},{2,2,0},{2,2,-2}},
 {{0,0,0},{4,0,0},{4,0,-2}},
 {{0,-2,0},{2,-2,0},{2,-2,-2}}};
pwr[x_,y_]:=If[x==0 && y==0, 1, x^y];
bern[n_,i_,u_]:=Binomial[n,i]pwr[u,i]pwr[1-u,n-i]
bzSurf[p_]:={Sum[p[[i+1,j+1,1]]bern[n,i,u]bern[n,j,w],
 {i,0,n,1}, {j,0,n,1}],
 Sum[p[[i+1,j+1,2]]bern[n,i,u]bern[n,j,w],
 {i,0,n,1}, {j,0,n,1}],
 Sum[p[[i+1,j+1,3]]bern[n,i,u]bern[n,j,w],
 {i,0,n,1}, {j,0,n,1}]};
patch[s_]:=
ParametricPlot3D[bzSurf[s],{u,0,1,.1}, {w,0.02,.98,.1}];
g3=Graphics3D[{AbsolutePointSize[3],
 Table[Point[p1[[i,j]]],{i,1,n+1},{j,1,n+1}]}]
g4=Graphics3D[{AbsolutePointSize[3],
 Table[Point[p2[[i,j]]],{i,1,n+1},{j,1,n+1}]}]
Show[patch[p1],patch[p2],g3,g4,
 DisplayFunction->$DisplayFunction]
```

Figure 6.26: Two Bézier Surface Patches.

big, but is easy to solve with appropriate mathematical software. A general equation in this set is

$$\mathbf{P}(u_k, w_l) = \mathbf{B}_m(u_k)\,\mathbf{P}\,\mathbf{B}_n(w_l) = \mathbf{Q}_{kl} \quad \text{for} \quad k = 0, 1, \ldots, m \quad \text{and} \quad l = 0, 1, \ldots, n.$$

Example: We choose $m = 3$ and $n = 2$. The system of equations becomes

$$\left[(1-u_k)^3, 3u_k(1-u_k)^2, 3u_k^2(1-u_k), u_k^3\right]\begin{bmatrix} \mathbf{P}_{00} & \mathbf{P}_{01} & \mathbf{P}_{02} \\ \mathbf{P}_{10} & \mathbf{P}_{11} & \mathbf{P}_{12} \\ \mathbf{P}_{20} & \mathbf{P}_{21} & \mathbf{P}_{22} \\ \mathbf{P}_{30} & \mathbf{P}_{31} & \mathbf{P}_{32} \end{bmatrix}\begin{bmatrix} (1-w_l)^2 \\ 2w_l(1-w_l) \\ w_l^2 \end{bmatrix} = \mathbf{Q}_{kl},$$

for $k = 0, 1, 2, 3$ and $l = 0, 1, 2$. This is a system of 12 equations in the 12 unknowns $\mathbf{P}_{ij}$. In most cases the u_k values can be equally spaced between 0 and 1 (in our case 0, 0.25, 0.5, 0.75, and 1), and the same for the w_l values (in our case, 0, 0.5, and 1).

6.22 Rational Bézier Surfaces

Section 6.15 describe the rational Bézier curve. The principle of this type of curve can be extended to surfaces, and this section discusses the rational rectangular Bézier surface patch. This type of surface is expressed by

$$\mathbf{P}(u, w) = \frac{\sum_{i=0}^{n}\sum_{j=0}^{m} w_{ij} B_{n,i}(u)\mathbf{P}_{ij} B_{m,j}(w)}{\sum_{k=0}^{n}\sum_{l=0}^{m} w_{kl} B_{n,k}(u) B_{m,l}(w)} \qquad 0 \le u, w \le 1. \qquad (6.36)$$

When all the weights w_{ij} are set to 1, Equation (6.36) reduces to the original rectangular Bézier surface patch. The weights serve as additional parameters and provide fine, accurate control of the shape of the surface. Figure 6.27 shows how the surface patch of Figure 6.20 can be pulled toward the center point [point $(4, 1, 1)$] by assigning $w_{22} = 5$, while keeping the other weights set to 1.

Note that weights of 0 and negative weights can also be used, as long as the denominator of Equation (6.36) is not zero.

◇ **Exercise 6.23:** Use the code of Figure 6.27 to construct a closed rational Bézier surface patch based on a grid of 2×4 control points.

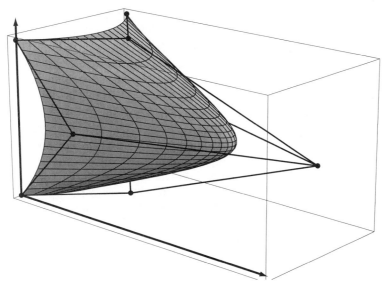

```
(* A Rational Bezier Surface *)
Clear[pwr,bern,spnts,n,m,wt,bzSurf,cpnts,patch,vlines,hlines,axes];
<<:Graphics:ParametricPlot3D.m
spnts={{{0,0,0},{1,0,1},{0,0,2}},
 {{1,1,0},{4,1,1},{1,1,2}}, {{0,2,0},{1,2,1},{0,2,2}}};
m=Length[spnts[[1]]]-1; n=Length[Transpose[spnts][[1]]]-1;
wt=Table[1, {i,1,n+1},{j,1,m+1}];
wt[[2,2]]=5;
pwr[x_,y_]:=If[x==0 && y==0, 1, x^y];
bern[n_,i_,u_]:=Binomial[n,i]pwr[u,i]pwr[1-u,n-i]
bzSurf[u_,w_]:=
 Sum[wt[[i+1,j+1]]spnts[[i+1,j+1]]bern[n,i,u]bern[m,j,w], {i,0,n}, {j,0,m}]/
 Sum[wt[[i+1,j+1]]bern[n,i,u]bern[m,j,w], {i,0,n}, {j,0,m}];
patch=ParametricPlot3D[bzSurf[u,w],{u,0,1}, {w,0,1},
 Compiled->False, DisplayFunction->Identity];
cpnts=Graphics3D[{AbsolutePointSize[4], (* control points *)
 Table[Point[spnts[[i,j]]], {i,1,n+1},{j,1,m+1}]}];
vlines=Graphics3D[{AbsoluteThickness[1], (* control polygon *)
 Table[Line[{spnts[[i,j]],spnts[[i+1,j]]}], {i,1,n}, {j,1,m+1}]}];
hlines=Graphics3D[{AbsoluteThickness[1],
 Table[Line[{spnts[[i,j]],spnts[[i,j+1]]}], {i,1,n+1}, {j,1,m}]}];
maxx=Max[Flatten[Table[Part[spnts[[i,j]], 1], {i,1,n+1}, {j,1,m+1}]]];
maxy=Max[Flatten[Table[Part[spnts[[i,j]], 2], {i,1,n+1}, {j,1,m+1}]]];
maxz=Max[Flatten[Table[Part[spnts[[i,j]], 3], {i,1,n+1}, {j,1,m+1}]]];
axes=Graphics3D[{AbsoluteThickness[1.5], (* the coordinate axes *)
 Line[{{0,0,maxz},{0,0,0},{maxx,0,0},{0,0,0},{0,maxy,0}}]}];
Show[cpnts,hlines,vlines,axes,patch, PlotRange->All, DefaultFont->{"cmr10",10},
 DisplayFunction->$DisplayFunction, ViewPoint->{2.783, -3.090, 1.243}];
```

Figure 6.27: A Rational Bézier Surface Patch.

6.23 Triangular Bézier Surfaces

The first surface to be derived with Bézier methods was the triangular patch, not the rectangular. It was developed in 1959 by de Casteljau at Citroën. The triangular Bézier patch, and its properties, is the topic of this section, but it should be noted that the ideas and techniques described here can be extended to Bézier surface patches with any number of edges. [DeRose and Loop 89] discusses one approach, termed *S-patch*, to this problem.

The triangular Bézier patch is based on control points $\mathbf{P}_{ijk}$ arranged in a roughly triangular shape. Each control point is three-dimensional and is assigned three indexes ijk such that $0 \le i, j, k \le n$ and $i + j + k = n$. The value of n is selected by the user depending on how large and complex the patch should be and how many points are given. Generally, a large n allows for a finer control of surface details but involves more computations. The following convention is used here. The first index, i, corresponds to the left side of the triangle, the second index, j, corresponds to the base, and the third index, k, corresponds to the right side. The indexing convention for $n = 1, 2, 3$, and 4 are shown in Figure 6.28. There are $n + 1$ points on each side of the triangle and because of the way the points are arranged there is a total of $\frac{1}{2}(n + 1)(n + 2)$ control points:

$$
\begin{array}{cccc}
& & & \mathbf{P}_{040} \\
& & \mathbf{P}_{030} & \mathbf{P}_{031}\,\mathbf{P}_{130} \\
& \mathbf{P}_{020} & \mathbf{P}_{021}\,\mathbf{P}_{120} & \mathbf{P}_{022}\,\mathbf{P}_{121}\,\mathbf{P}_{220} \\
\mathbf{P}_{010} & \mathbf{P}_{011}\,\mathbf{P}_{110} & \mathbf{P}_{012}\,\mathbf{P}_{111}\,\mathbf{P}_{210} & \mathbf{P}_{013}\,\mathbf{P}_{112}\,\mathbf{P}_{211}\,\mathbf{P}_{310} \\
\mathbf{P}_{001}\,\mathbf{P}_{100} & \mathbf{P}_{002}\,\mathbf{P}_{101}\,\mathbf{P}_{200} & \mathbf{P}_{003}\,\mathbf{P}_{102}\,\mathbf{P}_{201}\,\mathbf{P}_{300} & \mathbf{P}_{004}\,\mathbf{P}_{103}\,\mathbf{P}_{202}\,\mathbf{P}_{301}\,\mathbf{P}_{400} \\
\end{array}
$$

Figure 6.28: Control Points for Four Triangular Bézier Patches.

The surface patch itself is defined by the trinomial theorem [Equation (6.3)] as

$$
\mathbf{P}(u, v, w) = \sum_{i+j+k=n} \mathbf{P}_{ijk} \frac{n!}{i!\,j!\,k!} u^i v^j w^k = \sum_{i+j+k=n} \mathbf{P}_{ijk} B^n_{ijk}(u, v, w), \tag{6.37}
$$

where $u + v + w = 1$. Note that even though $\mathbf{P}(u, v, w)$ seems to depend on three parameters, it only depends on two since their sum is constant. The quantities

$$
B^n_{ijk}(u, v, w) = \frac{n!}{i!\,j!\,k!} u^i v^j w^k
$$

are the Bernstein polynomials in two variables (bivariate). They are listed here for $n = 1, 2, 3$, and 4

$$
\begin{array}{cccc}
& & & v^4 \\
& & v^3 & 4v^3w \; 4uv^3 \\
& v^2 & 3v^2w \; 3uv^2 & 6v^2w^2 \; 12uv^2w \; 6u^2v^2 \\
v & 2vw \; 2uv & 3vw^2 \; 6uvw \; 3u^2v & 4vw^3 \; 12uvw^2 \; 12u^2vw \; 4u^3v \\
w \; u & w^2 \; 2uw \; u^2 & w^3 \; 3uw^2 \; 3u^2w \; u^3 & w^4 \; 4uw^3 \; 6u^2w^2 \; 4u^3w \; u^4 \\
\end{array}
$$

The three boundary curves are obtained from Equation (6.37) by setting each of the three parameters in turn to zero. Setting, for example, $u = 0$ causes all terms of Equation (6.37) except those with $i = 0$ to vanish. The result is

$$\mathbf{P}(0, v, w) = \sum_{j+k=n} \mathbf{P}_{0jk} \frac{n!}{j!\, k!} v^j w^k, \quad \text{where} \quad v + w = 1. \tag{6.38}$$

Since $v + w = 1$, Equation (6.38) can be written

$$\mathbf{P}(v) = \sum_{j+k=n} \mathbf{P}_{0jk} \frac{n!}{j!\, k!} v^j (1-v)^k = \sum_{j=0}^{n} \mathbf{P}_{0j,n-j} \frac{n!}{j!\, (n-j)!} v^j (1-v)^{n-j}, \tag{6.39}$$

and this is a Bézier curve.

Example: We illustrate the case $n = 2$. There should be three control points on each side of the triangle, for a total of $\frac{1}{2}(2+1)(2+2) = 6$ points. We select simple coordinates:

$$(1, 3, 1)$$
$$(0.5, 1, 0)\,(1.5, 1, 0)$$
$$(0, 0, 0)\,(1, 0, -1)\,(2, 0, 0)$$

Note that four points have $z = 0$ and are therefore on the same plane. It is only the other two points, with $z = \pm 1$, that cause this surface to be nonflat.

The expression of the surface is

$$\begin{aligned}
\mathbf{P}(u, v, w) &= \sum_{i+j+k=2} \mathbf{P}_{ijk} \frac{n!}{i!\, j!\, k!} u^i v^j w^k \\
&= \mathbf{P}_{002} \frac{2!}{0!\, 0!\, 2!} w^2 + \mathbf{P}_{101} \frac{2!}{1!\, 0!\, 1!} uw + \mathbf{P}_{200} \frac{2!}{2!\, 0!\, 0!} u^2 \\
&\quad + \mathbf{P}_{011} \frac{2!}{0!\, 1!\, 1!} vw + \mathbf{P}_{110} \frac{2!}{1!\, 1!\, 0!} uv + \mathbf{P}_{020} \frac{2!}{0!\, 2!\, 0!} v^2 \\
&= (0, 0, 0) w^2 + (1, 0, -1) 2uw + (2, 0, 0) u^2 \\
&\quad + (0.5, 1, 0) 2vw + (1.5, 1, 0) 2uv + (1, 3, 1) v^2 \\
&= (2uw + 2u^2 + vw + 3uv + v^2, 2vw + 2uv + 3v^2, -2uw + v^2).
\end{aligned}$$

It is now easy to verify that the following special values of u, v, and w produce the three corner points:

u	v	w	point
0	0	1	(0,0,0)
0	1	0	(1,3,1)
1	0	0	(2,0,0)

But the most important feature of this triangular surface patch is the way it is displayed as a wireframe. The principle is to display this surface as a mesh of *three* families of curves (compare this with the two families in the case of a rectangular surface patch).

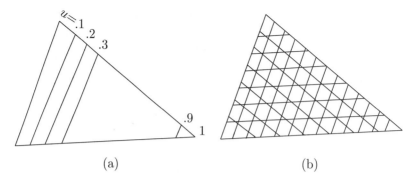

Figure 6.29: (a) Lines in the u Direction. (b) The Complete Surface Patch.

Each family consists of curves that are roughly parallel to one side of the triangle (Figure 6.29a,b).

⬦ **Exercise 6.24:** Write pseudo-code to draw the three families of curves.

A triangle of points can be stored in a one-dimensional array in computer memory. A simple way of doing this is to store the top point $\mathbf{P}_{0n0}$ at the beginning of the array, followed by a short segment consisting of the two points $\mathbf{P}_{0,n-1,1}$ and $\mathbf{P}_{1,n-1,0}$ of the next row down, followed by a longer segment with three points, and so on, ending with a segment with the $n+1$ points $\mathbf{P}_{00n}$, $\mathbf{P}_{1,0,n-1}$, through $\mathbf{P}_{n00}$ of the bottom row of the triangle. A direct check verifies that the points $\mathbf{P}_{ijk}$ of triangle row j, where $0 \leq j \leq n$, start at location $j(j+1)/2+1$ of the array, so they can be indexed by $j(j+1)/2+1+i$.

Figure 6.30 lists *Mathematica* code to compute one point on such a surface patch. Note that j is incremented from 0 to n (from the bottom to the top of the triangle), so the first iteration needs the points in the last segment of the array and the last iteration needs the single point at the start of the array. This is why the index to array `pnts` depends on j as $(n-j)(n-j+1)/2+1$ instead of as $j(j+1)/2+1$.

```
(* Triangular Bezier surface patch *)
pnts={{3,3,0}, {2,2,0},{4,2,1}, {1,1,0},{3,1,1},{5,1,2},
 {0,0,0},{2,0,1},{4,0,2},{6,0,3}};
B[i_,j_,k_]:=(n!/(i! j! k!))u^i v^j w^k;
n=3; u=1/6; v=2/6; w=3/6; Tsrpt={0,0,0};
indx:=(n-j)(n-j+1)/2+1+i;
Do[{k=n-i-j, Tsrpt=Tsrpt+B[i,j,k] pnts[[indx]]}, {j,0,n}, {i,0,n-j}];
Tsrpt
```

Figure 6.30: Code for One Point in a Triangular Bézier Patch.

Figure 6.31 shows a triangular Bézier surface patch for $n = 3$. Note how the wireframe consists of three sets of curves and how the curves remain roughly parallel and don't converge toward the three corners. (This should be compared with the triangular Coons patch of Figure 3.14 and with the lofted sweep surface of Figure 9.3. Each of these surfaces is displayed as two families of curves and has one dark corner as a result.) The control points and control polygon are also shown. The *Mathematica* code for this type

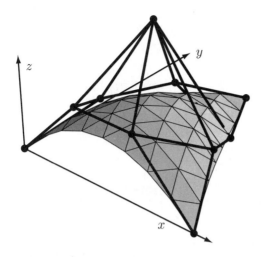

```
(* Triangular Bezier patch by Garry Helzer *)
rules=Solve[{u{a1,b1}+v{a2,b2}+w{a3,b3}=={x,y},u+v+w==1},{u,v,w}]
BarycentricCoordinates[Polygon[{{a1_,b1_},{a2_,b2_},{a3_,b3_}}]] \
[{x_,y_}]={u,v,w}/.rules//Flatten
Subdivide[l_]:=1/. Polygon[{p_,q_,r_}] :> Polygon /@ \
({{p+p,p+q,p+r},{p+q,q+q,q+r},{p+r,q+r,r+r},{p+q,q+r,r+p}}/2)
Transform[F_][L_]:= L /. Polygon[l_] :> Polygon[F /@ l]
P[L_][{u_,v_,w_}]:=
Module[{x,y,z,n=(Sqrt[8Length[L]+1]-3)/2},
 ((List @@ Expand[(x+y+z)^n]) /. {x->u,y->v,z->w}).L]
Param[T_,L_][{x_,y_}]:=With[{p=BarycentricCoordinates[T][{x, y}]},P[L][p]]
```

Run the code below in a separate cell

```
(* Triangular bezier patch for n=3 *)
T=Polygon[{{1, 0}, {0, 1}, {0, 0}}];
L={P300,P210,P120,P030, P201,P111,P021, P102,P012, P003} \
={{3,0,0},{2.5,1,.5},{2,2,0},{1.5,3,0},
 {2,0,1},{1.5,1,2},{1,2,.5}, {1,0,1},{.5,1,.5}, {0,0,0}};
SubT=Nest[Subdivide, T, 3];
Patch=Transform[Param[T, L]][SubT];
cpts={PointSize[0.02], Point/@L};
coord={AbsoluteThickness[1],
Line/@{{{0,0,0},{3.2,0,0}},{{0,0,0},{0,3.4,0}},{{0,0,0},{0,0,1.3}}}};
cpolygon={AbsoluteThickness[2],
Line[{P300,P210,P120,P030,P021,P012,P003,P102,P201,P300}],
Line[{P012,P102,P111,P120,P021,P111,P201,P210,P111,P012}]};
Show[Graphics3D[{cpolygon,cpts,coord,Patch}], Boxed->False, PlotRange->All,
ViewPoint->{2.620, -3.176, 2.236}];
```

Figure 6.31: A Triangular Bézier Surface Patch For $n = 3$.

When an object is digitized mechanically, the result is a large set of points. Such a set can be converted to a set of triangles by the Delaunay triangulation algorithm. This method produces a collection of edges that satisfy the following property: For each edge we can find a circle containing the edge's endpoints but not containing any other points.

of surface is due to Garry Helzer and it works by recursively subdividing the triangular patch into subtriangles. Figure C.2 shows two triangular Bézier patches for $n = 2$ and $n = 4$.

6.23.1 Scaffolding Construction

The scaffolding construction (or de Casteljau algorithm) of Section 6.6 can be directly extended to triangular Bézier patches. The bivariate Bernstein polynomials that are the basis of this type of surface are given by Equation (6.3), rewritten here

$$B_{i,j,k}^n(u,v,w) = \sum_{\substack{i,j,k \geq 0}}^{i+j+k=n} \frac{(i+j+k)!}{i!j!k!} u^i v^j w^k = \sum_{\substack{i,j,k \geq 0}}^{i+j+k=n} \frac{n!}{i!j!k!} u^i v^j w^k. \quad (6.3)$$

Direct checking verifies that these polynomials satisfy the recursion relation

$$B_{i,j,k}^n(u,v,w) = u B_{i-1,jk}^{n-1}(u,v,w) + v B_{i,j-1,k}^{n-1}(u,v,w) + w B_{i,j,k-1}^{n-1}(u,v,w), \quad (6.40)$$

and this relation is the basis of the de Casteljau algorithm for the triangular Bézier patch.

The algorithm starts with the original control points $\mathbf{P}_{ijk}$ which are labeled $\mathbf{P}_{ijk}^0$. The user selects a triplet (u,v,w) where $u + v + w = 1$ and performs the following step n times to compute intermediate points $\mathbf{P}_{i,j,k}^r$ for $r = 1, \ldots, n$ and $i + j + k = n - r$

$$\mathbf{P}_{i,j,k}^r = u \mathbf{P}_{i+1,j,k}^{r-1} + v \mathbf{P}_{i,j+1,k}^{r-1} + w \mathbf{P}_{i,j,k+1}^{r-1}.$$

The last step produces the single point $\mathbf{P}_{000}^n$ that's also the point produced by the selected triplet (u,v,w) on the triangular Bézier patch.

The algorithm is illustrated here for $n = 3$. Figure 6.28 shows the 10 control points. Assuming that the user has selected appropriate values for the parameter triplet (u,v,w), the first step of the algorithm produces the six intermediate points for $n = 2$ (Figure 6.32)

$$\begin{aligned}
\mathbf{P}_{002}^1 &= u \mathbf{P}_{102}^0 + v \mathbf{P}_{012}^0 + w \mathbf{P}_{003}^0, & \mathbf{P}_{101}^1 &= u \mathbf{P}_{201}^0 + v \mathbf{P}_{111}^0 + w \mathbf{P}_{102}^0, \\
\mathbf{P}_{200}^1 &= u \mathbf{P}_{300}^0 + v \mathbf{P}_{210}^0 + w \mathbf{P}_{201}^0, & \mathbf{P}_{011}^1 &= u \mathbf{P}_{111}^0 + v \mathbf{P}_{021}^0 + w \mathbf{P}_{012}^0, \\
\mathbf{P}_{110}^1 &= u \mathbf{P}_{210}^0 + v \mathbf{P}_{120}^0 + w \mathbf{P}_{111}^0, & \mathbf{P}_{020}^1 &= u \mathbf{P}_{120}^0 + v \mathbf{P}_{030}^0 + w \mathbf{P}_{021}^0.
\end{aligned}$$

The second step produces the three intermediate points for $n = 1$

$$\begin{aligned}
\mathbf{P}_{001}^2 &= u \mathbf{P}_{101}^1 + v \mathbf{P}_{011}^1 + w \mathbf{P}_{002}^1, \\
\mathbf{P}_{100}^2 &= u \mathbf{P}_{200}^1 + v \mathbf{P}_{110}^1 + w \mathbf{P}_{101}^1, \\
\mathbf{P}_{010}^2 &= u \mathbf{P}_{110}^1 + v \mathbf{P}_{020}^1 + w \mathbf{P}_{011}^1.
\end{aligned}$$

And the third step produces the single point

$$\mathbf{P}_{000}^3 = u \mathbf{P}_{100}^2 + v \mathbf{P}_{010}^2 + w \mathbf{P}_{001}^2.$$

This is the point that corresponds to the particular triplet (u,v,w) on the triangular patch defined by the 10 original control points.

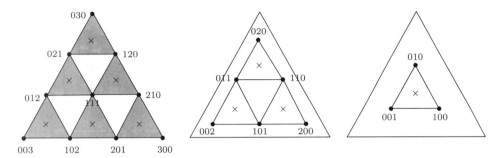

Figure 6.32: Scaffolding in a Triangular Bézier Patch.

⋄ **Exercise 6.25:** Illustrate this algorithm for $n = 4$. Start with the 15 original control points and list the four steps of the scaffolding. The final result should be the single point $\mathbf{P}_{000}^4$. Assume that the user has selected appropriate values for the parameter triplet (u, v, w),

⋄ **Exercise 6.26:** Assuming the values $u = 1/6$, $v = 2/6$, and $w = 3/6$, and the 10 control points

$$(3, 3, 0)$$
$$(2, 2, 0) \, (4, 2, 1)$$
$$(1, 1, 0) \, (3, 1, 1) \, (5, 1, 2)$$
$$(0, 0, 0) \, (2, 0, 1) \, (4, 0, 2) \, (6, 0, 3)$$

apply the de Casteljau algorithm to compute point $\mathbf{P}_{000}^3$, then use Equation (6.37) to compute surface point $\mathbf{P}(1/6, 2/6/3/6)$ and show that the two points are identical.

It can be shown that a general intermediate point $\mathbf{P}_{i,j,k}^r(u, v, w)$ obtained in the scaffolding process can be computed directly from the control points without having to go through the intermediate steps of the scaffolding construction, as follows

$$\mathbf{P}_{ijk}^r(u, v, w) = \sum_{a+b+c=r} B_{abc}^r(u, v, w) \mathbf{P}_{i+a, j+b, k+c}.$$

Example: For $n = 3$ and $r = 1$, point $\mathbf{P}_{002}^1$ is computed directly from the control points as the sum

$$\mathbf{P}_{002}^1 = \sum_{a+b+c=1} B_{abc}^1(u, v, w) \mathbf{P}_{0+a, 0+b, 2+c} = u\mathbf{P}_{102} + v\mathbf{P}_{012} + w\mathbf{P}_{003}.$$

For $n = 3$ and $r = 2$, point $\mathbf{P}_{001}^2$ is computed directly as the sum

$$\mathbf{P}_{001}^2 = \sum_{a+b+c=2} B_{abc}^2(u, v, w) \mathbf{P}_{0+a, 0+b, 1+c}$$
$$= v^2\mathbf{P}_{021} + 2vw\mathbf{P}_{012} + 2uv\mathbf{P}_{111} + w^2\mathbf{P}_{003} + 2uw\mathbf{P}_{102} + u^2\mathbf{P}_{201}.$$

⋄ **Exercise 6.27:** For $n = 4$, compute intermediate points $\mathbf{P}_{001}^3$ and $\mathbf{P}_{111}^1$ directly from the control points.

6.23.2 Subdivision

A triangular Bézier patch can be subdivided into three triangular Bézier patches by a
process similar to the one described in Section 6.8 for the Bézier curve. New control
points for the three new patches are computed in two steps. First, all the intermediate
points generated in the scaffolding steps are computed, then the original interior control
points are deleted. We illustrate this process first for $n = 3$ and $n = 4$, then for the
general case.

A triangular Bézier patch for $n = 3$ is defined by 10 control points, of which nine
are exterior. The user first selects the point inside the surface patch where the three
new triangles will meet. This is done by selecting a barycentric triplet (u, v, w). The
user then executes three steps of the scaffolding process to generate $6 + 3 + 1 = 10$ new
intermediate points. The new points are added to the nine exterior control points and
the single interior point $\mathbf{P}_{111}$ is deleted. The resulting 19 points are divided into three
overlapping sets of 10 points each (Figure 6.33) that define three adjacent triangular
Bézier patches inside the original patch.

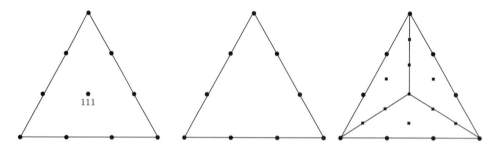

Figure 6.33: Subdividing the Triangular Bézier Patch for $n = 3$.

A triangular Bézier patch for $n = 4$ is defined by 15 control points, of which 12
are exterior. The user selects a barycentric triplet (u, v, w) and executes four steps of
the scaffolding process to generate $9 + 6 + 3 + 1 = 19$ new intermediate points. The
new points are added to the 12 exterior control points and the three interior points are
deleted. The resulting 31 points are divided into three overlapping sets of 15 points each
that define three adjacent triangular Bézier patches inside the original patch.

◇ **Exercise 6.28:** Draw a diagram for this case, similar to Figure 6.33.

In general, a triangular Bézier patch is defined by $\frac{1}{2}(n+1)(n+2)$ control points, of
which $1 + 2 + \underbrace{2 + \cdots + 2}_{n-2} + (n+1) = 3n$ points are exterior. The scaffolding construction
is then performed, creating $3(n-1)$ points in step 1, $3(n-2)$ points in step 2, and so
on, down to $3[n - (n-1)] = 3$ points in step $n-1$ and one point in step n, for a total
of $\frac{3n}{2}(n-1) + 1$ points. For $n = 3$ through 7, these numbers are 10, 19, 31, 46, and 64.
(Note that there are no interior points for $n = 1$ and $n = 2$.) These new points, added
to the original exterior points, provide $\frac{3n}{2}(n-1) + 1 + 3n = \frac{3n}{2}(n+1) + 1$ points. For
$n = 3$ through 7, these numbers are 19, 31, 46, 64, and 84. These numbers are enough

to construct three adjacent triangular Bézier patches defined by $\frac{1}{2}(n+1)(n+2)$ control points each.

The user always starts a subdivision by selecting a surface point $\mathbf{P}(u, v, w)$ where the three new triangular patches will meet. A special case occurs if this point is located on an edge of the original triangular patch (i.e., if one of u, v, or w is zero). In such a case, the original triangle is subdivided into two, instead of three triangular patches. This may be useful in cases where only a few extra points are required to reshape the surface.

6.23.3 Degree Elevation

Section 6.9 describes how to elevate the degree of a Bézier curve. This section employs the same ideas to elevate the degree of a triangular Bézier patch. Given a triangular patch of order n defined by $\frac{1}{2}(n+1)(n+2)$ control points $\mathbf{P}_{ijk}$, it is easy to compute a new set of control points $\mathbf{Q}_{ijk}$ that represent the same surface as a triangular patch of order $n+1$. The basic relation is

$$\sum_{\substack{i,j,k \geq 0}}^{i+j+k=n} \mathbf{P}_{ijk} B_{i,j,k}^n(u, v, w) = \sum_{i+j+k=n+1} \mathbf{Q}_{ijk} B_{i,j,k}^{n+1}(u, v, w).$$

It can be shown, employing methods similar to those of Section 6.9, that the new points $\mathbf{Q}_{ijk}$ are obtained from the original control points $\mathbf{P}_{ijk}$ by

$$\mathbf{Q}_{ijk} = \frac{1}{n+1}\left[i\mathbf{P}_{i-1,j,k} + j\mathbf{P}_{i,j-1,k} + k\mathbf{P}_{i,j,k-1}\right].$$

Example: We elevate the degree of a triangular Bézier patch from $n = 2$ to $n = 3$. The 10 new control points are obtained from the six original ones by

$$\mathbf{Q}_{003} = \mathbf{P}_{002}, \quad \mathbf{Q}_{102} = \tfrac{1}{3}(\mathbf{P}_{002} + 2\mathbf{P}_{101}), \quad \mathbf{Q}_{201} = \tfrac{1}{3}(2\mathbf{P}_{101} + \mathbf{P}_{200}), \quad \mathbf{Q}_{300} = \mathbf{P}_{200},$$
$$\mathbf{Q}_{012} = \tfrac{1}{3}(\mathbf{P}_{002} + 2\mathbf{P}_{011}), \quad \mathbf{Q}_{111} = \tfrac{1}{3}(\mathbf{P}_{011} + \mathbf{P}_{101} + \mathbf{P}_{110}), \quad \mathbf{Q}_{210} = \tfrac{1}{3}(2\mathbf{P}_{110} + \mathbf{P}_{200}),$$
$$\mathbf{Q}_{021} = \tfrac{1}{3}(2\mathbf{P}_{011} + \mathbf{P}_{020}), \quad \mathbf{Q}_{120} = \tfrac{1}{3}(\mathbf{P}_{020} + 2\mathbf{P}_{110}), \quad \mathbf{Q}_{030} = \mathbf{P}_{020}.$$

It is possible to elevate the degree of a patch repeatedly. Each degree elevation increases the number of control points and moves them closer to the actual surface. At the limit, the number of control points approaches infinity and the net of points approaches the surface patch.

6.24 Joining Triangular Bézier Patches

The triangular Bézier surface patch is used in cases where a large surface happens to be easier to break up into triangular patches than into rectangular ones. It is therefore important to discover the conditions for smooth joining of these surface patches. The conditions should be expressed in terms of constraints on the control points.

These constraints are developed here for cubic surface patches, but the principles are the same for higher-degree patches. The idea is to calculate three vectors that are tangent to the surface at the common boundary curve. Intuitively, the condition for a smooth join is that these vectors be coplanar (although they can have different magnitudes). We proceed in three steps:

Step 1. Figure 6.34 shows two triangular Bézier cubic patches, $\mathbf{P}(u, v, w)$ and $\mathbf{Q}(u, v, w)$, joined at the common boundary curve $\mathbf{P}(0, v, w) = \mathbf{Q}(0, v, w)$. Equation (6.39) shows how the boundary curves can be expressed as Bézier curves. Based on this equation, our common boundary curve can be written

$$\mathbf{P}(v) = \sum_{j+k=3} \frac{3!}{j!\,k!} v^j (1-v)^{3-j} \mathbf{P}_{0jk}.$$

This is easy to differentiate with respect to v and the result is

$$\frac{d\mathbf{P}(v)}{dv} = 3v^2(\mathbf{P}_{030} - \mathbf{P}_{021}) + 6v(1-v)(\mathbf{P}_{021} - \mathbf{P}_{012}) + 3(1-v)^2(\mathbf{P}_{012} - \mathbf{P}_{003})$$
$$= 3v^2\mathbf{B}_3 + 6v(1-v)\mathbf{B}_2 + (1-v)^2\mathbf{B}_1, \tag{6.41}$$

where each of the $\mathbf{B}_i$ vectors is defined as the difference of two control points. They can be seen in the figure as thick arrows going from $\mathbf{P}_{003}$ to $\mathbf{P}_{030}$.

Step 2. Another vector is computed that's tangent to the patch $\mathbf{P}(u, v, w)$ along the common boundary. This is done by calculating the tangent vector to the surface in the u direction and substituting $u = 0$. We first write the expression for the surface patch without the parameter w (it can be eliminated because $w = 1 - u - v$):

$$\mathbf{P}(u, v) = \sum_{i+j+k=3} \frac{3!}{i!\,j!\,k!} u^i v^j (1-u-v)^k \mathbf{P}_{ijk}.$$

This is easy to differentiate with respect to u and it yields

$$\left. \frac{\partial \mathbf{P}(u, v)}{\partial u} \right|_{u=0} = 3v^2(\mathbf{P}_{120} - \mathbf{P}_{021}) + 6v(1-v)(\mathbf{P}_{111} - \mathbf{P}_{012})$$
$$+ 3(1-v)^2(\mathbf{P}_{102} - \mathbf{P}_{003}) \tag{6.42}$$
$$= 3v^2\mathbf{A}_3 + 6v(1-v)\mathbf{A}_2 + 3(1-v)^2\mathbf{A}_1,$$

where each of the $\mathbf{A}_i$ vectors is again defined as the difference of two control points. They can be seen in the figure as thick arrows going, for example, from $\mathbf{P}_{003}$ to $\mathbf{P}_{102}$.

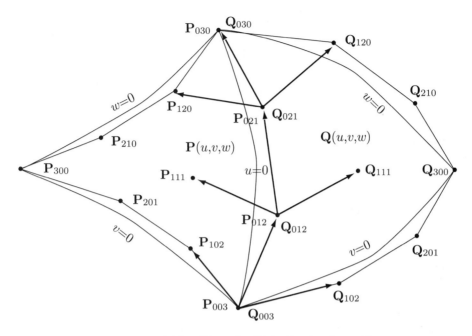

Figure 6.34: Joining Triangular Bézier Patches Smoothly.

Step 3. The third vector is the tangent to the other surface patch $\mathbf{Q}(u, v, w)$ along the common boundary. It is expressed as

$$
\begin{aligned}
\left.\frac{\partial \mathbf{Q}(u, v)}{\partial u}\right|_{u=0} &= 3v^2(\mathbf{Q}_{120} - \mathbf{Q}_{021}) + 6v(1-v)(\mathbf{Q}_{111} - \mathbf{Q}_{012}) \\
&\quad + 3(1-v)^2(\mathbf{Q}_{102} - \mathbf{Q}_{003}) \\
&= 3v^2\mathbf{C}_3 + 6v(1-v)\mathbf{C}_2 + 3(1-v)^2\mathbf{C}_1,
\end{aligned}
\tag{6.43}
$$

where each of the $\mathbf{C}_i$ vectors is again defined as the difference of two control points. They can be seen in the figure as thick arrows going, for example, from $\mathbf{Q}_{003}$ to $\mathbf{Q}_{102}$.

The condition for smooth joining is that the vectors defined by Equations (6.41) through (6.43) be coplanar for any value of v. This can be expressed as

$$
\begin{aligned}
3v^2\mathbf{B}_3 &+ 6v(1-v)\mathbf{B}_2 + (1-v)^2\mathbf{B}_1 \\
&= \alpha(3v^2\mathbf{A}_3 + 6v(1-v)\mathbf{A}_2 + 3(1-v)^2\mathbf{A}_1) \\
&\quad + \beta(3v^2\mathbf{C}_3 + 6v(1-v)\mathbf{C}_2 + 3(1-v)^2\mathbf{C}_1),
\end{aligned}
\tag{6.44}
$$

or, equivalently,

$$
v^2(\mathbf{B}_3 - \alpha\mathbf{A}_3 - \beta\mathbf{C}_3) + 2v(1-v)(\mathbf{B}_2 - \alpha\mathbf{A}_2 - \beta\mathbf{C}_2) + (1-v)^2(\mathbf{B}_1 - \alpha\mathbf{A}_1 - \beta\mathbf{C}_1) = 0.
$$

Since this should hold for any value of v, it can be written as the set of three equations:

$$\mathbf{B}_1 = \alpha\mathbf{A}_1 + \beta\mathbf{C}_1,$$
$$\mathbf{B}_2 = \alpha\mathbf{A}_2 + \beta\mathbf{C}_2, \qquad (6.45)$$
$$\mathbf{B}_3 = \alpha\mathbf{A}_3 + \beta\mathbf{C}_3.$$

Each of the three sets of vectors $\mathbf{B}_i$, $\mathbf{A}_i$, and $\mathbf{C}_i$ ($i = 1, 2, 3$) should therefore be coplanar. This condition can be expressed for the control points by saying that each of the three quadrilaterals given by

$$\mathbf{P}_{003} = \mathbf{Q}_{003}, \quad \mathbf{P}_{102}, \quad \mathbf{P}_{012} = \mathbf{Q}_{012}, \quad \mathbf{Q}_{102},$$
$$\mathbf{P}_{012} = \mathbf{Q}_{012}, \quad \mathbf{P}_{111}, \quad \mathbf{P}_{021} = \mathbf{Q}_{021}, \quad \mathbf{Q}_{111},$$
$$\mathbf{P}_{021} = \mathbf{Q}_{021}, \quad \mathbf{P}_{120}, \quad \mathbf{P}_{030} = \mathbf{Q}_{030}, \quad \mathbf{Q}_{120},$$

should be planar. In the special case $\alpha = \beta = 1$, each quadrilateral should be a square. Otherwise, each should have the same ratio of height to width.

The condition for such a set of three vectors to be coplanar is simple to derive. Figure 6.35 shows a quadrilateral with four corner points $\mathbf{A}$, $\mathbf{B}$, $\mathbf{C}$, and $\mathbf{D}$. Two dashed segments are shown, connecting $\mathbf{A}$ to $\mathbf{B}$ and $\mathbf{C}$ to $\mathbf{D}$. The condition for a flat quadrilateral (four coplanar corners) is that the two segments intersect. The first segment can be expressed parametrically as $(1 - u)\mathbf{A} + u\mathbf{B}$ and the second segment can be similarly expressed as $(1 - w)\mathbf{C} + w\mathbf{D}$. If there exist u and w in the interval $[0, 1]$ such that $(1 - u)\mathbf{A} + u\mathbf{B} = (1 - w)\mathbf{C} + w\mathbf{D}$, then the quadrilateral is flat.

Figure 6.35: A Quadrilateral.

6.24.1 Joining Rectangular and Triangular Bézier Patches

A smooth joining of a rectangular and a triangular surface patches, both of order n, may be useful in many practical applications. Figure 6.36a shows the numbering of the control points for the case $n = 4$. Points $\mathbf{Q}_{ijk}$ define the triangular patch and points $\mathbf{P}_{ij}$ define the rectangular patch. There are four pairs (in general, n pairs) of identical points. The problem of joining surface patches of such different topologies can be greatly simplified by elevating the degree (Section 6.23.3) of the two rightmost columns of control points of the triangular patch. The column of four points $\mathbf{Q}_{0jk}$ where $j + k = 3$ is transformed to five points $\mathbf{R}_{0jk}$ where $j + k = 4$, and the column of three points $\mathbf{Q}_{1jk}$ where $1 + j + k = 3$ is transformed to four points $\mathbf{R}_{1jk}$ where $1 + j + k = 4$. Figure 6.36b shows the new points and how, together with the column of four points $\mathbf{P}_{10}$ through $\mathbf{P}_{13}$, they create four quadrilaterals. The condition for smooth joining of the patches is that each quadrilateral be flat.

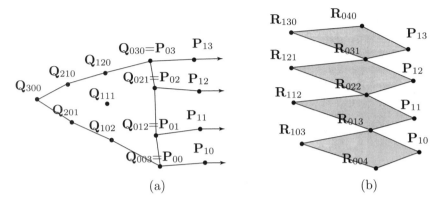

Figure 6.36: Smooth Joining of Triangular and Rectangular Bézier Surface Patches.

In general, there are $n + 1$ such quadrilaterals, and each condition can be written explicitly, as an equation, in terms of some of the points $\mathbf{P}_{1i}$, $\mathbf{Q}_{0jk}$, and $\mathbf{Q}_{1jk}$. A general equation is

$$(1 - \alpha)\mathbf{R}_{0,i,n-i} + \alpha\mathbf{R}_{0,i+1,n-i} = (1 - \beta)\mathbf{R}_{1,i,n-i} + \beta\mathbf{P}_{1,i}, \quad \text{for} \quad i = 0, 1, \ldots, n.$$

When the $\mathbf{R}_{ijk}$ points are expressed in terms of the original $\mathbf{Q}_{ijk}$ points, this relation becomes

$$\frac{1 - \alpha}{n}\left[i\mathbf{Q}_{0,i-1,n-i} + (n - i)\mathbf{Q}_{0,i,n-i}\right] + \frac{\alpha}{n}[(i + 1)\mathbf{Q}_{0,i,n-i} + (n - i)\mathbf{Q}_{0,i+1,n-i-1}]$$

$$= \frac{\beta}{n}\left[\mathbf{Q}_{0,i,n-i} + i\mathbf{Q}_{1,i-1,n-i} + (n - i)\mathbf{Q}_{1,i,n-i-1}\right] + \beta\mathbf{P}_{1i}.$$

Note that the quantities α and β in these equations should be indexed by i. In general, each quadrilateral has its own α_i and β_i, but the surface designer can start by guessing values for these $2(n + 1)$ quantities, then use them as parameters and vary them (while still keeping each quadrilateral flat), until the surface is molded to the desired shape.

If the rectangular patch is given and the triangular patch has to be designed and manipulated to connect smoothly to it, then the n points $\mathbf{Q}_{1jk}$ (the column to the left of the common boundary) are the unknowns. Conversely, if we start from the triangular patch and want to select control points for the rectangular patch, then the unknowns are the $n + 1$ control points $\mathbf{P}_{1i}$ (the column to the right of the common boundary). [Liu and Hoschek 89] has a detailed analysis of the conditions for smooth connection of various types of Bézier surface patches.

6.25 Reparametrizing the Bézier Surface

We illustrate the method described here by applying it to the bicubic Bézier surface patch. The expression for this patch is given by Equations (6.32) and (6.31):

$$\mathbf{P}(u,w) = \sum_{i=0}^{3}\sum_{j=0}^{3} B_{3,i}(u)\mathbf{P}_{i,j}B_{3,j}(w)$$

$$= \sum_{i=0}^{3}\sum_{j=0}^{3}(u^3, u^2, u, 1)\mathbf{M}\mathbf{P}\mathbf{M}^{-1}(w^3, w^2, w, 1)^T,$$

where $\mathbf{M}$ is the basis matrix

$$\mathbf{M} = \begin{pmatrix} -1 & 3 & -3 & 1 \\ 3 & -6 & 3 & 0 \\ -3 & 3 & 0 & 0 \\ 1 & 0 & 0 & 0 \end{pmatrix}$$

and $\mathbf{P}$ is the 4×4 matrix of control points

$$\begin{pmatrix} \mathbf{P}_{3,0} & \mathbf{P}_{3,1} & \mathbf{P}_{3,2} & \mathbf{P}_{3,3} \\ \mathbf{P}_{2,0} & \mathbf{P}_{2,1} & \mathbf{P}_{2,2} & \mathbf{P}_{2,3} \\ \mathbf{P}_{1,0} & \mathbf{P}_{1,1} & \mathbf{P}_{1,2} & \mathbf{P}_{1,3} \\ \mathbf{P}_{0,0} & \mathbf{P}_{0,1} & \mathbf{P}_{0,2} & \mathbf{P}_{0,3} \end{pmatrix}.$$

This surface patch can be reparametrized with the method of Section 6.10. We select part of patch $\mathbf{P}(u,w)$, e.g., the part where u varies from a to b, and define it as a new patch $\mathbf{Q}(u,w)$ where both u and w vary in the range $[0,1]$. The method discussed here shows how to obtain the control points $\mathbf{Q}_{ij}$ of patch $\mathbf{Q}(u,w)$ as functions of a, b and points $\mathbf{P}_{ij}$.

> B-splines are the defacto standard that drives today's sophisticated computer graphics applications. This method is also responsible for the developments that have transformed computer-aided geometric design from the era of hand-built models and manual measurements to fast computations and three-dimensional renderings.

Suppose that we want to reparametrize the "left" part of $\mathbf{P}(u,w)$, i.e., the part where $0 \le u \le 0.5$. Applying the methods of Section 6.10, we select $a = 0$, $b = 0.5$ and can write

$$\mathbf{P}(u/2, w) = (u^3, u^2, u, 1)\mathbf{M}\mathbf{B}\mathbf{P}\mathbf{M}^{-1}(w^3, w^2, w, 1)^T,$$

where $\mathbf{B}$ is given by Equation (6.22)

$$\mathbf{B} = \begin{pmatrix} (1-a)^3 & 3(a-1)^2a & 3(1-a)a^2 & a^3 \\ (a-1)^2(1-b) & (a-1)(-2a-b+3ab) & a(a+2b-3ab) & a^2b \\ (1-a)(-1+b)^2 & (b-1)(-a-2b+3ab) & b(2a+b-3ab) & ab^2 \\ (1-b)^3 & 3(b-1)^2b & 3(1-b)b^2 & b^3 \end{pmatrix}.$$

Exercise 6.17 shows that selecting $a = 0$ and $b = 0.5$ reduces matrix $\mathbf{B}$ to

$$
\mathbf{B} = \begin{pmatrix} 1 & 0 & 0 & 0 \\ \frac{1}{2} & \frac{1}{2} & 0 & 0 \\ \frac{1}{4} & \frac{1}{2} & \frac{1}{4} & 0 \\ \frac{1}{8} & \frac{3}{8} & \frac{3}{8} & \frac{1}{8} \end{pmatrix}.
$$

The new control points for our surface patch are therefore

$$
\begin{pmatrix} \mathbf{Q}_{3,0} & \mathbf{Q}_{3,1} & \mathbf{Q}_{3,2} & \mathbf{Q}_{3,3} \\ \mathbf{Q}_{2,0} & \mathbf{Q}_{2,1} & \mathbf{Q}_{2,2} & \mathbf{Q}_{2,3} \\ \mathbf{Q}_{1,0} & \mathbf{Q}_{1,1} & \mathbf{Q}_{1,2} & \mathbf{Q}_{1,3} \\ \mathbf{Q}_{0,0} & \mathbf{Q}_{0,1} & \mathbf{Q}_{0,2} & \mathbf{Q}_{0,3} \end{pmatrix} = \begin{pmatrix} 1 & 0 & 0 & 0 \\ \frac{1}{2} & \frac{1}{2} & 0 & 0 \\ \frac{1}{4} & \frac{1}{2} & \frac{1}{4} & 0 \\ \frac{1}{8} & \frac{3}{8} & \frac{3}{8} & \frac{1}{8} \end{pmatrix} \begin{pmatrix} \mathbf{P}_{3,0} & \mathbf{P}_{3,1} & \mathbf{P}_{3,2} & \mathbf{P}_{3,3} \\ \mathbf{P}_{2,0} & \mathbf{P}_{2,1} & \mathbf{P}_{2,2} & \mathbf{P}_{2,3} \\ \mathbf{P}_{1,0} & \mathbf{P}_{1,1} & \mathbf{P}_{1,2} & \mathbf{P}_{1,3} \\ \mathbf{P}_{0,0} & \mathbf{P}_{0,1} & \mathbf{P}_{0,2} & \mathbf{P}_{0,3} \end{pmatrix}
$$

$$
= \begin{pmatrix} \mathbf{P}_{3,0} & \mathbf{P}_{3,1} \\ \frac{1}{2}\mathbf{P}_{3,0} + \frac{1}{2}\mathbf{P}_{2,0} & \frac{1}{2}\mathbf{P}_{3,1} + \frac{1}{2}\mathbf{P}_{2,1} \\ \frac{1}{4}\mathbf{P}_{3,0} + \frac{1}{2}\mathbf{P}_{2,0} + \frac{1}{4}\mathbf{P}_{1,0} & \frac{1}{4}\mathbf{P}_{3,1} + \frac{1}{2}\mathbf{P}_{2,1} + \frac{1}{4}\mathbf{P}_{1,1} \\ \frac{1}{8}\mathbf{P}_{3,0} + \frac{3}{8}\mathbf{P}_{2,0} + \frac{3}{8}\mathbf{P}_{1,0} + \frac{1}{8}\mathbf{P}_{0,0} & \frac{1}{8}\mathbf{P}_{3,1} + \frac{3}{8}\mathbf{P}_{2,1} + \frac{3}{8}\mathbf{P}_{1,1} + \frac{1}{8}\mathbf{P}_{1,0} \\ \mathbf{P}_{3,2} & \mathbf{P}_{3,3} \\ \frac{1}{2}\mathbf{P}_{3,2} + \frac{1}{2}\mathbf{P}_{2,2} & \frac{1}{2}\mathbf{P}_{3,3} + \frac{1}{2}\mathbf{P}_{2,3} \\ \frac{1}{4}\mathbf{P}_{3,2} + \frac{1}{2}\mathbf{P}_{2,2} + \frac{1}{4}\mathbf{P}_{1,2} & \frac{1}{4}\mathbf{P}_{3,3} + \frac{1}{2}\mathbf{P}_{2,3} + \frac{1}{4}\mathbf{P}_{1,3} \\ \frac{1}{8}\mathbf{P}_{3,2} + \frac{3}{8}\mathbf{P}_{2,2} + \frac{3}{8}\mathbf{P}_{1,2} + \frac{1}{8}\mathbf{P}_{2,0} & \frac{1}{8}\mathbf{P}_{3,3} + \frac{3}{8}\mathbf{P}_{2,3} + \frac{3}{8}\mathbf{P}_{1,3} + \frac{1}{8}\mathbf{P}_{3,0} \end{pmatrix}.
$$

In general, suppose we want to reparametrize that portion of patch $\mathbf{P}(u, w)$ where $a \leq u \leq b$ and $c \leq w \leq d$. We can write

$$
\mathbf{Q}(u, w)
$$
$$
= \mathbf{P}([b - a]u + a, [d - c]w + c)
$$
$$
= \left(([b-a]u+a)^3, ([b-a]u+a)^2, ([b-a]u+a), 1 \right) \mathbf{M} \cdot \mathbf{P} \cdot \mathbf{M}^{-1} \begin{pmatrix} ([d-c]w+c)^3 \\ ([d-c]w+c)^2 \\ [d-c]w+c \\ 1 \end{pmatrix}
$$
$$
= (u^3, u^2, u, 1)\mathbf{A}_{ab}\mathbf{M} \cdot \mathbf{P} \cdot \mathbf{M}^T \cdot \mathbf{A}_{cd}^T(w^3, w^2, w, 1)^T
$$
$$
= (u^3, u^2, u, 1)\mathbf{M}(\mathbf{M}^{-1} \cdot \mathbf{A}_{ab} \cdot \mathbf{M})\mathbf{P}(\mathbf{M}^T \cdot \mathbf{A}_{cd}^T \cdot (\mathbf{M}^T)^{-1})\mathbf{M}^T(w^3, w^2, w, 1)^T
$$
$$
= (u^3, u^2, u, 1)\mathbf{M} \cdot \mathbf{B}_{ab} \cdot \mathbf{P} \cdot \mathbf{B}_{cd}^T \cdot \mathbf{M}^T(w^3, w^2, w, 1)^T
$$
$$
= (u^3, u^2, u, 1)\mathbf{M} \cdot \mathbf{Q} \cdot \mathbf{M}^T(w^3, w^2, w, 1)^T, \tag{6.46}
$$

where $\mathbf{B}_{ab} = \mathbf{M}^{-1} \cdot \mathbf{A}_{ab} \cdot \mathbf{M}$, $\mathbf{B}_{cd}^T = \mathbf{M}^T \cdot \mathbf{A}_{cd}^T \cdot (\mathbf{M}^T)^{-1}$, $\mathbf{Q} = \mathbf{B}_{ab} \cdot \mathbf{P} \cdot \mathbf{B}_{cd}^T$, and

$$
\mathbf{A}_{ab} = \begin{pmatrix} (b-a)^3 & 0 & 0 & 0 \\ 3a(b-a)^2 & (b-a)^2 & 0 & 0 \\ 3a^2(b-a) & 2a(b-a) & b-a & 0 \\ a^3 & a^2 & a & 1 \end{pmatrix}.
$$

The elements of $\mathbf{Q}$ depend on a, b, c, and d, and the $\mathbf{P}_{ij}$'s and are quite complex. They can be produced by the following *Mathematica* code:

```
B={{(1 - a)^3, 3*(-1 + a)^2*a, 3*(1 - a)*a^2, a^3},
  {(-1 + a)^2*(1 - b), (-1 + a)*(-2*a - b + 3*a*b),
   a*(a + 2*b - 3*a*b),
   a^2*b}, {(1 - a)*(-1 + b)^2, (-1 + b)*(-a - 2*b + 3*a*b),
   b*(2*a + b - 3*a*b), a*b^2},
  {(1 - b)^3, 3*(-1 + b)^2*b, 3*(1 - b)*b^2, b^3}};
TB={{(1 - c)^3, (-1 + c)^2*(1 - d), (1 - c)*(-1 + d)^2,
   (1 - d)^3},
  {3*(-1 + c)^2*c, (-1 + c)*(-2*c - d + 3*c*d),
   (-1 + d)*(-c - 2*d + 3*c*d), 3*(-1 + d)^2*d},
  {3*(1 - c)*c^2, c*(c + 2*d - 3*c*d), d*(2*c + d - 3*c*d),
   3*(1 - d)*d^2},
  {c^3, c^2*d, c*d^2, d^3}};
P={{P30,P31,P32,P33},{P20,P21,P22,P23},
  {P10,P11,P12,P13},{P00,P01,P02,P03}};
Q=Simplify[B.P.TB]
```

6.26 The Gregory Patch

John A. Gregory developed this method to extend the Coons surface patch. The Gregory method, however, becomes very practical when it is applied to extend the bicubic Bézier patch. Recall that such a patch is based on $4 \times 4 = 16$ control points (Figure 6.37a). We can divide the 16 points into two groups: the interior points, consisting of the four points $\mathbf{P}_{11}$, $\mathbf{P}_{12}$, $\mathbf{P}_{21}$, and $\mathbf{P}_{22}$, and the boundary points, consisting of the remaining 12 points. Experience shows that there are too few interior points to fine-tune the shape of the patch. Moving point $\mathbf{P}_{11}$, for example, affects both the direction from $\mathbf{P}_{01}$ to $\mathbf{P}_{11}$, and the direction from $\mathbf{P}_{10}$ to $\mathbf{P}_{11}$.

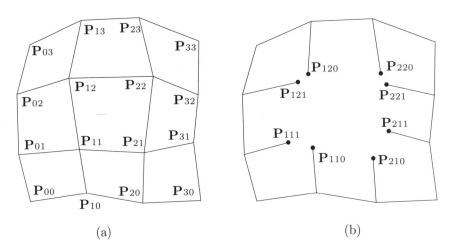

(a)   (b)

Figure 6.37: (a) A Bicubic Bézier Patch. (b) A Gregory Patch.

The idea in the Gregory patch is to split each of the four interior points into two points. Hence, instead of point $\mathbf{P}_{11}$, for example, there should be two points $\mathbf{P}_{110}$ and $\mathbf{P}_{111}$, both in the vicinity of the original $\mathbf{P}_{11}$. Moving $\mathbf{P}_{110}$ affects the shape of the patch only in the direction from $\mathbf{P}_{10}$ to $\mathbf{P}_{110}$. The shape of the patch around point $\mathbf{P}_{01}$ is not affected (at least, not significantly). Thus, the bicubic Gregory patch is defined by 20 points (Figure 6.37b), eight interior points and 12 boundary points. Points $\mathbf{P}_{110}$ and $\mathbf{P}_{111}$ can initially be set equal to $\mathbf{P}_{11}$, then moved interactively in different directions to obtain the right shape of the surface.

To calculate the surface, we first define 16 new points $\mathbf{Q}_{ij}$, then use Equation (6.31) with the new points as control points and with $n = m = 3$. Twelve of the $\mathbf{Q}$ points are boundary points and are identical to the boundary $\mathbf{P}$ points. The remaining four $\mathbf{Q}$ points are interior and each is calculated from a pair of interior $\mathbf{P}$ points. Their definitions are the following

$$\mathbf{Q}_{11}(u, w) = \frac{u\mathbf{P}_{110} + w\mathbf{P}_{111}}{u + w}, \qquad \mathbf{Q}_{21}(u, w) = \frac{(1 - u)\mathbf{P}_{210} + w\mathbf{P}_{211}}{1 - u + w},$$
$$\mathbf{Q}_{12}(u, w) = \frac{u\mathbf{P}_{120} + (1 - w)\mathbf{P}_{121}}{u + 1 - w}, \qquad \mathbf{Q}_{22}(u, w) = \frac{(1 - u)\mathbf{P}_{220} + (1 - w)\mathbf{P}_{221}}{1 - u + 1 - w}.$$

Note that $\mathbf{Q}_{11}(u, w)$ is a barycentric sum of two $\mathbf{P}$ points, so it is well defined. Even though u and w are independent and each is varied from 0 to 1 independently of the other, the sum is always a point on the straight segment connecting $\mathbf{P}_{110}$ to $\mathbf{P}_{111}$. The same is true for the other three interior $\mathbf{Q}$ points.

After calculating the new points, the Gregory patch is defined as the bicubic Bézier patch

$$\mathbf{P}(u, w) = \sum_{i=0}^{3} \sum_{j=0}^{3} B_{3,i}(w)\mathbf{Q}_{i,j}B_{3,j}(u).$$

(Note that four of the 16 points $\mathbf{Q}_{i,j}$ depend on the parameters u and w.)

6.26.1 The Gregory Tangent Vectors

The first derivatives of the Gregory patch are more complex than those of the bicubic Bézier patch, because four of the control points depend on the parameters u and w. The derivatives are

$$\frac{\partial \mathbf{P}(u, w)}{\partial u}$$
$$= \sum_{i=0}^{3} \sum_{j=0}^{3} \frac{d\, B_{3,i}(u)}{du} B_{3,j}(w)\mathbf{Q}_{i,j}(u, w) + \sum_{i=0}^{3} \sum_{j=0}^{3} B_{3,i}(u)B_{3,j}(w)\frac{\partial \mathbf{Q}_{i,j}(u, w)}{\partial u},$$
$$\frac{\partial \mathbf{P}(u, w)}{\partial w}$$
$$= \sum_{i=0}^{3} \sum_{j=0}^{3} B_{3,i}(u)\frac{d\, B_{3,j}(w)}{dw}\mathbf{Q}_{i,j}(u, w) + \sum_{i=0}^{3} \sum_{j=0}^{3} B_{3,i}(u)B_{3,j}(w)\frac{\partial \mathbf{Q}_{i,j}(u, w)}{\partial w}.$$

Each derivative is the sum of two similar terms, each of which has the same format as a derivative of the bicubic Bézier patch. Therefore, only one procedure is needed to calculate the derivatives numerically. This procedure is called twice for each partial derivative. The second call involves the derivatives of the control points, which are shown here.

The 12 boundary $\mathbf{Q}$ points don't depend on u or w, so their derivatives are zero. The eight derivatives of the four interior points are

$$\frac{\partial \mathbf{Q}_{11}(u, w)}{\partial u} = \frac{w(\mathbf{P}_{110} - \mathbf{P}_{111})}{(u + w)^2}, \qquad \frac{\partial \mathbf{Q}_{11}(u, w)}{\partial w} = \frac{u(\mathbf{P}_{110} - \mathbf{P}_{111})}{(u + w)^2},$$

$$\frac{\partial \mathbf{Q}_{21}(u, w)}{\partial u} = \frac{w(\mathbf{P}_{210} - \mathbf{P}_{211})}{(1 - u + w)^2}, \qquad \frac{\partial \mathbf{Q}_{21}(u, w)}{\partial w} = \frac{(1 - u)(\mathbf{P}_{210} - \mathbf{P}_{211})}{(1 - u + w)^2},$$

$$\frac{\partial \mathbf{Q}_{12}(u, w)}{\partial u} = \frac{(1 - w)(\mathbf{P}_{120} - \mathbf{P}_{121})}{(u + 1 - w)^2}, \qquad \frac{\partial \mathbf{Q}_{12}(u, w)}{\partial w} = \frac{u(\mathbf{P}_{120} - \mathbf{P}_{121})}{(u + 1 - w)^2},$$

$$\frac{\partial \mathbf{Q}_{22}(u, w)}{\partial u} = \frac{(1 - w)(\mathbf{P}_{220} - \mathbf{P}_{221})}{(1 - u + 1 - w)^2}, \qquad \frac{\partial \mathbf{Q}_{22}(u, w)}{\partial w} = \frac{(1 - u)(\mathbf{P}_{220} - \mathbf{P}_{221})}{(1 - u + 1 - w)^2}.$$

After the first derivatives (the tangent vectors) have been calculated numerically at a point, they are used to numerically calculate the normal vector at the point.

It is interesting to observe that the Bernshteĭn polynomial of degree 1, i.e., the function $z(t) = (1 - t)z_1 + t z_2$, is precisely the mediation operator $t[z_1, z_2]$ that we discussed in the previous chapter.

Donald Knuth, *The MetafontBook* (1986)

7
B-Spline Approximation

B-spline methods for curves and surfaces were first proposed in the 1940s but were seriously developed only in the 1970s, by several researchers, most notably R. Riesenfeld. They have been studied extensively, have been considerably extended since the 1970s, and much is currently known about them. The designation "B" stands for Basis, so the full name of this approach to curve and surface design is the basis spline. This chapter discusses the important types of B-spline curves and surfaces, including the most versatile one, the nonuniform rational B-spline (NURBS, Section 7.14).

The B-spline curve overcomes the main disadvantages of the Bézier curve which are (1) the degree of the Bézier curve depends on the number of control points, (2) it offers only global control, and (3) individual segments are easy to connect with C^1 continuity, but C^2 is difficult to obtain. The B-spline curve features local control and any desired degree of continuity. To obtain C^n continuity, the individual spline segments have to be polynomials of degree n. The B-spline curve is an approximating curve and is therefore defined by control points. However, in addition to the control points, the user has to specify the values of certain quantities called "knots." They are real numbers that offer additional control over the shape of the curve. The basic approach taken in the first part of this chapter ignores the knots, but they are introduced in Section 7.8 and their effect on the curve is explored.

There are several types of B-splines. In the *uniform* (also called periodic) B-spline (Sections 7.1 and 7.2), the knot values are uniformly spaced and all the weight functions have the same shape and are shifted with respect to each other. In the *nonuniform* B-spline (Section 7.11), the knots are specified by the user and the weight functions are generally different. There is also an *open uniform* B-spline (Section 7.10), where the knots are not uniform but are specified in a simple way. In a *rational* B-spline (Section 7.14), the weight functions are in the form of a ratio of two polynomials. In a *nonrational* B-spline, they are polynomials in t. The B-spline is an approximating curve based on control points, but there is also an *interpolating* version that passes through the points (Section 7.7). Section 7.4 shows how tension can be added to the B-spline.

B-splines are mathematically more sophisticated than other types of splines, so we start with a gentle introduction. We first use basic assumptions to derive the expressions for the quadratic and cubic uniform B-splines directly and without mentioning knots. We then show how to extend the derivations to uniform B-splines of any order. Following this, we discuss a different, recursive formulation of the weight functions of the uniform, open uniform, and nonuniform B-splines.

7.1 The Quadratic Uniform B-Spline

We start with the quadratic uniform B-spline. We assume that $n + 1$ control points, $\mathbf{P}_0, \mathbf{P}_1, \ldots, \mathbf{P}_n$, are given and we want to construct a spline curve where each segment $\mathbf{P}_i(t)$ is a quadratic parametric polynomial based on three points, $\mathbf{P}_{i-1}$, $\mathbf{P}_i$, and $\mathbf{P}_{i+1}$. We require that the segments connect with C^1 continuity (only cubic and higher-degree polynomial segments can have C^2 or higher continuities) and that the entire curve has local control. To achieve all this, we have to give up something and we elect to give up the requirement that a segment will pass through its first and last control points. We denote the start and end points of segment $\mathbf{P}_i(t)$ by $\mathbf{K}_i$ and $\mathbf{K}_{i+1}$, respectively and we call them *joint points*, or just *joints*. These points are still unknown and will have to be determined. Figure 7.1a shows two quadratic segments $\mathbf{P}_1(t)$ and $\mathbf{P}_2(t)$ defined by the four control points $\mathbf{P}_0$, $\mathbf{P}_1$, $\mathbf{P}_2$, and $\mathbf{P}_3$. The first segment goes from joint $\mathbf{K}_1$ to joint $\mathbf{K}_2$ and the second segment goes from joint $\mathbf{K}_2$ to joint $\mathbf{K}_3$, where the joints are drawn tentatively and will have to be determined and redrawn. Note that each segment is defined by three control points, so its control polygon has two edges. The first spline segment is defined only by $\mathbf{P}_0$, $\mathbf{P}_1$, and $\mathbf{P}_2$, so any changes in $\mathbf{P}_3$ will not affect it. This is how local control is achieved in a B-spline.

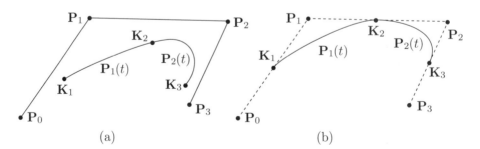

(a)   (b)

Figure 7.1: The Quadratic Uniform B-Spline.

We use the usual notation for the two segments

$$\mathbf{P}_i(t) = (t^2, t, 1)\mathbf{M} \begin{pmatrix} \mathbf{P}_{i-1} \\ \mathbf{P}_i \\ \mathbf{P}_{i+1} \end{pmatrix}, \quad i = 1, 2, \tag{7.1}$$

where $\mathbf{M}$ is the 3×3 basis matrix whose nine elements have to be calculated. We define three functions $a(t)$, $b(t)$, and $c(t)$ by:

$$(t^2, t, 1)\mathbf{M} = (t^2, t, 1) \begin{pmatrix} a_2 & b_2 & c_2 \\ a_1 & b_1 & c_1 \\ a_0 & b_0 & c_0 \end{pmatrix}$$

$$= (a_2 t^2 + a_1 t + a_0, b_2 t^2 + b_1 t + b_0, c_2 t^2 + c_1 t + c_0)$$

$$= \big(a(t), b(t), c(t)\big). \tag{7.2}$$

The nine elements of $\mathbf{M}$ are determined from the following three requirements:

1. The two segments should meet at a common joint and their tangent vectors should be equal at that point. This is expressed as

$$\mathbf{P}_1(1) = \mathbf{P}_2(0), \quad \mathbf{P}_1^t(1) = \mathbf{P}_2^t(0) \tag{7.3}$$

and produces the explicit equations (where a dot indicates differentiation with respect to t)

$$a(1)\mathbf{P}_0 + b(1)\mathbf{P}_1 + c(1)\mathbf{P}_2 = a(0)\mathbf{P}_1 + b(0)\mathbf{P}_2 + c(0)\mathbf{P}_3,$$
$$\dot{a}(1)\mathbf{P}_0 + \dot{b}(1)\mathbf{P}_1 + \dot{c}(1)\mathbf{P}_2 = \dot{a}(0)\mathbf{P}_1 + \dot{b}(0)\mathbf{P}_2 + \dot{c}(0)\mathbf{P}_3.$$

Since the control points $\mathbf{P}_i$ are arbitrary and can be any points, we can rewrite these two equations in the form

$$\begin{aligned} a(1) &= 0, & \dot{a}(1) &= 0, & \text{for } \mathbf{P}_0, \\ b(1) &= a(0), & \dot{b}(1) &= \dot{a}(0), & \text{for } \mathbf{P}_1, \\ c(1) &= b(0), & \dot{c}(1) &= \dot{b}(0), & \text{for } \mathbf{P}_2, \\ 0 &= c(0), & 0 &= \dot{c}(0), & \text{for } \mathbf{P}_3. \end{aligned}$$

Using the notation of Equation (7.2), this can be written

$$\begin{aligned} a_2 + a_1 + a_0 &= 0, & 2a_2 + a_1 &= 0, \\ b_2 + b_1 + b_0 &= a_0, & 2b_2 + b_1 &= 0, \\ c_2 + c_1 + c_0 &= b_0, & 2c_2 + c_1 &= 0, \\ 0 &= c_0, & 0 &= c_1. \end{aligned} \tag{7.4}$$

This requirement produces eight equations for the nine unknown matrix elements.

2. The entire curve should be independent of the particular coordinate system used, which implies that the weight functions of each segment should be barycentric, i.e., $a(t) + b(t) + c(t) \equiv 1$. This condition can be written explicitly as

$$a_2 + b_2 + c_2 = 0, \quad a_1 + b_1 + c_1 = 0, \quad a_0 + b_0 + c_0 = 1, \tag{7.5}$$

and these add three more equations.

We now have 11 equations for the nine unknowns, but it is easy to show that only nine of the 11 are independent. The sum of the first two of Equations (7.5) equals the sum of the three equations in the right column of Equation (7.4). Taking this into account, the equations can be solved uniquely, yielding

$$a_2 = 1/2, \quad a_1 = -1, \quad a_0 = 1/2,$$
$$b_2 = -1, \quad b_1 = 1, \quad b_0 = 1/2,$$
$$c_2 = 1/2, \quad c_1 = 0, \quad c_0 = 0.$$

The general quadratic B-spline segment, Equation (7.1), can now be written as

$$
\mathbf{P}_i(t) = \frac{1}{2}(t^2, t, 1)
\begin{pmatrix}
1 & -2 & 1 \\
-2 & 2 & 0 \\
1 & 1 & 0
\end{pmatrix}
\begin{pmatrix}
\mathbf{P}_{i-1} \\
\mathbf{P}_i \\
\mathbf{P}_{i+1}
\end{pmatrix}
$$
$$
= \frac{1}{2}(t^2 - 2t + 1)\mathbf{P}_{i-1} + \frac{1}{2}(-2t^2 + 2t + 1)\mathbf{P}_i + \frac{t^2}{2}\mathbf{P}_{i+1}, \quad i = 1, 2.
$$

(7.6)

We are now in a position to determine the start and end points, $\mathbf{K}_i$ and $\mathbf{K}_{i+1}$ of segment i. They are

$$
\mathbf{K}_i = \mathbf{P}_i(0) = \frac{1}{2}(\mathbf{P}_{i-1} + \mathbf{P}_i), \quad \mathbf{K}_{i+1} = \mathbf{P}_i(1) = \frac{1}{2}(\mathbf{P}_i + \mathbf{P}_{i+1}).
$$

Thus, the quadratic spline segment starts in the middle of the straight segment $\mathbf{P}_{i-1}\mathbf{P}_i$ and ends at the middle of the straight segment $\mathbf{P}_i\mathbf{P}_{i+1}$, as shown in Figure 7.1b.

The tangent vector of the general quadratic B-spline segment is easily obtained from Equation (7.6). It is

$$
\mathbf{P}_i^t(t) = \frac{1}{2}(2t, 1, 0)
\begin{bmatrix}
1 & -2 & 1 \\
-2 & 2 & 0 \\
1 & 1 & 0
\end{bmatrix}
\begin{bmatrix}
\mathbf{P}_{i-1} \\
\mathbf{P}_i \\
\mathbf{P}_{i+1}
\end{bmatrix}
= (t-1)\mathbf{P}_{i-1} + (-2t+1)\mathbf{P}_i + t\mathbf{P}_{i+1}. \quad (7.7)
$$

The tangent vectors at both ends of the segment are therefore

$$
\mathbf{P}^t(0) = \mathbf{P}_i - \mathbf{P}_{i-1}, \quad \mathbf{P}^t(1) = \mathbf{P}_{i+1} - \mathbf{P}_i,
$$

i.e., each of them points in the direction of one of the edges of the control polygon of the spline segment.

Since a quadratic spline segment is a polynomial of degree 2, we require continuity of the first derivative only. It is easy to show that the second derivative of our segment is $\mathbf{P}_{i-1} - 2\mathbf{P}_i + \mathbf{P}_{i+1}$. It is constant for a segment but is different for different segments.

Equation (8.4) of Section 8.2 shows a relation between the quadratic B-spline and Bézier curves. A similar relation between the corresponding cubic curves is illustrated in Section 7.5.

Example: Given the four control points $\mathbf{P}_0 = (1,0)$, $\mathbf{P}_1 = (1,1)$, $\mathbf{P}_2 = (2,1)$, and $\mathbf{P}_3 = (2,0)$ (Figure 7.2), the first quadratic spline segment is obtained from Equation (7.6)

$$\mathbf{P}_1(t) = \frac{1}{2}(t^2, t, 1) \begin{pmatrix} 1 & -2 & 1 \\ -2 & 2 & 0 \\ 1 & 1 & 0 \end{pmatrix} \begin{pmatrix} \mathbf{P}_0 \\ \mathbf{P}_1 \\ \mathbf{P}_2 \end{pmatrix}$$

$$= \frac{1}{2}(t^2 - 2t + 1)(1,0) + \frac{1}{2}(-2t^2 + 2t + 1)(1,1) + \frac{t^2}{2}(2,1)$$

$$= (t^2/2 + 1, -t^2/2 + t + 1/2).$$

It starts at joint $\mathbf{K}_1 = \mathbf{P}_1(0) = (1, \frac{1}{2})$ and ends at joint $\mathbf{K}_2 = \mathbf{P}_1(1) = (\frac{3}{2}, 1)$.

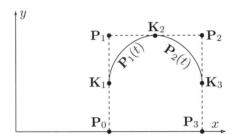

Figure 7.2: A Quadratic Uniform B-Spline Example.

The tangent vector of this segment is obtained from Equation (7.7)

$$\mathbf{P}_1^t(t) = \frac{1}{2}(2t, 1, 0) \begin{pmatrix} 1 & -2 & 1 \\ -2 & 2 & 0 \\ 1 & 1 & 0 \end{pmatrix} \begin{pmatrix} \mathbf{P}_0 \\ \mathbf{P}_1 \\ \mathbf{P}_2 \end{pmatrix}$$

$$= (t-1)(1,0) + (-2t+1)(1,1) + t(2,1)$$

$$= (t, 1-t).$$

Thus, the first segment starts going in direction $\mathbf{P}_1^t(0) = (0,1)$ (straight up) and ends going in direction $\mathbf{P}_1^t(1) = (1,0)$ (to the right).

◇ **Exercise 7.1:** Calculate the second segment, its tangent vector, and joint point $\mathbf{K}_3$.

Closed Quadratic B-Splines: Closed curves are sometimes needed and a closed B-spline curve is easy to construct. Given the usual $n+1$ control points, we extend them cyclically to obtain the $n+3$ points

$$\mathbf{P}_n, \ \mathbf{P}_0, \ \mathbf{P}_1, \ \mathbf{P}_2, \ \dots, \ \mathbf{P}_{n-1}, \ \mathbf{P}_n, \ \mathbf{P}_0$$

and compute the curve by applying Equation (7.6) to the $n+1$ geometry vectors

$$\begin{pmatrix} \mathbf{P}_n \\ \mathbf{P}_0 \\ \mathbf{P}_1 \end{pmatrix} \ \begin{pmatrix} \mathbf{P}_0 \\ \mathbf{P}_1 \\ \mathbf{P}_2 \end{pmatrix} \ \begin{pmatrix} \mathbf{P}_1 \\ \mathbf{P}_2 \\ \mathbf{P}_3 \end{pmatrix} \cdots \begin{pmatrix} \mathbf{P}_{n-2} \\ \mathbf{P}_{n-1} \\ \mathbf{P}_n \end{pmatrix} \ \begin{pmatrix} \mathbf{P}_{n-1} \\ \mathbf{P}_n \\ \mathbf{P}_0 \end{pmatrix}.$$

Example: Given the four control points $\mathbf{P}_0 = (1,0)$, $\mathbf{P}_1 = (1,1)$, $\mathbf{P}_2 = (2,1)$, and $\mathbf{P}_3 = (2,0)$ of the previous example, it is easy to close the curve by calculating the two additional segments

$$
\mathbf{P}_0(t) = \frac{1}{2}(t^2, t, 1)\begin{pmatrix} 1 & -2 & 1 \\ -2 & 2 & 0 \\ 1 & 1 & 0 \end{pmatrix}\begin{pmatrix} \mathbf{P}_3 \\ \mathbf{P}_0 \\ \mathbf{P}_1 \end{pmatrix}
$$

$$
= \frac{1}{2}(t^2 - 2t + 1)(2,0) + \frac{1}{2}(-2t^2 + 2t + 1)(1,0) + \frac{t^2}{2}(1,1)
$$

$$
= (t^2/2 - t + 3/2, t^2/2).
$$

$$
\mathbf{P}_3(t) = \frac{1}{2}(t^2, t, 1)\begin{pmatrix} 1 & -2 & 1 \\ -2 & 2 & 0 \\ 1 & 1 & 0 \end{pmatrix}\begin{pmatrix} \mathbf{P}_2 \\ \mathbf{P}_3 \\ \mathbf{P}_0 \end{pmatrix}
$$

$$
= \frac{1}{2}(t^2 - 2t + 1)(2,1) + \frac{1}{2}(-2t^2 + 2t + 1)(2,0) + \frac{t^2}{2}(1,0)
$$

$$
= (-t^2/2 + 2, t^2/2 - t + 1/2).
$$

The four segments connect the four joint points $(1, 1/2)$, $(3/2, 1)$, $(2, 1/2)$, $(3/2, 0)$ and back to $(1, 1/2)$.

The **B** stands for "basis".

7.2 The Cubic Uniform B-Spline

This curve is again defined by $n + 1$ control points and it consists of spline segments $\mathbf{P}_i(t)$, each a PC defined by four control points $\mathbf{P}_{i-1}$, $\mathbf{P}_i$, $\mathbf{P}_{i+1}$, and $\mathbf{P}_{i+2}$. The general form of segment i is therefore

$$
\mathbf{P}_i(t) = (t^3, t^2, t, 1)\mathbf{M}\begin{pmatrix} \mathbf{P}_{i-1} \\ \mathbf{P}_i \\ \mathbf{P}_{i+1} \\ \mathbf{P}_{i+2} \end{pmatrix}, \tag{7.8}
$$

where $\mathbf{M}$ is a 4×4 matrix whose 16 elements have to be determined by translating the constraints on the curve into 16 equations and solving them. The constraints are (1) two segments should meet with C^2 continuity and (2) the entire curve should be independent of the particular coordinate system. As in the quadratic case, we give up the requirement that a segment $\mathbf{P}_i(t)$ starts and ends at control points, and we denote its extreme points by $\mathbf{K}_i$ and $\mathbf{K}_{i+1}$. These joints can be computed as soon as the expression for the segment is derived. Figure 7.3a shows a tentative design for two cubic segments.

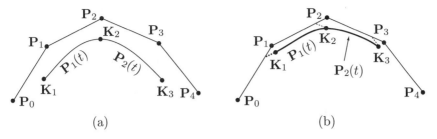

Figure 7.3: The Cubic Uniform B-Spline.

We start the derivation by writing

$$(t^3, t^2, t, 1)\mathbf{M} = (t^3, t^2, t, 1) \begin{pmatrix} a_3 & b_3 & c_3 & d_3 \\ a_2 & b_2 & c_2 & d_2 \\ a_1 & b_1 & c_1 & d_1 \\ a_0 & b_0 & c_0 & d_0 \end{pmatrix}$$

$$= (a_3 t^3 + a_2 t^2 + a_1 t + a_0, b_3 t^3 + b_2 t^2 + b_1 t + b_0,$$

$$c_3 t^3 + c_2 t^2 + c_1 t + c_0, d_3 t^3 + d_2 t^2 + d_1 t + d_0)$$

$$= \big(a(t), b(t), c(t), d(t)\big).$$

The first three constraints are expressed by

$$\mathbf{P}_1(1) = \mathbf{P}_2(0), \quad \mathbf{P}_1^t(1) = \mathbf{P}_2^t(0), \quad \mathbf{P}_1^{tt}(1) = \mathbf{P}_2^{tt}(0),$$

or, explicitly

$$a(1)\mathbf{P}_0 + b(1)\mathbf{P}_1 + c(1)\mathbf{P}_2 + d(1)\mathbf{P}_3 = a(0)\mathbf{P}_1 + b(0)\mathbf{P}_2 + c(0)\mathbf{P}_3 + d(0)\mathbf{P}_4,$$

$$\dot{a}(1)\mathbf{P}_0 + \dot{b}(1)\mathbf{P}_1 + \dot{c}(1)\mathbf{P}_2 + \dot{d}(1)\mathbf{P}_3 = \dot{a}(0)\mathbf{P}_1 + \dot{b}(0)\mathbf{P}_2 + \dot{c}(0)\mathbf{P}_3 + \dot{d}(0)\mathbf{P}_4,$$

$$\ddot{a}(1)\mathbf{P}_0 + \ddot{b}(1)\mathbf{P}_1 + \ddot{c}(1)\mathbf{P}_2 + \ddot{d}(1)\mathbf{P}_3 = \ddot{a}(0)\mathbf{P}_1 + \ddot{b}(0)\mathbf{P}_2 + \ddot{c}(0)\mathbf{P}_3 + \ddot{d}(0)\mathbf{P}_4.$$

Using the definitions of $a(t)$ and its relatives, this can be written explicitly as

$$\begin{aligned} a_3 + a_2 + a_1 + a_0 = 0, \quad & 3a_3 + 2a_2 + a_1 = 0, \quad & 6a_3 + 2a_2 = 0, \\ b_3 + b_2 + b_1 + b_0 = a_0, \quad & 3b_3 + 2b_2 + b_1 = a_1, \quad & 6b_3 + 2b_2 = 2a_2, \\ c_3 + c_2 + c_1 + c_0 = b_0, \quad & 3c_3 + 2c_2 + c_1 = b_1, \quad & 6c_3 + 2c_2 = 2b_2, \\ d_3 + d_2 + d_1 + d_0 = c_0, \quad & 3d_3 + 2d_2 + d_1 = c_1, \quad & 6d_3 + 2d_2 = 2c_2, \\ & 0 = d_0, \quad 0 = d_1, \quad 0 = 2d_2. \end{aligned}$$

(7.9)

These are 15 equations for the 16 unknowns.

We already know from the quadratic case that the weight functions of each segment should be barycentric, i.e., $a(t) + b(t) + c(t) + d(t) \equiv 1$. This condition can be written explicitly as

$$\begin{aligned} a_3 + b_3 + c_3 + d_3 = 0, \quad & a_2 + b_2 + c_2 + d_2 = 0, \\ a_1 + b_1 + c_1 + d_1 = 0, \quad & a_0 + b_0 + c_0 + d_0 = 1, \end{aligned}$$

(7.10)

and they add four more equations. We now have 19 equations, but only 16 of them are independent, since the first three equations of Equation (7.10) can be obtained by summing the first four equations of the left column of Equation (7.9). The system of equations can therefore be uniquely solved and the solutions are

$$a_3 = -1/6, \quad a_2 = 1/2, \quad a_1 = -1/2, \quad a_0 = 1/6,$$
$$b_3 = 1/2, \quad b_2 = -1, \quad b_1 = 0, \quad b_0 = 2/3,$$
$$c_3 = -1/2, \quad c_2 = 1/2, \quad c_1 = 1/2, \quad c_0 = 1/6,$$
$$d_3 = 1/6, \quad d_2 = 0, \quad d_1 = 0, \quad d_0 = 0.$$

The cubic B-spline segment can now be expressed as

$$
\begin{aligned}
\mathbf{P}_i(t) &= \frac{1}{6}(t^3, t^2, t, 1)
\begin{pmatrix}
-1 & 3 & -3 & 1 \\
3 & -6 & 3 & 0 \\
-3 & 0 & 3 & 0 \\
1 & 4 & 1 & 0
\end{pmatrix}
\begin{pmatrix}
\mathbf{P}_{i-1} \\
\mathbf{P}_i \\
\mathbf{P}_{i+1} \\
\mathbf{P}_{i+2}
\end{pmatrix} \\
&= \frac{1}{6}(-t^3 + 3t^2 - 3t + 1)\mathbf{P}_{i-1} + \frac{1}{6}(3t^3 - 6t^2 + 4)\mathbf{P}_i \\
&\quad + \frac{1}{6}(-3t^3 + 3t^2 + 3t + 1)\mathbf{P}_{i+1} + \frac{t^3}{6}\mathbf{P}_{i+2}.
\end{aligned}
\tag{7.11}
$$

The two extreme points are therefore

$$\mathbf{K}_i = \mathbf{P}_i(0) = \frac{1}{6}(\mathbf{P}_{i-1} + 4\mathbf{P}_i + \mathbf{P}_{i+1}), \text{ and } \mathbf{K}_{i+1} = \mathbf{P}_i(1) = \frac{1}{6}(\mathbf{P}_i + 4\mathbf{P}_{i+1} + \mathbf{P}_{i+2}).$$

In order to interpret them geometrically, we write them as

$$
\begin{aligned}
\mathbf{K}_i &= \left(\frac{1}{6}\mathbf{P}_{i-1} + \frac{5}{6}\mathbf{P}_i\right) + \frac{1}{6}\left(\mathbf{P}_{i+1} - \mathbf{P}_i\right), \\
\mathbf{K}_{i+1} &= \left(\frac{1}{6}\mathbf{P}_i + \frac{5}{6}\mathbf{P}_{i+1}\right) + \frac{1}{6}\left(\mathbf{P}_{i+2} - \mathbf{P}_{i+1}\right).
\end{aligned}
\tag{7.12}
$$

Point $\mathbf{K}_i$ is the sum of the point $(\frac{1}{6}\mathbf{P}_{i-1} + \frac{5}{6}\mathbf{P}_i)$ and one-sixth of the vector $(\mathbf{P}_{i+1} - \mathbf{P}_i)$. Point $\mathbf{K}_{i+1}$ has a similar interpretation. Both are shown in Figure 7.3b.

⋄ **Exercise 7.2:** Show another way to interpret $\mathbf{P}_i(0)$ and $\mathbf{P}_i(1)$ geometrically.

Users, especially those familiar with Bézier curves, find it counterintuitive that the B-spline curve does not start and end at its terminal control points. This "inconvenient" feature can be modified—and the curve made to start and end at its extreme points—by adding two *phantom* endpoints, $\mathbf{P}_{-1}$ and $\mathbf{P}_{n+1}$, at both ends of the curve, and placing those points at locations that would force the curve to start at $\mathbf{P}_0$ and end at $\mathbf{P}_n$. The calculation of this case is simple. The first segment starts at $\frac{1}{6}[\mathbf{P}_{-1} + 4\mathbf{P}_0 + \mathbf{P}_1]$. This value will equal $\mathbf{P}_0$ if we select $\mathbf{P}_{-1} = 2\mathbf{P}_0 - \mathbf{P}_1$. Similarly, the last segment ends at $\frac{1}{6}[\mathbf{P}_{n-1} + 4\mathbf{P}_n + \mathbf{P}_{n+1}]$ and this value equals $\mathbf{P}_n$ if we select $\mathbf{P}_{n+1} = 2\mathbf{P}_n - \mathbf{P}_{n-1}$.

Adding phantom points adds two segments to the curve, but this has the advantage that the tangents at the start and the end of the curve have known directions. The former is in the direction from $\mathbf{P}_0$ to $\mathbf{P}_1$ and the latter is from $\mathbf{P}_{n-1}$ to $\mathbf{P}_n$ (same as the end tangents of a Bézier curve). The tangent vector at the start of the first segment is $\frac{1}{2}\mathbf{P}_{-1} + \frac{1}{2}\mathbf{P}_1 = \mathbf{P}_1 - \mathbf{P}_0$, and similarly for the end tangent of the last segment.

The tangent vector of the general cubic B-spline segment is

$$\mathbf{P}_i^t(t) = \frac{1}{6}(-3t^2 + 6t - 3)\mathbf{P}_{i-1} + \frac{1}{6}(9t^2 - 12t)\mathbf{P}_i + \frac{1}{6}(-9t^2 + 6t + 3)\mathbf{P}_{i+1} + \frac{t^2}{2}\mathbf{P}_{i+2}.$$

As a result, the extreme tangent vectors are

$$\mathbf{P}_i^t(0) = \frac{1}{2}(\mathbf{P}_{i+1} - \mathbf{P}_{i-1}), \quad \mathbf{P}_i^t(1) = \frac{1}{2}(\mathbf{P}_{i+2} - \mathbf{P}_i). \tag{7.13}$$

They have simple geometric interpretations.

The second derivative of the cubic segment is

$$\mathbf{P}_i^{tt}(t) = \frac{1}{6}(-6t + 6)\mathbf{P}_{i-1} + \frac{1}{6}(18t - 12)\mathbf{P}_i + \frac{1}{6}(-18t + 6)\mathbf{P}_{i+1} + t\mathbf{P}_{i+2},$$

and it's easy to see that $\mathbf{P}_i^{tt}(1) = \mathbf{P}_{i+1}^{tt}(0) = \mathbf{P}_i - 2\mathbf{P}_{i+1} + \mathbf{P}_{i+2}$, which proves the C^2 continuity of this curve.

Example: We select the five points $\mathbf{P}_0 = (0,0)$, $\mathbf{P}_1 = (0,1)$, $\mathbf{P}_2 = (1,1)$, $\mathbf{P}_3 = (2,1)$, and $\mathbf{P}_4 = (2,0)$. They have simple, integer coordinates to simplify the computations. We use these points to construct two cubic B-spline segments. The first one is given by Equation (7.11)

$$\mathbf{P}_1(t) = \frac{1}{6}(-t^3 + 3t^2 - 3t + 1)(0,0) + \frac{1}{6}(3t^3 - 6t^2 + 4)(0,1)$$
$$+ \frac{1}{6}(-3t^3 + 3t^2 + 3t + 1)(1,1) + \frac{t^3}{6}(2,1)$$
$$= (-t^3/6 + t^2/2 + t/2 + 1/6, t^3/6 - t^2/2 + t/2 + 5/6).$$

It starts at joint $\mathbf{K}_1 = \mathbf{P}_1(0) = (1/6, 5/6)$ and ends at joint $\mathbf{K}_2 = \mathbf{P}_1(1) = (1,1)$. Notice that these joint points can be verified from Equation (7.12). The tangent vector of this segment is

$$\mathbf{P}_1^t(t) = \frac{1}{6}(-3t^2 + 6t - 3)(0,0) + \frac{1}{6}(9t^2 - 12t)(0,1)$$
$$+ \frac{1}{6}(-9t^2 + 6t + 3)(1,1) + \frac{t^2}{2}(2,1)$$
$$= (-t^2/2 + t + 1/2, t^2/2 - t + 1/2).$$

The two extreme tangents are $\mathbf{P}_1^t(0) = (1/2, 1/2)$ and $\mathbf{P}_1^t(1) = (1,0)$. These can also be verified by Equation (7.13). Figure 7.4 shows this segment and its successor (the dashed curves).

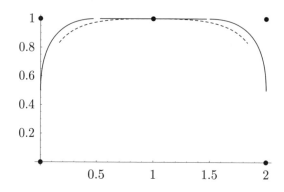

```
(* B-spline example of 2 cubic segs and 3 quadr segs for 5 points *)
Clear[Pt,T,t,M3,comb,a,g1,g2,g3];
Pt={{0,0},{0,1},{1,1},{2,1},{2,0}};
(* first, 2 cubic segments (dashed) *)
T[t_]:={t^3,t^2,t,1};
M3={{-1,3,-3,1},{3,-6,3,0},{-3,0,3,0},{1,4,1,0}}/6;
comb[i_]:=(T[t].M3)[[i]] Pt[[i+a]];
g1=Graphics[{PointSize[.02], Point/@Pt}];
a=0;
g2=ParametricPlot[comb[1]+comb[2]+comb[3]+comb[4], {t,0,.95},
 Compiled->False, PlotRange->All, DisplayFunction->Identity,
 PlotStyle->AbsoluteDashing[{2,2}]];
a=1;
g3=ParametricPlot[comb[1]+comb[2]+comb[3]+comb[4], {t,0.05,1},
 Compiled->False, PlotRange->All, DisplayFunction->Identity,
 PlotStyle->AbsoluteDashing[{2,2}]];
(* Now the 3 quadratic segments (solid) *)
T[t_]:={t^2,t,1};
M2={{1,-2,1},{-2,2,0},{1,1,0}}/2;
comb[i_]:=(T[t].M2)[[i]] Pt[[i+a]];
a=0;
g4=ParametricPlot[comb[1]+comb[2]+comb[3], {t,0,.97},
 Compiled->False, PlotRange->All, DisplayFunction->Identity];
a=1;
g5=ParametricPlot[comb[1]+comb[2]+comb[3], {t,0.03,.97},
 Compiled->False, PlotRange->All, DisplayFunction->Identity];
a=2;
g6=ParametricPlot[comb[1]+comb[2]+comb[3], {t,0,1},
 Compiled->False, PlotRange->All, DisplayFunction->Identity];
Show[g2,g3,g4,g5,g6,g1, PlotRange->All, DefaultFont->{"cmr10", 10},
 DisplayFunction->$DisplayFunction];
```

Figure 7.4: Two Cubic (Dashed) and Three Quadratic (Solid) Segments of a B-spline.

⋄ **Exercise 7.3:** Calculate the second spline segment $\mathbf{P}_2(t)$, its tangent vector, and joint $\mathbf{K}_3$.

⋄ **Exercise 7.4:** Use the five control points of the example above to construct the three segments and determine the four joints of the *quadratic* uniform B-spline defined by the points.

Exercise 7.4 shows that the same $n + 1$ control points can be used to construct a quadratic or a cubic B-spline curve (or a B-spline curve of any order up to $n + 1$). This is in contrast to the Bézier curve whose order is determined by the number of control points. This is also the reason why both n and the degree of the polynomials that make up the spline segments are needed to identify a B-spline. In practice, we use n and k (the *order*) to identify a B-spline. The order is simply the degree plus 1. Thus, a B-spline defined by five control points $\mathbf{P}_0$ through $\mathbf{P}_4$ can be of order 2 (linear, with four segments), order 3 (quadratic, with three segments), order 4 (cubic, with two segments), or order 5, (quintic, with one segment).

Figure 7.5a,b,c shows how a Bézier curve, a cubic B-spline, and a quadratic B-spline, respectively, are attracted to their control polygons. We already know that these three types of curves don't have the same endpoints, so this figure is only qualitative. It only shows how the various types of curves are attracted to their control points.

Collinear Points: Segment $\mathbf{P}_2(t)$ of Exercise 7.4 depends on points $\mathbf{P}_1$, $\mathbf{P}_2$, and $\mathbf{P}_3$ that are located on the line $y = 1$. This is why this segment is horizontal (and therefore straight). We conclude that the B-spline can consist of curved and straight segments connected with any desired continuity. All that's necessary in order to have a straight segment is to have enough collinear control points. In the case of a quadratic B-spline, three collinear points will result in a straight segment that will connect to its neighbors (curved or straight) with C^1 continuity. In the case of a cubic B-spline, four collinear points will result in a straight segment that will connect to its neighbors (curved or straight) with C^2 continuity, and similarly for higher-degree uniform B-splines.

A Closed Cubic B-Spline Curve: closing a cubic B-spline is similar to closing a quadratic curve. Given a set of $n+1$ control points, we extend them cyclically to obtain the $n + 4$ points

$$\mathbf{P}_n, \ \mathbf{P}_0, \ \mathbf{P}_1, \ \mathbf{P}_2, \ \ldots, \ \mathbf{P}_{n-1}, \ \mathbf{P}_n, \ \mathbf{P}_0, \ \mathbf{P}_1,$$

and compute the curve by applying Equation (7.11) to the $n + 1$ geometry vectors

$$\begin{pmatrix} \mathbf{P}_n \\ \mathbf{P}_0 \\ \mathbf{P}_1 \\ \mathbf{P}_2 \end{pmatrix} \ \begin{pmatrix} \mathbf{P}_0 \\ \mathbf{P}_1 \\ \mathbf{P}_2 \\ \mathbf{P}_3 \end{pmatrix} \ \begin{pmatrix} \mathbf{P}_1 \\ \mathbf{P}_2 \\ \mathbf{P}_3 \\ \mathbf{P}_4 \end{pmatrix} \ \cdots \ \begin{pmatrix} \mathbf{P}_{n-2} \\ \mathbf{P}_{n-1} \\ \mathbf{P}_n \\ \mathbf{P}_0 \end{pmatrix} \ \begin{pmatrix} \mathbf{P}_{n-1} \\ \mathbf{P}_n \\ \mathbf{P}_0 \\ \mathbf{P}_1 \end{pmatrix}.$$

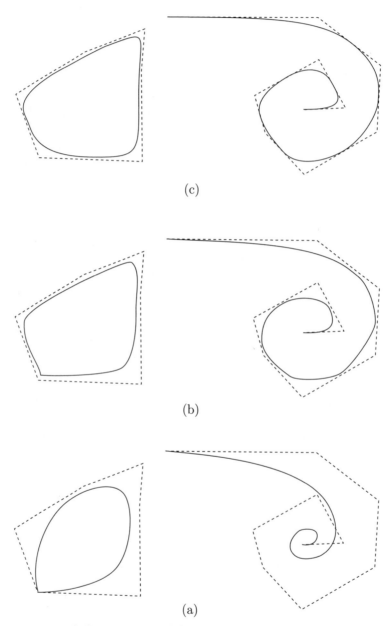

Figure 7.5: A Comparison of (a) Bézier, (b) Cubic B-Spline, and (c) Quadratic B-Spline Curves.

7.3 Multiple Control Points

It is possible to have several identical control points and a set of identical points is referred to as a multiple point. We use the uniform cubic B-spline [Equation (7.11)] as an example, but higher-degree uniform B-splines behave similarly.

We start with a double control point. Consider the cubic segment $\mathbf{P}_1(t)$ defined by the four control points $\mathbf{P}_0$, $\mathbf{P}_1 = \mathbf{P}_2$, and $\mathbf{P}_3$. Its expression is

$$\mathbf{P}_1(t) = \frac{1}{6}(-t^3 + 3t^2 - 3t + 1)\mathbf{P}_0 + \frac{1}{6}(-3t^2 + 3t + 5)\mathbf{P}_1 + \frac{t^3}{6}\mathbf{P}_3,$$

which implies
$$\mathbf{P}_1(0) = \frac{1}{6}\mathbf{P}_0 + \frac{5}{6}\mathbf{P}_1, \quad \mathbf{P}_1(1) = \frac{5}{6}\mathbf{P}_1 + \frac{1}{6}\mathbf{P}_3.$$

This segment therefore starts and ends at the same points as the general cubic segment and also has the same extreme tangent vectors. The difference is that it is strongly attracted to the double point.

Next, we consider a triple point. The five control points $\mathbf{P}_0$, $\mathbf{P}_1 = \mathbf{P}_2 = \mathbf{P}_3$, and $\mathbf{P}_4$ define the two cubic segments

$$\begin{aligned}
\mathbf{P}_1(t) &= \frac{1}{6}(-t^3 + 3t^2 - 3t + 1)\mathbf{P}_0 + \frac{1}{6}(t^3 - 3t^2 + 3t + 5)\mathbf{P}_1 \\
&= (1 - u)\mathbf{P}_0 + u\mathbf{P}_1, \quad \text{for } u = (t^3 - 3t^2 + 3t + 5)/6, \\
\mathbf{P}_2(t) &= \frac{1}{6}(-t^3 + 6)\mathbf{P}_1 + \frac{t^3}{6}\mathbf{P}_4 \\
&= (1 - w)\mathbf{P}_1 + w\mathbf{P}_4, \quad \text{for } w = t^3/6.
\end{aligned}$$

The parameter substitutions above show that these segments are straight (Figure 7.6). The extreme points of the two segments are

$$\begin{aligned}
\mathbf{P}_1(0) &= \frac{1}{6}\mathbf{P}_0 + \frac{5}{6}\mathbf{P}_1, \quad \mathbf{P}_1(1) = \mathbf{P}_1, \\
\mathbf{P}_2(0) &= \mathbf{P}_1, \quad \mathbf{P}_2(1) = \frac{5}{6}\mathbf{P}_1 + \frac{1}{6}\mathbf{P}_4,
\end{aligned}$$

showing that the segments meet at the triple control point.

In general, a cubic segment is attracted to a double control point and passes through a triple control point. A degree-4 segment is attracted to double and triple control points and passes through quadruple points, and similarly for higher-degree uniform segments.

The tangent vectors of the two cubic segments are

$$\begin{aligned}
\mathbf{P}_1^t(t) &= \frac{1}{6}(-3t^2 + 6t - 3)\mathbf{P}_0 + \frac{1}{6}(3t^2 - 6t + 3)\mathbf{P}_1, \\
\mathbf{P}_2^t(t) &= -\frac{t^2}{2}\mathbf{P}_1 + \frac{t^2}{2}\mathbf{P}_4,
\end{aligned}$$

yielding the extreme directions

$$\mathbf{P}_1^t(0) = \frac{1}{2}(\mathbf{P}_1 - \mathbf{P}_0), \quad \mathbf{P}_1^t(1) = 0 \cdot \mathbf{P}_0 + 0 \cdot \mathbf{P}_1 = (0,0),$$

$$\mathbf{P}_2^t(0) = (0,0), \quad \mathbf{P}_2^t(1) = \frac{1}{2}(\mathbf{P}_4 - \mathbf{P}_1).$$

Thus, the first segment starts in the direction from $\mathbf{P}_0$ to the triple point $\mathbf{P}_1$. The second segment ends going in the direction from $\mathbf{P}_1$ to $\mathbf{P}_4$. However, at the triple point, both tangents are indefinite, suggesting a cusp. It turns out that the two segments are straight lines (Figure 7.6).

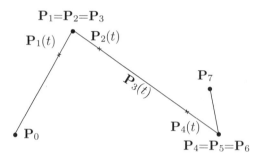

Figure 7.6: A Triple Point.

⋄ **Exercise 7.5:** Given the eight control points $\mathbf{P}_0$, $\mathbf{P}_1 = \mathbf{P}_2 = \mathbf{P}_3$, $\mathbf{P}_4 = \mathbf{P}_5 = \mathbf{P}_6$, and $\mathbf{P}_7$, calculate the two cubic segments $\mathbf{P}_3(t)$ and $\mathbf{P}_4(t)$ and their start and end points (Figure 7.6).

⋄ **Exercise 7.6:** Show that a cubic B-spline segment passes through its first control point if it is a triple point.

As a corollary, we deduce that a uniform cubic B-spline curve where every control point is triple is a polyline.

Example: We consider the case where both terminal points are triple and there are two other points in between. The total number of control points is eight and they satisfy $\mathbf{P}_0 = \mathbf{P}_1 = \mathbf{P}_2$ and $\mathbf{P}_5 = \mathbf{P}_6 = \mathbf{P}_7$. The five cubic spline segments are

$$\mathbf{P}_1(t) = \frac{1}{6}(-t^3 + 6)\mathbf{P}_0 + \frac{t^3}{6}\mathbf{P}_3,$$

$$\mathbf{P}_2(t) = \frac{1}{6}(2t^3 - 3t^2 - 3t + 5)\mathbf{P}_0 + \frac{1}{6}(-3t^3 + 3t^2 + 3t + 1)\mathbf{P}_3 + \frac{t^3}{6}\mathbf{P}_4,$$

$$\mathbf{P}_3(t) = \frac{1}{6}(-t^3 + 3t^2 - 3t + 1)\mathbf{P}_0 + \frac{1}{6}(3t^3 - 6t^2 + 4)\mathbf{P}_3$$

$$+ \frac{1}{6}(-3t^3 + 3t^2 + 3t + 1)\mathbf{P}_4 + \frac{t^3}{6}\mathbf{P}_5, \tag{7.14}$$

$$\mathbf{P}_4(t) = \frac{1}{6}(-t^3 + 3t^2 - 3t + 1)\mathbf{P}_3 + \frac{1}{6}(3t^3 - 6t^2 + 4)\mathbf{P}_4$$

$$+ \frac{1}{6}(-2t^3 + 3t^2 + 3t + 1)\mathbf{P}_5,$$

$$\mathbf{P}_5(t) = \frac{1}{6}(-t^3 + 3t^2 - 3t + 1)\mathbf{P}_4 + \frac{1}{6}(t^3 - 3t^2 + 3t + 5)\mathbf{P}_5.$$

It is easy to see that they satisfy $\mathbf{P}_1(0) = \mathbf{P}_0$ and $\mathbf{P}_5(1) = \mathbf{P}_5$ and that they meet at the four points

$$\frac{5}{6}\mathbf{P}_0 + \frac{1}{6}\mathbf{P}_3, \quad \frac{1}{6}\mathbf{P}_0 + \frac{4}{6}\mathbf{P}_3 + \frac{1}{6}\mathbf{P}_4, \quad \frac{1}{6}\mathbf{P}_3 + \frac{4}{6}\mathbf{P}_4 + \frac{1}{6}\mathbf{P}_5, \quad \text{and} \quad \frac{1}{6}\mathbf{P}_4 + \frac{5}{6}\mathbf{P}_5.$$

If we want to keep the two extreme points as triples, we can edit this curve only by moving the two interior points $\mathbf{P}_3$ and $\mathbf{P}_4$. Moving $\mathbf{P}_4$ affects the last four segments, and moving $\mathbf{P}_3$ affects the first four segments. This type of curve is therefore similar to a Bézier curve in that it starts and ends at its extreme control points and it features only limited local control.

◇ **Exercise 7.7:** Given the eight control points $\mathbf{P}_0 = \mathbf{P}_1 = \mathbf{P}_2 = (1,0)$, $\mathbf{P}_3 = (2,1)$, $\mathbf{P}_4 = (4,0)$, and $\mathbf{P}_5 = \mathbf{P}_6 = \mathbf{P}_7 = (4,1)$, use Equation (7.14) to calculate the cubic uniform B-spline curve defined by these points and compare it to the Bézier curve defined by the points.

7.4 Cubic B-Splines with Tension

Adding a tension parameter to the uniform cubic B-spline is similar to tension in the cardinal spline (Section 5.4). We use Hermite interpolation [Equation (4.7)] to calculate a PC segment that starts and ends at the same points as a cubic B-spline and whose extreme tangent vectors point in the same directions as those of the cubic B-spline, but whose magnitudes are controlled by a tension parameter s. Substituting $\frac{1}{6}\mathbf{P}_0 + \frac{4}{6}\mathbf{P}_1 + \frac{1}{6}\mathbf{P}_2$ and $\frac{1}{6}\mathbf{P}_1 + \frac{4}{6}\mathbf{P}_2 + \frac{1}{6}\mathbf{P}_3$ for the terminal points and $s(\mathbf{P}_2 - \mathbf{P}_0)$ and $s(\mathbf{P}_3 - \mathbf{P}_1)$ for the extreme tangents, we write Equation (4.7) and manipulate it such that it ends up looking like a uniform cubic B-spline segment, Equation (7.11).

$$
\mathbf{P}(t) = (t^3, t^2, t, 1)
\begin{pmatrix}
2 & -2 & 1 & 1 \\
-3 & 3 & -2 & -1 \\
0 & 0 & 1 & 0 \\
1 & 0 & 0 & 0
\end{pmatrix}
\begin{pmatrix}
\frac{1}{6}\mathbf{P}_0 + \frac{4}{6}\mathbf{P}_1 + \frac{1}{6}\mathbf{P}_2 \\
\frac{1}{6}\mathbf{P}_1 + \frac{4}{6}\mathbf{P}_2 + \frac{1}{6}\mathbf{P}_3 \\
s(\mathbf{P}_2 - \mathbf{P}_0) \\
s(\mathbf{P}_3 - \mathbf{P}_1)
\end{pmatrix}
$$

$$
= \frac{1}{6}\Big[\big(t^3(2-s) + t^2(2s-3) - st + 1\big)\mathbf{P}_0 + \big(t^3(6-s) + t^2(s-9) + 4\big)\mathbf{P}_1
$$

$$
+ \big(t^3(s-6) + t^2(9-2s) + st + 1\big)\mathbf{P}_2 + \big(t^3(s-2) + t^2(3-s)\big)\mathbf{P}_3 \Big]
$$

$$
= \frac{1}{6}(t^3, t^2, t, 1)
\begin{pmatrix}
2-s & 6-s & s-6 & s-2 \\
2s-3 & s-9 & 9-2s & 3-s \\
-s & 0 & s & 0 \\
1 & 4 & 1 & 0
\end{pmatrix}
\begin{pmatrix}
\mathbf{P}_0 \\
\mathbf{P}_1 \\
\mathbf{P}_2 \\
\mathbf{P}_3
\end{pmatrix}. \tag{7.15}
$$

A quick check verifies that Equation (7.15) reduces to the uniform cubic B-spline segment, Equation (7.11), for $s = 3$. This value is therefore considered the "neutral" or "standard" value of the tension parameter s. Since s controls the length of the tangent vectors, small values of s should produce the effects of higher tension and, in the extreme, the value $s = 0$ should result in indefinite tangent vectors and in the spline segment becoming a straight line. To show this, we rewrite Equation (7.15) for $s = 0$:

$$
\begin{aligned}
\mathbf{P}(t) &= \frac{1}{6}(t^3, t^2, t, 1)
\begin{pmatrix}
2 & 6 & -6 & -2 \\
-3 & -9 & 9 & 3 \\
0 & 0 & 0 & 0 \\
1 & 4 & 1 & 0
\end{pmatrix}
\begin{pmatrix}
\mathbf{P}_0 \\
\mathbf{P}_1 \\
\mathbf{P}_2 \\
\mathbf{P}_3
\end{pmatrix} \\
&= \frac{1}{6}(2t^3 - 3t^2 + 1)\mathbf{P}_0 + \frac{1}{6}(6t^3 - 9t^2 + 4)\mathbf{P}_1 \\
&\quad + \frac{1}{6}(-6t^3 + 9t^2 + 1)\mathbf{P}_2 + \frac{1}{6}(-2t^3 + 3t^2)\mathbf{P}_3.
\end{aligned}
$$

Substituting $T = 3t^2 - 2t^3$ for the parameter t changes the above expression to the form

$$
\mathbf{P}(T) = \frac{1}{6}(-\mathbf{P}_0 - 3\mathbf{P}_1 + 3\mathbf{P}_2 + \mathbf{P}_3)T + \frac{1}{6}(\mathbf{P}_0 + 4\mathbf{P}_1 + \mathbf{P}_2),
$$

which is a straight line from $\mathbf{P}(0) = \frac{1}{6}(\mathbf{P}_0 + 4\mathbf{P}_1 + \mathbf{P}_2)$ to $\mathbf{P}(1) = \frac{1}{6}(\mathbf{P}_1 + 4\mathbf{P}_2 + \mathbf{P}_3)$.

The tangent vector of Equation (7.15) is

$$
\begin{aligned}
\mathbf{P}^t(t) &= \frac{1}{6}(3t^2, 2t, 1, 0)
\begin{pmatrix}
2-s & 6-s & s-6 & s-2 \\
2s-3 & s-9 & 9-2s & 3-s \\
-s & 0 & s & 0 \\
1 & 4 & 1 & 0
\end{pmatrix}
\begin{pmatrix}
\mathbf{P}_0 \\
\mathbf{P}_1 \\
\mathbf{P}_2 \\
\mathbf{P}_3
\end{pmatrix} \\
&= \frac{1}{6}\Big[\big(3t^2(2-s) + 2t(2s-3) - s\big)\mathbf{P}_0 + \big(3t^2(6-s) + 2t(s-9)\big)\mathbf{P}_1 \\
&\quad + \big(3t^2(s-6) + 2t(9-2s) + s\big)\mathbf{P}_2 + \big(3t^2(s-2) + 2t(3-s)\big)\mathbf{P}_3\Big].
\end{aligned}
\tag{7.16}
$$

The extreme tangents are

$$
\mathbf{P}^t(0) = \frac{s}{6}(\mathbf{P}_2 - \mathbf{P}_0) \quad \text{and} \quad \mathbf{P}^t(1) = \frac{s}{6}(\mathbf{P}_3 - \mathbf{P}_1).
$$

Substituting $s = 0$ in Equation (7.16) yields the tangent vector for the case of infinite tension

$$
\begin{aligned}
\mathbf{P}^t(t) &= \frac{1}{6}\Big[6(t^2 - t)\mathbf{P}_0 + 18(t^2 - t)\mathbf{P}_1 - 18(t^2 - t)\mathbf{P}_2 - 6(t^2 - t)\mathbf{P}_3\Big] \\
&= (t^2 - t)(\mathbf{P}_0 + 3\mathbf{P}_1 - 3\mathbf{P}_2 - \mathbf{P}_3).
\end{aligned}
\tag{7.17}
$$

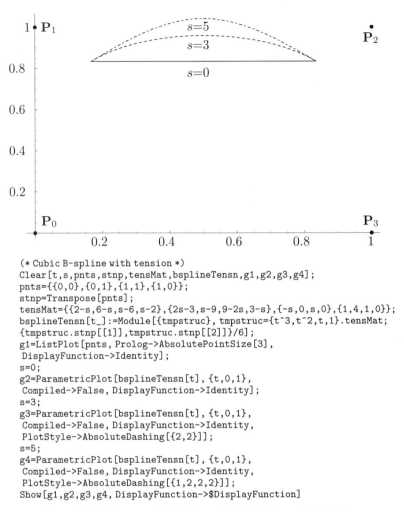

```
(* Cubic B-spline with tension *)
Clear[t,s,pnts,stnp,tensMat,bsplineTensn,g1,g2,g3,g4];
pnts={{0,0},{0,1},{1,1},{1,0}};
stnp=Transpose[pnts];
tensMat={{2-s,6-s,s-6,s-2},{2s-3,s-9,9-2s,3-s},{-s,0,s,0},{1,4,1,0}};
bsplineTensn[t_]:=Module[{tmpstruc}, tmpstruc={t^3,t^2,t,1}.tensMat;
{tmpstruc.stnp[[1]],tmpstruc.stnp[[2]]}/6];
g1=ListPlot[pnts, Prolog->AbsolutePointSize[3],
 DisplayFunction->Identity];
s=0;
g2=ParametricPlot[bsplineTensn[t], {t,0,1},
 Compiled->False, DisplayFunction->Identity];
s=3;
g3=ParametricPlot[bsplineTensn[t], {t,0,1},
 Compiled->False, DisplayFunction->Identity,
 PlotStyle->AbsoluteDashing[{2,2}]];
s=5;
g4=ParametricPlot[bsplineTensn[t], {t,0,1},
 Compiled->False, DisplayFunction->Identity,
 PlotStyle->AbsoluteDashing[{1,2,2,2}]];
Show[g1,g2,g3,g4, DisplayFunction->$DisplayFunction]
```

Figure 7.7: Figure and Code for a Cubic B-Spline with Tension.

⬦ **Exercise 7.8:** Since the spline segment is a straight line in this case, its tangent vector should always point in the same direction. Use Equation (7.17) to show that this is so.

Figure 7.7 illustrates the effect of tension on a cubic B-spline. Three curves are shown, corresponding to s values of 0, 3, and 5.

See also Section 6.11 for a discussion of cubic Bézier curves with tension.

> Sex alleviates tension and love causes it.
> —Woody Allen (as Andrew) in *A Midsummer Night's Sex Comedy* (1982)

7.5 Cubic B-Spline and Bézier Curves

Given a cubic B-spline segment $\mathbf{P}(t)$ based on the four control points $\mathbf{P}_0$, $\mathbf{P}_1$, $\mathbf{P}_2$, and $\mathbf{P}_3$, it is easy to find four control points $\mathbf{Q}_0$, $\mathbf{Q}_1$, $\mathbf{Q}_2$, and $\mathbf{Q}_3$ such that the Bézier curve $\mathbf{Q}(t)$ defined by them will have the same shape as $\mathbf{P}(t)$. This is done by equating the matrices of Equation (7.11) that define $\mathbf{P}(t)$ to those of Equation (6.8) that define $\mathbf{Q}(t)$:

$$
\begin{pmatrix}
-1 & 3 & -3 & 1 \\
3 & -6 & 3 & 0 \\
-3 & 0 & 3 & 0 \\
1 & 4 & 1 & 0
\end{pmatrix}
\begin{pmatrix}
\mathbf{P}_0 \\ \mathbf{P}_1 \\ \mathbf{P}_2 \\ \mathbf{P}_3
\end{pmatrix}
=
\begin{pmatrix}
-1 & 3 & -3 & 1 \\
3 & -6 & 3 & 0 \\
-3 & 3 & 0 & 0 \\
1 & 0 & 0 & 0
\end{pmatrix}
\begin{pmatrix}
\mathbf{Q}_0 \\ \mathbf{Q}_1 \\ \mathbf{Q}_2 \\ \mathbf{Q}_3
\end{pmatrix}.
$$

The solutions are

$$
\mathbf{Q}_0 = \frac{1}{6}\left(\mathbf{P}_0 + 4\mathbf{P}_1 + \mathbf{P}_2\right),
$$

$$
\mathbf{Q}_1 = \frac{1}{6}\left(4\mathbf{P}_1 + 2\mathbf{P}_2\right),
$$

$$
\mathbf{Q}_2 = \frac{1}{6}\left(2\mathbf{P}_1 + 4\mathbf{P}_2\right),
$$

$$
\mathbf{Q}_3 = \frac{1}{6}\left(\mathbf{P}_1 + 4\mathbf{P}_2 + \mathbf{P}_3\right).
$$

Equation (8.4) of Section 8.2 shows a similar relation between the quadratic B-spline and Bézier curves.

7.6 Higher-Degree Uniform B-Splines

The methods of Sections 7.1 and 7.2 can be employed to construct uniform B-splines of higher degrees. It can be shown (see, for example, [Yamaguchi 88], p. 329) that the degree-n uniform B-spline segment is given by

$$
\mathbf{P}_i(t) = (t^n, \ldots, t^2, t, 1)\mathbf{M}
\begin{pmatrix}
\mathbf{P}_{i-1} \\
\mathbf{P}_i \\
\mathbf{P}_{i+1} \\
\vdots \\
\mathbf{P}_{i+n-1}
\end{pmatrix},
$$

where the elements m_{ij} of the basis matrix $\mathbf{M}$ are

$$
m_{ij} = \frac{1}{n!}\binom{n}{i}\sum_{k=j}^{n}(n-k)^i(-1)^{k-j}\binom{n+1}{k-j}.
$$

Figure 7.8 shows a few examples of these matrices.

$$\mathbf{M}_1 = \frac{1}{1!} \begin{pmatrix} -1 & 1 \\ 1 & 0 \end{pmatrix}$$

$$\mathbf{M}_2 = \frac{1}{2!} \begin{pmatrix} 1 & -2 & 1 \\ -2 & 2 & 0 \\ 1 & 1 & 0 \end{pmatrix}$$

$$\mathbf{M}_3 = \frac{1}{3!} \begin{pmatrix} -1 & 3 & -3 & 1 \\ 3 & -6 & 3 & 0 \\ -3 & 0 & 3 & 0 \\ 1 & 4 & 1 & 0 \end{pmatrix}$$

$$\mathbf{M}_4 = \frac{1}{4!} \begin{pmatrix} 1 & -4 & 6 & -4 & 1 \\ -4 & 12 & -12 & 4 & 0 \\ 6 & -6 & -6 & 6 & 0 \\ -4 & -12 & 12 & 4 & 0 \\ 1 & 11 & 11 & 1 & 0 \end{pmatrix}$$

$$\mathbf{M}_5 = \frac{1}{5!} \begin{pmatrix} -1 & 5 & -10 & 10 & -5 & 1 \\ 5 & -20 & 30 & -20 & 5 & 0 \\ -10 & 20 & 0 & -20 & 10 & 0 \\ 10 & 20 & -60 & 20 & 10 & 0 \\ -5 & -50 & 0 & 50 & 5 & 0 \\ 1 & 26 & 66 & 26 & 1 & 0 \end{pmatrix}$$

$$\mathbf{M}_6 = \frac{1}{6!} \begin{pmatrix} 1 & -6 & 15 & -20 & 15 & -6 & 1 \\ -6 & 30 & -60 & 60 & -30 & 6 & 0 \\ 15 & -45 & 30 & 30 & -45 & 15 & 0 \\ -20 & -20 & 160 & -160 & 20 & 20 & 0 \\ 15 & 135 & -150 & -150 & 135 & 15 & 0 \\ -6 & -150 & -240 & 240 & 150 & 6 & 0 \\ 1 & 57 & 302 & 302 & 57 & 1 & 0 \end{pmatrix}$$

Figure 7.8: Some Basis Matrices for Uniform B-Splines.

7.7 Interpolating B-Splines

The B-spline is an approximating curve. Its shape is determined by the control points $\mathbf{P}_i$, but the curve itself does not pass through those points. Instead, it passes through the joints $\mathbf{K}_i$. In our notation so far, we have assumed that the cubic uniform B-spline is based on $n + 1$ control points and passes through $n - 1$ joint points. The number of control points for the cubic curve is therefore always two more than the number of joints.

> One person's constant is another person's variable.
>
> —Susan Gerhart

This section deals with the opposite problem. We show how to employ B-splines to construct an interpolating cubic spline curve that passes through a set of $n + 1$ given data points $\mathbf{K}_0, \mathbf{K}_1, \dots, \mathbf{K}_n$. The curve must consist of n segments and the idea is to use the $\mathbf{K}_i$ points to calculate a new set of points $\mathbf{P}_i$, then use the new points as the control points of a cubic uniform B-spline curve. To obtain n cubic segments, we need $n + 3$ points and we denote them by $\mathbf{P}_{-1}$ through $\mathbf{P}_{n+1}$.

Using $\mathbf{P}_i$ as our control points, Equation (7.11) shows that the general segment $\mathbf{P}_i(t)$ terminates at $\mathbf{P}_i(1) = \frac{1}{6}[\mathbf{P}_{i-2} + 4\mathbf{P}_{i-1} + \mathbf{P}_i]$. We require that the segment ends at point $\mathbf{K}_{i-1}$, which produces the equation $\frac{1}{6}[\mathbf{P}_{i-2} + 4\mathbf{P}_{i-1} + \mathbf{P}_i] = \mathbf{K}_{i-1}$. When this equation is repeated for $0 \le i \le n$, we get a system of $n + 1$ equations with the $\mathbf{P}_i$s as the unknowns. However, there are $n + 3$ unknowns ($\mathbf{P}_{-1}$ through $\mathbf{P}_{n+1}$), so we need two more equations.

The required equations are obtained by considering the tangent vectors of the interpolating curve at its two ends. We denote the tangent at the start by $\mathbf{T}_1$. It is given by $\mathbf{T}_1 = \frac{1}{2}(\mathbf{P}_1 - \mathbf{P}_{-1})$, so it points in the direction from $\mathbf{P}_{-1}$ to $\mathbf{P}_1$; similarly for the end tangent $\mathbf{T}_n = \frac{1}{2}(\mathbf{P}_{n+1} - \mathbf{P}_{n-1})$. After these two relations are included, the resulting system of $n + 3$ equations is

$$
n+3\left\{ \frac{1}{6} \begin{pmatrix} -3 & 0 & 3 & 0 & \dots & 0 & 0 & 0 \\ 1 & 4 & 1 & 0 & \dots & 0 & 0 & 0 \\ 0 & 1 & 4 & 1 & \dots & 0 & 0 & 0 \\ \vdots & & & & & & & \vdots \\ 0 & 0 & 0 & 0 & \dots & 4 & 1 & 0 \\ 0 & 0 & 0 & 0 & \dots & 1 & 4 & 1 \\ 0 & 0 & 0 & 0 & \dots & -3 & 0 & 3 \end{pmatrix} \begin{pmatrix} \mathbf{P}_{-1} \\ \mathbf{P}_0 \\ \mathbf{P}_1 \\ \vdots \\ \mathbf{P}_{n-1} \\ \mathbf{P}_n \\ \mathbf{P}_{n+1} \end{pmatrix} = \begin{pmatrix} \mathbf{T}_1 \\ \mathbf{K}_0 \\ \mathbf{K}_1 \\ \vdots \\ \mathbf{K}_{n-1} \\ \mathbf{K}_n \\ \mathbf{T}_n \end{pmatrix} \right. . \tag{7.18}
$$

$$\underbrace{\hphantom{XXXXXXXXXXXXX}}_{n+3}$$

The user specifies the values of the two extreme tangents $\mathbf{T}_1$ and $\mathbf{T}_n$, the equations are solved, and the $\mathbf{P}_i$ points are then used in the usual way to calculate a cubic uniform B-spline that passes through the original points $\mathbf{K}_i$. This process should be compared to the similar computation of the cubic spline, Section 5.1. Specifically, Equation (7.18) should be compared with Equation (5.7).

Notice that the coefficient matrix of Equation (7.18) is not diagonally dominant because of the four ± 3's. We can, however, modify it slightly by writing the system of equations in the form

$$
n+3\left\{\frac{1}{6}\begin{pmatrix} -3/2 & 0 & 3/2 & 0 & \dots & 0 & 0 & 0 \\ 1 & 4 & 1 & 0 & \dots & 0 & 0 & 0 \\ 0 & 1 & 4 & 1 & \dots & 0 & 0 & 0 \\ & \vdots & & & & & \vdots & \\ 0 & 0 & 0 & 0 & \dots & 4 & 1 & 0 \\ 0 & 0 & 0 & 0 & \dots & 1 & 4 & 1 \\ 0 & 0 & 0 & 0 & \dots & -3/2 & 0 & 3/2 \end{pmatrix}}_{n+3} \begin{pmatrix} \mathbf{P}_{-1} \\ \mathbf{P}_0 \\ \mathbf{P}_1 \\ \vdots \\ \mathbf{P}_{n-1} \\ \mathbf{P}_n \\ \mathbf{P}_{n+1} \end{pmatrix} = \begin{pmatrix} \mathbf{T}_1/2 \\ \mathbf{K}_0 \\ \mathbf{K}_1 \\ \vdots \\ \mathbf{K}_{n-1} \\ \mathbf{K}_n \\ \mathbf{T}_n/2 \end{pmatrix}. \quad (7.19)
$$

The coefficient matrix of Equation (7.19) is columnwise diagonally dominant and is therefore nonsingular. Thus, this system of equations has a unique solution, but this system is mathematically identical to Equation (7.18), so that system of equations also has a unique solution.

Example: This is the opposite of the example on page 259. We start with $\mathbf{K}_0 = (1/6, 5/6)$, $\mathbf{K}_1 = (1, 1)$, $\mathbf{K}_2 = (11/6, 5/6)$, and the two extreme tangents $\mathbf{T}_1 = (1/2, 1/2)$ and $\mathbf{T}_2 = (1/2, -1/2)$, and set up the 5×5 system of equations

$$
\frac{1}{6}\begin{pmatrix} -3 & 0 & 3 & 0 & 0 \\ 1 & 4 & 1 & 0 & 0 \\ 0 & 1 & 4 & 1 & 0 \\ 0 & 0 & 1 & 4 & 1 \\ 0 & 0 & -3 & 0 & 3 \end{pmatrix} \begin{pmatrix} \mathbf{P}_{-1} \\ \mathbf{P}_0 \\ \mathbf{P}_1 \\ \mathbf{P}_2 \\ \mathbf{P}_3 \end{pmatrix} = \begin{pmatrix} (1/2, 1/2) \\ (1/6, 5/6) \\ (1, 1) \\ (11/6, 5/6) \\ (1/2, -1/2) \end{pmatrix}.
$$

This is easy to solve and the solutions are $\mathbf{P}_{-1} = (0, 0)$, $\mathbf{P}_0 = (0, 1)$, $\mathbf{P}_1 = (1, 1)$, $\mathbf{P}_2 = (2, 1)$, and $\mathbf{P}_3 = (2, 0)$, identical to the original control points of the above-mentioned example.

7.8 A Knot Vector-Based Approach

The knot vector approach to the uniform B-spline curve assumes that the curve is a weighted sum, $\mathbf{P}(t) = \sum_{i=0}^{n} \mathbf{P}_i B_{n,i}(t)$ of the control points with unknown weight functions that have to be determined. The method is similar to that used in deriving the Bézier curve (Section 6.2). The cubic uniform B-spline is used here as an example, but this approach can be applied to B-splines of any order. We assume that five control points are given—so that five weight functions, $B_{4,0}(t)$ through $B_{4,4}(t)$ are required—and that the curve will consist of two cubic segments. In this approach we assume that each spline segment is traced when the parameter t varies over an interval of one unit, from an integer value u to the next integer $u+1$. The u values are called the *knots* of the B-spline. Since they are the integers $0, 1, 2, \dots$, they are uniformly distributed, hence the name *uniform* B-spline. To trace out a two-segment spline curve, t should vary in the interval $[0, 2]$.

The guiding principle is that each weight function should be a cubic polynomial, should have a maximum at the vicinity of "its" control point, and should drop to zero when away from the point. A general weight function should therefore have the bell shape shown in Figure 7.9a. To derive such a function, we write it as the union of four parts, $b_0(t)$, $b_1(t)$, $b_2(t)$, and $b_3(t)$, each a simple cubic polynomial, and each defined over one unit of t. Figure 7.9b shows how each weight $B_{4,i}(t)$ is defined over a range of five knots and is zero elsewhere

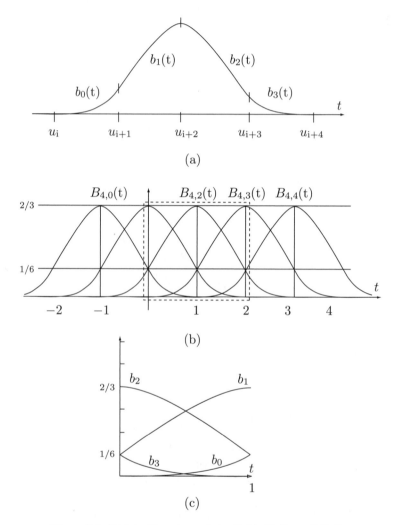

Figure 7.9: Weight Functions of the Cubic Uniform B-Spline.

The following considerations are employed to set up equations to calculate the $b_i(t)$ functions:

1. They should be barycentric.
2. They should provide C^2 continuity at the three points where they join.
3. $b_0(t)$ and its first two derivatives should be zero at the start point $b_0(0)$.
4. $b_3(t)$ and its first two derivatives should be zero at the end point $b_3(1)$.

We adopt the notation $b_i(t) = A_i t^3 + B_i t^2 + C_i t + D_i$. The conditions above yield the following equations:

1. The single equation $B_{4,0}(0) + B_{4,1}(0) + B_{4,2}(0) + B_{4,3}(0) = 1$. This is a special case of condition 1. We see later that the $b_i(t)$ functions resulting from our equations are, in fact, barycentric.

2. Condition 2 yields the nine equations

$$
\begin{aligned}
b_0(1) = b_1(0), \quad & \dot{b}_0(1) = \dot{b}_1(0), \quad & \ddot{b}_0(1) = \ddot{b}_1(0), \\
b_1(1) = b_2(0), \quad & \dot{b}_1(1) = \dot{b}_2(0), \quad & \ddot{b}_1(1) = \ddot{b}_2(0), \\
b_2(1) = b_3(0), \quad & \dot{b}_2(1) = \dot{b}_3(0), \quad & \ddot{b}_2(1) = \ddot{b}_3(0).
\end{aligned}
\tag{7.20}
$$

The first two derivatives of $b_i(t)$ are

$$
\frac{db_i(t)}{dt} = \dot{b}_i(t) = 3A_i t^2 + 2B_i t + C_i, \qquad \frac{d^2 b_i(t)}{dt^2} = \ddot{b}_i(t) = 6A_i t + 2B_i,
$$

so the nine equations above can be written explicitly as

$$
\begin{aligned}
A_0 + B_0 + C_0 + D_0 = D_1, \quad & 3A_0 + 2B_0 + C_0 = C_1, \quad & 6A_0 + 2B_0 = 2B_1, \\
A_1 + B_1 + C_1 + D_1 = D_2, \quad & 3A_1 + 2B_1 + C_1 = C_2, \quad & 6A_1 + 2B_1 = 2B_2, \\
A_2 + B_2 + C_2 + D_2 = D_3, \quad & 3A_2 + 2B_2 + C_2 = C_3, \quad & 6A_2 + 2B_2 = 2B_3.
\end{aligned}
$$

3. Condition 3 yields the three equations

$$
D_0 = 0, \quad C_0 = 0, \quad 2B_0 = 0.
$$

4. Condition 4 yields the three equations

$$
A_3 + B_3 + C_3 + D_3 = 0, \quad 3A_3 + 2B_3 + C_3 = 0, \quad 6A_3 + 2B_3 = 0.
$$

Thus, we end up with 16 equations that are easy to solve. Their solutions are

$$
\begin{aligned}
b_0(t) = \frac{1}{6} t^3, \quad & b_1(t) = \frac{1}{6}(1 + 3t + 3t^2 - 3t^3), \\
b_2(t) = \frac{1}{6}(4 - 6t^2 + 3t^3), \quad & b_3(t) = \frac{1}{6}(1 - 3t + 3t^2 - t^3).
\end{aligned}
\tag{7.21}
$$

The proof that the $b_i(t)$ functions are barycentric is now trivial. Figure 7.9c shows the shapes of the four weights.

Now that the weight functions are known, the entire curve can be expressed as the weighted sum $\mathbf{P}(t) = \sum_{i=0}^{n} \mathbf{P}_i B_{4,i}(t)$, where the weights all look the same and are shifted with respect to each other by using different ranges for t. Each weight $B_{4,i}(t)$ is nonzero only in the (open) interval (u_{i-3}, u_{i+1}) (Figure 7.9b).

Each curve segment $\mathbf{P}_i(t)$ can now be expressed as the barycentric sum of the four weighted points $\mathbf{P}_{i-3}$ through $\mathbf{P}_i$ (or, alternatively, as a linear combination of the $B_{4,i}(t)$ functions), $\mathbf{P}_i(t) = \sum_{j=-3}^{0} \mathbf{P}_{i+j} B_{4,i+j}(t)$, where $u_i \le t < u_{i+1}$. The next (crucial) step is to realize that in the range $u_i \leq t < u_{i+1}$, only component b_3 of $B_{4,i-3}$ is nonzero and similarly for the other three weights (see the dashed box of Figure 7.9b). The segment can therefore be written

$$
\begin{aligned}
\mathbf{P}_i(t) &= \sum_{j=3}^{0} \mathbf{P}_{i-j} b_j(t) \\
&= \frac{1}{6}\mathbf{P}_{i-3}(-t^3 + 3t^2 - 3t + 1) + \frac{1}{6}\mathbf{P}_{i-2}(3t^3 - 6t^2 + 4) \\
&\quad + \frac{1}{6}\mathbf{P}_{i-1}(-3t^3 + 3t^2 + 3t + 1) + \frac{1}{6}\mathbf{P}_i t^3 \\
&= \frac{1}{6}(t^3, t^2, t, 1)
\begin{pmatrix}
-1 & 3 & -3 & 1 \\
3 & -6 & 3 & 0 \\
-3 & 0 & 3 & 0 \\
1 & 4 & 1 & 0
\end{pmatrix}
\begin{pmatrix}
\mathbf{P}_{i-3} \\
\mathbf{P}_{i-2} \\
\mathbf{P}_{i-1} \\
\mathbf{P}_i
\end{pmatrix},
\end{aligned}
\tag{7.22}
$$

an expression identical (except for the choice of index i) to Equation (7.11). This approach to deriving the weight functions can be generalized for the nonuniform B-spline.

The dashed box of Figure 7.9b illustrates how the $B_{4,i}(t)$ weight functions blend the five control points in the two spline segments. The first weight, $B_{4,0}(t)$, goes down from $1/6$ to 0 when t varies from 0 to 1. Thus, the first control point $\mathbf{P}_0$ starts by contributing $1/6$ of its value to the curve, then decreases its contribution until it disappears at $t = 1$. This is why $\mathbf{P}_0$ does not contribute to the second segment. The second weight, $B_{4,1}(t)$, starts at $2/3$ (when $t = 0$), goes down to $1/6$ for $t = 1$, then all the way to 0 when t reaches 2. This is how the second control point $\mathbf{P}_1$ participates in the blend that generates the first two spline segments. Notice how the weight functions have their maxima at integer values of t, how only three weights are nonzero at these values, and how there are four nonzero weights for any other values of t.

Figure 7.10a shows the weight functions for the linear uniform B-spline. Each has the form of a hat, going from 0 to 1 and back to 0. They also have their maxima at integer values of t. The weight functions of the quadratic B-spline are shown in Figure 7.10b. Notice how each varies from 0 to $3/4$, how they meet at a height of $1/2$, and how their maxima are at half-integer values of t. The first weight, $B_{3,0}(t)$, drops from $1/2$ to 0 for the first spline segment (i.e., when t varies in the interval $[0, 1]$) and remains zero for the second and subsequent segments. The second weight, $B_{3,1}(t)$, climbs from $1/2$ to 1, then drops back to $1/2$ for the first segment. For the second segment, this weight goes down from $1/2$ to 0. These diagrams provide a clear understanding of how the control points are blended by the uniform B-spline.

The general B-spline weight functions are normally denoted by $N_{ik}(t)$ and can be defined recursively. Before delving into this topic, however, we show how the uniform B-spline curve itself can be defined recursively, similar to the recursive definition of the Bézier curve [Equation (6.11)]. Given a set of $n + 1$ control points $\mathbf{P}_0$ through $\mathbf{P}_n$ and a uniform knot vector $(t_0, t_1, \ldots, t_{n+k})$ (a set of equally-spaced $n + k + 1$ nondecreasing real numbers), the B-spline of order k is defined as

$$
\mathbf{P}(t) = \mathbf{P}_l^{(k-1)}(t), \quad \text{where} \quad t_l \leq t < t_{l+1}
\tag{7.23}
$$

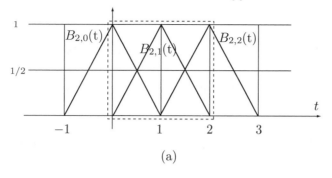

(a)

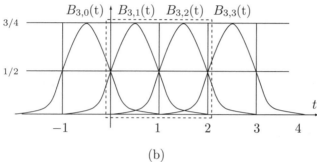

(b)

Figure 7.10: Weight Functions of the Linear and the Quadratic B-Splines.

and where the quantities $\mathbf{P}_i^{(j)}(t)$ are defined recursively by

$$\mathbf{P}_i^{(j)}(t) = \begin{cases} \mathbf{P}_i, & \text{for } j = 0, \\ (1 - T_{ij})\mathbf{P}_{i-1}^{(j-1)}(t) + T_{ij}\mathbf{P}_i^{(j-1)}(t), & \text{for } j > 0, \end{cases}$$

and

$$T_{ij} = \frac{t - t_i}{t_{i+k-j} - t_i}.$$

Figure 7.11 is a pyramid that illustrates how the quantities $\mathbf{P}_l^{(k-1)}(t)$ are constructed recursively. Each $\mathbf{P}_i^{(j)}(t)$ in the figure is constructed as a barycentric sum of the two quantities immediately to its left. Equation (7.23) is the *geometric* definition of the uniform B-spline.

We now turn to the *algebraic* (or analytical) definition of the general (uniform and nonuniform) B-spline curve. It is defined as the weighted sum

$$\mathbf{P}(t) = \sum_{i=0}^{n} \mathbf{P}_i N_{ik}(t),$$

where the weight functions $N_{ik}(t)$ are defined recursively by

$$N_{i1}(t) = \begin{cases} 1, & \text{if } t \in [t_i, t_{i+1}), \\ 0, & \text{otherwise,} \end{cases} \tag{7.24}$$

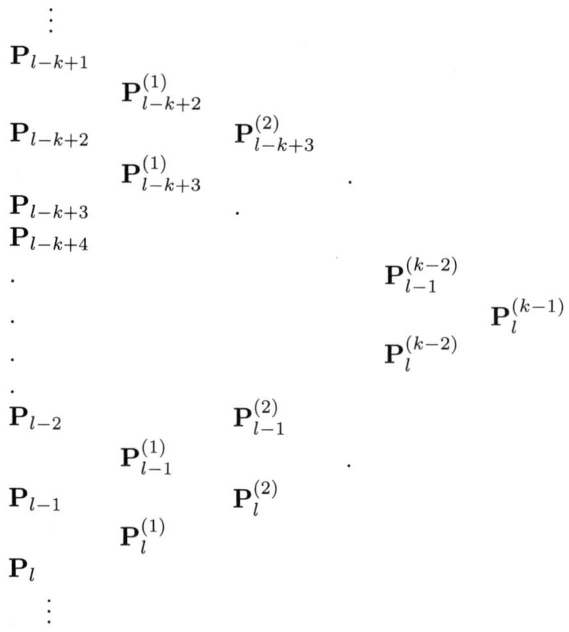

$$\vdots$$

$\mathbf{P}_{l-k+1}$

$\quad\quad\quad \mathbf{P}^{(1)}_{l-k+2}$

$\quad\quad\quad\quad\quad\quad \mathbf{P}^{(2)}_{l-k+3}$

$\mathbf{P}_{l-k+2}$

$\quad\quad\quad \mathbf{P}^{(1)}_{l-k+3}$

$\mathbf{P}_{l-k+3}$

$\mathbf{P}_{l-k+4}$

$\quad\quad\quad\quad\quad\quad\quad\quad\quad \mathbf{P}^{(k-2)}_{l-1}$

$\quad\quad\quad\quad\quad\quad\quad\quad\quad\quad\quad\quad \mathbf{P}^{(k-1)}_{l}$

$\quad\quad\quad\quad\quad\quad\quad\quad\quad \mathbf{P}^{(k-2)}_{l}$

$\mathbf{P}_{l-2}$

$\quad\quad\quad\quad\quad\quad \mathbf{P}^{(2)}_{l-1}$

$\quad\quad\quad \mathbf{P}^{(1)}_{l-1}$

$\mathbf{P}_{l-1}$

$\quad\quad\quad\quad\quad\quad \mathbf{P}^{(2)}_{l}$

$\quad\quad\quad \mathbf{P}^{(1)}_{l}$

$\mathbf{P}_{l}$

$$\vdots$$

Figure 7.11: Recursive Construction of $\mathbf{P}^{(k-1)}_{l}(t)$.

(note how the interval starts at t_i but does not reach t_{i+1}; such an interval is closed on the left and open on the right) and

$$N_{ik}(t) = \frac{t - t_i}{t_{i+k-1} - t_i} N_{i,k-1}(t) + \frac{t_{i+k} - t}{t_{i+k} - t_{i+1}} N_{i+1,k-1}(t), \quad \text{where} \quad 0 \le i \le n. \quad (7.25)$$

The weights $N_{ik}(t)$ may be tedious to calculate in the general case, where the knots t_i can be any, but are easy to calculate in the special case where the knot vector is the uniform sequence $(0, 1, \ldots, n + k)$, i.e., when $t_i = i$. Here are examples for the first few values of k.

For $k = 1$, the weight functions are defined by

$$N_{i1}(t) = \begin{cases} 1, & \text{if } t \in [i, i+1), \\ 0, & \text{otherwise.} \end{cases} \quad (7.26)$$

This results in the "step" functions shown in Figure 7.12. Notice how each step is closed on the left and open on the right and how $N_{i1}(t)$ is nonzero only in the interval $[i, i+1)$ (this interval is its *support*). It is also clear that each of them is a shifted version of its predecessor, so we can express any of them as a shifted version of the first one and write $N_{i1}(t) = N_{01}(t - i)$.

For $k = 2$, the weight functions can be calculated for any i from Equation (7.25)

$$N_{02}(t) = \frac{t - t_0}{t_1 - t_0} N_{01}(t) + \frac{t_2 - t}{t_2 - t_1} N_{11}(t)$$

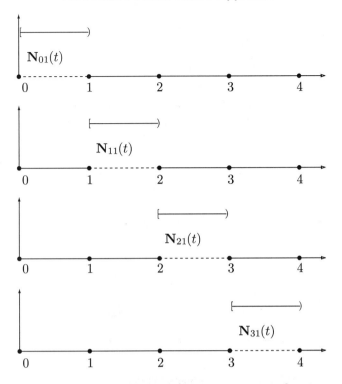

Figure 7.12: Uniform B-Spline Weight Functions for $k = 1$.

$$= tN_{01}(t) + (2 - t)N_{11}(t)$$

$$= \begin{cases} t, & \text{when } 0 \le t < 1, \\ 2 - t, & \text{when } 1 \le t < 2, \\ 0, & \text{otherwise,} \end{cases}$$

$$N_{12}(t) = \frac{t - t_1}{t_2 - t_1} N_{11}(t) + \frac{t_3 - t}{t_3 - t_2} N_{21}(t)$$

$$= (t - 1)N_{11}(t) + (3 - t)N_{21}(t)$$

$$= \begin{cases} t - 1, & \text{when } 1 \le t < 2, \\ 3 - t, & \text{when } 2 \le t < 3, \\ 0, & \text{otherwise,} \end{cases}$$

$$N_{22}(t) = \frac{t - t_2}{t_3 - t_2} N_{21}(t) + \frac{t_4 - t}{t_4 - t_3} N_{31}(t)$$

$$= (t - 2)N_{21}(t) + (4 - t)N_{31}(t)$$

$$= \begin{cases} t - 2, & \text{when } 2 \le t < 3, \\ 4 - t, & \text{when } 3 \le t < 4, \\ 0, & \text{otherwise.} \end{cases}$$

The hat-shaped functions are shown in Figure 7.13. Notice how $N_{i2}(t)$ spans the interval $[i, i+2)$. It is also obvious that each of them is a shifted version of its predecessor, so we can express any of them as a shifted version of the first one and write $N_{i2}(t) = N_{02}(t-i)$.

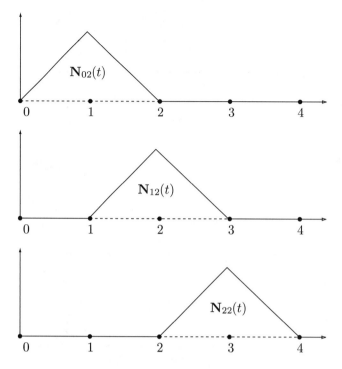

Figure 7.13: Uniform B-Spline Weight Functions for $k = 2$.

For $k = 3$, the calculations are similar:

$$
\begin{aligned}
N_{03}(t) &= \frac{t - t_0}{t_2 - t_0} N_{02}(t) + \frac{t_3 - t}{t_3 - t_1} N_{12}(t) \\
&= \frac{t}{2} N_{02}(t) + \frac{3 - t}{2} N_{12}(t) \\
&= \begin{cases}
t^2/2, & \text{when } 0 \le t < 1, \\
\frac{t^2}{2}(2 - t) + \frac{3-t}{2}(t - 1), & \text{when } 1 \le t < 2, \\
(3 - t)^2/2, & \text{when } 2 \le t < 3, \\
0, & \text{otherwise,}
\end{cases} \\
&= \begin{cases}
t^2/2, & \text{when } 0 \le t < 1, \\
(-2t^2 + 6t - 3)/2, & \text{when } 1 \le t < 2, \\
(3 - t)^2/2, & \text{when } 2 \le t < 3, \\
0, & \text{otherwise,}
\end{cases} \\
N_{13}(t) &= \frac{t - t_1}{t_3 - t_1} N_{12}(t) + \frac{t_4 - t}{t_4 - t_2} N_{22}(t) \\
&= \frac{t - 1}{2} N_{12}(t) + \frac{4 - t}{2} N_{22}(t)
\end{aligned}
$$

$$= \begin{cases} (t-1)^2/2, & \text{when } 1 \le t < 2, \\ (-2t^2 + 10t - 11)/2, & \text{when } 2 \le t < 3, \\ (4-t)^2/2, & \text{when } 3 \le t < 4, \\ 0, & \text{otherwise.} \end{cases}$$

Each of these curves (Figure 7.14) is a spline whose three segments are quadratic polynomials (i.e., parabolic arcs) joined smoothly at the knots. Notice again that the support of $N_{i3}(t)$ is the interval $[i, i+3)$ and that they are shifted versions of each other, allowing us to write $N_{i3}(t) = N_{03}(t-i)$.

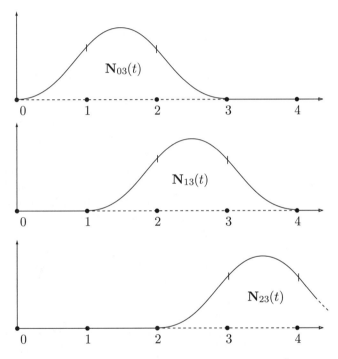

Figure 7.14: Uniform B-Spline Weight Functions for $k = 3$.

⋄ **Exercise 7.9:** How can we show that the various $N_{i3}(t)$ are shifted versions of each other?

In general, the support of $N_{ik}(t)$ is the interval $[i, i+k)$ and $N_{ik}(t) = N_{0k}(t-i)$. Figure 7.15 shows how a general weight function $\mathbf{N}_{ik}(t)$ is constructed recursively. Each $\mathbf{N}_{ij}(t)$ function in this triangle is constructed as a weighted sum of the two functions immediately to its left.

The geometric and algebraic definitions of the B-spline look different but it can be shown that they are identical. The proof of this is called the Cox–DeBoor (or DeBoor–Cox) formula [DeBoor 72].

$\mathbf{N}_{i,1}$
$\mathbf{N}_{i+1,1}$.

.   .

. $\mathbf{N}_{i,k-2}$

                                              $\mathbf{N}_{i,k-1}$
                          $\mathbf{N}_{i+1,k-2}$ $\mathbf{N}_{i,k}$
                                        $\mathbf{N}_{i+1,k-1}$
. $\mathbf{N}_{i+2,k-2}$

.

.   .

$\mathbf{N}_{i+k-2,1}$.
$\mathbf{N}_{i+k-1,1}$

Figure 7.15: Recursive Construction of $\mathbf{N}_{i,k}(t)$.

7.9 Recursive Definitions of the B-Spline

The order k of the B-spline curve is an integer in the interval $[2, n+1]$ (it is possible to have $k = 1$, but the curve degenerates in this case to just a plot of the control points). Each blending function $N_{ik}(t)$ has support over k intervals $[t_i, t_{i+k-1})$ and is zero outside its support. The knot vector $(t_0, t_1, \ldots, t_{n+k})$ consists of $n + k + 1$ nondecreasing real numbers t_i. These values define $n + k$ subintervals $[t_i, t_{i+1})$. The two extreme values t_0 and t_n are selected based on the values of n and k. Any terms of the form $0/0$ or $x/0$ in the calculation of the blending functions are assumed to be zero. Editing the B-spline curve can be done by (1) adding, moving, or deleting control points without changing the order k, (2) changing the order k without modifying the control points, and (3) increasing the size of the knot vector. The knot vector contains $n + k + 1$ values, so increasing its size implies that either n or k should be increased. Here are a few more properties of the curve:

1. Plotting the B-spline curve is done by varying the parameter t over the range of knot values $[t_{k-1}, t_{n+1})$.

2. Each segment of the curve (between two consecutive knot values) depends on k control points. This is why the curve has local control and it also implies that the maximum value of k is the number $n + 1$ of control points.

3. Any control point participates in at most k segments.

4. The curve lies inside the convex hull defined by at most k control points. This means that the curve passes close to the control points, a feature that makes it easy for a designer to place these points in order to obtain the right curve shape.

5. The blending functions $N_{ik}(t)$ are barycentric for any t in the interval $[t_{k-1}, t_{n+1})$. They are also nonnegative and, except for $k = 1$, each has one maximum.

6. The curve and its first $k - 1$ derivatives are continuous over the entire range (except that nonuniform B-splines can have discontinuities, see Figure 7.19d).

7. The entire curve can be affinely transformed by transforming the control points, then redrawing the curve from the new points.

One important difference between the B-spline and the Bézier curve is the use of a *knot vector*. This feature (which has already been mentioned) consists of a nondecreasing sequence of real numbers called *knots*. The knot vector adds flexibility to the curve and provides better control of its shape, but its use requires experience. There are three common ways to select the values in the knot vector, namely uniform, open uniform, and nonuniform. In a uniform B-spline the knot values are equally spaced. An example is $(-2, -1.5, -0.5, 0, 0.5, 1, 1.5)$, but more typical examples are a vector with normalized values between 0 and 1 $(0, 0.2, 0.4, 0.6, 0.8, 1)$ or a vector with integer values $(0, 1, 2, 3, 4, 5, 6)$. Figure 7.16 lists *Mathematica* code to calculate, print, and plot the weight functions for any set of knots.

```
(* B-spline weight functions printed and plotted *)
Clear[bspl,knt,i,k,n,t,p]
bspl[i_,k_,t_]:=If[knt[[i+k]]==knt[[i+1]],0, (* 0<=i<=n *)
 bspl[i,k-1,t] (t-knt[[i+1]])/(knt[[i+k]]-knt[[i+1]])] \
 +If[knt[[i+1+k]]==knt[[i+2]],0,
 bspl[i+1,k-1,t] (knt[[i+1+k]]-t)/(knt[[i+1+k]]-knt[[i+2]])];
bspl[i_,1,t_]:=If[knt[[i+1]]<=t<knt[[i+2]], 1, 0];
n=4; k=3; (* Note: 0<=k<=n *)
(* knt=Table[i, {i,0,n+k}]; *) (* knots for the uniform case *)
knt={0,0,0,1,2,3,3,3}; (* knots for the NONuniform case *)
(* Show the weight functions *)
Do[Print["N(",i,",",k,",",t,")=",Simplify[bspl[i,k,t]]], {i,0,n}]
(* Plot them. Plots are separated using .97 instead of 1 *)
Do[p[i+1]=Plot[bspl[i,k,t], {t,k-.97,n+.97},
DisplayFunction->Identity], {i,0,n}]
Show[Table[p[i+1], {i,0,n}], Ticks->None,
DisplayFunction->$DisplayFunction]
```

Figure 7.16: Code for the B-Spline Weight Functions.

7.10 Open Uniform B-Splines

The open uniform B-spline is obtained when the knot vector is uniform except at its two ends, where knot values are repeated k times. The following are simple examples:

For $n = 3$ and $k = 2$, there are $n + k + 1 = 6$ knots, e.g., $(0, 0, 1, 2, 3, 3)$.

For $n = 4$ and $k = 4$, there are $n + k + 1 = 9$ knots, e.g., $(0, 0, 0, 0, 1, 2, 2, 2, 2)$.

For $n = 3$ and $k = 2$, there are $n + k + 1 = 6$ knots, e.g., $(0, 0, 0.33, 0.67, 1, 1)$.

For $n = 4$ and $k = 4$, there are $n + k + 1 = 9$ knots, e.g., $(0, 0, 0, 0, 0.5, 1, 1, 1, 1)$.

(Notice how the last two examples are normalized.) In general, given values for n and k, we can generate an integer open knot vector by setting

$$t_i = \begin{cases} 0, & \text{for } 0 \leq i < k, \\ i - k + 1, & \text{for } k \leq i \leq n, \\ n - k + 2, & \text{for } n < i \leq n + k, \end{cases} \qquad \text{for} \quad 0 \leq i \leq n + k. \qquad (7.27)$$

An open uniform B-spline curve starts at $\mathbf{P}_0$ and ends at $\mathbf{P}_n$. This feature makes it easy to generate closed curves of this type. The two extreme tangents of this curve point in the directions from $\mathbf{P}_0$ to $\mathbf{P}_1$ and from $\mathbf{P}_{n-1}$ to $\mathbf{P}_n$, respectively. This is why open uniform B-spline curves are similar to Bézier curves. In fact, when $k = n+1$ (i.e., when the degree of the polynomials is n), these curves have knot vectors of the form $(0, 0, \ldots, 0, 1, 1, \ldots, 1)$ and they reduce to Bézier curves.

Example: (1) Five control points $\mathbf{P}_0$ through $\mathbf{P}_4$ are given, implying that $n = 4$. We select order 3 (i.e., segments that are polynomials of degree 2) and use Equation (7.27) to construct the knot sequence $(0, 0, 0, 1, 2, 3, 3, 3)$. The parameter t varies from $t_{k-1} = t_2 = 0$ to $t_{n+1} = t_5 = 3$, so our curve will consist of three segments. Each of the blending functions $N_{i3}(t)$ (where $0 \le i \le n$) is nonzero over three subintervals of t and is calculated from Equations (7.24) and (7.25). The result is

$$
\begin{aligned}
N_{03}(t) &= (1-t)^2, & 0 \le t < 1, \\[4pt]
N_{13}(t) &= \frac{1}{2}\begin{cases} -3t^2 + 4t, & 0 \le t < 1, \\ (2-t)^2, & 1 \le t < 2, \end{cases} \\[4pt]
N_{23}(t) &= \frac{1}{2}\begin{cases} t^2, & 0 \le t < 1, \\ -2t^2 + 6t - 3, & 1 \le t < 2, \\ (3-t)^2, & 2 \le t < 3, \end{cases} \\[4pt]
N_{33}(t) &= \frac{1}{2}\begin{cases} (t-1)^2, & 1 \le t < 2, \\ -3t^2 + 14t - 15, & 2 \le t < 3, \end{cases} \\[4pt]
N_{43}(t) &= (t-2)^2, & 2 \le t < 3,
\end{aligned}
$$

so the three spline segments are

$$
\begin{aligned}
\mathbf{P}_1(t) &= (1-t)^2\mathbf{P}_0 + \tfrac{1}{2}t(4-3t)\mathbf{P}_1 + \tfrac{1}{2}t^2\mathbf{P}_2, & 0 \le t < 1, \\
\mathbf{P}_2(t) &= \tfrac{1}{2}(2-t)^2\mathbf{P}_1 + \tfrac{1}{2}\big[t(2-t)+(t-1)(3-t)\big]\mathbf{P}_2 + \tfrac{1}{2}(t-1)^2\mathbf{P}_3, & 1 \le t < 2, \\
\mathbf{P}_3(t) &= \tfrac{1}{2}(3-t)^2\mathbf{P}_2 + \tfrac{1}{2}(3-t)(3t-5)\mathbf{P}_3 + (t-2)^2\mathbf{P}_4, & 2 \le t < 3.
\end{aligned}
$$

It is now easy to calculate where each segment starts and ends:

$$
\begin{aligned}
\mathbf{P}_1(0) &= \mathbf{P}_0, & \mathbf{P}_1(1) &= (\mathbf{P}_1 + \mathbf{P}_2)/2, \\
\mathbf{P}_2(1) &= (\mathbf{P}_1 + \mathbf{P}_2)/2, & \mathbf{P}_2(2) &= (\mathbf{P}_2 + \mathbf{P}_3)/2, \\
\mathbf{P}_3(2) &= (\mathbf{P}_2 + \mathbf{P}_3)/2, & \mathbf{P}_3(3) &= \mathbf{P}_4,
\end{aligned}
$$

Figure 7.17 shows a typical example of the three segments (with intentional gaps between them).

⋄ **Exercise 7.10:** Show that the three spline segments provide C^1 continuity at the two interior points $\mathbf{P}_1(1) = \mathbf{P}_2(1)$ and $\mathbf{P}_2(2) = \mathbf{P}_3(2)$.

Example: (2) We again choose five control points but this time we select $k = n + 1 = 5$. The curve will therefore consist of degree-4 polynomial segments. Such a segment requires five points (it has five coefficients, so five equations are needed), which is why we will end up with just one segment. Equation (7.27) is again used to construct the knot vector $(0, 0, 0, 0, 0, 1, 1, 1, 1, 1)$. The parameter t varies from $t_{k-1} = t_4 = 0$ to

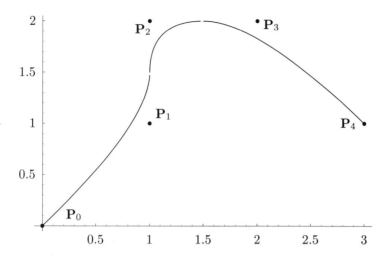

```
(* Plot a B-spline curve. Can also print the weight functions *)
Clear[bspl,knt,i,k,n,t,p,g1,g2,pnt] (* First the weight functions *)
bspl[i_,k_,t_]:=If[knt[[i+k]]==knt[[i+1]],0, (* 0<=i<=n *)
 bspl[i,k-1,t] (t-knt[[i+1]])/(knt[[i+k]]-knt[[i+1]])] \
 +If[knt[[i+1+k]]==knt[[i+2]],0,
 bspl[i+1,k-1,t] (knt[[i+1+k]]-t)/(knt[[i+1+k]]-knt[[i+2]])];
bspl[i_,1,t_]:=If[knt[[i+1]]<=t<knt[[i+2]], 1, 0];
n=4; k=3; (* Note: 0<=k<=n *)
(* knt=Table[i, {i,0,n+k}]; knots for the uniform case *)
knt={0,0,0,1,2,3,3,3}; (* knots for the open-unif or non-uniform cases *)
(* Do[Print[bspl[i,k,t]], {i,0,n}] Display the weight functions *)
pnt={{0,0},{1,1},{1,2},{2,2},{3,1}}; (* test for n+1=5 control points *)
p[t_]:=Sum[pnt[[i+1]] bspl[i,k,t], {i,0,n}] (* The curve as a weighted sum *)
g1=ListPlot[pnt, Prolog->AbsolutePointSize[3], DisplayFunction->Identity];
g2=ParametricPlot[p[t], {t,0,.97}, Compiled->False, DisplayFunction->Identity];
g3=ParametricPlot[p[t], {t,1,1.97}, Compiled->False, DisplayFunction->Identity];
g4=ParametricPlot[p[t], {t,2,3}, Compiled->False, DisplayFunction->Identity];
Show[g1,g2,g3,g4, PlotRange->All, DisplayFunction->$DisplayFunction,
 DefaultFont->{"cmr10", 10}];
```

Figure 7.17: An Open Uniform B-Spline.

$t_{n+1} = t_5 = 1$, showing again that the curve will consist of one segment. This should be a Bézier curve, because $k = n + 1$.

The calculation of the blending functions $N_{i5}(t)$ (where $0 \leq i \leq n$) is shown here in detail. We start with the nine functions $N_{i1}(t)$ that are calculated from Equation (7.24)

$$N_{01} = 1 \text{ when } t_0 \leq t < t_1, N_{11} = 1 \text{ when } t_1 \leq t < t_2, \ldots, N_{81} = 1 \text{ when } t_8 \leq t < t_9.$$

Since $t_0 = t_1 = t_2 = t_3 = t_4 = 0$ and $t_5 = t_6 = t_7 = t_8 = t_9 = 1$, we conclude that

$$N_{41} = 1 \quad \text{when} \quad t \in [t_4, t_5) = [0, 1),$$

and the other eight functions $N_{i1}(t)$ are zero. The next step is to calculate the eight functions $N_{i2}(t)$ from Equation (7.25):

$$N_{02}(t) = \frac{t - t_0}{t_1 - t_0} N_{01} + \frac{t_2 - t}{t_2 - t_1} N_{11} = 0,$$

$$N_{12}(t) = \frac{t - t_1}{t_2 - t_1} N_{11} + \frac{t_3 - t}{t_3 - t_2} N_{21} = 0,$$

$$N_{22}(t) = \frac{t - t_2}{t_3 - t_2} N_{21} + \frac{t_4 - t}{t_4 - t_3} N_{31} = 0,$$

$$N_{32}(t) = \frac{t - t_3}{t_4 - t_3} N_{31} + \frac{t_5 - t}{t_5 - t_4} N_{41} = 0 + (1 - t),$$

$$N_{42}(t) = \frac{t - t_4}{t_5 - t_4} N_{41} + \frac{t_6 - t}{t_6 - t_5} N_{51} = t + 0,$$

$$N_{52}(t) = \frac{t - t_5}{t_6 - t_5} N_{51} + \frac{t_7 - t}{t_7 - t_6} N_{61} = 0,$$

$$N_{62}(t) = \frac{t - t_6}{t_7 - t_6} N_{61} + \frac{t_8 - t}{t_8 - t_7} N_{71} = 0,$$

$$N_{72}(t) = \frac{t - t_7}{t_8 - t_7} N_{71} + \frac{t_9 - t}{t_9 - t_8} N_{81} = 0.$$

Only $N_{32}(t)$ and $N_{42}(t)$ are nonzero. The seven functions $N_{i3}(t)$ are calculated similarly:

$$N_{03}(t) = \frac{t - t_0}{t_2 - t_0} N_{02} + \frac{t_3 - t}{t_3 - t_1} N_{12} = 0,$$

$$N_{13}(t) = \frac{t - t_1}{t_3 - t_1} N_{12} + \frac{t_4 - t}{t_4 - t_2} N_{22} = 0,$$

$$N_{23}(t) = \frac{t - t_2}{t_4 - t_2} N_{22} + \frac{t_5 - t}{t_5 - t_3} N_{32} = 0 + (1 - t)^2,$$

$$N_{33}(t) = \frac{t - t_3}{t_5 - t_3} N_{32} + \frac{t_6 - t}{t_6 - t_4} N_{42} = t(1 - t) + (1 - t)t,$$

$$N_{43}(t) = \frac{t - t_4}{t_6 - t_4} N_{42} + \frac{t_7 - t}{t_7 - t_5} N_{52} = t^2 + 0,$$

$$N_{53}(t) = \frac{t - t_5}{t_7 - t_5} N_{52} + \frac{t_8 - t}{t_8 - t_6} N_{62} = 0,$$

$$N_{63}(t) = \frac{t - t_6}{t_8 - t_6} N_{62} + \frac{t_9 - t}{t_9 - t_7} N_{72} = 0.$$

Three of the seven functions are nonzero. The six functions $N_{i4}(t)$ are

$$N_{04}(t) = \frac{t - t_0}{t_3 - t_0} N_{03} + \frac{t_4 - t}{t_4 - t_1} N_{13} = 0,$$

$$N_{14}(t) = \frac{t - t_1}{t_4 - t_1} N_{13} + \frac{t_5 - t}{t_5 - t_2} N_{23} = 0 + (1 - t)^3,$$

$$N_{24}(t) = \frac{t - t_2}{t_5 - t_2} N_{23} + \frac{t_6 - t}{t_6 - t_3} N_{33} = t(1 - t)^2 + 2t(1 - t)^2,$$

$$N_{34}(t) = \frac{t - t_3}{t_6 - t_3} N_{33} + \frac{t_7 - t}{t_7 - t_4} N_{43} = 2t^2(1 - t) + (1 - t)t^2,$$

$$N_{44}(t) = \frac{t - t_4}{t_7 - t_4} N_{43} + \frac{t_8 - t}{t_8 - t_5} N_{53} = t^3,$$

$$N_{54}(t) = \frac{t - t_5}{t_8 - t_5} N_{53} + \frac{t_9 - t}{t_9 - t_6} N_{63} = 0.$$

Four of them are nonzero. The last step is the calculation of the five functions $N_{i5}(t)$:

$$N_{05}(t) = \frac{t - t_0}{t_4 - t_0} N_{04} + \frac{t_5 - t}{t_5 - t_1} N_{14} = (1 - t)^4,$$

$$N_{15}(t) = \frac{t - t_1}{t_5 - t_1} N_{14} + \frac{t_6 - t}{t_6 - t_2} N_{24} = t(1 - t)^3 + 3t(1 - t)^3,$$

$$N_{25}(t) = \frac{t - t_2}{t_6 - t_2} N_{24} + \frac{t_7 - t}{t_7 - t_3} N_{34} = 3t^2(1 - t)^2 + 3t^2(1 - t)^2,$$

$$N_{35}(t) = \frac{t - t_3}{t_7 - t_3} N_{34} + \frac{t_8 - t}{t_8 - t_4} N_{44} = 3t^3(1 - t) + (1 - t)t^3,$$

$$N_{45}(t) = \frac{t - t_4}{t_8 - t_4} N_{44} + \frac{t_9 - t}{t_9 - t_5} N_{54} = t^4.$$

All five are nonzero and they should look familiar (they are the Bernstein polynomials for $n = 4$). The curve consists of the single segment

$$\mathbf{P}(t) = \sum_{i=0}^{4} N_{i5}(t)\mathbf{P}_i$$
$$= (1 - t)^4 \mathbf{P}_0 + 4t(1 - t)^3 \mathbf{P}_1 + 6t^2(1 - t)^2 \mathbf{P}_2 + 4t^3(1 - t)\mathbf{P}_3 + t^4 \mathbf{P}_4,$$

which is the Bézier curve defined by the five points. The B-spline curve is again shown to be more general than the Bézier curve, since it contains the latter as a special case.

It is the multiplicity of knot values that causes the open B-spline to start and end at its extreme control points. This is easy to understand when we realize that every subinterval $[t_i, t_{i+1})$ of knots corresponds to one segment $\mathbf{P}_i(t)$ of the B-spline. When $t_i = t_{i+1}$, that segment reduces to a point. The result is that each repeat of a knot value decreases the continuity at a joint point by 1. Consider, for example, the open B-spline of order $k = 4$. The individual spline segments are degree-3 (cubic) polynomials that have C^2 continuity at their joint points. If knot t_i has multiplicity 2 (i.e., $t_i = t_{i+1}$), then segment $\mathbf{P}_i(t)$ reduces to a point and segments $\mathbf{P}_{i-1}(t)$ and $\mathbf{P}_{i+1}(t)$ meet at a joint point with C^1 continuity. If knot t_i has multiplicity 3 ($t_i = t_{i+1} = t_{i+2}$), then segments $\mathbf{P}_i(t)$ and $\mathbf{P}_{i+1}(t)$ reduce to points and segments $\mathbf{P}_{i-1}(t)$ and $\mathbf{P}_{i+2}(t)$ meet at a joint point (which in this case is a control point) with C^0 continuity. If the first knot has multiplicity 4 ($t_0 = t_1 = t_2 = t_3$), then segments $\mathbf{P}_0(t)$, $\mathbf{P}_1(t)$, and $\mathbf{P}_2(t)$ reduce to points and segment $\mathbf{P}_3(t)$ starts at that point with no continuity.

7.11 Nonuniform B-Splines

The nonuniform B-spline is more general than the uniform or open B-splines, although it is not the most general type of this curve. It is obtained when the knot values are not equally spaced. The only requirement is that the knots be nondecreasing. Adjusting the knot values (as well as having multiple values) is a feature that helps fine-tune the shape of the curve. Multiple knots can be used to pull the curve in a certain direction and to create a cusp or even a discontinuity at a join point. Nonuniform B-splines can get complex, so we limit the discussion in this section to order-4 (i.e., degree-3) nonuniform B-splines. This is not a serious limitation, as this type is the most commonly used and it makes it easier to understand the properties and behavior of the nonuniform B-spline.

In the case of order-4 nonuniform B-splines, the knot vector contains values from t_0 to t_{n+4} (there are four more knots than control points), so the minimum number of knots is eight (since the minimum number of control points is four) and the parameter t varies, in this case, from $t_{k-1} = t_3$ to $t_{n+1} = t_4$. Spline segment $\mathbf{P}_i(t)$ depends on control points $\mathbf{P}_{i-3}$, $\mathbf{P}_{i-2}$, $\mathbf{P}_{i-1}$, and $\mathbf{P}_i$ and its expression is

$$\mathbf{P}_i(t) = N_{i-3,4}(t)\mathbf{P}_{i-3} + N_{i-2,4}(t)\mathbf{P}_{i-2} + N_{i-1,4}(t)\mathbf{P}_{i-1} + N_{i,4}(t)\mathbf{P}_i,$$

where $3 \le i \le n$ and $t_i \le t \le t_{i+1}$. There are $n - 2$ segments denoted by $\mathbf{P}_3(t)$ through $\mathbf{P}_n(t)$. When $n = 3$ (four control points), the curve consists of just one segment. When knot t_i has multiplicity 2 (i.e., $t_i = t_{i+1}$), segment $\mathbf{P}_i(t)$ reduces to a point. As has been mentioned earlier, it is this feature that makes the nonuniform B-spline so flexible, powerful, and therefore useful in practical work.

The weight functions are defined recursively by Equations (7.24) and (7.25) but go up to N_{i4} only:

$$N_{i1}(t) = \begin{cases} 1, & \text{if } t \in [t_i, t_{i+1}), \\ 0, & \text{otherwise}, \end{cases}$$

$$N_{i2}(t) = \frac{t - t_i}{t_{i+1} - t_i} N_{i,1}(t) + \frac{t_{i+2} - t}{t_{i+2} - t_{i+1}} N_{i+1,1}(t),$$

$$N_{i3}(t) = \frac{t - t_i}{t_{i+2} - t_i} N_{i,2}(t) + \frac{t_{i+3} - t}{t_{i+3} - t_{i+1}} N_{i+1,2}(t), \qquad (7.28)$$

$$N_{i4}(t) = \frac{t - t_i}{t_{i+3} - t_i} N_{i,3}(t) + \frac{t_{i+4} - t}{t_{i+4} - t_{i+1}} N_{i+1,3}(t).$$

The first set, $N_{i1}(t)$, are horizontal segments. The second set, $N_{i2}(t)$, are straight lines. The third set are quadratic polynomials and the fourth set, $N_{i4}(t)$, are cubic polynomials. Each cubic segment is defined by four control points and lies in the convex hull defined by the points. Thus, segment $\mathbf{P}_i(t)$ is defined by points $\mathbf{P}_{i-3}$, $\mathbf{P}_{i-2}$, $\mathbf{P}_{i-1}$, and $\mathbf{P}_i$, while segment $\mathbf{P}_{i+1}(t)$ is defined by points $\mathbf{P}_{i-2}$, $\mathbf{P}_{i-1}$, $\mathbf{P}_i$, and $\mathbf{P}_{i+1}$.

Figure 7.19 illustrates the effect of knot multiplicities using $n = 7$ (i.e., eight points) as an example. The knot vector should contain $n + k + 1 = 7 + 4 + 1 = 12$ values and t should vary from $t_{k-1} = t_3$ to $t_{n+1} = t_8$, a total of five subintervals. The four parts of the figure show cubic B-spline curves constructed with the knot vectors

$$(-3, -2, -1, 0, 1, 2, 3, 4, 5, 6, 7, 8), \quad (-3, -2, -1, 0, 1, 1, 2, 3, 4, 5, 6, 7),$$

```
(* 8-Point Nonuniform Cubic B-Spline Example. Five Segments *)
Clear[g,Q,pts,seg];
P0={0,0}; P1={0,1}; P2={1,1}; P3={1,0}; P4={2,0}; P5={2.75,1}; P6={3,1}; P7={3,0};
pts=Graphics[{PointSize[.01], Point/@{P0,P1,P2,P3,P4,P5,P6,P7}}];
seg={AbsoluteDashing[{2,2}], Line[{P1,P2,P3}], Line[{P4,P5,P6,P7}]};
Q[t_]:={((1-t)^3 P0 +(3t^3-6t^2+4) P1 +(-3t^3+3t^2+3t+1) P2 +t^3 P3)/6,
((2-t)^3 P1 +(3t^3-15t^2+21t-5) P2 +(-3t^3+12t^2-12t+4) P3 +(t-1)^3 P4)/6,
((3-t)^3 P2 +(3t^3-24t^2+60t-44) P3 +(-3t^3+21t^2-45t+31) P4 +(t-2)^3 P5)/6,
((4-t)^3 P3 +(3t^3-33t^2+117t-131) P4 +(-3t^3+30t^2-96t+100) P5 +(t-3)^3 P6)/6,
((5-t)^3 P4 +(3t^3-42t^2+192t-284) P5 +(-3t^3+39t^2-165t+229) P6 +(t-4)^3 P7)/6};
g=Table[ParametricPlot[Q[t][[i]], {t,i-1,0.97i},
Compiled->False, DisplayFunction->Identity], {i,1,5}];
Show[g, pts, Graphics[seg], PlotRange->All, DefaultFont->{"cmr10", 10},
DisplayFunction->$DisplayFunction, AspectRatio->Automatic];
```

For the four segments of part (b), the only difference is

```
Q[t_]:={(1-t)^3/6 P0 +(11t^3-15t^2-3t+7)/12 P1+(-5t^3+3t^2+3t+1)/4 P2 +t^3/2 P3,
(2-t)^3/2 P2 +(5t^3-27t^2+45t-21)/4 P3+(-11t^3+51t^2-69t+29)/12 P4 +(t-1)^3/6 P5,
(3-t)^3/4 P3 +(7t^3-57t^2+147t-115)/12 P4+(-3t^3+21t^2-45t+31)/6 P5 +(t-2)^3/6 P6,
((4-t)^3 P4 +(3t^3-33t^2+117t-131)P5+(-3t^3+30t^2-96t+100)P6 +(t-3)^3 P7)/6};
g=Table[ParametricPlot[Q[t][[i]], {t,i-1,0.97i},
Compiled->False, DisplayFunction->Identity], {i,1,4}];
```

For the three segments of part (c), the only difference is

```
Q[t_]:={(1-t)^3 P0 /6+(11t^3-15t^2-3t+7)P1 /12+(-7t^3+3t^2+3t+1)P2 /4+t^3 P3,
(2-t)^3 P3+(7t^3-39t^2+69t-37)P4 /4+(-11t^3+51t^2-69t+29) P5 /12+(t-1)^3 P6 /6,
(3-t)^3 P4 /4+(7t^3-57t^2+147t-115)P5 /12+(-3t^3+21t^2-45t+31) P6 /6+(t-2)^3 P7 /6};
g=Table[ParametricPlot[Q[t][[i]], {t,i-1,0.97i},
Compiled->False, DisplayFunction->Identity], {i,1,3}];
```

For the two segments of part (d), the only difference is

```
Q[t_]:={(1-t)^3P0 /6 +(11t^3-15t^2-3t+7)P1 /12+(-7t^3+3t^2+3t+1)P2 /4 +t^3 P3,
(2-t)^3 P4 +(7t^3-39t^2+69t-37)P5 /4+(-11t^3+51t^2-69t+29)P6 /12+(t-1)^3P7 /6};
g=Table[ParametricPlot[Q[t][[i]], {t,i-1,0.97i},
Compiled->False, DisplayFunction->Identity], {i,1,2}];
```

Figure 7.18: Code for an 8-Point Nonuniform B-Spline Example, Figure 7.19.

$$(-3, -2, -1, 0, 1, 1, 1, 2, 3, 4, 5, 6), \quad (-3, -2, -1, 0, 1, 1, 1, 1, 2, 3, 4, 5),$$

respectively. Notice that only six knots, t_3 through t_8, are really important. The rest are distinct and uniform but less important, since only some of them are used in calculating the blending functions.

In Figure 7.19a, all knots have multiplicity 1, each segment is defined by four points, and adjacent segments share three points. The first segment, $\mathbf{P}_3(t)$, is defined by points $\mathbf{P}_0$, $\mathbf{P}_1$, $\mathbf{P}_2$, and $\mathbf{P}_3$, while the last segment, $\mathbf{P}_7(t)$, is defined by points $\mathbf{P}_4$, $\mathbf{P}_5$, $\mathbf{P}_6$, and $\mathbf{P}_7$. The five segments join with C^2 continuity. In Figure 7.19b, we set $t_4 = t_5$, thereby reducing segment $\mathbf{P}_4(t)$ to zero length, causing segments $\mathbf{P}_3(t)$ and $\mathbf{P}_5(t)$ to meet at join $t_4 = t_5$. However, these segments share just two control points, $\mathbf{P}_2$ and $\mathbf{P}_3$, so they have less "in common" and, consequently, join with only C^1 continuity. In Figure 7.19c, we set $t_4 = t_5 = t_6$, thereby reducing segments $\mathbf{P}_4(t)$ and $\mathbf{P}_5(t)$ to zero length and causing segments $\mathbf{P}_3(t)$ and $\mathbf{P}_6(t)$ to meet. These segments share just one control point, namely $\mathbf{P}_3$, so they meet at this point, with C^0 continuity. In Figure 7.19d, we set

$t_4 = t_5 = t_6 = t_7$, so now we have three zero-length segments, namely $\mathbf{P}_4(t)$, $\mathbf{P}_5(t)$, and $\mathbf{P}_6(t)$. Segments $\mathbf{P}_3(t)$ and $\mathbf{P}_7(t)$ now have to meet, but they don't have any common control points. The result is a *discontinuity* (a break) in the curve between points $\mathbf{P}_3$ and $\mathbf{P}_4$.

Figure 7.18 lists the code for Figure 7.19.

Example: This long example is divided into two parts.

Part a. In this part, we calculate the blending functions and spline segments of the curve of Figure 7.19a, where the knot vector is the uniform sequence

$$(-3, -2, -1, 0, 1, 2, 3, 4, 5, 6, 7, 8).$$

The calculations are done bearing in mind that t varies from $t_3 = 0$ to $t_8 = 5$. We need to calculate all the functions $N_{i4}(t)$ that are nonzero in the five subintervals $[0, 1)$, $[1, 2)$, $[2, 3)$, $[3, 4)$, and $[4, 5)$. Four blending functions are used to construct each of the five spline segments, so segment $\mathbf{P}_3(t)$ is defined by functions $N_{04}(t)$ through $N_{34}(t)$, segment $\mathbf{P}_4(t)$ is defined by functions $N_{14}(t)$ through $N_{44}(t)$, and segment $\mathbf{P}_7(t)$ is defined by functions $N_{44}(t)$ through $N_{74}(t)$. The first step is to calculate N_{i1}:

$$N_{31} = 1 \quad \text{for} \quad t \in [0, 1), \quad N_{41} = 1 \quad \text{for} \quad t \in [1, 2),$$
$$N_{51} = 1 \quad \text{for} \quad t \in [2, 3), \quad N_{61} = 1 \quad \text{for} \quad t \in [3, 4), \quad N_{71} = 1 \quad \text{for} \quad t \in [4, 5),$$

and N_{01}, N_{11}, N_{21}, N_{81}, N_{91}, $N_{10,1}$, and $N_{11,1}$ are zero in the range $0 \le t < 5$.

Step 2 is to calculate functions N_{i2} that are nonzero for $0 \le t < 5$:

$$N_{02}(t) = \frac{t - t_0}{t_1 - t_0} N_{01} + \frac{t_2 - t}{t_2 - t_1} N_{11} = 0,$$

$$N_{12}(t) = \frac{t - t_1}{t_2 - t_1} N_{11} + \frac{t_3 - t}{t_3 - t_2} N_{21} = 0,$$

$$N_{22}(t) = \frac{t - t_2}{t_3 - t_2} N_{21} + \frac{t_4 - t}{t_4 - t_3} N_{31} = (1 - t) \quad \text{for} \quad t \in [0, 1),$$

$$N_{32}(t) = \frac{t - t_3}{t_4 - t_3} N_{31} + \frac{t_5 - t}{t_5 - t_4} N_{41} = \begin{cases} t & \text{for } t \in [0, 1), \\ 2 - t & \text{for } t \in [1, 2), \end{cases}$$

$$N_{42}(t) = \frac{t - t_4}{t_5 - t_4} N_{41} + \frac{t_6 - t}{t_6 - t_5} N_{51} = \begin{cases} t - 1 & \text{for } t \in [1, 2), \\ 3 - t & \text{for } t \in [2, 3), \end{cases}$$

$$N_{52}(t) = \frac{t - t_5}{t_6 - t_5} N_{51} + \frac{t_7 - t}{t_7 - t_6} N_{61} = \begin{cases} t - 2 & \text{for } t \in [2, 3), \\ 4 - t & \text{for } t \in [3, 4), \end{cases}$$

$$N_{62}(t) = \frac{t - t_6}{t_7 - t_6} N_{61} + \frac{t_8 - t}{t_8 - t_7} N_{71} = \begin{cases} t - 3 & \text{for } t \in [3, 4), \\ 5 - t & \text{for } t \in [4, 5), \end{cases}$$

$$N_{72}(t) = \frac{t - t_7}{t_8 - t_7} N_{71} + \frac{t_9 - t}{t_9 - t_8} N_{81} = t - 4 \quad \text{for} \quad t \in [4, 5).$$

This step terminates at $N_{72}(t)$ since $N_{82}(t)$ and its successors are zero for $0 \le t < 5$.

Step 3 requires the calculation of several functions N_{i3}:

$$N_{03}(t) = \frac{t - t_0}{t_2 - t_0} N_{02} + \frac{t_3 - t}{t_3 - t_1} N_{12} = 0,$$

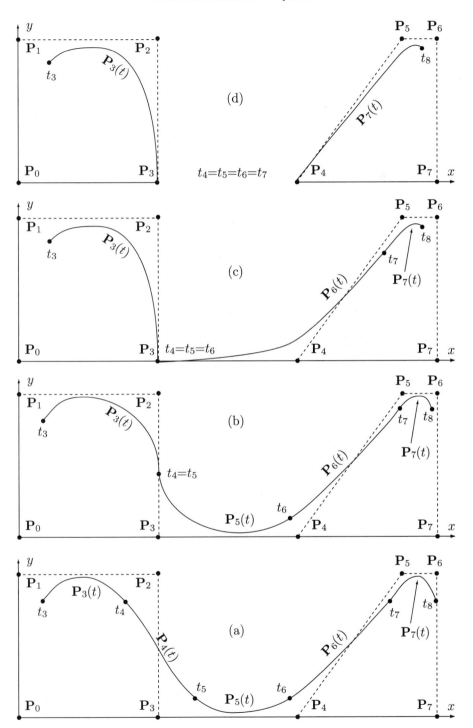

Figure 7.19: An Eight-Point Nonuniform B-Spline Curve with Multiple Knots.

$$N_{13}(t) = \frac{t-t_1}{t_3-t_1}N_{12} + \frac{t_4-t}{t_4-t_2}N_{22} = \frac{1}{2}(1-t)^2 \qquad\qquad \text{for } t \in [0,1),$$

$$N_{23}(t) = \frac{t-t_2}{t_4-t_2}N_{22} + \frac{t_5-t}{t_5-t_3}N_{32} = \frac{1}{2}\begin{cases}(-2t^2+2t+1) & \text{for } t \in [0,1), \\ (2-t)^2 & \text{for } t \in [1,2),\end{cases}$$

$$N_{33}(t) = \frac{t-t_3}{t_5-t_3}N_{32} + \frac{t_6-t}{t_6-t_4}N_{42} = \frac{1}{2}\begin{cases}t^2 & \text{for } t \in [0,1), \\ (-2t^2+6t-3) & \text{for } t \in [1,2), \\ (3-t)^2 & \text{for } t \in [2,3),\end{cases}$$

$$N_{43}(t) = \frac{t-t_4}{t_6-t_4}N_{42} + \frac{t_7-t}{t_7-t_5}N_{52} = \frac{1}{2}\begin{cases}(t-1)^2 & \text{for } t \in [1,2), \\ (-2t^2+10t-11) & \text{for } t \in [2,3), \\ (4-t)^2 & \text{for } t \in [3,4),\end{cases}$$

$$N_{53}(t) = \frac{t-t_5}{t_7-t_5}N_{52} + \frac{t_8-t}{t_8-t_6}N_{62} = \frac{1}{2}\begin{cases}(t-2)^2 & \text{for } t \in [2,3), \\ (-2t^2+14t-23) & \text{for } t \in [3,4), \\ (5-t)^2 & \text{for } t \in [4,5),\end{cases}$$

$$N_{63}(t) = \frac{t-t_6}{t_8-t_6}N_{62} + \frac{t_9-t}{t_9-t_7}N_{72} = \frac{1}{2}\begin{cases}(t-3)^2 & \text{for } t \in [3,4), \\ (-2t^2+18t-39) & \text{for } t \in [4,5),\end{cases}$$

$$N_{73}(t) = \frac{t-t_7}{t_9-t_7}N_{72} + \frac{t_{10}-t}{t_{10}-t_8}N_{82} = \frac{1}{2}(t-4)^2 \qquad\qquad \text{for } t \in [4,5).$$

We stop at N_{73} since N_{83} and its successors are zero for $0 \le t < 5$.

The last step involves the calculation of eight functions N_{i4}:

$$N_{04}(t) = \frac{t-t_0}{t_3-t_0}N_{03} + \frac{t_4-t}{t_4-t_1}N_{13} = \frac{1}{6}(1-t)^3 \qquad\qquad \text{for } t \in [0,1),$$

$$N_{14}(t) = \frac{t-t_1}{t_4-t_1}N_{13} + \frac{t_5-t}{t_5-t_2}N_{23} = \frac{1}{6}\begin{cases}(3t^3-6t^2+4) & \text{for } t \in [0,1), \\ (2-t)^3 & \text{for } t \in [1,2),\end{cases}$$

$$N_{24}(t) = \frac{t-t_2}{t_5-t_2}N_{23} + \frac{t_6-t}{t_6-t_3}N_{33} = \frac{1}{6}\begin{cases}(-3t^3+3t^2+3t+1) & \text{for } t \in [0,1), \\ (3t^3-15t^2+21t-5) & \text{for } t \in [1,2), \\ (3-t)^3 & \text{for } t \in [2,3),\end{cases}$$

$$N_{34}(t) = \frac{t-t_3}{t_6-t_3}N_{33} + \frac{t_7-t}{t_7-t_4}N_{43} = \frac{1}{6}\begin{cases}t^3 & \text{for } t \in [0,1), \\ (-3t^3+12t^2-12t+4) & \text{for } t \in [1,2), \\ (3t^3-24t^2+60t-44) & \text{for } t \in [2,3), \\ (4-t)^3 & \text{for } t \in [3,4),\end{cases}$$

$$N_{44}(t) = \frac{t-t_4}{t_7-t_4}N_{43} + \frac{t_8-t}{t_8-t_5}N_{53} = \frac{1}{6}\begin{cases}(t-1)^3 & \text{for } t \in [1,2), \\ (-3t^3+21t^2-45t+31) & \text{for } t \in [2,3), \\ (3t^3-33t^2+117t-131) & \text{for } t \in [3,4), \\ (5-t)^3 & \text{for } t \in [4,5),\end{cases}$$

$$N_{54}(t) = \frac{t-t_5}{t_8-t_5}N_{53} + \frac{t_9-t}{t_9-t_6}N_{63} = \frac{1}{6}\begin{cases}(t-2)^3 & \text{for } t \in [2,3), \\ (-3t^3+30t^2-96t+100) & \text{for } t \in [3,4), \\ (3t^3-42t^2+192t-284) & \text{for } t \in [4,5),\end{cases}$$

$$N_{64}(t) = \frac{t-t_6}{t_9-t_6}N_{63} + \frac{t_{10}-t}{t_{10}-t_7}N_{73} = \frac{1}{6}\begin{cases}(t-3)^3 & \text{for } t \in [3,4), \\ (-3t^3+39t^2-165t+229) & \text{for } t \in [4,5),\end{cases}$$

$$N_{74}(t) = \frac{t - t_7}{t_{10} - t_7} N_{73} + \frac{t_{11} - t}{t_{11} - t_8} N_{83} = \frac{1}{6}(t - 4)^3 \qquad\qquad \text{for } t \in [4, 5).$$

A careful study of this last group shows that N_{84} and its successors are zero for $0 \leq t < 5$.

The last group of blending functions can now be used to construct the five spline segments:

$$\begin{aligned}
\mathbf{P}_3(t) &= N_{04}(t)\mathbf{P}_0 + N_{14}(t)\mathbf{P}_1 + N_{24}(t)\mathbf{P}_2 + N_{34}(t)\mathbf{P}_3 & t \in [0, 1) \\
&= \frac{1}{6}\big[(1 - t)^3 \mathbf{P}_0 + (3t^3 - 6t^2 + 4)\mathbf{P}_1 \\
&\quad + (-3t^3 + 3t^2 + 3t + 1)\mathbf{P}_2 + t^3 \mathbf{P}_3\big], \\
\mathbf{P}_4(t) &= N_{14}(t)\mathbf{P}_1 + N_{24}(t)\mathbf{P}_2 + N_{34}(t)\mathbf{P}_3 + N_{44}(t)\mathbf{P}_4 & t \in [1, 2) \\
&= \frac{1}{6}\big[(2 - t)^3 \mathbf{P}_1 + (3t^3 - 15t^2 + 21t - 5)\mathbf{P}_2 \\
&\quad + (-3t^3 + 12t^2 - 12t + 4)\mathbf{P}_3 + (t - 1)^3 \mathbf{P}_4\big], \\
\mathbf{P}_5(t) &= N_{24}(t)\mathbf{P}_2 + N_{34}(t)\mathbf{P}_3 + N_{44}(t)\mathbf{P}_4 + N_{54}(t)\mathbf{P}_5 & t \in [2, 3) \\
&= \frac{1}{6}\big[(3 - t)^3 \mathbf{P}_2 + (3t^3 - 24t^2 + 60t - 44)\mathbf{P}_3 \\
&\quad + (-3t^3 + 21t^2 - 45t + 31)\mathbf{P}_4 + (t - 2)^3 \mathbf{P}_5\big], \\
\mathbf{P}_6(t) &= N_{34}(t)\mathbf{P}_3 + N_{44}(t)\mathbf{P}_4 + N_{54}(t)\mathbf{P}_5 + N_{64}(t)\mathbf{P}_6 & t \in [3, 4) \\
&= \frac{1}{6}\big[(4 - t)^3 \mathbf{P}_3 + (3t^3 - 33t^2 + 117t - 131)\mathbf{P}_4 \\
&\quad + (-3t^3 + 30t^2 - 96t + 100)\mathbf{P}_5 + (t - 3)^3 \mathbf{P}_6\big], \\
\mathbf{P}_7(t) &= N_{44}(t)\mathbf{P}_4 + N_{54}(t)\mathbf{P}_5 + N_{64}(t)\mathbf{P}_6 + N_{74}(t)\mathbf{P}_7 & t \in [4, 5) \\
&= \frac{1}{6}\big[(5 - t)^3 \mathbf{P}_4 + (3t^3 - 42t^2 + 192t - 284)\mathbf{P}_5 \\
&\quad + (-3t^3 + 39t^2 - 165t + 229)\mathbf{P}_6 + (t - 4)^3 \mathbf{P}_7\big].
\end{aligned}$$

A direct check verifies that each segment has barycentric weights. The entire curve starts at $\mathbf{P}_3(0) = (\mathbf{P}_0 + 4\mathbf{P}_1 + \mathbf{P}_2)/6$ and ends at $\mathbf{P}_7(5) = (\mathbf{P}_5 + 4\mathbf{P}_6 + \mathbf{P}_7)/6$. The four joint points between the segments are

$$\mathbf{P}_3(1) = \mathbf{P}_4(1) = (\mathbf{P}_1 + 4\mathbf{P}_2 + \mathbf{P}_3)/6, \quad \mathbf{P}_4(2) = \mathbf{P}_5(2) = (\mathbf{P}_2 + 4\mathbf{P}_3 + \mathbf{P}_4)/6,$$
$$\mathbf{P}_5(3) = \mathbf{P}_6(3) = (\mathbf{P}_3 + 4\mathbf{P}_4 + \mathbf{P}_5)/6, \quad \mathbf{P}_6(4) = \mathbf{P}_7(4) = (\mathbf{P}_4 + 4\mathbf{P}_5 + \mathbf{P}_6)/6.$$

The coordinates of the control points of Figure 7.19a are $\mathbf{P}_0 = (0, 0)$, $\mathbf{P}_1 = (0, 1)$, $\mathbf{P}_2 = (1, 1)$, $\mathbf{P}_3 = (1, 0)$, $\mathbf{P}_4 = (2, 0)$, $\mathbf{P}_5 = (2.75, 1)$, $\mathbf{P}_6 = (3, 1)$, and $\mathbf{P}_7 = (3, 0)$. The curve therefore starts at $(1/6, 5/6)$, ends at $(2.96, 5/6)$, and passes through the joins $(5/6, 5/6)$, $(7/6, 1/6)$, $(1.96, 1/6)$, and $(2.67, 5/6)$.

Figure 7.20 lists the code that computes the weight functions for this case. This code is general and can also compute B-spline weight functions for the uniform and open uniform cases.

Part b: To continue the example, we now calculate the blending functions and spline segments of the curve of Figure 7.19b where the knot vector is the nonuniform

```
(* Compute the nonuniform weight functions for the 8-point example that follows *)
Clear[bspl,knt]
bspl[i_,k_,t_]:=If[knt[[i+k]]==knt[[i+1]],0, (* 0<=i<=n *)
 bspl[i,k-1,t] (t-knt[[i+1]])/(knt[[i+k]]-knt[[i+1]])] \
 +If[knt[[i+1+k]]==knt[[i+2]],0,
 bspl[i+1,k-1,t] (knt[[i+1+k]]-t)/(knt[[i+1+k]]-knt[[i+2]])];
bspl[i_,1,t_]:=If[knt[[i+1]]<=t<knt[[i+2]], 1, 0];
n=4; k=4; (* Note: 0<=k<=n *)
knt={-3,-2,-1,0,1,2,3,4,5,6,7,8}; (* knots for nonuniform case *)
bspl[i,k,t] (* assign a value to i *)
```

Figure 7.20: Eight-Point Nonuniform B-Spline Example; Code for Blending Functions.

$(-3, -2, -1, 0, 1, 1, 2, 3, 4, 5, 6, 7)$. Notice that we now have $t_4 = t_5 = 1$, resulting in different blending functions and different spline segments.

It is important to realize that t varies in this case from $t_3 = 0$ to $t_8 = 4$. The five intervals of t for the five spline segments are $[0, 1)$, $[1, 1)$, $[1, 2)$, $[2, 3)$, and $[3, 4)$. The second segment $\mathbf{P}_4(t)$ has now been reduced to a single point.

The first step is to calculate N_{i1}:

$$N_{31} = 1 \text{ for } t \in [0, 1), \quad N_{41} = 1 \text{ for } t \in [1, 1),$$
$$N_{51} = 1 \text{ for } t \in [1, 2), \quad N_{61} = 1 \text{ for } t \in [2, 3), \quad N_{71} = 1 \text{ for } t \in [3, 4),$$

and N_{01}, N_{11}, N_{21}, N_{81}, N_{91}, $N_{10,1}$, and $N_{11,1}$ are zero in the range $0 \le t < 4$.

Step 2 is to calculate functions N_{i2} that are nonzero for $0 \le t < 4$:

$$N_{02}(t) = \frac{t - t_0}{t_1 - t_0} N_{01} + \frac{t_2 - t}{t_2 - t_1} N_{11} = 0,$$

$$N_{12}(t) = \frac{t - t_1}{t_2 - t_1} N_{11} + \frac{t_3 - t}{t_3 - t_2} N_{21} = 0,$$

$$N_{22}(t) = \frac{t - t_2}{t_3 - t_2} N_{21} + \frac{t_4 - t}{t_4 - t_3} N_{31} = (1 - t) \quad \text{for } t \in [0, 1),$$

$$N_{32}(t) = \frac{t - t_3}{t_4 - t_3} N_{31} + \frac{t_5 - t}{t_5 - t_4} N_{41} = t \qquad \text{for } t \in [0, 1),$$

$$N_{42}(t) = \frac{t - t_4}{t_5 - t_4} N_{41} + \frac{t_6 - t}{t_6 - t_5} N_{51} = 2 - t \quad \text{for } t \in [1, 2),$$

$$N_{52}(t) = \frac{t - t_5}{t_6 - t_5} N_{51} + \frac{t_7 - t}{t_7 - t_6} N_{61} = \begin{cases} t - 1 & \text{for } t \in [1, 2), \\ 3 - t & \text{for } t \in [2, 3), \end{cases}$$

$$N_{62}(t) = \frac{t - t_6}{t_7 - t_6} N_{61} + \frac{t_8 - t}{t_8 - t_7} N_{71} = \begin{cases} t - 2 & \text{for } t \in [2, 3), \\ 4 - t & \text{for } t \in [3, 4), \end{cases}$$

$$N_{72}(t) = \frac{t - t_7}{t_8 - t_7} N_{71} + \frac{t_9 - t}{t_9 - t_8} N_{81} = t - 4 \quad \text{for } t \in [3, 4).$$

This step terminates at $N_{72}(t)$ since $N_{82}(t)$ and its successors are zero for $0 \le t < 4$.

Step 3 requires the calculation of several functions N_{i3}:

$$N_{03}(t) = \frac{t - t_0}{t_2 - t_0} N_{02} + \frac{t_3 - t}{t_3 - t_1} N_{12} = 0,$$

$$N_{13}(t) = \frac{t - t_1}{t_3 - t_1} N_{12} + \frac{t_4 - t}{t_4 - t_2} N_{22} = \frac{1}{2}(1-t)^2 \qquad \text{for } t \in [0,1),$$

$$N_{23}(t) = \frac{t - t_2}{t_4 - t_2} N_{22} + \frac{t_5 - t}{t_5 - t_3} N_{32} = \frac{1}{2}(-3t^2 + 2t + 1) \qquad \text{for } t \in [0,1),$$

$$N_{33}(t) = \frac{t - t_3}{t_5 - t_3} N_{32} + \frac{t_6 - t}{t_6 - t_4} N_{42} = \begin{cases} t^2 & \text{for } t \in [0,1), \\ (2-t)^2 & \text{for } t \in [1,2), \end{cases}$$

$$N_{43}(t) = \frac{t - t_4}{t_6 - t_4} N_{42} + \frac{t_7 - t}{t_7 - t_5} N_{52} = \frac{1}{2} \begin{cases} (-3t^2 + 10t - 7) & \text{for } t \in [1,2), \\ (3-t)^2 & \text{for } t \in [2,3), \end{cases}$$

$$N_{53}(t) = \frac{t - t_5}{t_7 - t_5} N_{52} + \frac{t_8 - t}{t_8 - t_6} N_{62} = \frac{1}{2} \begin{cases} (t-1)^2 & \text{for } t \in [1,2), \\ (-2t^2 + 10t - 11) & \text{for } t \in [2,3), \\ (4-t)^2 & \text{for } t \in [3,4), \end{cases}$$

$$N_{63}(t) = \frac{t - t_6}{t_8 - t_6} N_{62} + \frac{t_9 - t}{t_9 - t_7} N_{72} = \frac{1}{2} \begin{cases} (t-2)^2 & \text{for } t \in [2,3), \\ (-2t^2 + 14t - 23) & \text{for } t \in [3,4), \end{cases}$$

$$N_{73}(t) = \frac{t - t_7}{t_9 - t_7} N_{72} + \frac{t_{10} - t}{t_{10} - t_8} N_{82} = \frac{1}{2}(t-3)^2 \qquad \text{for } t \in [3,4).$$

Here we stop at N_{73} since N_{83} and its successors are zero for $0 \le t < 4$.

The last step involves the calculation of eight functions N_{i4}:

$$N_{04}(t) = \frac{t - t_0}{t_3 - t_0} N_{03} + \frac{t_4 - t}{t_4 - t_1} N_{13} = \frac{1}{6}(1-t)^3 \qquad \text{for } t \in [0,1),$$

$$N_{14}(t) = \frac{t - t_1}{t_4 - t_1} N_{13} + \frac{t_5 - t}{t_5 - t_2} N_{23} = \frac{1}{12}(11t^3 - 15t^2 - 3t + 7) \qquad \text{for } t \in [0,1),$$

$$N_{24}(t) = \frac{t - t_2}{t_5 - t_2} N_{23} + \frac{t_6 - t}{t_6 - t_3} N_{33} = \begin{cases} \frac{1}{4}(-5t^3 + 3t^2 + 3t + 1) & \text{for } t \in [0,1), \\ \frac{1}{2}(2-t)^3 & \text{for } t \in [1,2), \end{cases}$$

$$N_{34}(t) = \frac{t - t_3}{t_6 - t_3} N_{33} + \frac{t_7 - t}{t_7 - t_4} N_{43} = \begin{cases} \frac{1}{2}t^3 & \text{for } t \in [0,1), \\ \frac{1}{4}(5t^3 - 27t^2 + 45t - 21) & \text{for } t \in [1,2), \\ \frac{1}{4}(3-t)^3 & \text{for } t \in [2,3), \end{cases}$$

$$N_{44}(t) = \frac{t - t_4}{t_7 - t_4} N_{43} + \frac{t_8 - t}{t_8 - t_5} N_{53} = \begin{cases} \frac{1}{12}(-11t^3 + 51t^2 - 69t + 29) & \text{for } t \in [1,2), \\ \frac{1}{12}(7t^3 - 57t^2 + 147t - 115) & \text{for } t \in [2,3), \\ \frac{1}{6}(4-t)^3 & \text{for } t \in [3,4), \end{cases}$$

$$N_{54}(t) = \frac{t - t_5}{t_8 - t_5} N_{53} + \frac{t_9 - t}{t_9 - t_6} N_{63} = \frac{1}{6} \begin{cases} (t-1)^3 & \text{for } t \in [1,2), \\ (-3t^3 + 21t^2 - 45t + 31) & \text{for } t \in [2,3), \\ (3t^3 - 33t^2 + 117t - 131) & \text{for } t \in [3,4), \end{cases}$$

$$N_{64}(t) = \frac{t - t_6}{t_9 - t_6} N_{63} + \frac{t_{10} - t}{t_{10} - t_7} N_{73} = \frac{1}{6} \begin{cases} (t-2)^3 & \text{for } t \in [2,3), \\ (-3t^3 + 30t^2 - 96t + 100) & \text{for } t \in [3,4), \end{cases}$$

$$N_{74}(t) = \frac{t - t_7}{t_{10} - t_7} N_{73} + \frac{t_{11} - t}{t_{11} - t_8} N_{83} = \frac{1}{6}(t-3)^3 \qquad \text{for } t \in [3,4).$$

This group of blending functions can now be used to construct the five spline segments

$$\mathbf{P}_3(t) = N_{04}(t)\mathbf{P}_0 + N_{14}(t)\mathbf{P}_1 + N_{24}(t)\mathbf{P}_2 + N_{34}(t)\mathbf{P}_3 \qquad t \in [0,1)$$

$$= \frac{1}{6}(1-t)^3\mathbf{P}_0 + \frac{1}{12}(11t^3 - 15t^2 - 3t + 7)\mathbf{P}_1$$

$$+ \frac{1}{4}(-5t^3 + 3t^2 + 3t + 1)\mathbf{P}_2 + \frac{1}{2}t^3\mathbf{P}_3,$$

$$\mathbf{P}_4(t) = N_{14}(1)\mathbf{P}_1 + N_{24}(1)\mathbf{P}_2 + N_{34}(1)\mathbf{P}_3 + N_{44}(1)\mathbf{P}_4 \qquad t \in [1,1)$$

$$= 0\mathbf{P}_1 + \frac{1}{2}\mathbf{P}_2 + \frac{1}{2}\mathbf{P}_3 + 0\mathbf{P}_4 = (\mathbf{P}_2 + \mathbf{P}_3)/2 \quad \text{(a point)},$$

$$\mathbf{P}_5(t) = N_{24}(t)\mathbf{P}_2 + N_{34}(t)\mathbf{P}_3 + N_{44}(t)\mathbf{P}_4 + N_{54}(t)\mathbf{P}_5 \qquad t \in [1,2)$$

$$= \frac{1}{2}(2-t)^3\mathbf{P}_2 + \frac{1}{4}(5t^3 - 27t^2 + 45t - 21)\mathbf{P}_3$$

$$+ \frac{1}{12}(-11t^3 + 51t^2 - 69t + 29)\mathbf{P}_4 + \frac{1}{6}(t-1)^3\mathbf{P}_5,$$

$$\mathbf{P}_6(t) = N_{34}(t)\mathbf{P}_3 + N_{44}(t)\mathbf{P}_4 + N_{54}(t)\mathbf{P}_5 + N_{64}(t)\mathbf{P}_6 \qquad t \in [2,3)$$

$$= \frac{1}{4}(3-t)^3\mathbf{P}_3 + \frac{1}{12}(7t^3 - 57t^2 + 147t - 115)\mathbf{P}_4$$

$$+ \frac{1}{6}(-3t^3 + 21t^2 - 45t + 31)\mathbf{P}_5 + \frac{1}{6}(t-2)^3\mathbf{P}_6,$$

$$\mathbf{P}_7(t) = N_{44}(t)\mathbf{P}_4 + N_{54}(t)\mathbf{P}_5 + N_{64}(t)\mathbf{P}_6 + N_{74}(t)\mathbf{P}_7 \qquad t \in [3,4)$$

$$= \frac{1}{6}\Big[(4-t)^3\mathbf{P}_4 + (3t^3 - 33t^2 + 117t - 131)\mathbf{P}_5$$

$$+ (-3t^3 + 30t^2 - 96t + 100)\mathbf{P}_6 + (t-3)^3\mathbf{P}_7\Big].$$

A direct check verifies that each segment has barycentric weights. The entire curve starts at $\mathbf{P}_3(0) = (2\mathbf{P}_0 + 7\mathbf{P}_1 + 3\mathbf{P}_2)/12$ and ends at $\mathbf{P}_7(4) = (\mathbf{P}_5 + 4\mathbf{P}_6 + \mathbf{P}_7)/6$. The three joint points between the segments are

$$\mathbf{P}_3(1) = \mathbf{P}_5(1) = (\mathbf{P}_2 + \mathbf{P}_3)/2, \quad \mathbf{P}_5(2) = \mathbf{P}_6(2) = (3\mathbf{P}_3 + 7\mathbf{P}_4 + 2\mathbf{P}_5)/12,$$
$$\mathbf{P}_6(3) = \mathbf{P}_7(3) = (\mathbf{P}_4 + 4\mathbf{P}_5 + \mathbf{P}_6)/6.$$

(End of example.)

⋄ **Exercise 7.11:** Calculate the blending functions and spline segments for the curves of Figure 7.19c,d.

This example illustrates the power and flexibility of the nonuniform B-spline. Other curve methods make it possible to control the shape of a curve by moving control points, by subdividing the curve and adding points, and by repeating certain points. The nonuniform B-spline method can employ all these operations but can also fine-tune the curve by changing the values of knots and by using multiple knots.

7.12 Matrix Form of the Nonuniform B-Spline

The Cox–DeBoor recursive formula, Equations (7.24) and (7.25), is general and can be used to calculate the blending functions of the uniform, open, and nonuniform B-splines. However, it is complex and slow to calculate. Explicit, matrix-based expressions for the B-spline are simpler and faster to use. Such expressions have been derived for the uniform quadratic B-spline in Section 7.1 [Equation (7.6)] and for the uniform cubic B-spline in Section 7.2 [Equation (7.11)]. Similar expressions are derived in this section for the linear, quadratic, and cubic *nonuniform* B-splines. We temporarily use the notation u instead of t for the parameter and u_i instead of t_i for the knots.

For the linear case, where $k = 2$, the Cox–DeBoor formula becomes

$$N_{i2} = \frac{u - u_i}{u_{i+1} - u_i} N_{i1}(u) + \frac{u_{i+2} - u}{u_{i+2} - u_{i+2}} N_{i+1,1}(u)$$

$$= \begin{cases} \dfrac{u - u_i}{u_{i+1} - u_i} & \text{for } u \in [u_i, u_{i+1}), \\ \dfrac{u_{i+2} - u}{u_{i+2} - u_{i+1}} & \text{for } u \in [u_{i+1}, u_{i+2}), \\ 0 & \text{otherwise.} \end{cases} \tag{7.29}$$

For $i = 0$, this becomes

$$N_{02} = \begin{cases} \dfrac{u - u_0}{u_1 - u_0} & \text{for } u \in [u_0, u_1), \\ \dfrac{u_2 - u}{u_2 - u_1} & \text{for } u \in [u_1, u_2), \\ 0 & \text{otherwise.} \end{cases} \tag{7.30}$$

The other blending function N_{12} is easily obtained from Equation (7.30) by incrementing all the indices.

Blending function N_{02} is zero over the subinterval $[u_2, u_3)$ and blending function N_{12} is zero over $[u_0, u_1)$. It is therefore only over the interval $[u_1, u_2)$ that both these functions are nonzero, so the parameter u should vary from u_1 to u_2. Over this interval, we have

$$N_{02}(u) = \frac{u_2 - u}{u_2 - u_1}, \quad N_{12}(u) = \frac{u_3 - u}{u_3 - u_2}. \tag{7.31}$$

To derive the expression for the linear spline, we denote $\Delta = u_2 - u_1$ and define the parameter t by

$$t = \frac{u - u_1}{\Delta} = \frac{u - u_1}{u_2 - u_1}.$$

Notice that $u = u_1 \rightarrow t = 0$ and $u = u_2 \rightarrow t = 1$. Also, $u - u_1 = t\Delta$ and $u - u_2 = \Delta(t - 1)$. Substituting this in Equation (7.31) yields the matrix expression for the linear nonuniform B-spline

$$\mathbf{P}(t) = (t, 1) \begin{pmatrix} -1 & 1 \\ 1 & 0 \end{pmatrix} \begin{pmatrix} \mathbf{P}_0 \\ \mathbf{P}_1 \end{pmatrix}. \tag{7.32}$$

When t varies from 0 to 1, this becomes the straight line from $\mathbf{P}_0$ to $\mathbf{P}_1$. The nonuniform linear B-spline does not depend on Δ, so it is identical to the uniform linear B-spline.

> When you get an 8 on the midterm, there ain't a curve in the world that can save you.
>
> —Unknown

Next, we derive the matrix form of the quadratic case. Applying the Cox–DeBoor formula to Equation (7.30), we get the first quadratic blending function N_{03}:

$$
N_{03}(u) = \begin{cases}
\dfrac{u - u_0}{u_2 - u_0} \cdot \dfrac{u - u_0}{u_1 - u_0} & \text{for } u \in [u_0, u_1), \\[2mm]
\dfrac{u - u_0}{u_2 - u_0} \cdot \dfrac{u_2 - u}{u_2 - u_1} + \dfrac{u_3 - u}{u_3 - u_1} \cdot \dfrac{u - u_1}{u_2 - u_1} & \text{for } u \in [u_1, u_2), \\[2mm]
\dfrac{u_3 - u}{u_3 - u_1} \cdot \dfrac{u_3 - u}{u_3 - u_2} & \text{for } u \in [u_2, u_3), \\[2mm]
0 & \text{otherwise.}
\end{cases}
\tag{7.33}
$$

Functions N_{13} and N_{23} are obtained from Equation (7.33) by incrementing all the indices. When this is done, we observe that each of the three blending functions N_{i3} is zero over different intervals and it is only over subinterval $[u_2, u_3)$ that all three are nonzero, and their values are

$$
\begin{aligned}
N_{03}(u) &= \frac{u_3 - u}{u_3 - u_1} \cdot \frac{u_3 - u}{u_3 - u_2}, \\[2mm]
N_{13}(u) &= \frac{u - u_1}{u_3 - u_1} \cdot \frac{u_3 - u}{u_3 - u_2} + \frac{u_4 - u}{u_4 - u_2} \cdot \frac{u - u_2}{u_3 - u_2}, \\[2mm]
N_{23}(u) &= \frac{u - u_2}{u_4 - u_2} \cdot \frac{u - u_2}{u_3 - u_2}.
\end{aligned}
\tag{7.34}
$$

Since the knot vector is nonuniform, the differences between consecutive knots may be different and we denote them

$$
\Delta_1 = u_2 - u_1, \quad \Delta_2 = u_3 - u_2, \quad \Delta_3 = u_4 - u_3.
$$

We also define $t = (u - u_2)/\Delta_2$, which implies

$$
\begin{aligned}
u - u_1 &= t\Delta_2 + \Delta_1, \\
u - u_2 &= t\Delta_2, \\
u - u_3 &= (t - 1)\Delta_2, \\
u - u_4 &= t\Delta_2 - (\Delta_2 + \Delta_3).
\end{aligned}
\tag{7.35}
$$

Equations (7.34) and (7.35) yield the matrix form of the nonuniform quadratic B-spline

$$
\mathbf{P}(t) = (t^2, t, 1) \begin{pmatrix} a & -a-b & b \\ -2a & 2a & 0 \\ a & 1-a & 0 \end{pmatrix} \begin{pmatrix} \mathbf{P}_0 \\ \mathbf{P}_1 \\ \mathbf{P}_2 \end{pmatrix},
\tag{7.36}
$$

where

$$
a = \frac{\Delta_2}{\Delta_1 + \Delta_2}, \qquad b = \frac{\Delta_2}{\Delta_2 + \Delta_3},
$$

and t varies from 0 to 1 (note that $u = u_2 \rightarrow t = 0$ and $u = u_3 \rightarrow t = 1$).

B-splines were known to and studied by Nikolai Lobachevsky whose major contribution to mathematics is perhaps the so-called non-Euclidean (hyperbolic) geometry in the late eighteenth century. The modern version described here was developed, in the late 1970s, by C. DeBoor, M. Cox and L. Mansfield. Note that their algorithm is a generalization of de Casteljau's scaffolding method.

The next example derives the matrix form of the nonuniform cubic B-spline. We apply the Cox–DeBoor formula to Equation (7.33) to obtain the first of the four blending functions N_{i4}:

$$
N_{04}(u) = \begin{cases}
\dfrac{u - u_0}{u_3 - u_0} \cdot \dfrac{u - u_0}{u_2 - u_0} \cdot \dfrac{u - u_0}{u_1 - u_0} & \text{for } u \in [u_0, u_1), \\[2ex]
\begin{aligned}
&\dfrac{u - u_0}{u_3 - u_0} \cdot \dfrac{u - u_0}{u_2 - u_0} \cdot \dfrac{u_2 - u}{u_2 - u_1} \\
+ &\dfrac{u - u_0}{u_3 - u_0} \cdot \dfrac{u_3 - u}{u_3 - u_1} \cdot \dfrac{u - u_1}{u_2 - u_1} \\
+ &\dfrac{u_4 - u}{u_4 - u_1} \cdot \dfrac{u - u_1}{u_3 - u_1} \cdot \dfrac{u - u_1}{u_2 - u_1}
\end{aligned} & \text{for } u \in [u_1, u_2), \\[4ex]
\begin{aligned}
&\dfrac{u - u_0}{u_3 - u_0} \cdot \dfrac{u_3 - u}{u_3 - u_1} \cdot \dfrac{u_3 - u}{u_3 - u_2} \\
+ &\dfrac{u_4 - u}{u_4 - u_1} \cdot \dfrac{u - u_1}{u_3 - u_1} \cdot \dfrac{u_3 - u}{u_3 - u_2} \\
+ &\dfrac{u_4 - u}{u_4 - u_1} \cdot \dfrac{u_4 - u}{u_4 - u_2} \cdot \dfrac{u - u_2}{u_3 - u_2}
\end{aligned} & \text{for } u \in [u_2, u_3), \\[4ex]
\dfrac{u_4 - u}{u_4 - u_1} \cdot \dfrac{u_4 - u}{u_4 - u_2} \cdot \dfrac{u_4 - u}{u_4 - u_3} & \text{for } u \in [u_3, u_4), \\[2ex]
0 & \text{otherwise.}
\end{cases}
\tag{7.37}
$$

The remaining three blending functions N_{14}, N_{24}, and N_{34} are obtained from Equation (7.37) by incrementing all the indices. When this is done we observe, as before, that each of the four blending functions N_{i4} is zero over different intervals and it is only over subinterval $[u_3, u_4)$ that all four are nonzero. Their values are

$$
\begin{aligned}
N_{04}(u) ={}& \frac{u_4 - u}{u_4 - u_1} \cdot \frac{u_4 - u}{u_4 - u_2} \cdot \frac{u_4 - u}{u_4 - u_3}, \\
N_{14}(u) ={}& \frac{u - u_1}{u_4 - u_1} \cdot \frac{u_4 - u}{u_4 - u_2} \cdot \frac{u_4 - u}{u_4 - u_3} + \frac{u_5 - u}{u_5 - u_2} \cdot \frac{u - u_2}{u_4 - u_2} \cdot \frac{u_4 - u}{u_4 - u_3} \\
& + \frac{u_5 - u}{u_5 - u_2} \cdot \frac{u_5 - u}{u_5 - u_3} \cdot \frac{u - u_3}{u_4 - u_3}, \\
N_{24}(u) ={}& \frac{u - u_2}{u_5 - u_2} \cdot \frac{u - u_2}{u_4 - u_2} \cdot \frac{u_4 - u}{u_4 - u_3} + \frac{u - u_2}{u_5 - u_2} \cdot \frac{u_5 - u}{u_5 - u_3} \cdot \frac{u - u_3}{u_4 - u_3} \\
& + \frac{u_6 - u}{u_6 - u_3} \cdot \frac{u - u_3}{u_5 - u_3} \cdot \frac{u - u_3}{u_4 - u_3}, \\
N_{34}(u) ={}& \frac{u - u_3}{u_6 - u_3} \cdot \frac{u - u_3}{u_5 - u_3} \cdot \frac{u - u_3}{u_4 - u_3}.
\end{aligned}
\tag{7.38}
$$

Since the knot vector is nonuniform, the differences between consecutive knots may be different and we denote them by

$$\Delta_1 = u_2 - u_1, \quad \Delta_2 = u_3 - u_2, \quad \Delta_3 = u_4 - u_3,$$
$$\Delta_4 = u_5 - u_4, \quad \Delta_5 = u_6 - u_5, \quad t = (u - u_3)/\Delta_3.$$

This implies

$$
\begin{aligned}
u - u_1 &= t\Delta_3 + (\Delta_1 + \Delta_2), \\
u - u_2 &= t\Delta_3 + \Delta_2, \\
u - u_3 &= t\Delta_3, \\
u - u_4 &= (t-1)\Delta_3, \\
u - u_5 &= t\Delta_3 - (\Delta_3 + \Delta_4), \\
u - u_6 &= t\Delta_3 - (\Delta_3 + \Delta_4 + \Delta_5).
\end{aligned}
\tag{7.39}
$$

Equations (7.38) and (7.39) yield the matrix form of the nonuniform cubic B-spline:

$$
\mathbf{P}(t) = (t^3, t^2, t, 1)
\begin{pmatrix}
-a & a+b+c & -b-c-d & d \\
3a & -3a-3b & 3b & 0 \\
-3a & 3a-3e & 3e & 0 \\
a & 1-a-f & f & 0
\end{pmatrix}
\begin{pmatrix}
\mathbf{P}_0 \\
\mathbf{P}_1 \\
\mathbf{P}_2 \\
\mathbf{P}_3
\end{pmatrix},
\tag{7.40}
$$

where

$$
a = \frac{\Delta_3^2}{(\Delta_1 + \Delta_2 + \Delta_3)(\Delta_2 + \Delta_3)}, \qquad
d = \frac{\Delta_3^2}{(\Delta_3 + \Delta_4 + \Delta_5)(\Delta_4 + \Delta_5)},
$$
$$
b = \frac{\Delta_3^2}{(\Delta_2 + \Delta_3 + \Delta_4)(\Delta_2 + \Delta_3)}, \qquad
e = \frac{\Delta_2 \Delta_3}{(\Delta_2 + \Delta_3 + \Delta_4)(\Delta_2 + \Delta_3)},
$$
$$
c = \frac{\Delta_3^2}{(\Delta_2 + \Delta_3 + \Delta_4)(\Delta_3 + \Delta_4)}, \qquad
f = \frac{\Delta_2^2}{(\Delta_2 + \Delta_3 + \Delta_4)(\Delta_2 + \Delta_3)}.
$$

The quantities Δ_i are defined as differences of knot values $u_{i+1} - u_i$ and a good choice for those differences is the chord lengths between points. However, a cubic spline segment requires five Δ_i's, but there are only three chords between the four points defining it. In general, a B-spline curve is defined by $n+1$ points, having n chords between them, but $n+2$ differences Δ_i are required. A standard technique is to select

$$\Delta_1 = \Delta_2 = |\mathbf{P}_1 - \mathbf{P}_0|, \qquad \Delta_{n+1} = \Delta_{n+2} = |\mathbf{P}_n - \mathbf{P}_{n-1}|,$$

and $\Delta_i = |\mathbf{P}_{i-1} - \mathbf{P}_{i-2}|$ for $i = 3, 4, \ldots, n$.

The last topic discussed in this section is the relation between the quadratic uniform and quadratic nonuniform B-splines. Given three control points $\mathbf{Q}_0$, $\mathbf{Q}_1$, and $\mathbf{Q}_2$, the uniform quadratic B-spline $\mathbf{Q}(t)$ defined by them is given by Equation (7.6)

$$
\mathbf{Q}(t) = \frac{1}{2}(t^2, t, 1)
\begin{pmatrix}
1 & -2 & 1 \\
-2 & 2 & 0 \\
1 & 1 & 0
\end{pmatrix}
\begin{pmatrix}
\mathbf{Q}_0 \\
\mathbf{Q}_1 \\
\mathbf{Q}_2
\end{pmatrix}.
\tag{7.6}
$$

The nonuniform quadratic B-spline defined by three control points $\mathbf{P}_0$, $\mathbf{P}_1$, and $\mathbf{P}_2$ is given by Equation (7.36). If we require the two curves to be identical for any value of the parameter t, we obtain the equation

$$\frac{1}{2}\begin{pmatrix} 1 & -2 & 1 \\ -2 & 2 & 0 \\ 1 & 1 & 0 \end{pmatrix}\begin{pmatrix} \mathbf{Q}_0 \\ \mathbf{Q}_1 \\ \mathbf{Q}_2 \end{pmatrix} = \begin{pmatrix} a & -a-b & b \\ -2a & 2a & 0 \\ a & 1-a & 0 \end{pmatrix}\begin{pmatrix} \mathbf{P}_0 \\ \mathbf{P}_1 \\ \mathbf{P}_2 \end{pmatrix}.$$

This is a system of three equations where we assume that the unknowns are the $\mathbf{Q}_i$'s. The solutions are

$$\mathbf{Q}_0 = 2a\mathbf{P}_0 + (1-2a)\mathbf{P}_1, \quad \mathbf{Q}_1 = \mathbf{P}_1, \quad \text{and} \quad \mathbf{Q}_2 = (1-2b)\mathbf{P}_1 + 2b\mathbf{P}_2.$$

To see the geometrical interpretation of these relations, we write

$$\mathbf{Q}_0 = 2a\mathbf{P}_0 + (1-2a)\mathbf{P}_1 = 2a\mathbf{P}_0 + 2(1-a)\mathbf{P}_1 - \mathbf{P}_1 = 2\mathbf{P}(0) - \mathbf{P}_1 = 2\mathbf{Q}(0) - \mathbf{Q}_1,$$

which implies $\mathbf{Q}_0 - \mathbf{Q}(0) = \mathbf{Q}(0) - \mathbf{Q}_1$. The distance between $\mathbf{Q}_0$ and $\mathbf{Q}(0)$ equals the distance between $\mathbf{Q}(0)$ and $\mathbf{Q}_1$, and a similar relation among $\mathbf{Q}_1$, $\mathbf{Q}(1)$, and $\mathbf{Q}_2$.

The conclusion is that a group of three points $\mathbf{P}_0$, $\mathbf{P}_1$, and $\mathbf{P}_2$ defining a single quadratic nonuniform B-spline segment $\mathbf{P}(t)$ can be replaced by a group of three points $\mathbf{Q}_0$, $\mathbf{Q}_1$, and $\mathbf{Q}_2$ defining a single quadratic *uniform* B-spline segment $\mathbf{Q}(t)$ identical to $\mathbf{P}(t)$. However, given a set of $n+1$ control points $\mathbf{P}_i$ for a nonuniform B-spline curve, they cannot, in general, be replaced by a set of $n+1$ points $\mathbf{Q}_i$ that produce an identical uniform B-spline curve.

7.13 Subdividing the B-spline Curve

The B-spline curve is easy to manipulate by moving the control points and varying the knots. Still, if the curve is based on too few points, it may "refuse" to get the right shape, no matter what. More control points can be added, in such a case, by subdividing the curve, a process similar to subdividing the Bézier curve (Section 6.8). The method described here is called the Oslo algorithm and the discussion follows [Cohen et al. 80] and [Prautzsch 84].

(Control points can also be added by raising the degree of the B-spline curve, similar to the degree elevation of the Bézier curve, Section 6.9. This operation is discussed in [Cohen et al. 85].)

The idea behind subdividing a curve is that there are many (even infinitely many) sets of control points that produce the same B-spline curve. Normally, we are interested in the smallest number of control points that will produce a given curve, but if we cannot get the right shape with the original $n+1$ control points, we need to find a set of $n+2$ points that will produce *the same curve*, then move the new points around, attempting to bring the curve to the desired shape.

Given a set of $n+1$ control points $\mathbf{P}_i$ and a knot vector $(t_0, t_1, \ldots, t_{n+k})$, we start the subdivision process by inserting several new knots, thereby obtaining a new knot vector $(u_0, u_1, \ldots, u_{m+k})$ where $m > n$. The new, subdivided curve is based on the $m + 1$ control points $\mathbf{Q}_j$ defined by the Oslo algorithm as

$$\mathbf{Q}_j = \sum_{i=0}^{n} a_{ij}^k \mathbf{P}_i, \quad \text{where} \quad 0 \leq i \leq n \quad \text{and} \quad 0 \leq j \leq m,$$

where the coefficients a_{ij}^k are defined recursively by a relation similar to the Cox–DeBoor formula

$$a_{ij}^1 = \begin{cases} 1, & t_i \leq u_j < t_{i+1}, \\ 0, & \text{otherwise}, \end{cases} \tag{7.41}$$

$$a_{ij}^k = \frac{u_{j+k-1} - t_i}{t_{i+k-1} - t_i} a_{ij}^{k-1} + \frac{t_{i+k} - u_{j+k-1}}{t_{i+k} - t_{i+1}} a_{i+1,j}^{k-1}. \tag{7.42}$$

This relation guarantees that $\sum_{i}^{n} a_{ij}^k = 1$, for $0 \leq j \leq m$.

If the original knot vector is uniform, inserting a single knot will convert it to a nonuniform vector. However, an open knot vector can sometimes remain open after inserting new knots, as the following example shows. Suppose that we have the open vector $(0, 0, 0, 1, 2, 2, 2)$, where t varies from 0 to 2. This corresponds to a two-segment curve and we want to subdivide both segments. We first multiply each knot by 2, obtaining the vector $(0, 0, 0, 2, 4, 4, 4)$ that produces the same curve when $0 \leq t < 4$. Next, we insert knots 1 and 3 to obtain the knot vector $(0, 0, 0, 1, 2, 3, 4, 4, 4)$. This vector is still open and it corresponds to the four segments $[0, 1)$, $[1, 2)$, $[2, 3)$, and $[3, 4)$.

Example: We assume four control points and quadratic segments (i.e., $k = 3$). We already know that each segment is defined by three points, so two segments are needed for this curve. The knot vector is assumed to be uniform and it goes from $t_0 = 0$ to $t_{n+k} = t_6 = 6$. The parameter t varies from $t_{k-1} = t_2 = 2$ to $t_{n+1} = t_4 = 4$; two subintervals. This again shows that the curve consists of two spline segments, the first for the subinterval $[t_2, t_3)$ and the second for $[t_3, t_4)$. We decide to subdivide the first segment. This segment is defined by points $\mathbf{P}_0$, $\mathbf{P}_1$, and $\mathbf{P}_2$ (notice that $n = 2$ for this subdivision), so the subdivision process should produce four points, $\mathbf{Q}_0$, $\mathbf{Q}_1$, $\mathbf{Q}_2$, and $\mathbf{Q}_3$ (this implies $m = 3$), such that the two quadratic segments defined by them will have the same shape as the segment being subdivided.

To perform the subdivision, we need to insert a new knot between $t_2 = 2$ and $t_3 = 3$. We (somewhat arbitrarily) select its value to be 2.5. The new knot vector is

$$(u_0, u_1, u_2, u_3, u_4, u_5, u_6, u_7) = (0, 1, 2, 2.5, 3, 4, 5, 6),$$

and it is nonuniform. The calculation of the a_{ij}^k coefficients is done by varying i from 0 to $n = 2$ and varying j from 0 to $m = 3$. It requires three steps, for $k = 1, 2, 3$ (notice that this k is not the same as the order of the B-spline).

Step 1: We use Equation (7.41). A direct comparison of the t_i and u_i knots shows that the only nonzero a_{ij}^1 coefficients are a_{00}^1, a_{11}^1, a_{22}^1, and a_{23}^1. Each has a value of 1.

Step 2: We calculate a_{ij}^2 for $j = 0, 1, 2, 3$ from Equation (7.42). For each value of j, we stop when we get coefficients that add up to 1. The nonzero coefficients are

$$a_{00}^2 = \frac{u_1 - t_0}{t_1 - t_0} a_{00}^1 + \frac{t_2 - u_1}{t_2 - t_1} a_{10}^1 = \frac{1 - 0}{1 - 0} \cdot 1 = 1,$$

$$a_{11}^2 = \frac{u_2 - t_1}{t_2 - t_1} a_{11}^1 + \frac{t_3 - u_2}{t_3 - t_2} a_{21}^1 = \frac{2 - 1}{2 - 1} \cdot 1 = 1,$$

$$a_{12}^2 = \frac{u_3 - t_1}{t_2 - t_1} a_{12}^1 + \frac{t_3 - u_3}{t_3 - t_2} a_{22}^1 = \frac{3 - 2.5}{3 - 2} \cdot 1 = 1/2,$$

$$a_{22}^2 = \frac{u_3 - t_2}{t_3 - t_2} a_{22}^1 + \frac{t_4 - u_3}{t_4 - t_3} a_{32}^1 = \frac{2.5 - 2}{3 - 2} \cdot 1 = 1/2,$$

$$a_{23}^2 = \frac{u_4 - t_2}{t_3 - t_2} a_{23}^1 + \frac{t_4 - u_4}{t_4 - t_3} a_{33}^1 = \frac{3 - 2}{3 - 2} \cdot 1 = 1.$$

Step 3: The coefficients of step 2 are used to calculate a_{ij}^3:

$$a_{00}^3 = \frac{u_2 - t_0}{t_2 - t_0} a_{00}^2 + \frac{t_3 - u_2}{t_3 - t_1} a_{10}^2 = \frac{2 - 0}{2 - 0} \cdot 1 = 1,$$

$$a_{01}^3 = \frac{u_3 - t_0}{t_2 - t_0} a_{01}^2 + \frac{t_3 - u_3}{t_3 - t_1} a_{11}^2 = \frac{3 - 2.5}{3 - 1} \cdot 1 = 1/4,$$

$$a_{11}^3 = \frac{u_3 - t_1}{t_3 - t_1} a_{11}^2 + \frac{t_4 - u_3}{t_4 - t_2} a_{21}^2 = \frac{2.5 - 1}{3 - 1} \cdot 1 = 3/4,$$

$$a_{12}^3 = \frac{u_4 - t_1}{t_3 - t_1} a_{12}^2 + \frac{t_4 - u_4}{t_4 - t_2} a_{22}^2 = \frac{3 - 1}{3 - 1} \cdot \frac{1}{2} + \frac{4 - 3}{4 - 2} \cdot \frac{1}{2} = 3/4,$$

$$a_{22}^3 = \frac{u_4 - t_2}{t_4 - t_2} a_{22}^2 + \frac{t_5 - u_4}{t_5 - t_3} a_{32}^2 = \frac{3 - 2}{4 - 2} \cdot \frac{1}{2} = 1/4,$$

$$a_{23}^3 = \frac{u_5 - t_2}{t_4 - t_2} a_{23}^2 + \frac{t_5 - u_5}{t_5 - t_3} a_{33}^2 = \frac{4 - 2}{4 - 2} \cdot 1 = 1.$$

The four new control points can now be calculated. They are

$$\mathbf{Q}_0 = \sum_{i=0}^{3} a_{i0}^3 \mathbf{P}_i = a_{00}^3 \mathbf{P}_0 = \mathbf{P}_0,$$

$$\mathbf{Q}_1 = \sum_{i=0}^{3} a_{i1}^3 \mathbf{P}_i = a_{01}^3 \mathbf{P}_0 + a_{11}^3 \mathbf{P}_1 = \frac{1}{4} \mathbf{P}_0 + \frac{3}{4} \mathbf{P}_1,$$

$$\mathbf{Q}_2 = \sum_{i=0}^{3} a_{i2}^3 \mathbf{P}_i = a_{12}^3 \mathbf{P}_1 + a_{22}^3 \mathbf{P}_2 = \frac{3}{4} \mathbf{P}_1 + \frac{1}{4} \mathbf{P}_2,$$

$$\mathbf{Q}_3 = \sum_{i=0}^{3} a_{i3}^3 \mathbf{P}_i = a_{23}^3 \mathbf{P}_2 = \mathbf{P}_2.$$

The two quadratic B-spline segments defined by $\mathbf{Q}_0 \mathbf{Q}_1 \mathbf{Q}_2$ and $\mathbf{Q}_2 \mathbf{Q}_3 \mathbf{Q}_4$ have the same shape as the original segment defined by $\mathbf{P}_0 \mathbf{P}_1 \mathbf{P}_2$, but they are easier to modify since they are based on four points.

7.14 Nonuniform Rational B-Splines (NURBS)

The use of a knot vector is one reason why the B-spline curve is more general than the Bézier and other curve methods. The $n+k+1$ knots can be used as parameters and can be varied by the user/designer to obtain the desired shape of the curve. The rational B-spline, described in this section, employs an additional set of $n+1$ parameters w_i, called *weights*, to add even greater flexibility to the curve. In addition to this feature, the rational B-spline has several more important advantages as follows:

1. It makes it possible to create curves that are true conic sections. It is well known that a polynomial cannot represent a circle. More generally, it cannot represent arbitrary conic sections. It is easy to show that the Bézier and B-spline curves can represent approximate circles (Appendix B). If precise circles or conic sections are needed, then rational curves are the natural choice.

2. It is invariant under perspective projections. We know that curves that are barycentric sums are invariant under affine transformations. If we want to rotate, scale, shear, or translate such a curve, we can apply the transformation to the control points and use the transformed points to draw the transformed curve. There is no need to apply the transformation to every pixel on the curve. However, if we want to project a space (three-dimensional) curve in perspective on a two-dimensional output device, we have to individually project every pixel on the curve. With a rational curve, we can (perspective) project the control points and use the projected, two-dimensional points to calculate the projected curve.

3. It reduces to the nonrational B-spline when all the weights w_i are set to 1. This means that a software package for rational B-splines can be used to generate nonrational B-splines (uniform, open, and nonuniform). This also implies that the nonuniform rational B-spline (NURBS for short) is the most general parametric curve. It can take many shapes and can easily be reduced to simpler forms. Because of this, NURBS is today the defacto standard for curve design. Three excellent references to NURBS are [Farin 99], [Piegl 97], and [Rogers 01].

Perhaps the best way to introduce rational B-splines (and rational curves in general) is by means of homogeneous coordinates. This method starts by adding an extra dimension to points, so a two-dimensional point becomes a triplet (x, y, w) and a three-dimensional point becomes a 4-tuple (x, y, z, w). After transforming or manipulating the point, it is projected back to its original number of dimensions by dividing its coordinates by w. Given four-dimensional control points $\mathbf{Q}_i = (x_i, y_i, z_i, w_i)$, where we assume for convenience that the w_i coordinates are nonnegative, we can define a (nonrational) B-spline curve as

$$\mathbf{P}_{\mathrm{nr}}(t) = \sum_{i=0}^{n} \mathbf{Q}_i N_{ik}(t).$$

From this we get the rational B-spline $\mathbf{P}_{\mathrm{r}}(t)$ by isolating that part of $\mathbf{P}_{\mathrm{nr}}(t)$ that depends on the fourth coordinates w_i and dividing by this part.

$$\mathbf{P}_{\mathrm{r}}(t) = \frac{\sum_{i=0}^{n} \mathbf{P}_i w_i N_{ik}(t)}{\sum_{i=0}^{n} w_i N_{ik}(t)} = \sum_{i=0}^{n} \mathbf{P}_i R_{ik}(t), \tag{7.43}$$

where $\mathbf{P}_i = (x_i, y_i, z_i)$ are three-dimensional control points and $R_{ik}(t)$ are the new, rational blending functions defined by

$$R_{ik}(t) = \frac{w_i N_{ik}(t)}{\sum_{i=0}^n w_i N_{ik}(t)}. \tag{7.44}$$

This type of curve has most of the properties of the nonrational B-spline. The following should be mentioned in particular:

1. The new blending functions $R_{ik}(t)$ are nonnegative and barycentric.

2. The curve reduces to the nonrational curve when all the weights w_i equal 1 [this is a direct consequence of Equation (7.44)].

3. Since the rational curve is the four-dimensional generalization of the nonrational B-spline, the algorithms for curve subdivision and degree elevation of the B-spline can be used for the rational version. They simply have to be executed on the four-dimensional control points (x_i, y_i, z_i, w_i).

So much for the definition of the rational B-spline. The main question is how to select values for the weights in order to modify the shape of the curve in a predictable way. In order to isolate the effect of one weight on the curve, we first observe that Equation (7.43) implies that when $w_k = 0$, point $\mathbf{P}_k$ has no effect on the curve. To see how increasing the value of a weight affects the curve, we select an index $0 \le k \le n$ and divide Equation (7.43) by w_k

$$\mathbf{P}_\mathrm{r}(t) = \frac{\sum_{i=0,i\ne k}^n \mathbf{P}_i \frac{w_i}{w_k} N_{ik}(t) + \mathbf{P}_k N_{kk}(t)}{\sum_{i=0,i\ne k}^n \frac{w_i}{w_k} N_{ik}(t) + N_{kk}(t)}.$$

It is easy to see that as w_k grows without limit, the result approaches point $\mathbf{P}_k$. We therefore conclude that those curve segments that are affected by $\mathbf{P}_k$ will approach this point as weight w_k grows.

The rest of this section describes two approaches to understanding the weights and their effects on the curve. The first approach is to set all weights $w_i = 1$, then change the value of one of them and see how it affects the blending functions. The second approach is to derive specific sets of weights that will produce B-spline curves that are conic sections. The first approach is illustrated by a detailed example.

Example: This is an extension of the open B-spline example on page 282. We assume $n = 4$ (five control points), select order $k = 3$ (quadratic polynomial segments), and the knot vector $(0, 0, 0, 1, 2, 3, 3, 3)$. The parameter t varies from $t_{k-1} = t_2 = 0$ to $t_{n+1} = t_5 = 3$, so our curve consists of three segments. The nonrational blending functions $N_{i3}(t)$ are

$$N_{03}(t) = (1-t)^2, \qquad\qquad 0 \le t < 1,$$

$$N_{13}(t) = \frac{1}{2} \begin{cases} t(4 - 3t), & 0 \le t < 1, \\ (2-t)^2, & 1 \le t < 2, \end{cases}$$

$$N_{23}(t) = \frac{1}{2} \begin{cases} t^2, & 0 \le t < 1, \\ (-2t^2 + 6t - 3), & 1 \le t < 2, \\ (3-t)^2, & 2 \le t < 3, \end{cases}$$

$$N_{33}(t) = \frac{1}{2} \begin{cases} (t-1)^2, & 1 \le t < 2, \\ (-3t^2 + 14t - 15), & 2 \le t < 3, \end{cases}$$

$$N_{43}(t) = (t-2)^2, \qquad\qquad\qquad 2 \le t < 3.$$

Before we can calculate the rational blending functions, we have to select values for the five weights. We choose $(1, 1, w_2, 1, 1)$, where w_2 will later be assigned several different values. The result is

$$R_{03}(t) = \frac{w_0 N_{03}(t)}{\sum_{i=0}^{4} w_i N_{i3}(t)} = \frac{(1-t)^2}{(1-t)^2 + t(4-3t)/2 + w_2 t^2/2}, \qquad t \in [0, 1),$$

$$R_{13}(t) = \frac{w_1 N_{13}(t)}{\sum_{i=0}^{4} w_i N_{i3}(t)} = \begin{cases} \frac{t(4-3t)/2}{(1-t)^2 + t(4-3t)/2 + w_2 t^2/2}, & t \in [0, 1) \\ \frac{(2-t)^2/2}{(2-t)^2/2 + w_2(-2t^2 + 6t - 3)/2 + (t-1)^2/2}, & t \in [1, 2), \end{cases}$$

$$R_{23}(t) = \frac{w_2 N_{23}(t)}{\sum_{i=0}^{4} w_i N_{i3}(t)} = \begin{cases} \frac{w_2 t^2/2}{(1-t)^2 + t(4-3t)/2 + w_2 t^2/2}, & t \in [0, 1) \\ \frac{w_2(-2t^2 + 6t - 3)/2}{(2-t)^2/2 + w_2(-2t^2 + 6t - 3)/2 + (t-1)^2/2}, & t \in [1, 2) \\ \frac{w_2(3-t)^2/2}{w_2(3-t)^2/2 + (-3t^2 + 14t - 15)/2 + (t-2)^2}, & t \in [2, 3), \end{cases}$$

$$R_{33}(t) = \frac{w_3 N_{33}(t)}{\sum_{i=0}^{4} w_i N_{i3}(t)} = \begin{cases} \frac{(t-1)^2/2}{(2-t)^2/2 + w_2(-2t^2 + 6t - 3)/2 + (t-1)^2/2}, & t \in [1, 2) \\ \frac{(-3t^2 + 14t - 15)/2}{w_2(3-t)^2/2 + (-3t^2 + 14t - 15)/2 + (t-2)^2}, & t \in [2, 3), \end{cases}$$

$$R_{43}(t) = \frac{w_4 N_{43}(t)}{\sum_{i=0}^{4} w_i N_{i3}(t)} = \frac{(t-2)^2}{w_2(3-t)^2/2 + (-3t^2 + 14t - 15)/2 + (t-2)^2} \quad t \in [2, 3).$$

We next calculate the three spline segments for the four cases $w_2 = 0, 0.5, 1$, and 5. For $w_2 = 0$ the three segments are

$$\mathbf{P}_1(t) = \frac{(1-t)^2}{1 - t^2/2}\mathbf{P}_0 + \frac{(4-3t)t}{2-t^2}\mathbf{P}_1 + 0\mathbf{P}_2,$$

$$\mathbf{P}_2(t) = \frac{(2-t)^2}{5 - 6t + 2t^2}\mathbf{P}_1 + 0\mathbf{P}_2 + \frac{(t-1)^2}{5 - 6t + 2t^2}\mathbf{P}_3,$$

$$\mathbf{P}_3(t) = 0\mathbf{P}_2 + \frac{15 - 14t + 3t^2}{7 - 6t + t^2}\mathbf{P}_3 + \frac{2(-2+t)^2}{-7 + 6t - t^2}\mathbf{P}_4.$$

For $w_2 = 0.5$ they are

$$\mathbf{P}_1(t) = \frac{(1-t)^2}{1 - 0.25t^2}\mathbf{P}_0 + \frac{(4-3t)t}{2 - 0.5t^2}\mathbf{P}_1 + \frac{0.25t^2}{1 - 0.25t^2}\mathbf{P}_2,$$

$$\mathbf{P}_2(t) = \frac{(2-t)^2}{3.5 - 3t + t^2}\mathbf{P}_1 + \frac{0.25(-3 + 6t - 2t^2)}{1.75 - 1.5t + 0.5t^2}\mathbf{P}_2 + \frac{(t-1)^2}{3.5 - 0.5t^2}\mathbf{P}_3,$$

$$\mathbf{P}_3(t) = \frac{0.25(3-t)^2}{-1.25 + 1.5t - 0.25t^2}\mathbf{P}_2 + \frac{-15 + 14t - 3t^2}{-2.5 + 3.5t - 0.5t^2}\mathbf{P}_3 + \frac{(t-2)^2}{-1.25 + 1.5t - 0.25t^2}\mathbf{P}_4.$$

For $w_2 = 1$ we get

$$\mathbf{P}_1(t) = (1-t)^2\mathbf{P}_0 + \frac{(4-3t)t}{2}\mathbf{P}_1 + \frac{t^2}{2}\mathbf{P}_2,$$

$$\mathbf{P}_2(t) = \frac{(2-t)^2}{2}\mathbf{P}_1 + \frac{-3+6t-2t^2}{2}\mathbf{P}_2 + \frac{(t-1)^2}{2+6t-3t^2}\mathbf{P}_3,$$

$$\mathbf{P}_3(t) = \frac{(3-t)^2}{2}\mathbf{P}_2 + \frac{-15+14t-3t^2}{2}\mathbf{P}_3 + (t-2)^2\mathbf{P}_4.$$

Finally, for $w_2 = 5$ the segments are

$$\mathbf{P}_1(t) = \frac{(1-t)^2}{1+2t^2}\mathbf{P}_0 + \frac{(4-3t)t}{2+4t^2}\mathbf{P}_1 + \frac{5t^2}{2+4t^2}\mathbf{P}_2,$$

$$\mathbf{P}_2(t) = \frac{(2-t)^2}{-10+24t-8t^2}\mathbf{P}_1 + \frac{5(-3+6t-2t^2)}{-10+24t-8t^2}\mathbf{P}_2 + \frac{(t-1)^2}{-10+54t-23t^2}\mathbf{P}_3,$$

$$\mathbf{P}_3(t) = \frac{5(3-t)^2}{38-24t+4t^2}\mathbf{P}_2 + \frac{-15+14t-3t^2}{38-24t+4t^2}\mathbf{P}_3 + \frac{(t-2)^2}{19-12t+2t^2}\mathbf{P}_4.$$

They are plotted in Figure 7.21 for control points $\mathbf{P}_0 = (0,0)$, $\mathbf{P}_1 = (0,1)$, $\mathbf{P}_2 = (1,0)$, $\mathbf{P}_3 = (2,1)$, and $\mathbf{P}_4 = (2,0)$. It is easy to see how weight w_2 affects the shape of the curve by controlling the amount of "pull" that point $\mathbf{P}_2$ exerts on the curve. For $w_2 = 0$, point $\mathbf{P}_2$ has no effect. The curve is defined by the four remaining points and is identical to the control polygon of these points. As w_2 grows toward 5, the curve becomes more and more attracted to $\mathbf{P}_2$.

Now for the second approach. We are looking for specific sets of weights that will generate conic sections. Since the conics are described by quadratic equations and each is fully defined by means of three points, it makes sense to try rational B-splines of order $k = 3$ defined by three points (i.e., $n = 2$). The conic is easier to design if the B-spline curve starts and ends at control points, so it makes sense to use an open B-spline. Since we have selected $k = n + 1$, we know (from Section 7.10) that the open B-spline will be a Bézier curve. The knot vector for our curve is calculated by Equation (7.27) to be $(0,0,0,0,1,1,1,1)$. To simplify our task, we try the simple set of weights $(1, w_1, 1)$. Our problem is to find out for what values, if any, of w_1 we get precise conics.

There is no need to use the Cox–DeBoor recursive formula [Equation (7.25)] to calculate the blending functions because they are the quadratic Bernstein polynomials. The curve itself can easily be written

$$\mathbf{P}(t) = \frac{N_{03}(t)\mathbf{P}_0 + w_1 N_{13}(t)\mathbf{P}_1 + N_{23}(t)\mathbf{P}_2}{N_{03}(t) + w_1 N_{13}(t) + N_{23}(t)}$$

$$= \frac{(1-t)^2\mathbf{P}_0 + 2w_1 t(1-t)\mathbf{P}_1 + t^2\mathbf{P}_2}{(1-t)^2 + 2w_1 t(1-t) + t^2}. \tag{7.45}$$

$\diamond$ **Exercise 7.12:** Show that in the special case where $w_1 = 0$, the curve of Equation (7.45) reduces to the straight line between $\mathbf{P}_0$ and $\mathbf{P}_2$.

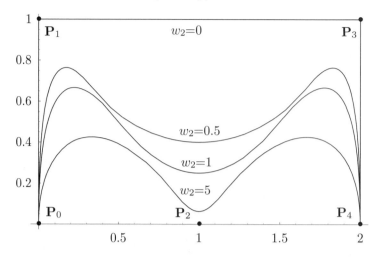

```
(* Rational B-spline example. w_2=0, .5, 1, 5 (Slow!) *)
Clear[bspl,knt,w,pnts,cur1,cur2,cur3,cur4,R] (* weight functions *)
bspl[i_,k_,t_]:=If[knt[[i+k]]==knt[[i+1]],0, (* 0<=i<=n *)
 bspl[i,k-1,t] (t-knt[[i+1]])/(knt[[i+k]]-knt[[i+1]])] \
 +If[knt[[i+1+k]]==knt[[i+2]],0,
 bspl[i+1,k-1,t] (knt[[i+1+k]]-t)/(knt[[i+1+k]]-knt[[i+2]])];
bspl[i_,1,t_]:=If[knt[[i+1]]<=t<knt[[i+2]], 1, 0];
R[i_,t_]:=(w[[i+1]] bspl[i,k,t])/Sum[w[[j+1]] bspl[j,k,t], {j,0,n}];
n=4; k=3; w={1,1,0,1,1}; (* weights *)
knt={0,0,0,1,2,3,3,3}; (* knots *)
pnts={{0,0}, {0,1}, {1,0}, {2,1}, {2,0}};
cur1=ParametricPlot[Sum[(R[i,t] pnts[[i+1]]), {i,0,n}], {t,0,3},
 PlotRange->All, DisplayFunction->Identity, Compiled->False];
w[[3]]=0.5;
cur2=ParametricPlot[Sum[(R[i,t] pnts[[i+1]]), {i,0,n}], {t,0,3},
 PlotRange->All, DisplayFunction->Identity, Compiled->False];
w[[3]]=1;
cur3=ParametricPlot[Sum[(R[i,t] pnts[[i+1]]), {i,0,n}], {t,0,3},
 PlotRange->All, DisplayFunction->Identity, Compiled->False];
w[[3]]=5;
cur4=ParametricPlot[Sum[(R[i,t] pnts[[i+1]]), {i,0,n}], {t,0,3},
 PlotRange->All, DisplayFunction->Identity, Compiled->False];
Show[cur1,cur2,cur3,cur4, PlotRange->All, DefaultFont->{"cmr10", 10},
DisplayFunction->$DisplayFunction];
```

Figure 7.21: Effects of Varying Weight w_2.

The midpoint $\mathbf{S}$ of the curve of Equation (7.45) is given by

$$\mathbf{S} = \mathbf{P}(0.5) = \frac{(\mathbf{P}_0 + \mathbf{P}_2)/2}{1 + w_1} + \frac{w_1 \mathbf{P}_1}{1 + w_1} = \frac{1}{1 + w_1}\mathbf{M} + \frac{w_1}{1 + w_1}\mathbf{P}_1 = (1-u)\mathbf{M} + u\mathbf{P}_1, \ \ (7.46)$$

where $\mathbf{M} = (\mathbf{P}_0 + \mathbf{P}_2)/2$ is the midpoint of $\mathbf{P}_0$ and $\mathbf{P}_2$ and $u \stackrel{\text{def}}{=} w_1/(1 + w_1)$. Thus, point $\mathbf{S}$, which is called the *shoulder point* of the curve moves along a straight line from $\mathbf{M}$ to $\mathbf{P}_1$ when w_1 varies from 0 to ∞ (or, equivalently, when u varies from 0 to 1).

Equation (7.46) also yields the relation

$$w_1 = \frac{\mathbf{M} - \mathbf{S}}{\mathbf{S} - \mathbf{P}_1},\tag{7.47}$$

which shows that w_1 is the ratio of two distances.

It can be shown (see, e.g., [Lee 86]) that the single weight w_1 determines the type of conic generated by Equation (7.45). Values in the range $(0, 1)$ generate an elliptic curve (with a circle as a special case). The value $w_1 = 1$ produces a parabolic curve, and values $w_1 > 1$ result in a hyperbolic curve. Figure 7.22 shows examples of these types of conics (notice that $\mathbf{S}$ is not necessarily the maximum point on these curves).

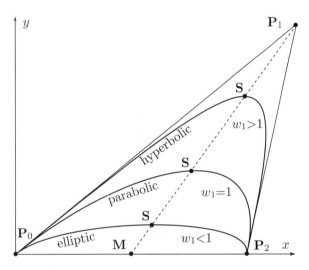

Figure 7.22: Conics Generated by Varying w_1.

A circle is formed when the three control points form an isosceles triangle. If we denote the base angle of this triangle by θ, it can be shown that a circular arc spanning 2θ degrees is obtained when $w_1 = \cos\theta$. The most common cases are $\theta = 60°$ and $\theta = 90°$. In the latter case (Figure 7.23b), a complete circle can easily be formed by using the symmetry of a circle and duplicating every point four times. In the former case (Figure 7.23a), a complete circle can be obtained by specifying six control points and calculating three spline segments.

Example: We are given the three points $\mathbf{P}_0 = (0, -1)R$, $\mathbf{P}_1 = (-1.732, -1)R$, and $\mathbf{P}_2 = (-0.866, 0.5)R$ of Figure 7.23a. Substituting these points in Equation (7.45) and setting w_1 to $\cos 60° = 0.5$ yields the $60°$ circular arc that goes from $\mathbf{P}_0$ to $\mathbf{P}_2$:

$$\begin{aligned}
\mathbf{P}(t) &= \frac{(1-t)^2\mathbf{P}_0 + 2w_1 t(1-t)\mathbf{P}_1 + t^2\mathbf{P}_2}{(1-t)^2 + 2w_1 t(1-t) + t^2} \\
&= R\frac{(1-t)^2(0, -1) + t(1-t)(-1.732, -1) + t^2(-0.866, 0.5)}{(1-t)^2 + t(1-t) + t^2}
\end{aligned}$$

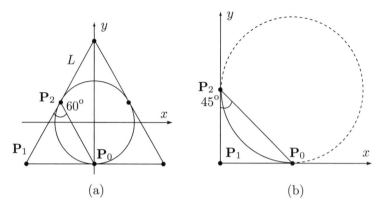

$$(a) \qquad\qquad (b)$$

```
(* One third of a circle done by rational B-spline *)
P0={0,-1}; P1={-1.732,-1}; P2={-0.866,0.5}; w1=0.5;
pnts=ListPlot[{P0,P1,P2}, Prolog->PointSize[.04], DisplayFunction->Identity];
axs={AbsoluteThickness[1], Line[{P0,P1,P2}]};
th=ParametricPlot[((1-t)^2 P0+2w1 t(1-t)P1+t^2 P2)/((1-t)^2+2w1 t(1-t)+t^2),
{t,0,1}, PlotRange->All, DisplayFunction->Identity, Compiled->False];
Show[Graphics[axs],th,pnts, PlotRange->All, DisplayFunction->$DisplayFunction];
```

Figure 7.23: Control Points for Circles.

$$= R\frac{(0.866t^2 - 1.732t, 0.5t^2 + t - 1)}{(1-t)^2 + t(1-t) + t^2}.$$

⋄ **Exercise 7.13:** Show how to figure out the coordinates of the three points from Figure 7.23a.

⋄ **Exercise 7.14:** Given the three points $\mathbf{P}_0 = (1,0)R$, $\mathbf{P}_1 = (0,0)$, and $\mathbf{P}_2 = (0,1)R$ of Figure 7.23b, calculate the quadratic rational B-spline segment defined by the points whose shape is a circular arc spanning $90°$.

7.15 Uniform B-Spline Surfaces

The uniform B-Spline surface patch is constructed as a Cartesian product of two uniform B-spline curves. The biquadratic B-spline surface patch, for example, is fully defined by nine control points and is constructed as the Cartesian product of Equation (7.6) with itself

$$
\mathbf{P}(u,w) = \left(\frac{1}{2}\right)^2 (u^2, u, 1)
\begin{pmatrix} 1 & -2 & 1 \\ -2 & 2 & 0 \\ 1 & 1 & 0 \end{pmatrix}
\begin{pmatrix} \mathbf{P}_{00} & \mathbf{P}_{01} & \mathbf{P}_{02} \\ \mathbf{P}_{10} & \mathbf{P}_{11} & \mathbf{P}_{12} \\ \mathbf{P}_{20} & \mathbf{P}_{21} & \mathbf{P}_{22} \end{pmatrix}
$$
$$
\times \begin{pmatrix} 1 & -2 & 1 \\ -2 & 2 & 0 \\ 1 & 1 & 0 \end{pmatrix}^T
\begin{pmatrix} w^2 \\ w \\ 1 \end{pmatrix}.
$$

(7.48)

Its four corner points are not the four extreme control points, but

$$
\mathbf{K}_{00} = \mathbf{P}(0,0) = \frac{1}{4}(\mathbf{P}_{00} + \mathbf{P}_{01} + \mathbf{P}_{10} + \mathbf{P}_{11}),
$$
$$
\mathbf{K}_{01} = \mathbf{P}(0,1) = \frac{1}{4}(\mathbf{P}_{01} + \mathbf{P}_{02} + \mathbf{P}_{11} + \mathbf{P}_{12}),
$$
$$
\mathbf{K}_{10} = \mathbf{P}(1,0) = \frac{1}{4}(\mathbf{P}_{10} + \mathbf{P}_{11} + \mathbf{P}_{20} + \mathbf{P}_{21}),
$$
$$
\mathbf{K}_{11} = \mathbf{P}(1,1) = \frac{1}{4}(\mathbf{P}_{11} + \mathbf{P}_{12} + \mathbf{P}_{21} + \mathbf{P}_{22}).
$$
(7.49)

Notice that corner point $\mathbf{K}_{00}$ can be written

$$
\mathbf{K}_{00} = \frac{1}{2}\left(\frac{\mathbf{P}_{00} + \mathbf{P}_{01}}{2} + \frac{\mathbf{P}_{10} + \mathbf{P}_{11}}{2}\right).
$$

This point is therefore located midway between points $(\mathbf{P}_{00} + \mathbf{P}_{01})/2$ and $(\mathbf{P}_{10} + \mathbf{P}_{11})/2$. Figure 7.24a shows this location, as well as the locations of the other three corner points, for the case where the control points are equally spaced.

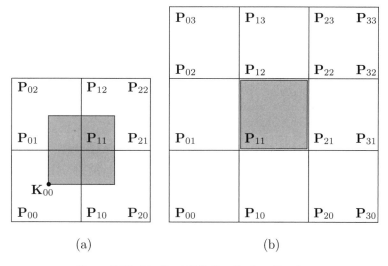

(a)   (b)

Figure 7.24: Idealized B-Spline Surface Patches.

Example: Given the nine points

$$
\mathbf{P}_{00} = (0,0,0), \quad \mathbf{P}_{01} = (0,1,0), \quad \mathbf{P}_{02} = (0,2,0),
$$
$$
\mathbf{P}_{10} = (1,0,0), \quad \mathbf{P}_{11} = (1,1,1), \quad \mathbf{P}_{12} = (1,2,0),
$$
$$
\mathbf{P}_{20} = (2,0,0), \quad \mathbf{P}_{21} = (2,1,0), \quad \mathbf{P}_{22} = (2,2,0),
$$

the biquadratic B-spline surface patch defined by them is given by the simple expression

$$\mathbf{P}(u, w) = (u + 1/2, w + 1/2, (-1 - 2u + 2u^2)(-1 - 2w + 2w^2)/4).$$

Its four corner points are

$$\mathbf{K}_{00} = \mathbf{P}(0, 0) = \left(\frac{1}{2}, \frac{1}{2}, \frac{1}{4}\right), \quad \mathbf{K}_{01} = \mathbf{P}(0, 1) = \left(\frac{1}{2}, \frac{3}{2}, \frac{1}{4}\right),$$

$$\mathbf{K}_{10} = \mathbf{P}(1, 0) = \left(\frac{3}{2}, \frac{1}{2}, \frac{1}{4}\right), \quad \mathbf{K}_{11} = \mathbf{P}(1, 1) = \left(\frac{3}{2}, \frac{3}{2}, \frac{1}{4}\right).$$

Figure 7.25 shows the relation between this surface and its control points.

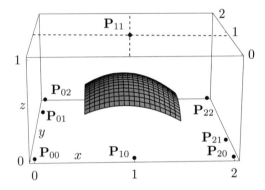

```
(* BiQuadratic B-spline Patch Example *)
<<:Graphics:ParametricPlot3D.m
Clear[T,Pnts,Q,comb,g1,g2];
T[t_]:={t^2,t,1};
Pnts={{{0,0,0},{0,1.5,0},{0,2,0}},{{1,0,0},{1,1,1},{1,2,0}},
  {{2,0,0},{2,0.5,0},{2,2,0}}};
Q={{1,-2,1},{-2,2,0},{1,1,0}};
g1=Graphics3D[{AbsolutePointSize[3], Table[Point[Pnts[[i,j]]],{i,1,3},{j,1,3}]}];
comb[i_]:=((1/4)T[u].Q.Pnts)[[i]] (Transpose[Q].T[w])[[i]]
g2=ParametricPlot3D[comb[1]+comb[2]+comb[3], {u,0,1},{w,0,1}, AspectRatio->Automatic,
  Ticks->{{0,1,2},{0,1,2},{0,1}}, Compiled->False, DisplayFunction->Identity];
Show[g2,g1, DisplayFunction->$DisplayFunction, ViewPoint->{-0.196, -4.177, 1.160},
  PlotRange->All, DefaultFont->{"cmr10", 10}];
```

Figure 7.25: A Biquadratic B-Spline Surface Patch.

◇ **Exercise 7.15:** Calculate the midpoint $\mathbf{P}(1/2, 1/2)$ of this patch.

From the dictionary

A line segment is a part of a line that is bounded by two end points. The midpoint of a segment is the unique point located at an equal distance from the two end points.

The bicubic B-spline patch is defined by a grid of 4×4 control points and is constructed as the Cartesian product of Equation (7.11) with itself

$$
\mathbf{P}(u, w) = \left(\frac{1}{6}\right)^2 (u^3, u^2, u, 1) \begin{pmatrix} -1 & 3 & -3 & 1 \\ 3 & -6 & 3 & 0 \\ -3 & 0 & 3 & 0 \\ 1 & 4 & 1 & 0 \end{pmatrix}
$$

$$
\times \begin{pmatrix} \mathbf{P}_{00} & \mathbf{P}_{01} & \mathbf{P}_{02} & \mathbf{P}_{03} \\ \mathbf{P}_{10} & \mathbf{P}_{11} & \mathbf{P}_{12} & \mathbf{P}_{13} \\ \mathbf{P}_{20} & \mathbf{P}_{21} & \mathbf{P}_{22} & \mathbf{P}_{23} \\ \mathbf{P}_{30} & \mathbf{P}_{31} & \mathbf{P}_{32} & \mathbf{P}_{33} \end{pmatrix} \begin{pmatrix} -1 & 3 & -3 & 1 \\ 3 & -6 & 3 & 0 \\ -3 & 0 & 3 & 0 \\ 1 & 4 & 1 & 0 \end{pmatrix}^T \begin{pmatrix} w^3 \\ w^2 \\ w \\ 1 \end{pmatrix} . \qquad (7.50)
$$

Its four corner points are

$$
\begin{aligned}
\mathbf{K}_{00} &= \mathbf{P}(0,0) \\
&= \frac{1}{36}(\mathbf{P}_{00} + \mathbf{P}_{02} + 4\mathbf{P}_{10} + 4\mathbf{P}_{12} + \mathbf{P}_{20} + 4\mathbf{P}_{01} + 16\mathbf{P}_{11} + 4\mathbf{P}_{21} + \mathbf{P}_{22}), \\
\mathbf{K}_{01} &= \mathbf{P}(0,1) \\
&= \frac{1}{36}(\mathbf{P}_{01} + 4\mathbf{P}_{02} + \mathbf{P}_{03} + 4\mathbf{P}_{11} + 16\mathbf{P}_{12} + 4\mathbf{P}_{13} + \mathbf{P}_{21} + 4\mathbf{P}_{22} + \mathbf{P}_{23}), \\
\mathbf{K}_{10} &= \mathbf{P}(1,0) \qquad\qquad\qquad\qquad\qquad\qquad\qquad\qquad\qquad\qquad (7.51) \\
&= \frac{1}{36}(\mathbf{P}_{10} + \mathbf{P}_{12} + 4\mathbf{P}_{20} + 4\mathbf{P}_{22} + \mathbf{P}_{30} + 4\mathbf{P}_{11} + 16\mathbf{P}_{21} + 4\mathbf{P}_{31} + \mathbf{P}_{32}), \\
\mathbf{K}_{11} &= \mathbf{P}(1,1) \\
&= \frac{1}{36}(\mathbf{P}_{11} + 4\mathbf{P}_{12} + \mathbf{P}_{13} + 4\mathbf{P}_{21} + 16\mathbf{P}_{22} + 4\mathbf{P}_{23} + \mathbf{P}_{31} + 4\mathbf{P}_{32} + \mathbf{P}_{33}).
\end{aligned}
$$

Each is a barycentric sum of nine control points. Notice that the first corner point can be rewritten in the form

$$
\mathbf{K}_{00} = \frac{1}{6}\left[\frac{1}{6}(\mathbf{P}_{00} + 4\mathbf{P}_{10} + \mathbf{P}_{20}) + \frac{4}{6}(\mathbf{P}_{01} + 4\mathbf{P}_{11} + \mathbf{P}_{21}) + \frac{1}{6}(\mathbf{P}_{02} + 4\mathbf{P}_{12} + \mathbf{P}_{22})\right]. \quad (7.52)
$$

This point is therefore the weighted sum of three points, each the weighted sum of three control points. Its precise location depends on the positions of the nine points involved.

$\diamond$ **Exercise 7.16:** What is the value of $\mathbf{K}_{00}$ for the special case where the control points are equally spaced?

The other three corner points can be expressed similarly. If all 16 points are equally spaced, the bicubic surface patch has its corners at the four control points $\mathbf{P}_{11}$, $\mathbf{P}_{21}$, $\mathbf{P}_{12}$, and $\mathbf{P}_{22}$ (Figure 7.24b shows an idealized diagram).

Large B-spline surfaces can be constructed from these bicubic patches by starting with a mesh of $(m + 1) \times (n + 1)$ control points $\mathbf{P}_{00}$ through $\mathbf{P}_{mn}$, dividing it into $(m - 2) \times (n - 2)$ overlapping groups of 4×4 points each, as in Figure 5.9 and applying Equation (7.50) to calculate a cubic patch for each group. The individual patches will not only connect at their joint points but will have C^2 continuity along their boundaries.

To show that the bicubic patches connect at the joints, we note how joint point $\mathbf{K}_{01}$ can be obtained from joint $\mathbf{K}_{00}$ by incrementing the second indices of the nine control points involved in their expressions [Equation (7.51)]. The same is true for joints $\mathbf{K}_{10}$ and $\mathbf{K}_{11}$. Similarly, joint point $\mathbf{K}_{10}$ can be obtained from $\mathbf{K}_{00}$ by incrementing the first index of each control point, and the same is true for joints $\mathbf{K}_{01}$ and $\mathbf{K}_{11}$.

To show first-order continuity we calculate, for example, the two tangent vectors $\mathbf{P}^u(u,0)$ and $\mathbf{P}^u(u,1)$ of boundary curves $\mathbf{P}(u,0)$ and $\mathbf{P}(u,1)$

$$
\begin{aligned}
\mathbf{P}^u(u,0) = \big(&-\mathbf{P}_{02} + \mathbf{P}_{20} + 4\mathbf{P}_{21} + \mathbf{P}_{22} - \mathbf{P}_{00}(u-1)^2 - 4\mathbf{P}_{01}(u-1)^2 \\
&+ 2\mathbf{P}_{02}u - 4\mathbf{P}_{10}u - 16\mathbf{P}_{11}u - 4\mathbf{P}_{12}u + 2\mathbf{P}_{20}u + 8\mathbf{P}_{21}u + 2\mathbf{P}_{22}u \\
&- \mathbf{P}_{02}u^2 + 3\mathbf{P}_{10}u^2 + 12\mathbf{P}_{11}u^2 + 3\mathbf{P}_{12}u^2 - 3\mathbf{P}_{20}u^2 - 12\mathbf{P}_{21}u^2 \\
&- 3\mathbf{P}_{22}u^2 + \mathbf{P}_{30}u^2 + 4\mathbf{P}_{31}u^2 + \mathbf{P}_{32}u^2\big)/12 \\
\mathbf{P}^u(u,1) = \big(&-\mathbf{P}_{03} + \mathbf{P}_{21} + 4\mathbf{P}_{22} + \mathbf{P}_{23} - \mathbf{P}_{01}(u-1)^2 - 4\mathbf{P}_{02}(u-1)^2 \\
&+ 2\mathbf{P}_{03}u - 4\mathbf{P}_{11}u - 16\mathbf{P}_{12}u - 4\mathbf{P}_{13}u + 2\mathbf{P}_{21}u + 8\mathbf{P}_{22}u + 2\mathbf{P}_{23}u \\
&- \mathbf{P}_{03}u^2 + 3\mathbf{P}_{11}u^2 + 12\mathbf{P}_{12}u^2 + 3\mathbf{P}_{13}u^2 - 3\mathbf{P}_{21}u^2 - 12\mathbf{P}_{22}u^2 \\
&- 3\mathbf{P}_{23}u^2 + \mathbf{P}_{31}u^2 + 4\mathbf{P}_{32}u^2 + \mathbf{P}_{33}u^2\big)/12
\end{aligned}
\tag{7.53}
$$

Equation (7.53) shows that tangent vector $\mathbf{P}^u(u,1)$ can be obtained from $\mathbf{P}^u(u,0)$ by incrementing the second index of every control point involved. Equation (7.54) illustrates the same property for the second derivatives, thereby showing second-order continuity:

$$
\begin{aligned}
\mathbf{P}^{uu}(u,0) = \big(&\mathbf{P}_{00} + \mathbf{P}_{02} - 2\mathbf{P}_{10} - 8\mathbf{P}_{11} - 2\mathbf{P}_{12} + \mathbf{P}_{20} + 4\mathbf{P}_{21} + \mathbf{P}_{22} \\
&- 4\mathbf{P}_{01}(u-1) - \mathbf{P}_{00}u - \mathbf{P}_{02}u + 3\mathbf{P}_{10}u + 12\mathbf{P}_{11}u + 3\mathbf{P}_{12}u \\
&- 3\mathbf{P}_{20}u - 12\mathbf{P}_{21}u - 3\mathbf{P}_{22}u + \mathbf{P}_{30}u + 4\mathbf{P}_{31}u + \mathbf{P}_{32}u\big)/6 \\
\mathbf{P}^{uu}(u,1) = \big(&\mathbf{P}_{01} + \mathbf{P}_{03} - 2\mathbf{P}_{11} - 8\mathbf{P}_{12} - 2\mathbf{P}_{13} + \mathbf{P}_{21} + 4\mathbf{P}_{22} + \mathbf{P}_{23} \\
&- 4\mathbf{P}_{02}(u-1) - \mathbf{P}_{01}u - \mathbf{P}_{03}u + 3\mathbf{P}_{11}u + 12\mathbf{P}_{12}u + 3\mathbf{P}_{13}u \\
&- 3\mathbf{P}_{21}u - 12\mathbf{P}_{22}u - 3\mathbf{P}_{23}u + \mathbf{P}_{31}u + 4\mathbf{P}_{32}u + \mathbf{P}_{33}u\big)/6.
\end{aligned}
\tag{7.54}
$$

7.16 Relation to Other Surfaces

This short section shows how the uniform bicubic B-spline surface patch can be expressed as either a bicubic Coons or a bicubic Bézier patch.

Bicubic Coons and B-Spline Patches. A bicubic B-spline surface patch can be written as a bicubic Coons patch. That patch [Equation (4.35), duplicated here] is defined in terms of four corner points, eight tangent vectors, and four twist vectors. These 16 quantities (the elements of matrix $\mathbf{C}$ below) can be expressed in terms of the 16 control points $\mathbf{P}_{ij}$ that define the B-spline patch. The idea is to equate the expression for the Coons surface

$$
\mathbf{Q}(u,w) = (u^3, u^2, u, 1)\mathbf{H}
\begin{pmatrix}
\mathbf{Q}_{00} & \mathbf{Q}_{01} & \mathbf{Q}_{00}^w & \mathbf{Q}_{01}^w \\
\mathbf{Q}_{10} & \mathbf{Q}_{11} & \mathbf{Q}_{10}^w & \mathbf{Q}_{11}^w \\
\mathbf{Q}_{00}^u & \mathbf{Q}_{01}^u & \mathbf{Q}_{00}^{uw} & \mathbf{Q}_{01}^{uw} \\
\mathbf{Q}_{10}^u & \mathbf{Q}_{11}^u & \mathbf{Q}_{10}^{uw} & \mathbf{Q}_{11}^{uw}
\end{pmatrix}
\mathbf{H}^T
\begin{pmatrix}
w^3 \\
w^2 \\
w \\
1
\end{pmatrix}
= \mathbf{UHCH}^T\mathbf{W}^T,
$$

(4.35)

with that of the B-spline surface, Equation (7.50), and solve for the 16 elements of matrix $\mathbf{C}$. This process is straightforward and the solutions are

$$
\mathbf{Q}_{00} = \frac{1}{6}\left(\frac{\mathbf{P}_{00}}{6} + \frac{4\mathbf{P}_{10}}{6} + \frac{\mathbf{P}_{20}}{6}\right) + \frac{4}{6}\left(\frac{\mathbf{P}_{01}}{6} + \frac{4\mathbf{P}_{11}}{6} + \frac{\mathbf{P}_{21}}{6}\right) + \frac{1}{6}\left(\frac{\mathbf{P}_{02}}{6} + \frac{4\mathbf{P}_{12}}{6} + \frac{\mathbf{P}_{22}}{6}\right),
$$

$$
\mathbf{Q}_{01} = \frac{1}{6}\left(\frac{\mathbf{P}_{01}}{6} + \frac{4\mathbf{P}_{11}}{6} + \frac{\mathbf{P}_{21}}{6}\right) + \frac{4}{6}\left(\frac{\mathbf{P}_{02}}{6} + \frac{4\mathbf{P}_{12}}{6} + \frac{\mathbf{P}_{22}}{6}\right) + \frac{1}{6}\left(\frac{\mathbf{P}_{03}}{6} + \frac{4\mathbf{P}_{13}}{6} + \frac{\mathbf{P}_{23}}{6}\right),
$$

$$
\mathbf{Q}_{10} = \frac{1}{6}\left(\frac{\mathbf{P}_{10}}{6} + \frac{4\mathbf{P}_{20}}{6} + \frac{\mathbf{P}_{30}}{6}\right) + \frac{4}{6}\left(\frac{\mathbf{P}_{11}}{6} + \frac{4\mathbf{P}_{21}}{6} + \frac{\mathbf{P}_{31}}{6}\right) + \frac{1}{6}\left(\frac{\mathbf{P}_{12}}{6} + \frac{4\mathbf{P}_{22}}{6} + \frac{\mathbf{P}_{32}}{6}\right),
$$

$$
\mathbf{Q}_{11} = \frac{1}{6}\left(\frac{\mathbf{P}_{11}}{6} + \frac{4\mathbf{P}_{21}}{6} + \frac{\mathbf{P}_{31}}{6}\right) + \frac{4}{6}\left(\frac{\mathbf{P}_{12}}{6} + \frac{4\mathbf{P}_{22}}{6} + \frac{\mathbf{P}_{32}}{6}\right) + \frac{1}{6}\left(\frac{\mathbf{P}_{13}}{6} + \frac{4\mathbf{P}_{23}}{6} + \frac{\mathbf{P}_{33}}{6}\right),
$$

$$
\mathbf{Q}_{00}^u = \frac{1}{6}\left(\frac{\mathbf{P}_{20} - \mathbf{P}_{00}}{2}\right) + \frac{4}{6}\left(\frac{\mathbf{P}_{21} - \mathbf{P}_{01}}{2}\right) + \frac{1}{6}\left(\frac{\mathbf{P}_{22} - \mathbf{P}_{02}}{2}\right),
$$

$$
\mathbf{Q}_{01}^u = \frac{1}{6}\left(\frac{\mathbf{P}_{21} - \mathbf{P}_{01}}{2}\right) + \frac{4}{6}\left(\frac{\mathbf{P}_{22} - \mathbf{P}_{02}}{2}\right) + \frac{1}{6}\left(\frac{\mathbf{P}_{23} - \mathbf{P}_{03}}{2}\right),
$$

$$
\mathbf{Q}_{10}^u = \frac{1}{6}\left(\frac{\mathbf{P}_{30} - \mathbf{P}_{10}}{2}\right) + \frac{4}{6}\left(\frac{\mathbf{P}_{31} - \mathbf{P}_{11}}{2}\right) + \frac{1}{6}\left(\frac{\mathbf{P}_{32} - \mathbf{P}_{12}}{2}\right),
$$

$$
\mathbf{Q}_{11}^u = \frac{1}{6}\left(\frac{\mathbf{P}_{31} - \mathbf{P}_{11}}{2}\right) + \frac{4}{6}\left(\frac{\mathbf{P}_{32} - \mathbf{P}_{12}}{2}\right) + \frac{1}{6}\left(\frac{\mathbf{P}_{33} - \mathbf{P}_{13}}{2}\right),
$$

$$
\mathbf{Q}_{00}^w = \frac{1}{2}\left(\frac{\mathbf{P}_{02}}{6} + \frac{4\mathbf{P}_{12}}{6} + \frac{\mathbf{P}_{22}}{6}\right) - \frac{1}{2}\left(\frac{\mathbf{P}_{00}}{6} + \frac{4\mathbf{P}_{10}}{6} + \frac{\mathbf{P}_{20}}{6}\right),
$$

$$
\mathbf{Q}_{01}^w = \frac{1}{2}\left(\frac{\mathbf{P}_{03}}{6} + \frac{4\mathbf{P}_{13}}{6} + \frac{\mathbf{P}_{23}}{6}\right) - \frac{1}{2}\left(\frac{\mathbf{P}_{01}}{6} + \frac{4\mathbf{P}_{11}}{6} + \frac{\mathbf{P}_{21}}{6}\right),
$$

$$
\mathbf{Q}_{10}^w = \frac{1}{2}\left(\frac{\mathbf{P}_{12}}{6} + \frac{4\mathbf{P}_{22}}{6} + \frac{\mathbf{P}_{32}}{6}\right) - \frac{1}{2}\left(\frac{\mathbf{P}_{10}}{6} + \frac{4\mathbf{P}_{20}}{6} + \frac{\mathbf{P}_{30}}{6}\right),
$$

$$
\mathbf{Q}_{11}^w = \frac{1}{2}\left(\frac{\mathbf{P}_{13}}{6} + \frac{4\mathbf{P}_{23}}{6} + \frac{\mathbf{P}_{33}}{6}\right) - \frac{1}{2}\left(\frac{\mathbf{P}_{11}}{6} + \frac{4\mathbf{P}_{21}}{6} + \frac{\mathbf{P}_{31}}{6}\right),
$$

$$
\mathbf{Q}_{00}^{uw} = \frac{1}{2}\left(\frac{\mathbf{P}_{22} - \mathbf{P}_{02}}{2}\right) - \frac{1}{2}\left(\frac{\mathbf{P}_{20} - \mathbf{P}_{00}}{2}\right),
$$

$$
\mathbf{Q}_{01}^{uw} = \frac{1}{2}\left(\frac{\mathbf{P}_{23} - \mathbf{P}_{03}}{2}\right) - \frac{1}{2}\left(\frac{\mathbf{P}_{21} - \mathbf{P}_{01}}{2}\right),
$$

$$
\mathbf{Q}_{10}^{uw} = \frac{1}{2}\left(\frac{\mathbf{P}_{32} - \mathbf{P}_{12}}{2}\right) - \frac{1}{2}\left(\frac{\mathbf{P}_{30} - \mathbf{P}_{10}}{2}\right),
$$

$$
\mathbf{Q}_{11}^{uw} = \frac{1}{2}\left(\frac{\mathbf{P}_{33} - \mathbf{P}_{13}}{2}\right) - \frac{1}{2}\left(\frac{\mathbf{P}_{31} - \mathbf{P}_{11}}{2}\right).
$$

Bézier and B-Spline Bicubic Patches. A bicubic B-spline surface patch can also be written in the form of a bicubic Bézier patch. The bicubic Bézier patch is fully defined by 16 control points $\mathbf{Q}_{ij}$ [the elements of matrix $\mathbf{P}$ of Equation (6.32)]. They can be expressed in terms of the 16 control points $\mathbf{P}_{ij}$ defining the B-spline patch. The idea is to equate the expressions for the bicubic Bézier and B-spline surface patches and solve for the elements of matrix $\mathbf{P}$. The solutions are

$$\mathbf{Q}_{00} = \frac{1}{6}\left(\frac{\mathbf{P}_{00}}{6} + \frac{4\mathbf{P}_{10}}{6} + \frac{\mathbf{P}_{20}}{6}\right) + \frac{4}{6}\left(\frac{\mathbf{P}_{01}}{6} + \frac{4\mathbf{P}_{11}}{6} + \frac{\mathbf{P}_{21}}{6}\right) + \frac{1}{6}\left(\frac{\mathbf{P}_{02}}{6} + \frac{4\mathbf{P}_{12}}{6} + \frac{\mathbf{P}_{22}}{6}\right),$$

$$\mathbf{Q}_{01} = \frac{4}{6}\left(\frac{\mathbf{P}_{01}}{6} + \frac{4\mathbf{P}_{11}}{6} + \frac{\mathbf{P}_{21}}{6}\right) + \frac{2}{6}\left(\frac{\mathbf{P}_{02}}{6} + \frac{4\mathbf{P}_{12}}{6} + \frac{\mathbf{P}_{32}}{6}\right),$$

$$\mathbf{Q}_{02} = \frac{2}{6}\left(\frac{\mathbf{P}_{01}}{6} + \frac{4\mathbf{P}_{11}}{6} + \frac{\mathbf{P}_{21}}{6}\right) + \frac{4}{6}\left(\frac{\mathbf{P}_{02}}{6} + \frac{4\mathbf{P}_{12}}{6} + \frac{\mathbf{P}_{32}}{6}\right),$$

$$\mathbf{Q}_{03} = \frac{1}{6}\left(\frac{\mathbf{P}_{01}}{6} + \frac{4\mathbf{P}_{11}}{6} + \frac{\mathbf{P}_{21}}{6}\right) + \frac{4}{6}\left(\frac{\mathbf{P}_{02}}{6} + \frac{4\mathbf{P}_{12}}{6} + \frac{\mathbf{P}_{22}}{6}\right) + \frac{1}{6}\left(\frac{\mathbf{P}_{03}}{6} + \frac{4\mathbf{P}_{13}}{6} + \frac{\mathbf{P}_{23}}{6}\right),$$

$$\mathbf{Q}_{10} = \frac{1}{6}\left(\frac{4\mathbf{P}_{10} + 2\mathbf{P}_{20}}{6}\right) + \frac{4}{6}\left(\frac{4\mathbf{P}_{11} + 2\mathbf{P}_{21}}{6}\right) + \frac{1}{6}\left(\frac{4\mathbf{P}_{12} + 2\mathbf{P}_{22}}{6}\right),$$

$$\mathbf{Q}_{11} = \frac{4}{6}\left(\frac{4\mathbf{P}_{11} + 2\mathbf{P}_{21}}{6}\right) + \frac{2}{6}\left(\frac{4\mathbf{P}_{12} + 2\mathbf{P}_{22}}{6}\right),$$

$$\mathbf{Q}_{12} = \frac{2}{6}\left(\frac{4\mathbf{P}_{11} + 2\mathbf{P}_{21}}{6}\right) + \frac{4}{6}\left(\frac{4\mathbf{P}_{12} + 2\mathbf{P}_{22}}{6}\right),$$

$$\mathbf{Q}_{13} = \frac{1}{6}\left(\frac{4\mathbf{P}_{11} + 2\mathbf{P}_{21}}{6}\right) + \frac{4}{6}\left(\frac{4\mathbf{P}_{12} + 2\mathbf{P}_{22}}{6}\right) + \frac{1}{6}\left(\frac{4\mathbf{P}_{13} + 2\mathbf{P}_{23}}{6}\right),$$

$$\mathbf{Q}_{20} = \frac{1}{6}\left(\frac{4\mathbf{P}_{10} + 2\mathbf{P}_{20}}{6}\right) + \frac{4}{6}\left(\frac{4\mathbf{P}_{11} + 2\mathbf{P}_{21}}{6}\right) + \frac{1}{6}\left(\frac{4\mathbf{P}_{12} + 2\mathbf{P}_{22}}{6}\right),$$

$$\mathbf{Q}_{21} = \frac{4}{6}\left(\frac{2\mathbf{P}_{11} + 4\mathbf{P}_{21}}{6}\right) + \frac{2}{6}\left(\frac{2\mathbf{P}_{12} + 4\mathbf{P}_{22}}{6}\right),$$

$$\mathbf{Q}_{22} = \frac{2}{6}\left(\frac{2\mathbf{P}_{11} + 4\mathbf{P}_{21}}{6}\right) + \frac{4}{6}\left(\frac{2\mathbf{P}_{12} + 4\mathbf{P}_{22}}{6}\right),$$

$$\mathbf{Q}_{23} = \frac{1}{6}\left(\frac{2\mathbf{P}_{11} + 4\mathbf{P}_{21}}{6}\right) + \frac{4}{6}\left(\frac{2\mathbf{P}_{12} + 4\mathbf{P}_{22}}{6}\right) + \frac{1}{6}\left(\frac{2\mathbf{P}_{13} + 4\mathbf{P}_{23}}{6}\right),$$

$$\mathbf{Q}_{30} = \frac{1}{6}\left(\frac{\mathbf{P}_{10}}{6} + \frac{4\mathbf{P}_{20}}{6} + \frac{\mathbf{P}_{30}}{6}\right) + \frac{4}{6}\left(\frac{\mathbf{P}_{11}}{6} + \frac{4\mathbf{P}_{21}}{6} + \frac{\mathbf{P}_{31}}{6}\right) + \frac{1}{6}\left(\frac{\mathbf{P}_{12}}{6} + \frac{4\mathbf{P}_{22}}{6} + \frac{\mathbf{P}_{32}}{6}\right),$$

$$\mathbf{Q}_{31} = \frac{4}{6}\left(\frac{\mathbf{P}_{11} + 4\mathbf{P}_{21} + \mathbf{P}_{31}}{6}\right) + \frac{2}{6}\left(\frac{\mathbf{P}_{12} + 4\mathbf{P}_{22} + \mathbf{P}_{32}}{6}\right),$$

$$\mathbf{Q}_{32} = \frac{2}{6}\left(\frac{\mathbf{P}_{11} + 4\mathbf{P}_{21} + \mathbf{P}_{31}}{6}\right) + \frac{4}{6}\left(\frac{\mathbf{P}_{12} + 4\mathbf{P}_{22} + \mathbf{P}_{32}}{6}\right),$$

$$\mathbf{Q}_{33} = \frac{1}{6}\left(\frac{\mathbf{P}_{11}}{6} + \frac{4\mathbf{P}_{21}}{6} + \frac{\mathbf{P}_{31}}{6}\right) + \frac{4}{6}\left(\frac{\mathbf{P}_{12}}{6} + \frac{4\mathbf{P}_{22}}{6} + \frac{\mathbf{P}_{32}}{6}\right) + \frac{1}{6}\left(\frac{\mathbf{P}_{13}}{6} + \frac{4\mathbf{P}_{23}}{6} + \frac{\mathbf{P}_{33}}{6}\right).$$

7.17 An Interpolating Bicubic Patch

The uniform bicubic B-spline surface patch is defined by 16 control points. A mesh of $(m+1) \times (n+1)$ control points can be used to calculate $(m-2) \times (n-2)$ such patches. Each patch has four corner points, but since the patches are connected, the total number of joint points is $(m-1) \times (n-1)$. This section shows how to solve the opposite problem, namely given a mesh of $(m-1) \times (n-1)$ data points $\mathbf{Q}_{1,1}$ through $\mathbf{Q}_{m-1,n-1}$, how to calculate the bicubic B-spline surface that passes through them.

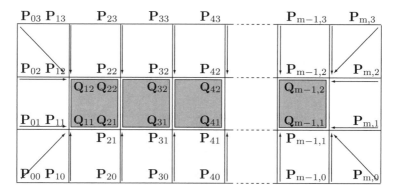

Figure 7.26: An Interpolating B-Spline Surface.

The given data points $\mathbf{Q}_{ij}$ are considered the joint points of the unknown surface and Equation (7.52) shows how they are related to the (yet unknown) control points $\mathbf{P}_{00}$ through $\mathbf{P}_{mn}$:

$$
\begin{aligned}
\mathbf{Q}_{ij} = \frac{1}{6}\Big[& \frac{1}{6}(\mathbf{P}_{i-1,j-1} + 4\mathbf{P}_{i,j-1} + \mathbf{P}_{i+1,j-1}) \\
& + \frac{4}{6}(\mathbf{P}_{i-1,j} + 4\mathbf{P}_{i,j} + \mathbf{P}_{i+1,j}) \\
& + \frac{1}{6}(\mathbf{P}_{i-1,j+1} + 4\mathbf{P}_{i,j+1} + \mathbf{P}_{i+1,j+1})\Big].
\end{aligned}
\tag{7.55}
$$

Equation (7.55) can be written $(m-1) \times (n-1)$ times, once for each given data point $\mathbf{Q}_{ij}$. The number of equations needed, however, is $(m+1) \times (n+1)$. We use the relation

$$(m+1) \times (n+1) = (m-1) \times (n-1) + 2m + 2n,$$

to figure out how many more equations are needed. The extra equations are obtained by the user specifying the vectors shown in Figure 7.26. There are $m-1$ vectors going from boundary control points $\mathbf{P}_{i,0}$ to the "bottom" data points $\mathbf{Q}_{i1}$. There are $m-1$ more such vectors going from the boundary control points $\mathbf{P}_{i,n+1}$ to the "top" data points $\mathbf{Q}_{i,n}$. In addition, there are $2(n-1)$ vectors going from the "left" and "right" boundary control points to the extreme data points $\mathbf{Q}_{1,j}$ and $\mathbf{Q}_{m-1,j}$. Finally, there are four vectors going from the four corner control points to the four corner data points.

Once all $2(n-1)+2(m-1)+4$ vectors have been specified, a system of $(m+1) \times (n+1)$ linear equations can be set and solved, to yield the control points.

If the surface should be closed along one dimension, some of the vectors don't have to be specified. For example, if the surface of Figure 7.26 should be closed in the vertical direction (i.e., if it should resemble a horizontal cylinder), then the bottom row of control points $\mathbf{P}_{i,0}$ should be duplicated and renamed $\mathbf{P}_{i,4}$, and the top row $\mathbf{P}_{i,3}$ should be duplicated and renamed $\mathbf{P}_{i,-1}$. Two extra rows of surface patches should be calculated, but every patch now has control points above and below it, so the $2(m-1)$ vertical vectors need not be specified by the user.

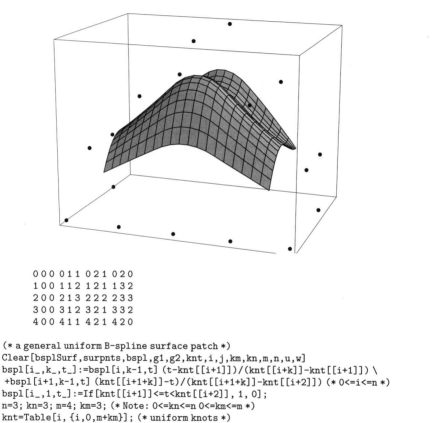

```
0 0 0  0 1 1  0 2 1  0 2 0
1 0 0  1 1 2  1 2 1  1 3 2
2 0 0  2 1 3  2 2 2  2 3 3
3 0 0  3 1 2  3 2 1  3 3 2
4 0 0  4 1 1  4 2 1  4 2 0
```

```
(* a general uniform B-spline surface patch *)
Clear[bsplSurf,surpnts,bspl,g1,g2,knt,i,j,km,kn,m,n,u,w]
bspl[i_,k_,t_]:=bspl[i,k-1,t] (t-knt[[i+1]])/(knt[[i+k]]-knt[[i+1]]) \
 +bspl[i+1,k-1,t] (knt[[i+1+k]]-t)/(knt[[i+1+k]]-knt[[i+2]]) (* 0<=i<=n *)
bspl[i_,1,t_]:=If[knt[[i+1]]<=t<knt[[i+2]], 1, 0];
n=3; kn=3; m=4; km=3; (* Note: 0<=kn<=n 0<=km<=m *)
knt=Table[i, {i,0,m+km}]; (* uniform knots *)
(* Input triplets from data file *)
surpnts=ReadList["surf.pnts", {Number,Number,Number}, RecordLists->True];
bsplSurf[u_,w_]:=Sum[Sum[surpnts[[i+1,j+1]]bspl[i,km,u],{i,0,m}]bspl[j,kn,w],{j,0,n}]
g1=Graphics3D[{AbsolutePointSize[3], Table[Point[surpnts[[i,j]]],{i,1,5},{j,1,4}]}];
g2=ParametricPlot3D[bsplSurf[u,w], {u,km-1,m+1},{w,kn-1,n+1},
DisplayFunction->Identity,
AspectRatio->Automatic, Compiled->False];
Show[g1,g2, PlotRange->All, DisplayFunction->$DisplayFunction,
DefaultFont->{"cmr10", 10}, ViewPoint->{1.389, -3.977, 1.042}];
```

Figure 7.27: A Quadratic-Cubic B-Spline Surface Patch.

7.18 The Quadratic-Cubic B-Spline Surface

This type of surface patch is defined by a 3×4 mesh of control points and its expression is a Cartesian product of the quadratic and cubic B-spline curves:

$$\mathbf{P}(u,w) = \left(\frac{1}{2}\right)\left(\frac{1}{6}\right)(u^2, u, 1)\begin{pmatrix} 1 & -2 & 1 \\ -2 & 2 & 0 \\ 1 & 1 & 0 \end{pmatrix}$$

$$\times \begin{pmatrix} \mathbf{P}_{00} & \mathbf{P}_{01} & \mathbf{P}_{02} & \mathbf{P}_{03} \\ \mathbf{P}_{10} & \mathbf{P}_{11} & \mathbf{P}_{12} & \mathbf{P}_{13} \\ \mathbf{P}_{20} & \mathbf{P}_{21} & \mathbf{P}_{22} & \mathbf{P}_{23} \end{pmatrix}\begin{pmatrix} -1 & 3 & -3 & 1 \\ 3 & -6 & 3 & 0 \\ -3 & 0 & 3 & 0 \\ 1 & 4 & 1 & 0 \end{pmatrix}^T\begin{pmatrix} w^3 \\ w^2 \\ w \\ 1 \end{pmatrix}.$$

Figure 7.27 is an example.

> The excellent mathematical and algorithmic properties, combined with successful industrial applications, have contributed to the enormous popularity of NURBS. NURBS play a role in the CAD/CAM/CAE world similar to that of the English language in science and business: "Want to talk business? Learn to talk NURBS".
>
> Les Piegl and Wayne Tiller, *The NURBS Book* (1996)

8
Subdivision Methods

8.1 Introduction

The Bézier curve can be constructed either as a weighted sum of control points or by the process of scaffolding. These are two very different approaches that lead to the same result. A third approach to curve and surface design, employing the process of *refinement* (also known as *subdivision* or *corner cutting*), is the topic of this chapter. Refinement is a general approach that can produce Bézier curves, B-spline curves, and other types of curves. Its main advantage is that it can easily be extended to surfaces.

8.2 Chaikin's Refinement Method

In 1974, George Chaikin came up with the idea of constructing a smooth curve from a small number of control points in several *refinement* steps. The principle of Chaikin's method is to start with a given set of control points $\mathbf{P}_i$, perform a computation that results in a new set of points $\mathbf{P}_i^1$, and repeat the process, producing more and more sets of points $\mathbf{P}_i^k$. Thus, the original control polygon is successively refined. Table 8.1 shows the notation used.

$$
\begin{array}{l}
\mathbf{P}_0, \ \mathbf{P}_1, \ \ldots, \ \mathbf{P}_n \\
\mathbf{P}_0^1, \ \mathbf{P}_1^1, \ \ldots, \ \mathbf{P}_{n_1}^1 \\
\mathbf{P}_0^2, \ \mathbf{P}_1^2, \ \ldots, \ \mathbf{P}_{n_2}^2 \\
\vdots \\
\mathbf{P}_0^k, \ \mathbf{P}_1^k, \ \ldots, \ \mathbf{P}_{n_k}^k
\end{array}
$$

Table 8.1: Refining Control Points.

Each point $\mathbf{P}_j^k$ is computed as a weighted sum of the points $\mathbf{P}_i^{k-1}$ of the previous iteration. Thus,

$$
\mathbf{P}_j^k = \sum_{i=0}^{n_{k-1}} a_{ijk}\mathbf{P}_i^{k-1} = (a_{0jk}, a_{1jk}, \ldots, a_{n_{k-1},jk}) \begin{pmatrix} \mathbf{P}_0^{k-1} \\ \mathbf{P}_1^{k-1} \\ \vdots \\ \mathbf{P}_{n_{k-1}}^{k-1} \end{pmatrix},
$$

where a_{ijk} are real coefficients. Notice that each iteration produces a different number $n_k + 1$ of points. If n_k gets smaller with k, then the number of points gets smaller and smaller until a single point is left. An example is the de Casteljau scaffolding construction, a process that produces one point of the Bézier curve. At the other extreme, n_k may get larger with k, producing more points in each iteration. We then stop after a few iterations and draw the curve by drawing straight segments between the points of the last iteration. An example of this case is the Chaikin algorithm, described in example (2).

Each iteration can be completely described by its coefficient matrix

$$
\begin{pmatrix} \mathbf{P}_0^k \\ \mathbf{P}_1^k \\ \vdots \\ \mathbf{P}_{n_k}^k \end{pmatrix} = \begin{pmatrix} a_{00k} & a_{10k} & \cdots & a_{n_{k-1},0k} \\ a_{01k} & a_{11k} & \cdots & a_{n_{k-1},1k} \\ \vdots & \vdots & & \vdots \\ a_{0,n_k,k} & a_{1,n_k,k} & \cdots & a_{n_{k-1},n_k,k} \end{pmatrix} \begin{pmatrix} \mathbf{P}_0^{k-1} \\ \mathbf{P}_1^{k-1} \\ \vdots \\ \mathbf{P}_{n_{k-1}}^{k-1} \end{pmatrix}
$$

$$
= \mathbf{M}_k \begin{pmatrix} \mathbf{P}_0^{k-1} \\ \mathbf{P}_1^{k-1} \\ \vdots \\ \mathbf{P}_{n_{k-1}}^{k-1} \end{pmatrix},
$$

(8.1)

where $\mathbf{M}_k$ has $n_k + 1$ rows and $n_{k-1} + 1$ columns. Since the number of iterations may be large, the number of coefficients a_{ijk} may be huge. In practice, this number is significantly reduced in three ways: (1) Using a rule of calculation where most of these coefficients are zero. (2) Using coefficients a_{ij} that are independent of k. (3) Using coefficients a_{ik} that are independent of j. Case 2 is called *uniform refinement* and case 3 is termed *stationary refinement*.

Example: (1) This is the de Casteljau scaffolding construction expressed as a refinement process. The rule of refinement is

$$
\mathbf{P}_j^{k+1} = 0.5(\mathbf{P}_j^k + \mathbf{P}_{j+1}^k),
$$

(8.2)

which implies that the a_i coefficients are independent of j and k (this is a stationary uniform refinement method) and are zero except for the two coefficients a_j and a_{j+1}. The a_{ijk}'s therefore depend on i only and are given by

$$
a_i = \begin{cases} 0.5, & i = j, j+1, \\ 0, & \text{otherwise.} \end{cases}
$$

Since $\mathbf{P}_j^k$ depends on $\mathbf{P}_j^{k-1}$ and $\mathbf{P}_{j+1}^{k-1}$, the largest value for j is $n_k - 1$. This means that each iteration reduces the number of points by 1 (Figure 8.2a). We start with the $n+1$ points $\mathbf{P}_0, \mathbf{P}_1, \ldots, \mathbf{P}_n$. The first iteration produces n points, the second iteration produces $n-1$ points, and so on, until iteration n produces one point. That point is located on the Bézier curve $\mathbf{P}(t)$ defined by the $n+1$ original control points. In fact, that point is $\mathbf{P}(0.5)$. If we generalize Equation (8.2) to $\mathbf{P}_j^{k+1} = (1-\alpha)\mathbf{P}_j^k + \alpha\mathbf{P}_{j+1}^k$, then the final point is $\mathbf{P}(\alpha)$. Matrix $\mathbf{M}_k$ of Equation (8.1) is

$$\mathbf{M}_k = \begin{pmatrix} 0.5 & 0.5 & 0 & 0 & \cdots & 0 \\ 0 & 0.5 & 0.5 & 0 & \cdots & 0 \\ 0 & 0 & 0.5 & 0.5 & \cdots & 0 \\ \vdots & \vdots & \vdots & \ddots & \ddots & \vdots \\ 0 & 0 & 0 & \cdots & 0.5 & 0.5 \end{pmatrix}.$$

It is independent of k and is of order $k \times (k+1)$.

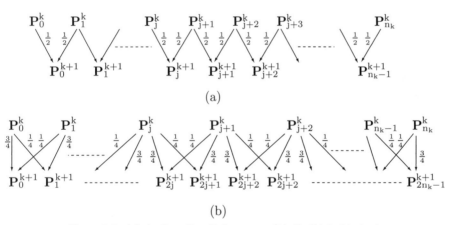

Figure 8.2: (a) de Casteljau Refinement. (b) Chaikin's Method.

Example: (2) We start with the $n+1$ control points $\mathbf{P}_0, \mathbf{P}_1, \ldots, \mathbf{P}_n$ and apply the rule of refinement

$$\mathbf{P}_{2j}^{k+1} = \frac{3}{4}\mathbf{P}_j^k + \frac{1}{4}\mathbf{P}_{j+1}^k, \qquad \mathbf{P}_{2j+1}^{k+1} = \frac{1}{4}\mathbf{P}_j^k + \frac{3}{4}\mathbf{P}_{j+1}^k. \tag{8.3}$$

(This is illustrated in Figure 8.2b.) The first iteration starts with the original $n+1$ points and produces the $2n$ points $\mathbf{P}_i^1$ shown in Table 8.3. Each subsequent iteration doubles the number of points and brings the points closer to the curve. After k iterations (where k depends on the required precision), the curve is displayed by drawing straight segments between the points produced in the last iteration.

This method is due to George Chaikin ([Chaikin 74] and [Riesenfeld 75]) and has a simple geometric interpretation, which is illustrated in Figure 8.5. Part (a) of the figure shows a control polygon made of five points. The rule of refinement is: take a segment

$$\mathbf{P}_0^1 = \tfrac{3}{4}\mathbf{P}_0 + \tfrac{1}{4}\mathbf{P}_1, \qquad \mathbf{P}_1^1 = \tfrac{1}{4}\mathbf{P}_0 + \tfrac{3}{4}\mathbf{P}_1,$$
$$\mathbf{P}_2^1 = \tfrac{3}{4}\mathbf{P}_1 + \tfrac{1}{4}\mathbf{P}_2, \qquad \mathbf{P}_3^1 = \tfrac{1}{4}\mathbf{P}_1 + \tfrac{3}{4}\mathbf{P}_2,$$
$$\mathbf{P}_4^1 = \tfrac{3}{4}\mathbf{P}_2 + \tfrac{1}{4}\mathbf{P}_3, \qquad \mathbf{P}_5^1 = \tfrac{1}{4}\mathbf{P}_2 + \tfrac{3}{4}\mathbf{P}_3,$$
$$\vdots \qquad\qquad\qquad \vdots$$
$$\mathbf{P}_{2n-2}^1 = \tfrac{3}{4}\mathbf{P}_{n-1} + \tfrac{1}{4}\mathbf{P}_n, \quad \mathbf{P}_{2n-1}^1 = \tfrac{1}{4}\mathbf{P}_{n-1} + \tfrac{3}{4}\mathbf{P}_n.$$

Table 8.3: First Iteration of Chaikin's Algorithm.

$\mathbf{P}_i\mathbf{P}_{i+1}$ of the control polygon and place two new points $\mathbf{Q}_i$ and $\mathbf{R}_i$ at distances from $\mathbf{P}_i$ of 1/4 and 3/4 the segment's size, respectively (Figure 8.5b, which justifies the term "corner cutting"). The new points are therefore given by

$$\mathbf{Q}_i = \frac{3}{4}\mathbf{P}_i + \frac{1}{4}\mathbf{P}_{i+1}, \quad \mathbf{R}_i = \frac{1}{4}\mathbf{P}_i + \frac{3}{4}\mathbf{P}_{i+1}.$$

This is repeated for all the polygon segments. If we start with $n+1$ control points defining a control polygon with n sides, we end up with $2n$ new points $\mathbf{Q}_i$ and $\mathbf{R}_i$. They should now be connected to form a new control polygon with $2n - 1$ sides. As this process is repeated (Figure 8.5c), the control polygons get closer to the smooth curve shown in Figure 8.5d. This figure also shows that the midpoint of any segment of the control polygon is a point on the Chaikin curve. In fact, the midpoint of any segment generated at any stage of the refinement is a point on the Chaikin curve.

◇ **Exercise 8.1:** Is this curve a Bézier curve?

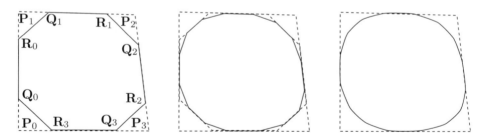

Figure 8.4: Chaikin's Algorithm for a Closed Curve.

This algorithm works for closed curves too. The only modification needed is to connect the last point $\mathbf{P}_n$ to the first one $\mathbf{P}_0$ and compute the two auxiliary points $\mathbf{Q}_n$ and $\mathbf{R}_n$. This can be done in a natural way if we copy $\mathbf{P}_0$ and name the duplicate $\mathbf{P}_{n+1}$. Figure 8.4 shows three instances in the construction of such a curve. Again, we see that the midpoint of any segment of the control polygon is a point on the closed Chaikin curve.

To identify the kind of curve that Chaikin's algorithm produces, let's consider the control polygon defined by the three points $\mathbf{P}_0$, $\mathbf{P}_1 = \mathbf{B}$, and $\mathbf{P}_2$ (Figure 8.6). Let $\mathbf{A}$ and $\mathbf{C}$ be the midpoints of segments $\mathbf{P}_0\mathbf{P}_1$ and $\mathbf{P}_1\mathbf{P}_2$, respectively, and let point $\mathbf{P}$ be

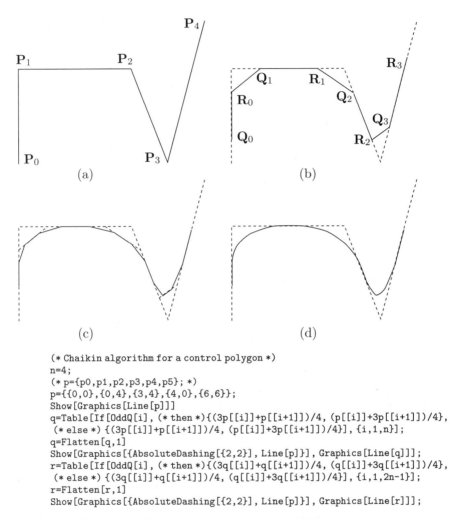

```
(* Chaikin algorithm for a control polygon *)
n=4;
(* p={p0,p1,p2,p3,p4,p5}; *)
p={{0,0},{0,4},{3,4},{4,0},{6,6}};
Show[Graphics[Line[p]]]
q=Table[If[OddQ[i], (* then *){(3p[[i]]+p[[i+1]])/4, (p[[i]]+3p[[i+1]])/4},
   (* else *) {(3p[[i]]+p[[i+1]])/4, (p[[i]]+3p[[i+1]])/4}}, {i,1,n}];
q=Flatten[q,1]
Show[Graphics[{AbsoluteDashing[{2,2}], Line[p]}], Graphics[Line[q]]];
r=Table[If[OddQ[i], (* then *){(3q[[i]]+q[[i+1]])/4, (q[[i]]+3q[[i+1]])/4},
   (* else *) {(3q[[i]]+q[[i+1]])/4, (q[[i]]+3q[[i+1]])/4}}, {i,1,2n-1}];
r=Flatten[r,1]
Show[Graphics[{AbsoluteDashing[{2,2}], Line[p]}], Graphics[Line[r]]];
```

Figure 8.5: Chaikin's Algorithm for a Control Polygon.

the midpoint of points $\mathbf{M}_{ab} = (\mathbf{A} + \mathbf{B})/2$ and $\mathbf{M}_{bc} = (\mathbf{B} + \mathbf{C})/2$. This point has the following properties:

1. It is located on the Bézier curve defined by points $\mathbf{A}$, $\mathbf{B}$, and $\mathbf{C}$ because it's been constructed using the de Casteljau scaffolding process.

2. It is located on the Chaikin curve defined by points $\mathbf{P}_0$, $\mathbf{P}_1$, and $\mathbf{P}_2$. This is because points $\mathbf{M}_{ab}$ and $\mathbf{M}_{bc}$ are the points constructed by the first step of Chaikin's algorithm and we already know that the midpoint of any Chaikin segment is a point on the Chaikin curve.

⋄ **Exercise 8.2:** Show that points $\mathbf{M}_{ab}$ and $\mathbf{M}_{bc}$ are the points constructed by the first step of Chaikin's algorithm.

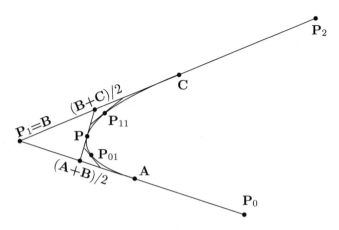

Figure 8.6: Points on the Chaikin Curve.

The second refinement step produces the two midpoints, $\mathbf{P}_{01}$ and $\mathbf{P}_{11}$ (Figure 8.6) using the recursive procedures

$$\mathbf{B} \leftarrow (\mathbf{A} + \mathbf{B})/2, \quad \mathbf{C} \leftarrow \mathbf{P}, \quad \mathbf{P}_{01} \leftarrow (\mathbf{A} + 2\mathbf{B} + \mathbf{C})/4,$$
$$\mathbf{A} \leftarrow \mathbf{P}, \quad \mathbf{B} \leftarrow (\mathbf{B} + \mathbf{C})/2, \quad \mathbf{P}_{11} \leftarrow (\mathbf{A} + 2\mathbf{B} + \mathbf{C})/4.$$

An argument similar to the previous one shows that these two points are also located on the quadratic Bézier curve defined by $\mathbf{A}$, $\mathbf{B}$, and $\mathbf{C}$ as well as on the Chaikin curve defined by $\mathbf{P}_0$, $\mathbf{P}_1$, and $\mathbf{P}_2$. Applying this argument to all the points generated by the refinement steps shows that they are located on both curves, which proves that the Chaikin curve defined by $\mathbf{P}_0$, $\mathbf{P}_1$, and $\mathbf{P}_2$ is identical to the quadratic Bézier curve defined by $\mathbf{A}$, $\mathbf{B}$, and $\mathbf{C}$. This Bézier curve is

$$\mathbf{P}(t) = (1 - t)^2 \mathbf{A} + 2t(1 - t)\mathbf{B} + t^2 \mathbf{C},$$

and it is easy to express in terms of the original control points $\mathbf{P}_i$

$$
\begin{aligned}
\mathbf{P}(t) &= (t^2, t, 1) \begin{pmatrix} 1 & -2 & 1 \\ -2 & 2 & 0 \\ 1 & 0 & 0 \end{pmatrix} \begin{pmatrix} \mathbf{A} \\ \mathbf{B} \\ \mathbf{C} \end{pmatrix} \\
&= (t^2, t, 1) \begin{pmatrix} 1 & -2 & 1 \\ -2 & 2 & 0 \\ 1 & 0 & 0 \end{pmatrix} \begin{pmatrix} (\mathbf{P}_0 + \mathbf{P}_1)/2 \\ \mathbf{P}_1 \\ (\mathbf{P}_1 + \mathbf{P}_2)/2 \end{pmatrix} \\
&= (t^2, t, 1) \begin{pmatrix} 1 & -2 & 1 \\ -2 & 2 & 0 \\ 1 & 0 & 0 \end{pmatrix} \begin{pmatrix} 1/2 & 1/2 & 0 \\ 0 & 1 & 0 \\ 0 & 1/2 & 1/2 \end{pmatrix} \begin{pmatrix} \mathbf{P}_0 \\ \mathbf{P}_1 \\ \mathbf{P}_2 \end{pmatrix} \\
&= \frac{1}{2}(t^2, t, 1) \begin{pmatrix} 1 & -2 & 1 \\ -2 & 2 & 0 \\ 1 & 1 & 0 \end{pmatrix} \begin{pmatrix} \mathbf{P}_0 \\ \mathbf{P}_1 \\ \mathbf{P}_2 \end{pmatrix}.
\end{aligned}
\tag{8.4}
$$

The result is the quadratic B-spline curve segment, Equation (7.6).

We therefore conclude that the curve produced by Chaikin's algorithm is not a new type of curve but the quadratic B-spline for points $\mathbf{P}_0$, $\mathbf{P}_1$, and $\mathbf{P}_2$. This fact lets us see the B-spline in a new light and it also shows a relation between the quadratic Bézier and B-spline curves.

$\diamond$ **Exercise 8.3:** (Easy). State this relation.

The Original Chaikin Algorithm

The description of Chaikin's algorithm in this section differs from that originally proposed by George Chaikin. Here is the original description of the method, as it appears in [Chaikin 74]. Start with four points $\mathbf{P}_1$ through $\mathbf{P}_4$ (Figure 8.7a). Points $\mathbf{P}_4$ and $\mathbf{P}_3$ are pushed into a stack and a new $\mathbf{P}_4 = (\mathbf{P}_2 + \mathbf{P}_3)/2$ is constructed. Points $\mathbf{P}_1$ and $\mathbf{P}_4$ are now compared. If their distance is greater than or equal to three pixels, then points $\mathbf{P}_2$ and $\mathbf{P}_3$ are recomputed according to

$$\mathbf{P}_3 = (\mathbf{P}_2 + \mathbf{P}_4)/2, \quad \mathbf{P}_2 = (\mathbf{P}_2 + \mathbf{P}_1)/2$$

(Figure 8.7b,c), points $\mathbf{P}_4$ and $\mathbf{P}_3$ are pushed into the stack, point $\mathbf{P}_4$ is recalculated, and the distance between $\mathbf{P}_1$ and $\mathbf{P}_4$ is checked. This is repeated until the distance becomes smaller than three pixels, in which case the short segment $\mathbf{P}_1\mathbf{P}_4$ is drawn, point $\mathbf{P}_4$ is renamed $\mathbf{P}_1$, the stack is popped twice and the resulting points are named $\mathbf{P}_2$ and $\mathbf{P}_4$, and the distance $\mathbf{P}_1\mathbf{P}_4$ is checked again. The process terminates when the stack is empty. Figure 8.7d is a flowchart of this algorithm.

$\diamond$ **Exercise 8.4:** Why compare the distance to three pixels and not to two?

8.3 Quadratic Uniform B-Spline by Subdivision

The uniform B-spline for a group of $n + 1$ control points can be constructed as a set of short segments, each a quadratic polynomial based on three control points. This section shows how Chaikin's algorithm (Section 8.2) can be applied to construct such a curve. We divide the original $n + 1$ control points into $n - 1$ overlapping groups of three points each, and use each group to calculate four new points. The groups are

$$\mathbf{P}_0\mathbf{P}_1\mathbf{P}_2, \quad \mathbf{P}_1\mathbf{P}_2\mathbf{P}_3, \ldots, \quad \mathbf{P}_{n-2}\mathbf{P}_{n-1}\mathbf{P}_n.$$

Subdividing the first group is done by

$$\begin{pmatrix} \mathbf{P}_0^1 \\ \mathbf{P}_1^1 \\ \mathbf{P}_2^1 \\ \mathbf{P}_3^1 \end{pmatrix} = \frac{1}{4}\begin{pmatrix} 3 & 1 & 0 \\ 1 & 3 & 0 \\ 0 & 3 & 1 \\ 0 & 1 & 3 \end{pmatrix}\begin{pmatrix} \mathbf{P}_0 \\ \mathbf{P}_1 \\ \mathbf{P}_2 \end{pmatrix} = \begin{pmatrix} \frac{3}{4}\mathbf{P}_0 + \frac{1}{4}\mathbf{P}_1 \\ \frac{1}{4}\mathbf{P}_0 + \frac{3}{4}\mathbf{P}_1 \\ \frac{3}{4}\mathbf{P}_1 + \frac{1}{4}\mathbf{P}_2 \\ \frac{1}{4}\mathbf{P}_1 + \frac{3}{4}\mathbf{P}_2 \end{pmatrix},$$

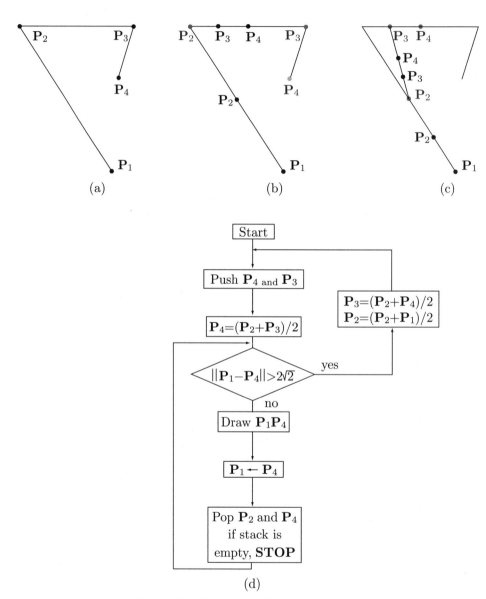

Figure 8.7: The Original Chaikin Algorithm.

and it yields the four new points $\mathbf{P}_0^1$, $\mathbf{P}_1^1$, $\mathbf{P}_2^1$, and $\mathbf{P}_3^1$. Subdividing the second group is done similarly and yields the four points $\mathbf{P}_2^1$, $\mathbf{P}_3^1$, $\mathbf{P}_4^1$, and $\mathbf{P}_5^1$, of which only the last two are new. Each subsequent group also yields two new points when subdivided. The process is then repeated on the $2n$ segments defined by the $2n+2$ new points $\mathbf{P}_i^1$, yielding $4n+4$ points $\mathbf{P}_i^2$. These points, in turn, define $4n+2$ segments. When the number of points is large enough, the curve can be drawn by connecting each pair of adjacent points with a straight segment.

It can be shown (see page 325) that the curve obtained this way is the quadratic uniform B-spline, Equation (7.6).

8.4 Cubic Uniform B-Spline by Subdivision

The approach to constructing cubic B-splines by subdivision is similar to that of Section 8.3. We show how Chaikin's methods (Section 8.2) can be applied to the construction of a cubic uniform B-spline for a set of $n+1$ control points $\mathbf{P}_i$. The points are divided into overlapping groups of four points each, and each group is used to calculate, by refinement, a PC that becomes a segment in the entire curve. These cubic segments have C^2 continuity. Since refinement is an iterative process, we denote the control points obtained in the kth subdivision step by $\mathbf{P}_i^k$. Thus, it makes sense to denote the original control points by $\mathbf{P}_i^0$. They are divided into the overlapping groups

$$\mathbf{P}_0^0\mathbf{P}_1^0\mathbf{P}_2^0\mathbf{P}_3^0, \quad \mathbf{P}_1^0\mathbf{P}_2^0\mathbf{P}_3^0\mathbf{P}_4^0, \ldots, \quad \mathbf{P}_{n-3}^0\mathbf{P}_{n-2}^0\mathbf{P}_{n-1}^0\mathbf{P}_n^0.$$

Figure 8.8a illustrates the refinement process that leads from the group of four control points $\mathbf{P}_0^0\mathbf{P}_1^0\mathbf{P}_2^0\mathbf{P}_3^0$ to a segment of a cubic uniform B-spline. The treatment for the other groups is similar. The figure shows the positions of the five iteration-1 points $\mathbf{P}_i^1$ and the seven points $\mathbf{P}_i^2$ resulting from iteration 2. The first refinement step computes the five points $\mathbf{P}_0^1\mathbf{P}_1^1\mathbf{P}_2^1\mathbf{P}_3^1\mathbf{P}_4^1$ as follows:

1. Each of the three points with even subscripts $\mathbf{P}_0^1\mathbf{P}_2^1\mathbf{P}_4^1$ (termed the *edge* points) is located at the center of a segment delimited by two of the original control points. Thus, $\mathbf{P}_0^1$ is located midway between $\mathbf{P}_0^0$ and $\mathbf{P}_1^0$.

2. Each of the two points with odd subscripts $\mathbf{P}_1^1$ and $\mathbf{P}_3^1$ (termed the *vertex* points) is located at the center of a segment whose endpoints are located at the centers of two segments delimited by two new edge points and one original control point. Thus, $\mathbf{P}_1^1$ is located at the center of the segment whose endpoints are located at the centers of the two segments delimited by the three points $\mathbf{P}_0^1$, $\mathbf{P}_1^0$ and $\mathbf{P}_2^1$.

The five points produced by the first refinement step can be expressed in terms of the four original control points by

$$\begin{pmatrix} \mathbf{P}_0^1 \\ \mathbf{P}_1^1 \\ \mathbf{P}_2^1 \\ \mathbf{P}_3^1 \\ \mathbf{P}_4^1 \end{pmatrix} = \begin{pmatrix} \frac{1}{2}(\mathbf{P}_0^0 + \mathbf{P}_1^0) \\ \frac{1}{8}(\mathbf{P}_0^0 + 6\mathbf{P}_1^0 + \mathbf{P}_2^0) \\ \frac{1}{2}(\mathbf{P}_1^0 + \mathbf{P}_2^0) \\ \frac{1}{8}(\mathbf{P}_1^0 + 6\mathbf{P}_2^0 + \mathbf{P}_3^0) \\ \frac{1}{2}(\mathbf{P}_2^0 + \mathbf{P}_3^0) \end{pmatrix} = \frac{1}{8} \begin{pmatrix} 4 & 4 & 0 & 0 \\ 1 & 6 & 1 & 0 \\ 0 & 4 & 4 & 0 \\ 0 & 1 & 6 & 1 \\ 0 & 0 & 4 & 4 \end{pmatrix} \begin{pmatrix} \mathbf{P}_0^0 \\ \mathbf{P}_1^0 \\ \mathbf{P}_2^0 \\ \mathbf{P}_3^0 \end{pmatrix}.$$

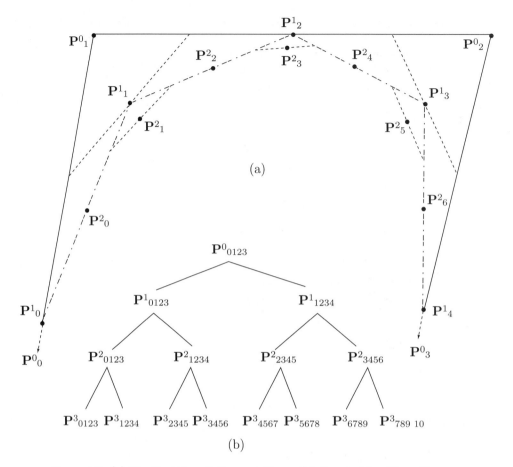

Figure 8.8: (a) The First Two Refinement Steps. (b) Groups After Three Steps.

Each of the new points $\mathbf{P}_i^1$ is computed from either two or three of the points $\mathbf{P}_j^0$. The five new points are then divided into two overlapping groups $\mathbf{P}_0^1\mathbf{P}_1^1\mathbf{P}_2^1\mathbf{P}_3^1$ and $\mathbf{P}_1^1\mathbf{P}_2^1\mathbf{P}_3^1\mathbf{P}_4^1$ of four points each, and the second subdivision step is applied to each group to produce five new points denoted by $\mathbf{P}_i^2$. Some of the $\mathbf{P}_i^2$ points, however, are identical, so this second step produces a total of seven distinct points. Figure 8.8b shows the points produced by the first three iterations of the refinement process and how each group of four points $\mathbf{P}_i^k$ produces two overlapping groups of four new points $\mathbf{P}_i^{k+1}$ each. The compact notation $\mathbf{P}_{0123}^3$ stands for a group of four points. It is easy to see that iteration k produces 2^k overlapping groups of four points each, for a total of $4 + (2^k - 1) = 3 + 2^k$ distinct points. Thus, iteration 0 (the original control points) consists of $3 + 2^0 = 4$ points, and iterations 1, 2, 3, and 4 produce 5, 7, 11, and 19 points, respectively.

Since each point produced in step k is computed from either two or three points of step $k - 1$, it is convenient to express a new triplet of points $\mathbf{P}_i^k\mathbf{P}_{i+1}^k\mathbf{P}_{i+2}^k$ as a function of a triplet $\mathbf{P}_j^{k-1}\mathbf{P}_{j+1}^{k-1}\mathbf{P}_{j+2}^{k-1}$. We illustrate this relation for $k = 1$

$$\begin{pmatrix} \mathbf{P}_0^1 \\ \mathbf{P}_1^1 \\ \mathbf{P}_2^1 \end{pmatrix} = \mathbf{A}\begin{pmatrix} \mathbf{P}_0^0 \\ \mathbf{P}_1^0 \\ \mathbf{P}_2^0 \end{pmatrix}, \quad \begin{pmatrix} \mathbf{P}_2^1 \\ \mathbf{P}_3^1 \\ \mathbf{P}_4^1 \end{pmatrix} = \mathbf{A}\begin{pmatrix} \mathbf{P}_1^0 \\ \mathbf{P}_2^0 \\ \mathbf{P}_3^0 \end{pmatrix}, \quad \text{where} \quad \mathbf{A} = \frac{1}{8}\begin{pmatrix} 4 & 4 & 0 \\ 1 & 6 & 1 \\ 0 & 4 & 4 \end{pmatrix},$$

or, using compact notation $\mathbf{P}_{012}^1 = \mathbf{A}\mathbf{P}_{012}^0$ and $\mathbf{P}_{234}^1 = \mathbf{A}\mathbf{P}_{123}^0$. In general $\mathbf{P}_{i\,i+1\,i+2}^1 = \mathbf{A}\mathbf{P}_{j\,j+1\,j+2}^0$ for even values of i and for $j = i, i-1$.

For $k = 2$, the computation of the seven points $\mathbf{P}_i^2$ can be summarized by the three overlapping triplets $\mathbf{P}_{012}^2 = \mathbf{A}\mathbf{P}_{012}^1$, $\mathbf{P}_{234}^2 = \mathbf{A}\mathbf{P}_{123}^1$, and $\mathbf{P}_{456}^2 = \mathbf{A}\mathbf{P}_{234}^1$, or in general $\mathbf{P}_{i\,i+1\,i+2}^2 = \mathbf{A}\mathbf{P}_{j\,j+1\,j+2}^1$, for even values of i and for $j = i$, $i-1$, and $i-2$. For $k = 3$, the calculation of the 11 points $\mathbf{P}_i^3$ is summarized by the five triplets $\mathbf{P}_{i\,i+1\,i+2}^3 = \mathbf{A}\mathbf{P}_{j\,j+1\,j+2}^2$, where i is even and $j = i, i-1, i-2$, and $i-3$. In general, the computation of the $3 + 2^k$ points of step k can be summarized by the $2^{k-1} + 1$ triplets $\mathbf{P}_{i\,i+1\,i+2}^k = \mathbf{A}\mathbf{P}_{j\,j+1\,j+2}^{k-1}$ where i is even and j goes through the values i, $i-1$ and so on, down to $i - (2^{k-1} - 1)$.

$\diamond$ **Exercise 8.5:** Write each of the nine triplets $\mathbf{P}_{i\,i+1\,i+2}^4$ (for even values of i) in terms of a triplet $\mathbf{P}_{j\,j+1\,j+2}^3$.

Because of the repeated use of matrix $\mathbf{A}$, most triplets produced in step k can be expressed in terms of triplets produced in earlier steps. For example, the trio of points $\mathbf{P}_{012}^3$ can be written as $\mathbf{A}\mathbf{P}_{012}^2 = \mathbf{A}^2\mathbf{P}_{012}^1 = \mathbf{A}^3\mathbf{P}_{012}^0$, the triplet $\mathbf{P}_{234}^3$ equals $\mathbf{A}\mathbf{P}_{123}^2$, and $\mathbf{P}_{456}^3$ can be written as $\mathbf{A}\mathbf{P}_{234}^2 = \mathbf{A}^2\mathbf{P}_{123}^1$. (Note that for the triplet on the left-hand side, the first subscript is always even, but the first subscript of the triplet on the right can be even or odd.) These relations point the way to moving forward from an earlier triplet to a later one. If we start, say, with the triplet $\mathbf{P}_{123}^1$, we can easily compute the triplets $\mathbf{P}_{234}^2$, $\mathbf{P}_{456}^3$, $\mathbf{P}_{89\,10}^4$, $\mathbf{P}_{16\,17\,18}^5$, and so on by multiplying the three points $\mathbf{P}_{123}^1$ by powers of $\mathbf{A}$. We can use this method to leapfrog across many recursion steps and proceed, in one step, from any triplet $\mathbf{P}_{i\,i+1\,i+2}^k$ to a triplet many subdivision steps later! In the limit, this can be written $\lim_{k\to\infty}\mathbf{P}_{i\,i+1\,i+2}^k = \mathbf{A}^\infty\mathbf{P}_{i\,i+1\,i+2}^k$, where $\mathbf{A}^\infty$ denotes $\lim_{k\to\infty}\mathbf{A}^k$. Any triplet $\mathbf{P}_{i\,i+1\,i+2}^k$ is an approximation to the ideal B-spline curve, but the limit $\lim_{k\to\infty}\mathbf{P}_{i\,i+1\,i+2}^k$ converges to a point on the actual curve.

The problem is therefore to calculate the limit of $\mathbf{A}^k$ as k approaches infinity, and this can easily be done with the help of the following theorem (see any text on matrices or linear algebra for the proof and for more information on eigenvalues and eigenvectors):

Theorem: If $\mathbf{A}$ is an $n \times n$ matrix for which there exist n linearly independent eigenvectors, then $\mathbf{A}$ has the form $\mathbf{Q}\Lambda\mathbf{Q}^{-1}$, where $\mathbf{Q}$ is the matrix whose columns are the n eigenvectors and Λ is the diagonal matrix whose diagonal elements are the eigenvalues of $\mathbf{A}$.

This theorem implies that $\mathbf{A}^2 = \mathbf{Q}\Lambda\mathbf{Q}^{-1}\mathbf{Q}\Lambda\mathbf{Q}^{-1} = \mathbf{Q}\Lambda^2\mathbf{Q}^{-1}$, and in general $\mathbf{A}^k = \mathbf{Q}\Lambda^k\mathbf{Q}^{-1}$. Following this theorem, we can write our matrix $\mathbf{A}$ (after its eigenvalues and a set of linearly independent eigenvectors have been computed with appropriate software) as

$$\mathbf{A} = \begin{pmatrix} 1 & -1 & 1 \\ -1/2 & 0 & 1 \\ 1 & 1 & 1 \end{pmatrix}\begin{pmatrix} 1/4 & 0 & 0 \\ 0 & 1/2 & 0 \\ 0 & 0 & 1 \end{pmatrix}\begin{pmatrix} 1/3 & -2/3 & 1/3 \\ -1/2 & 0 & 1/2 \\ 1/6 & 2/3 & 1/6 \end{pmatrix}.$$

Since matrix Λ is diagonal, we have

$$\lim_{k\to\infty} \Lambda^k = \lim_{k\to\infty} \begin{pmatrix} (1/4)^k & 0 & 0 \\ 0 & (1/2)^k & 0 \\ 0 & 0 & 1^k \end{pmatrix} = \begin{pmatrix} 0 & 0 & 0 \\ 0 & 0 & 0 \\ 0 & 0 & 1 \end{pmatrix}.$$

The limit $\mathbf{A}^\infty$ is therefore

$$\begin{pmatrix} 1 & -1 & 1 \\ -1/2 & 0 & 1 \\ 1 & 1 & 1 \end{pmatrix} \begin{pmatrix} 0 & 0 & 0 \\ 0 & 0 & 0 \\ 0 & 0 & 1 \end{pmatrix} \begin{pmatrix} 1/3 & -2/3 & 1/3 \\ -1/2 & 0 & 1/2 \\ 1/6 & 2/3 & 1/6 \end{pmatrix} = \frac{1}{6}\begin{pmatrix} 1 & 4 & 1 \\ 1 & 4 & 1 \\ 1 & 4 & 1 \end{pmatrix},$$

so we end up with the limits

$$\lim_{k\to\infty} \mathbf{P}^k_{i\,i+1\,i+2} = \frac{1}{6}\begin{pmatrix} 1 & 4 & 1 \\ 1 & 4 & 1 \\ 1 & 4 & 1 \end{pmatrix}\begin{pmatrix} \mathbf{P}^k_i \\ \mathbf{P}^k_{i+1} \\ \mathbf{P}^k_{i+2} \end{pmatrix} \overset{\text{def}}{=} \frac{1}{6}(1,4,1)\begin{pmatrix} \mathbf{P}^k_i \\ \mathbf{P}^k_{i+1} \\ \mathbf{P}^k_{i+2} \end{pmatrix}$$
$$= \frac{1}{6}(\mathbf{P}^k_i + 4\mathbf{P}^k_{i+1} + \mathbf{P}^k_{i+2}),$$

where k is any nonnegative integer. Notice that the three points of the triplet converge to the same point on the B-spline curve.

To summarize, we can (1) select four control points $\mathbf{P}^0_{0123}$, (2) select a value k and perform k refinement steps, (3) select a value i and a triplet $\mathbf{P}^k_{i\,i+1\,i+2}$, and (4) compute $(\mathbf{P}^k_i + 4\mathbf{P}^k_{i+1} + \mathbf{P}^k_{i+2})/6$. This will be a point on the cubic B-spline curve segment defined by the four original control points. To show that this is so, we can express each of the three points $\mathbf{P}^k_{i\,i+1\,i+2}$ in terms of the original control points $\mathbf{P}^0_{0123}$, and compare the result with the general cubic B-spline segment, Equation (7.11). Here are some examples.

Example: (1) We start with $k = 0$ and $i = 0$. The initial triplet is therefore $\mathbf{P}^0_{012}$.

$$\lim_{k\to\infty} \mathbf{P}^0_{012} = \frac{1}{6}(1,4,1)\begin{pmatrix} \mathbf{P}^0_0 \\ \mathbf{P}^0_1 \\ \mathbf{P}^0_2 \end{pmatrix} = \frac{1}{6}(\mathbf{P}^0_0 + 4\mathbf{P}^0_1 + \mathbf{P}^0_2),$$

which is the initial point $\mathbf{P}(0)$ of the B-spline segment, as can be seen from Equation (7.11).

Example: (2) The values $k = 0$ and $i = 1$ specify the triplet $\mathbf{P}^0_{123}$ (notice that i does not have to be even).

$$\lim_{k\to\infty} \mathbf{P}^0_{123} = \frac{1}{6}(1,4,1)\begin{pmatrix} \mathbf{P}^0_1 \\ \mathbf{P}^0_2 \\ \mathbf{P}^0_3 \end{pmatrix} = \frac{1}{6}(\mathbf{P}^0_1 + 4\mathbf{P}^0_2 + \mathbf{P}^0_3),$$

which is the final point $\mathbf{P}(1)$ of the B-spline segment, as can be seen from the same equation.

Example: (3) We perform one refinement step and select the triplet $\mathbf{P}^1_{123}$ specified by $k = 1$ and $i = 1$. When this triplet is expressed in terms of the control points $\mathbf{P}^0_i$, the result is

$$\lim_{k \to \infty} \mathbf{P}^1_{123} = \frac{1}{6}(\mathbf{P}^1_1 + 4\mathbf{P}^1_2 + \mathbf{P}^1_3)$$

$$= \frac{1}{6}\left(\frac{1}{8}(\mathbf{P}^0_0 + 6\mathbf{P}^0_1 + \mathbf{P}^0_2) + \frac{4}{2}(\mathbf{P}^0_1 + \mathbf{P}^0_2) + \frac{1}{8}(\mathbf{P}^0_1 + 6\mathbf{P}^0_2 + \mathbf{P}^0_3)\right)$$

$$= \frac{1}{48}(\mathbf{P}^0_0 + 23\mathbf{P}^0_1 + 23\mathbf{P}^0_2 + \mathbf{P}^0_3).$$

Equation (7.11) tells us that this is the midpoint $\mathbf{P}(1/2)$ of the curve segment.

⋄ **Exercise 8.6:** Select $k = 3$ and $i = 6$ and compute the point on the cubic B-spline curve segment obtained from these values at the limit of subdivision.

8.5 Biquadratic B-Spline Surface by Subdivision

The method of subdivision has been introduced in Section 8.2, where Chaikin's algorithm for curves is discussed. Generating the quadratic B-spline curve by subdivision is described in Section 8.3. This material should be reviewed before reading ahead. The technique of subdivision can be extended to surfaces that are defined by a mesh of control points. We use the biquadratic B-spline surface patch as an example. Such a patch is constructed by Equation (7.48) from a grid of 3×3 control points $\mathbf{P}_{ij}$. We denote this patch by BSP and the original points by $\mathbf{P}^0_{ij}$. The principle of constructing a BSP by subdivision is to find a way to subdivide the mesh of original points into a finer mesh with more points $\mathbf{P}^1_{ij}$ and, as a result, with more subpatches. If this is done right, the new control points $\mathbf{P}^1_{ij}$ will be closer to the ideal BSP surface than the original ones. When this process is repeated, it results in more and more control points $\mathbf{P}^k_{ij}$ that get closer and closer to the BSP. At the limit, we end up with infinitely many points that lie on the surface. In practice, we stop the subdivision process after a finite number k of steps and display the surface as a wireframe by connecting points $\mathbf{P}^k_{ij}$ with straight segments.

The refinement rule for a BSP $\mathbf{P}(u, w)$ employs reparametrization to calculate four new patches $\mathbf{Q}(u, w)$. The technique of reparametrization was introduced in Section 6.10 for curves and has been extended for Bézier surface patches in Section 6.25. It can easily be modified for the biquadratic B-spline surface by rewriting Equation (6.46) in the form

$$\mathbf{Q}(u, w) = \mathbf{P}([b - a]u + a, [d - c]w + c)$$

$$= (([b - a]u + a)^2, ([b - a]u + a), 1)\mathbf{M} \cdot \mathbf{P} \cdot \mathbf{M}^{-1}\begin{pmatrix} ([d - c]w + c)^2 \\ [d - c]w + c \\ 1 \end{pmatrix}$$

$$= (u^2, u, 1)\mathbf{A}_{ab}\mathbf{M} \cdot \mathbf{P} \cdot \mathbf{M}^T \cdot \mathbf{A}^T_{cd}(w^2, w, 1)^T$$

$$= (u^2, u, 1)\mathbf{M}(\mathbf{M}^{-1} \cdot \mathbf{A}_{ab} \cdot \mathbf{M})\mathbf{P}(\mathbf{M}^T \cdot \mathbf{A}^T_{cd} \cdot (\mathbf{M}^T)^{-1})\mathbf{M}^T(w^2, w, 1)^T$$

$$= (u^2, u, 1)\mathbf{M} \cdot \mathbf{B}_{ab} \cdot \mathbf{P} \cdot \mathbf{B}_{cd}^T \cdot \mathbf{M}^T (w^2, w, 1)^T$$
$$= (u^2, u, 1)\mathbf{M} \cdot \mathbf{Q} \cdot \mathbf{M}^T (w^2, w, 1)^T,$$

where

$$\mathbf{M} = \frac{1}{2} \begin{pmatrix} 1 & -2 & 1 \\ -2 & 2 & 0 \\ 1 & 1 & 0 \end{pmatrix}, \quad \mathbf{A}_{ab} = \begin{pmatrix} (b-a)^2 & 0 & 0 \\ 2a(b-a) & b-a & 0 \\ a^2 & a & 1 \end{pmatrix}$$

($\mathbf{M}$ is the basis matrix for the biquadratic B-spline surface),

$$\mathbf{P} = \begin{pmatrix} \mathbf{P}_{00} & \mathbf{P}_{01} & \mathbf{P}_{02} \\ \mathbf{P}_{10} & \mathbf{P}_{11} & \mathbf{P}_{12} \\ \mathbf{P}_{20} & \mathbf{P}_{21} & \mathbf{P}_{22} \end{pmatrix},$$

$$\mathbf{B}_{ab} = \mathbf{M}^{-1} \cdot \mathbf{A}_{ab} \cdot \mathbf{M}$$
$$= \begin{pmatrix} ((1-a)(1-2a+b))/2 & (1+3a-4a^2-b+2ab)/2 & a^2-(ab)/2 \\ 1/2-a/2-b/2+(ab)/2 & (1+a+b-2ab)/2 & (ab)/2 \\ ((1+a-2b)(1-b))/2 & (1-a+3b+2ab-4b^2)/2 & -(ab)/2+b^2 \end{pmatrix},$$

$$\mathbf{B}_{cd}^T = \mathbf{M}^T \cdot \mathbf{A}_{cd}^T \cdot (\mathbf{M}^T)^{-1}$$
$$= \begin{pmatrix} ((1-c)(1-2c+d))/2 & 1/2-c/2-d/2+(cd)/2 & ((1+c-2d)(1-d))/2 \\ (1+3c-4c^2-d+2cd)/2 & (1+c+d-2cd)/2 & (1-c+3d+2cd-4d^2)/2 \\ c^2-(cd)/2 & (cd)/2 & -(cd)/2+d^2 \end{pmatrix},$$

and

$$\mathbf{Q} = \mathbf{B}_{ab} \cdot \mathbf{P} \cdot \mathbf{B}_{cd}^T. \qquad (8.5)$$

The elements of $\mathbf{Q}$ depend on the four parameters a, b, c, and d, and on the $\mathbf{P}_{ij}$'s. Once the four parameters are known, matrix $\mathbf{Q}$ is easy to calculate symbolically with appropriate mathematical software.

The rule for subdividing a biquadratic B-spline surface patch $\mathbf{P}(u, w)$ is as follows: Call the original surface patch the subdivision step-0 surface. Use reparametrization to calculate the four step-1 surface patches defined by the following sets of parameters:

$$a = 0, \ b = 0.5, \ c = 0, \ d = 0.5, \quad a = 0.5, \ b = 1, \ c = 0, \ d = 0.5,$$
$$a = 0, \ b = 0.5, \ c = 0.5, \ d = 1, \quad a = 0.5, \ b = 1, \ c = 0.5, \ d = 1.$$

The basic idea is shown in idealized form in Figure 8.11. Each of the four new step-1 patches is defined by nine points, but some of the new points are identical, so the four patches are fully defined by 16 points $\mathbf{P}_{ij}^1$ for i and j from 00 to 33. The first of the four patches (Figure 8.11a) is constructed by setting $a = 0$, $b = 0.5$, $c = 0$, and $d = 0.5$ (it is a reparametrization of the "upper left" quadrant of the original, step-0 surface patch) and then applying Equation (8.5). The resulting nine control points $\mathbf{P}_{ij}^1$ are (from the code of Figure 8.9)

$$\mathbf{P}_{00}^1 = \frac{1}{16}(9\mathbf{P}_{00}^0 + 3\mathbf{P}_{10}^0 + 3\mathbf{P}_{01}^0 + \mathbf{P}_{11}^0), \quad \mathbf{P}_{01}^1 = \frac{1}{16}(3\mathbf{P}_{00}^0 + \mathbf{P}_{10}^0 + 9\mathbf{P}_{01}^0 + 3\mathbf{P}_{11}^0),$$

$$\mathbf{P}_{02}^1 = \frac{1}{16}(9\mathbf{P}_{01}^0 + 3\mathbf{P}_{11}^0 + 3\mathbf{P}_{02}^0 + \mathbf{P}_{12}^0), \quad \mathbf{P}_{10}^1 = \frac{1}{16}(3\mathbf{P}_{00}^0 + 9\mathbf{P}_{10}^0 + \mathbf{P}_{01}^0 + 3\mathbf{P}_{11}^0),$$

$$\mathbf{P}_{11}^1 = \frac{1}{16}(\mathbf{P}_{00}^0 + 3\mathbf{P}_{10}^0 + 3\mathbf{P}_{01}^0 + 9\mathbf{P}_{11}^0), \quad \mathbf{P}_{12}^1 = \frac{1}{16}(3\mathbf{P}_{01}^0 + 9\mathbf{P}_{11}^0 + \mathbf{P}_{02}^0 + 3\mathbf{P}_{12}^0),$$

$$\mathbf{P}_{20}^1 = \frac{1}{16}(9\mathbf{P}_{10}^0 + 3\mathbf{P}_{20}^0 + 3\mathbf{P}_{11}^0 + \mathbf{P}_{21}^0), \quad \mathbf{P}_{21}^1 = \frac{1}{16}(3\mathbf{P}_{10}^0 + \mathbf{P}_{20}^0 + 9\mathbf{P}_{11}^0 + 3\mathbf{P}_{21}^0),$$

$$\mathbf{P}_{22}^1 = \frac{1}{16}(9\mathbf{P}_{11}^0 + 3\mathbf{P}_{21}^0 + 3\mathbf{P}_{12}^0 + \mathbf{P}_{22}^0). \tag{8.6}$$

```
(* reparametrize biquadratic B-spline surface *)
Clear[a,b,c,d,A,B,TB,H,M,P,Q];
M={{1,-2,1},{-2,2,0},{1,1,0}}/2;
A={{(b-a)^2,0,0},{2a(b-a),b-a,0},{a^2,a,1}};
(* B=MatrixForm[Simplify[Inverse[M].A.M]] *)
B={{((1 - a)*(1 - 2*a + b))/2, (1 + 3*a - 4*a^2 - b + 2*a*b)/2,
  a^2 - (a*b)/2}, {1/2 - a/2 - b/2 + (a*b)/2, (1 + a + b - 2*a*b)/2,
  (a*b)/2}, {((1 + a - 2*b)*(1 - b))/2, (1 - a + 3*b + 2*a*b - 4*b^2)/2,
  -(a*b)/2 + b^2}};
TB={{((1 - c)*(1 - 2*c + d))/2, 1/2 - c/2 - d/2 + (c*d)/2,
  ((1 + c - 2*d)*(1 - d))/2},
  {(1 + 3*c - 4*c^2 - d + 2*c*d)/2, (1 + c + d - 2*c*d)/2,
  (1 - c + 3*d + 2*c*d - 4*d^2)/2},
  {c^2 - (c*d)/2, (c*d)/2, -(c*d)/2 + d^2}};
P={{P00,P01,P02},{P10,P11,P12},{P20,P21,P22}};
Q=Simplify[B.P.TB]
a=0; b=.5; c=0; d=.5; Q
```

Figure 8.9: Code for the Nine Control Points of the "Upper-Left" Patch.

These points can be interpreted geometrically in two ways as follows:

1. The original surface patch has four faces and each new control point is located on one of these faces. Such a point is a weighted sum (with weights 9/16, 3/16, 3/16, and 1/16) of the four points on its face. Figure 8.10 shows the four possible weight patterns.

2. Take the two points $\mathbf{P}_{00}^0$ and $\mathbf{P}_{10}^0$ and add them with weights 3/4 and 1/4. Do the same with $\mathbf{P}_{01}^0$ and $\mathbf{P}_{11}^0$. Add the two results also with weights 3/4 and 1/4, and call the resulting point $\mathbf{P}_{00}^1$. Each new point is therefore the sum of two quantities, each a sum of two points on the same edge, where all the sums use weights of 3/4 and 1/4. Recall that these are the weights used by the original Chaikin's algorithm.

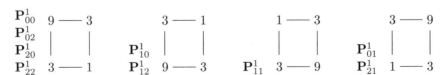

Figure 8.10: The Four Weight Patterns.

Using each of the other three sets of parameters to reparametrize the surface results in three more sets of nine more points each, only some of which are new. Figure 8.11b,c,d

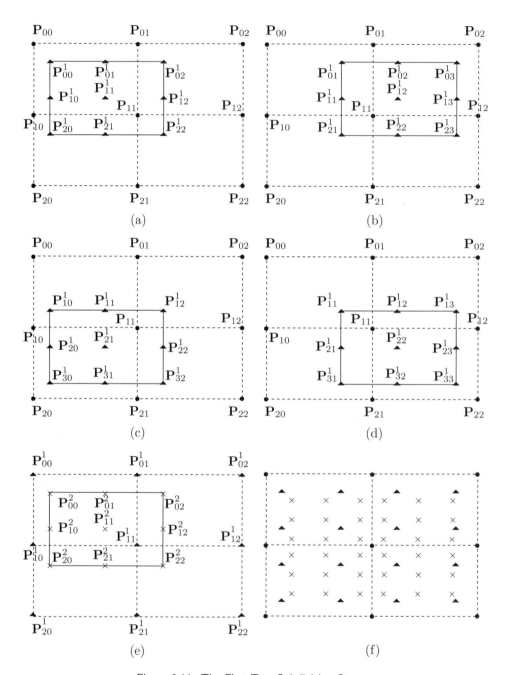

Figure 8.11: The First Two Subdivision Steps.

shows the points (as small triangles) for the sets

$$a = 0.5, \ b = 1, \ c = 0, \ d = 0.5, \quad \text{part (b)},$$
$$a = 0, \ b = 0.5, \ c = 0.5, \ d = 1, \quad \text{part (c)},$$
$$a = 0.5, \ b = 1, \ c = 0.5, \ d = 1, \quad \text{part (d)}.$$

The total number of points $\mathbf{P}_{ij}^1$ is $9 + 3 + 3 + 1 = 16$, enough for four new (step 1) biquadratic patches based on nine points each. We can either display the four surface patches or proceed to step 2.

In step 2 of the subdivision, the new mesh of 16 points is used to calculate $4 \times 9 = 36$ points $\mathbf{P}_{ij}^2$. Figure 8.11e shows nine of them, marked as $\times$, and Figure 8.11f shows all 36, enough points for $4 \times 4 = 16$ step-2 biquadratic patches based on nine points each.

This subdivision process is repeated several times, resulting in more and more points. When enough points have been obtained, the surface can be generated by connecting the points with short straight segments. It becomes a polygonal surface made of four-sided polygons (quadrilaterals).

An examination of all the parts of Figure 8.11 seems to suggest that the subdivision process produces smaller and smaller meshes of control points, thereby generating smaller and smaller surface patches. It is easy to show that this is not so. Let $\mathbf{Q}(u, w)$ denote the reparametrization of the "upper left" quadrant of the original surface patch $\mathbf{P}(u, w)$. These two patches are based on different meshes of control points, but we show that they have the same "upper left" corner point, i.e., $\mathbf{Q}(0, 0) = \mathbf{P}(0, 0)$. The corner point $\mathbf{P}(0, 0)$ of a general biquadratic B-spline surface patch $\mathbf{P}(u, w)$ is shown by Equation (7.49) to be

$$\mathbf{P}(0, 0) = \frac{1}{4}(\mathbf{P}_{00} + \mathbf{P}_{01} + \mathbf{P}_{10} + \mathbf{P}_{11}).$$

The corner point $\mathbf{Q}(0, 0)$ is therefore

$$
\begin{aligned}
\mathbf{Q}(0, 0) &= \frac{1}{4}(\mathbf{P}_{00}^1 + \mathbf{P}_{01}^1 + \mathbf{P}_{10}^1 + \mathbf{P}_{11}^1) \\
&= \frac{1}{4 \cdot 16}\big[(9\mathbf{P}_{00}^0 + 3\mathbf{P}_{10}^0 + 3\mathbf{P}_{01}^0 + \mathbf{P}_{11}^0) + (3\mathbf{P}_{00}^0 + \mathbf{P}_{10}^0 + 9\mathbf{P}_{01}^0 + 3\mathbf{P}_{11}^0) \\
&\quad + (3\mathbf{P}_{00}^0 + 9\mathbf{P}_{10}^0 + \mathbf{P}_{01}^0 + 3\mathbf{P}_{11}^0) + (\mathbf{P}_{00}^0 + 3\mathbf{P}_{10}^0 + 3\mathbf{P}_{01}^0 + 9\mathbf{P}_{11}^0) \\
&= \frac{1}{4}(\mathbf{P}_{00}^0 + \mathbf{P}_{01}^0 + \mathbf{P}_{10}^0 + \mathbf{P}_{11}^0)\big] \\
&= \mathbf{P}(0, 0).
\end{aligned}
$$

It turns out that even though consecutive steps of the subdivision process result in smaller meshes, those meshes converge to a limit and don't shrink indefinitely.

8.6 Bicubic B-Spline Surface by Subdivision

The technique used in this section to subdivide a bicubic B-spline surface patch is similar to the one used in Section 8.5 to subdivide the biquadratic B-spline surface.

The rule for subdividing a bicubic B-spline surface patch $\mathbf{P}(u, w)$ uses reparametrization to calculate four new patches $\mathbf{Q}(u, w)$. This is done by rewriting Equation (6.46) in the form

$$
\begin{aligned}
\mathbf{Q}&(u, w) \\
&= \mathbf{P}([b - a]u + a, [d - c]w + c) \\
&= \left(([b-a]u+a)^3, ([b-a]u+a)^2, ([b-a]u+a), 1\right)\mathbf{M} \cdot \mathbf{P} \cdot \mathbf{M}^{-1}\begin{pmatrix} ([d-c]w+c)^3 \\ ([d-c]w+c)^2 \\ [d-c]w+c \\ 1 \end{pmatrix} \\
&= (u^3, u^2, u, 1)\mathbf{A}_{ab}\mathbf{M} \cdot \mathbf{P} \cdot \mathbf{M}^T \cdot \mathbf{A}_{cd}^T(w^3, w^2, w, 1)^T \\
&= (u^3, u^2, u, 1)\mathbf{M}(\mathbf{M}^{-1} \cdot \mathbf{A}_{ab} \cdot \mathbf{M})\mathbf{P}(\mathbf{M}^T \cdot \mathbf{A}_{cd}^T \cdot (\mathbf{M}^T)^{-1})\mathbf{M}^T(w^3, w^2, w, 1)^T \\
&= (u^3, u^2, u, 1)\mathbf{M} \cdot \mathbf{B}_{ab} \cdot \mathbf{P} \cdot \mathbf{B}_{cd}^T \cdot \mathbf{M}^T(w^3, w^2, w, 1)^T \\
&= (u^3, u^2, u, 1)\mathbf{M} \cdot \mathbf{Q} \cdot \mathbf{M}^T(w^3, w^2, w, 1)^T,
\end{aligned} \tag{8.7}
$$

where

$$
\mathbf{M} = \frac{1}{6}\begin{pmatrix} -1 & 3 & -3 & 1 \\ 3 & -6 & 3 & 0 \\ -3 & 0 & 3 & 0 \\ 1 & 4 & 1 & 0 \end{pmatrix}, \quad \mathbf{A}_{ab} = \begin{pmatrix} (b-a)^2 & 0 & 0 \\ 2a(b-a) & b-a & 0 \\ a^2 & a & 1 \end{pmatrix}
$$

($\mathbf{M}$ is the basis matrix for the bicubic B-spline surface),

$$
\mathbf{P} = \begin{pmatrix} \mathbf{P}_{00} & \mathbf{P}_{01} & \mathbf{P}_{02} & \mathbf{P}_{03} \\ \mathbf{P}_{10} & \mathbf{P}_{11} & \mathbf{P}_{12} & \mathbf{P}_{13} \\ \mathbf{P}_{20} & \mathbf{P}_{21} & \mathbf{P}_{22} & \mathbf{P}_{23} \\ \mathbf{P}_{30} & \mathbf{P}_{31} & \mathbf{P}_{32} & \mathbf{P}_{33} \end{pmatrix},
$$

$$
\begin{aligned}
\mathbf{B}_{ab} &= \mathbf{M}^{-1} \cdot \mathbf{A}_{ab} \cdot \mathbf{M} \\
&= \left(\begin{array}{cc} ((1-a)(1-5a+6a^2+3b-7ab+2b^2))/6 & (4-22a^2+18a^3+20ab-21a^2b-4b^2+6ab^2)/6 \\ ((a-1)(-1+2a-2ab+b^2))/6 & (4-4a^2-4ab+6a^2b+2b^2-3ab^2)/6 \\ ((a-1)(1+a-2b)(b-1))/6 & (4+2a^2-4ab-3a^2b-4b^2+6ab^2)/6 \\ ((1-b)(1+3a+2a^2-5b-7ab+6b^2))/6 & (4-4a^2+20ab+6a^2b-22b^2-21ab^2+18b^3)/6 \end{array}\right. \\
& \qquad\qquad \left.\begin{array}{cc} 1/6+a+(11a^2)/6-3a^3-b/2-(5ab)/3+(7a^2b)/2+b^2/3-ab^2 & a^3-(7a^2b)/6+(ab^2)/3 \\ 1/6+a/2+a^2/3+(ab)/3-a^2b-b^2/6+(ab^2)/2 & (a(2a-b)b)/6 \\ 1/6-a^2/6+b/2+(ab)/3+(a^2b)/2+b^2/3-ab^2 & (ab(-a+2b))/6 \\ 1/6-a/2+a^2/3+b-(5ab)/3-a^2b+(11b^2)/6+(7ab^2)/2-3b^3 & (a^2b)/3-(7ab^2)/6+b^3 \end{array}\right),
\end{aligned}
$$

$$
\mathbf{B}_{cd}^T = \mathbf{M}^T \cdot \mathbf{A}_{cd}^T \cdot (\mathbf{M}^T)^{-1}
$$

$$
= \begin{pmatrix}
\begin{array}{l}
((1-c)(1-5c+6c^2+3d-7cd+2d^2))/6 \\
(4-22c^2+18c^3+20cd-21c^2d-4d^2+6cd^2)/6 \\
1/6+c+(11c^2)/6-3c^3-d/2-(5cd)/3+(7c^2d)/2+d^2/3-cd^2 \\
c^3-(7c^2d)/6+(cd^2)/3
\end{array}
&
\begin{array}{l}
((-1+c)(-1+2c-2cd+d^2))/6 \\
(4-4c^2-4cd+6c^2d+2d^2-3cd^2)/6 \\
1/6+c/2+c^2/3+(cd)/3-c^2d-d^2/6+(cd^2)/2 \\
(c(2c-d)d)/6
\end{array}
\\[2em]
\begin{array}{l}
((-1+c)(1+c-2d)(-1+d))/6 \\
(4+2c^2-4cd-3c^2d-4d^2+6cd^2)/6 \\
1/6-c^2/6+d/2+(cd)/3+(c^2d)/2+d^2/3-cd^2 \\
(cd(-c+2d))/6
\end{array}
&
\begin{array}{l}
((1-d)(1+3c+2c^2-5d-7cd+6d^2))/6 \\
(4-4c^2+20cd+6c^2d-22d^2-21cd^2+18d^3)/6 \\
1/6-c/2+c^2/3+d-(5cd)/3-c^2d+(11d^2)/6+(7cd^2)/2-3d^3 \\
(c^2d)/3-(7cd^2)/6+d^3
\end{array}
\end{pmatrix},
$$

and

$$\mathbf{Q} = \mathbf{B}_{ab} \cdot \mathbf{P} \cdot \mathbf{B}_{cd}^{T}. \tag{8.8}$$

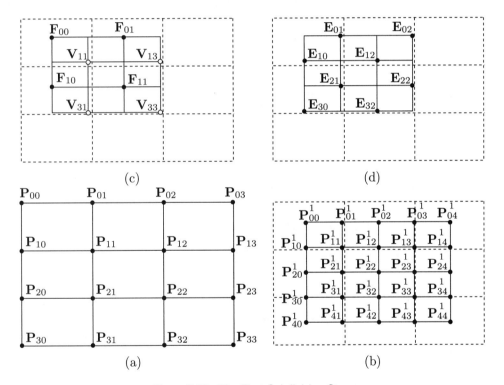

Figure 8.12: The First Subdivision Step.

The refinement rule for a bicubic B-spline patch $\mathbf{P}(u,w)$ is to use reparametrization to calculate the four surface patches defined by the following sets of parameters:

$$a = 0,\ b = 0.5,\ c = 0, d = 0.5, \quad a = 0.5,\ b = 1,\ c = 0,\ d = 0.5,$$
$$a = 0,\ b = 0.5,\ c = 0.5, d = 1, \quad a = 0.5,\ b = 1,\ c = 0.5,\ d = 1.$$

The basic idea is shown in idealized form in Figure 8.12a,b. Each of the new patches is defined by 16 points, but some of the new points are identical, so the four patches are fully defined by 25 points. The first of the four patches is constructed by setting $a = 0$, $b = 0.5$, $c = 0$, and $d = 0.5$ (this is a reparametrization of the "upper left" quadrant of the original surface patch) and applying Equation (8.8). The resulting 16 control points $\mathbf{P}_{ij}^1$ are (see Figure 8.13 for the computations)

$$\mathbf{P}_{00}^1 = \frac{1}{4}(\mathbf{P}_{00}^0 + \mathbf{P}_{10}^0 + \mathbf{P}_{01}^0 + \mathbf{P}_{11}^0),$$

$$\mathbf{P}_{01}^1 = \frac{1}{16}(\mathbf{P}_{00}^0 + \mathbf{P}_{10}^0 + 6(\mathbf{P}_{01}^0 + \mathbf{P}_{11}^0) + \mathbf{P}_{02}^0 + \mathbf{P}_{12}^0),$$

$$\mathbf{P}_{02}^1 = \frac{1}{4}(\mathbf{P}_{01}^0 + \mathbf{P}_{11}^0 + \mathbf{P}_{02}^0 + \mathbf{P}_{12}^0),$$

$$\mathbf{P}_{03}^1 = \frac{1}{16}(\mathbf{P}_{01}^0 + \mathbf{P}_{11}^0 + 6(\mathbf{P}_{02}^0 + \mathbf{P}_{12}^0) + \mathbf{P}_{03}^0 + \mathbf{P}_{13}^0),$$

$$\mathbf{P}_{01}^1 = \frac{1}{16}(\mathbf{P}_{00}^0 + \mathbf{P}_{01}^0 + 6(\mathbf{P}_{10}^0 + \mathbf{P}_{11}^0) + \mathbf{P}_{20}^0 + \mathbf{P}_{21}^0),$$

$$\mathbf{P}_{11}^1 = \frac{1}{64}(\mathbf{P}_{00}^0 + 6\mathbf{P}_{10}^0 + \mathbf{P}_{20}^0 + 6(\mathbf{P}_{01}^0 + 6\mathbf{P}_{11}^0 + \mathbf{P}_{21}^0) + \mathbf{P}_{02}^0 + 6\mathbf{P}_{12}^0 + \mathbf{P}_{22}^0),$$

$$\mathbf{P}_{12}^1 = \frac{1}{16}(\mathbf{P}_{01}^0 + \mathbf{P}_{12}^0 + 6(\mathbf{P}_{11}^0 + \mathbf{P}_{12}^0) + \mathbf{P}_{21}^0 + \mathbf{P}_{22}^0),$$

$$\mathbf{P}_{13}^1 = \frac{1}{64}(\mathbf{P}_{01}^0 + 6\mathbf{P}_{11}^0 + \mathbf{P}_{21}^0 + 6(\mathbf{P}_{02}^0 + 6\mathbf{P}_{12}^0 + \mathbf{P}_{22}^0) + \mathbf{P}_{03}^0 + 6\mathbf{P}_{13}^0 + \mathbf{P}_{23}^0),$$

$$\mathbf{P}_{20}^1 = \frac{1}{4}(\mathbf{P}_{10}^0 + \mathbf{P}_{20}^0 + \mathbf{P}_{11}^0 + \mathbf{P}_{21}^0), \qquad (8.9)$$

$$\mathbf{P}_{21}^1 = \frac{1}{16}(\mathbf{P}_{10}^0 + \mathbf{P}_{20}^0 + 6(\mathbf{P}_{11}^0 + \mathbf{P}_{21}^0) + \mathbf{P}_{12}^0 + \mathbf{P}_{22}^0),$$

$$\mathbf{P}_{22}^1 = \frac{1}{4}(\mathbf{P}_{11}^0 + \mathbf{P}_{21}^0 + \mathbf{P}_{12}^0 + \mathbf{P}_{22}^0),$$

$$\mathbf{P}_{23}^1 = \frac{1}{16}(\mathbf{P}_{11}^0 + \mathbf{P}_{21}^0 + 6(\mathbf{P}_{12}^0 + \mathbf{P}_{22}^0) + \mathbf{P}_{13}^0 + \mathbf{P}_{23}^0),$$

$$\mathbf{P}_{30}^1 = \frac{1}{16}(\mathbf{P}_{10}^0 + \mathbf{P}_{11}^0 + 6(\mathbf{P}_{20}^0 + \mathbf{P}_{21}^0) + \mathbf{P}_{30}^0 + \mathbf{P}_{31}^0),$$

$$\mathbf{P}_{31}^1 = \frac{1}{64}(\mathbf{P}_{10}^0 + 6\mathbf{P}_{20}^0 + \mathbf{P}_{30}^0 + 6(\mathbf{P}_{11}^0 + 6\mathbf{P}_{21}^0 + \mathbf{P}_{31}^0) + \mathbf{P}_{12}^0 + 6\mathbf{P}_{22}^0 + \mathbf{P}_{32}^0),$$

$$\mathbf{P}_{32}^1 = \frac{1}{16}(\mathbf{P}_{11}^0 + \mathbf{P}_{12}^0 + 6(\mathbf{P}_{21}^0 + \mathbf{P}_{22}^0) + \mathbf{P}_{31}^0 + \mathbf{P}_{32}^0),$$

$$\mathbf{P}_{33}^1 = \frac{1}{64}(\mathbf{P}_{11}^0 + 6\mathbf{P}_{21}^0 + \mathbf{P}_{31}^0 + 6(\mathbf{P}_{12}^0 + 6\mathbf{P}_{22}^0 + \mathbf{P}_{32}^0) + \mathbf{P}_{13}^0 + 6\mathbf{P}_{23}^0 + \mathbf{P}_{33}^0).$$

These points can be classified into face points, edge points, and vertex points. The four face points are (Figure 8.12c) $\mathbf{F}_{00} = \mathbf{P}_{00}^1$, $\mathbf{F}_{01} = \mathbf{P}_{02}^1$, $\mathbf{F}_{10} = \mathbf{P}_{20}^1$, and $\mathbf{F}_{11} = \mathbf{P}_{22}^1$. Each is the average of four corner points of one face of the original patch. The eight edge points are (Figure 8.12d)

$$\begin{aligned} \mathbf{E}_{01} &= \mathbf{P}_{01}^1, & \mathbf{E}_{02} &= \mathbf{P}_{03}^1, & \mathbf{E}_{10} &= \mathbf{P}_{10}^1, & \mathbf{E}_{12} &= \mathbf{P}_{12}^1, \\ \mathbf{E}_{21} &= \mathbf{P}_{21}^1, & \mathbf{E}_{22} &= \mathbf{P}_{23}^1, & \mathbf{E}_{30} &= \mathbf{P}_{30}^1, & \mathbf{E}_{32} &= \mathbf{P}_{32}^1. \end{aligned}$$

```
(* reparametrize bicubic B-spline surface *)
Clear[a,b,c,d,A,B,TB,H,M,P,Q];
M={{-1,3,-3,1},{3,-6,3,0},{-3,0,3,0},{1,4,1,0}}/6;
A={{(b-a)^3,0,0,0},{3a(b-a)^2,(b-a)^2,0,0},{3a^2(b-a),2a(b-a),b-a,0},{a^3,a^2,a,1}};
(*B=Simplify[Inverse[M].A.M] *)
B={{((1 - a)*(1 - 5*a + 6*a^2 + 3*b - 7*a*b + 2*b^2))/6,
  (4 - 22*a^2 + 18*a^3 + 20*a*b - 21*a^2*b - 4*b^2 + 6*a*b^2)/6,
  1/6 + a + (11*a^2)/6 - 3*a^3 - b/2 - (5*a*b)/3 + (7*a^2*b)/2 + b^2/3 -
  a*b^2, a^3 - (7*a^2*b)/6 + (a*b^2)/3},
 {((-1 + a)*(-1 + 2*a - 2*a*b + b^2))/6,
  (4 - 4*a^2 - 4*a*b + 6*a^2*b + 2*b^2 - 3*a*b^2)/6,
  1/6 + a/2 + a^2/3 + (a*b)/3 - a^2*b - b^2/6 + (a*b^2)/2,
  (a*(2*a - b)*b)/6}, {((-1 + a)*(1 + a - 2*b)*(-1 + b))/6,
  (4 + 2*a^2 - 4*a*b - 3*a^2*b - 4*b^2 + 6*a*b^2)/6,
  1/6 - a^2/6 + b/2 + (a*b)/3 + (a^2*b)/2 + b^2/3 - a*b^2,
  (a*b*(-a + 2*b))/6}, {((1 - b)*(1 + 3*a + 2*a^2 - 5*b - 7*a*b + 6*b^2))/
  6, (4 - 4*a^2 + 20*a*b + 6*a^2*b - 22*b^2 - 21*a*b^2 + 18*b^3)/6,
  1/6 - a/2 + a^2/3 + b - (5*a*b)/3 - a^2*b + (11*b^2)/6 + (7*a*b^2)/2 -
  3*b^3, (a^2*b)/3 - (7*a*b^2)/6 + b^3}};
TB={{((1 - a)*(1 - 5*a + 6*a^2 + 3*b - 7*a*b + 2*b^2))/6,
  ((-1 + a)*(-1 + 2*a - 2*a*b + b^2))/6,
  ((-1 + a)*(1 + a - 2*b)*(-1 + b))/6,
  ((1 - b)*(1 + 3*a + 2*a^2 - 5*b - 7*a*b + 6*b^2))/6},
 {(4 - 22*a^2 + 18*a^3 + 20*a*b - 21*a^2*b - 4*b^2 + 6*a*b^2)/6,
  (4 - 4*a^2 - 4*a*b + 6*a^2*b + 2*b^2 - 3*a*b^2)/6,
  (4 + 2*a^2 - 4*a*b - 3*a^2*b - 4*b^2 + 6*a*b^2)/6,
  (4 - 4*a^2 + 20*a*b + 6*a^2*b - 22*b^2 - 21*a*b^2 + 18*b^3)/6},
 {1/6 + a + (11*a^2)/6 - 3*a^3 - b/2 - (5*a*b)/3 + (7*a^2*b)/2 +
  b^2/3 - a*b^2, 1/6 + a/2 + a^2/3 + (a*b)/3 - a^2*b - b^2/6 +
  (a*b^2)/2, 1/6 - a^2/6 + b/2 + (a*b)/3 + (a^2*b)/2 + b^2/3 - a*b^2,
  1/6 - a/2 + a^2/3 + b - (5*a*b)/3 - a^2*b + (11*b^2)/6 + (7*a*b^2)/2 -
  3*b^3}, {a^3 - (7*a^2*b)/6 + (a*b^2)/3, (a*(2*a - b)*b)/6,
  (a*b*(-a + 2*b))/6, (a^2*b)/3 - (7*a*b^2)/6 + b^3}};
P={{P30,P31,P32,P33},{P20,P21,P22,P23},{P10,P11,P12,P13},{P00,P01,P02,P03}};
Q=Simplify[B.P.TB]
a=0; b=.5; c=0; d=.5; Q
```

Figure 8.13: Code for the 16 Control Points of the "Uper-Left" Patch.

Each is the average of two face points and the two points $\mathbf{P}_{ij}^0$ that are closest to it. The remaining four points are called vertex points. There is one vertex point for each interior vertex of the original mesh. The vertex points are shown in Figure 8.12c and they have the form $\mathbf{V} = (\mathbf{Q} + 2\mathbf{R} + \mathbf{S})/4$, where $\mathbf{S}$ is an interior vertex, $\mathbf{Q}$ is the average of the four face points located on the faces adjacent to $\mathbf{S}$, and $\mathbf{R}$ is the average of the midpoints of the four edges that meet at $\mathbf{S}$. As an example, consider the interior vertex $\mathbf{P}_{11}$ (Figure 8.14). If we denote $\mathbf{S} = \mathbf{P}_{11}$, then $\mathbf{Q}$ is the average of the four face points $\mathbf{F}_{00}$, $\mathbf{F}_{01}$, $\mathbf{F}_{10}$, and $\mathbf{F}_{11}$, and $\mathbf{R}$ is the average of the midpoints of the four edges $\mathbf{P}_{01}\mathbf{P}_{11}$, $\mathbf{P}_{10}\mathbf{P}_{11}$, $\mathbf{P}_{12}\mathbf{P}_{11}$, and $\mathbf{P}_{21}\mathbf{P}_{11}$ (the points labeled $\times$ in the figure). This interior vertex therefore corresponds to vertex point

$$\frac{1}{4}(\mathbf{Q} + 2\mathbf{R} + \mathbf{S})$$
$$= \frac{1}{4}(\mathbf{F}_{00} + \mathbf{F}_{01} + \mathbf{F}_{10} + \mathbf{F}_{11})$$

$$+ \frac{2}{4} \left(\frac{\mathbf{P}_{01} + \mathbf{P}_{11}}{2} + \frac{\mathbf{P}_{10} + \mathbf{P}_{11}}{2} + \frac{\mathbf{P}_{12} + \mathbf{P}_{11}}{2} + \frac{\mathbf{P}_{21} + \mathbf{P}_{11}}{2} \right) + \mathbf{P}_{11}$$

$$= \frac{1}{16} \left((\mathbf{P}_{00} + \mathbf{P}_{10} + \mathbf{P}_{01} + \mathbf{P}_{11}) + (\mathbf{P}_{10} + \mathbf{P}_{20} + \mathbf{P}_{11} + \mathbf{P}_{21}) \right.$$

$$\left. + (\mathbf{P}_{01} + \mathbf{P}_{11} + \mathbf{P}_{02} + \mathbf{P}_{12}) + (\mathbf{P}_{11} + \mathbf{P}_{21} + \mathbf{P}_{12} + \mathbf{P}_{22}) \right)$$

$$+ \frac{1}{4} \left(\mathbf{P}_{01} + \mathbf{P}_{10} + \mathbf{P}_{21} + \mathbf{P}_{12} + 4\mathbf{P}_{11} \right) + \mathbf{P}_{11}$$

$$= \frac{1}{16} (\mathbf{P}_{00} + 6\mathbf{P}_{10} + 6\mathbf{P}_{01} + 36\mathbf{P}_{11} + \mathbf{P}_{20} + 6\mathbf{P}_{21} + \mathbf{P}_{02} + 6\mathbf{P}_{12} + \mathbf{P}_{22})$$

$$= \mathbf{P}_{11}^1 .$$

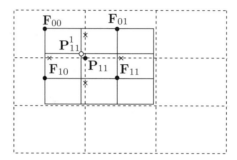

Figure 8.14: Constructing Vertex Point $\mathbf{P}_{11}^1$.

Here are the rules for calculating all 25 points $\mathbf{P}_{ij}^1$:

1. Construct one face point for each face of the original mesh. This point is the average of all the points defining the face.

2. Construct one edge point for each interior edge of the original mesh. This point is the average of the midpoint of the edge and the two face points of the faces adjacent to the edge.

3. Construct one vertex point for each interior vertex of the original mesh. This point is the average of (1) four face points, (2) four midpoints of edges, and (3) one interior vertex.

Since the original bicubic mesh consists of 9 faces, 12 interior edges, and 4 interior vertices, the first subdivision step results in 9 face points, 12 edge points, and 4 vertex points, a total of 25 points.

It should be noted that even though the mesh resulting from each subdivision is smaller than its predecessor, they don't shrink to a point but converge to a limit. All these meshes define the same bicubic B-spline surface.

⋄ **Exercise 8.7:** Equation (7.51) shows that the "top left" corner of a bicubic B-spline patch is given by

$$\mathbf{P}(0,0) = \frac{1}{36} (\mathbf{P}_{00} + \mathbf{P}_{02} + 4\mathbf{P}_{10} + 4\mathbf{P}_{12} + \mathbf{P}_{20} + 4\mathbf{P}_{01} + 16\mathbf{P}_{11} + 4\mathbf{P}_{21} + \mathbf{P}_{22}).$$

Show that this is still the same corner of the patch after one subdivision.

8.7 Polygonal Surfaces by Subdivision

Polygonal surfaces have been discussed in Section 2.2. Such a surface is normally obtained by measuring the coordinates of points on an object, either manually or with a three-dimensional digitizer. The designer then selects a set of points and the software connects those points with straight segments, resulting in a polygon. This is how the original mesh of points is converted to a set of polygons. The only condition is that the polygons be flat. The entire polygonal surface can then be shaded using Gouraud or Phong shading [Salomon 99]. If the result is not smooth enough, it can be improved by subdividing the original mesh of points, which is why the subdivision of polygonal surfaces is important.

8.8 Doo Sabin Surfaces

The method described in this section is due to Donald Doo and Malcolm Sabin [Doo and Sabin 78]. They observed that the method used in Section 8.5 to subdivide a biquadratic B-spline surface patch generates each new point $\mathbf{P}_{ij}^1$ as a weighted sum of four points: a vertex point, two edge points, and a face point. For example, Equation (8.6) gives point $\mathbf{P}_{00}^1$ as

$$\mathbf{P}_{00}^1 = \frac{1}{16}(9\mathbf{P}_{00}^0 + 3\mathbf{P}_{10}^0 + 3\mathbf{P}_{01}^0 + \mathbf{P}_{11}^0),$$

so we write it in the form

$$\mathbf{P}_{00}^1 = \frac{1}{16}(9\mathbf{P}_{00}^0 + 3\mathbf{P}_{10}^0 + 3\mathbf{P}_{01}^0 + \mathbf{P}_{11}^0)$$

$$= \frac{1}{16}\left(4\mathbf{P}_{00}^0 + 2(\mathbf{P}_{00}^0 + \mathbf{P}_{01}^0) + 2(\mathbf{P}_{00}^0 + \mathbf{P}_{10}^0) + (\mathbf{P}_{00}^0 + \mathbf{P}_{01}^0 + \mathbf{P}_{10}^0 + \mathbf{P}_{11}^0)\right)$$

$$= \frac{1}{4}\left(4\mathbf{P}_{00}^0 + (\mathbf{P}_{00}^0 + \mathbf{P}_{01}^0)/2 + (\mathbf{P}_{00}^0 + \mathbf{P}_{10}^0)/2 + (\mathbf{P}_{00}^0 + \mathbf{P}_{01}^0 + \mathbf{P}_{10}^0 + \mathbf{P}_{11}^0)/4\right)$$

$$= \frac{1}{4}\left(4\mathbf{V} + \mathbf{E}_1 + \mathbf{E}_2 + \mathbf{F}\right),$$

where $\mathbf{V}$ is the vertex point $\mathbf{P}_{00}^0$, $\mathbf{E}_1$ is the average of $\mathbf{P}_{00}^0$ and $\mathbf{P}_{01}^0$ (i.e., it is located midway between them), $\mathbf{E}_2$ is the average of $\mathbf{P}_{00}^0$ and $\mathbf{P}_{10}^0$, and $\mathbf{F}$ is the average of the four corners of the polygon being subdivided.

The idea of Doo and Sabin is to subdivide a mesh of points that consists of any polygons, not just quadrilaterals, by performing the following steps:

1. Consider a vertex $\mathbf{P}_i^0$ on the original mesh (Figure 8.15). It is located on a certain face F (and perhaps on other faces as well) and is at the intersection of two edges, $E1$ and $E2$ (two of the edges that form F). Create a new point $\mathbf{P}_i^1$ as a weighted average of $\mathbf{P}_i^0$, the two edge points adjacent to $\mathbf{P}_i^0$ (i.e., the center points of $E1$ and $E2$), and the face point that's the average of all the vertices forming F. Repeat this for every vertex $\mathbf{P}_i^0$. See Figure 8.16a for an example.

2. Consider face F again. It now contains some new points $\mathbf{P}_i^1$. Connect them so that they form a new polygon. This polygon will become a face in the new, refined surface. Repeat for all faces F (Figure 8.16b).

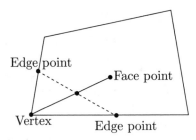

Figure 8.15: Edge and Face Points.

3. Consider again a vertex $\mathbf{P}_i^0$ on the original mesh. Such a vertex is normally common to several faces. For each of those faces, find the new point that's nearest $\mathbf{P}_i^0$. Connect those points to each other to form a new polygon. This polygon will also become a face in the new, subdivided surface. Repeat for all vertices $\mathbf{P}_i^0$ (Figure 8.16c).

4. Consider an edge of the original mesh of points. There will normally be two faces adjacent to this edge and they will have new points $\mathbf{P}_i^1$. Connect the new points around the edge to form a new polygon. This polygon will also become a face in the new, subdivided surface. Repeat this step for all edges (Figure 8.16d).

Notice that the new mesh may contain all kinds of polygons, not just triangles or quadrilaterals.

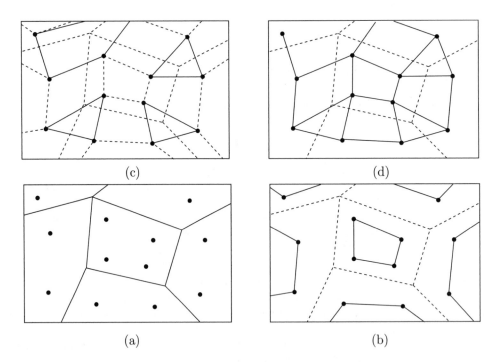

Figure 8.16: The First Doo–Sabin Subdivision Step.

8.9 Catmull–Clark Surfaces

The method described here is due to Edwin Catmull and Jim Clark [Catmull and Clark 78] and is an extension of the method of Section 8.6 to arbitrary polygonal surfaces. We have seen that subdividing a bicubic B-spline surface patch generates each new point $\mathbf{P}_{ij}^1$ as either a face point, an edge point, or a vertex point. A Catmull–Clark surface patch starts with an arbitrary polygonal surface and subdivides it by generating new face, edge, and vertex points and connecting them in a simple way. The rules for generating the points are the following:

1. A face point is calculated for each face of the original mesh. The point is simply the average of all the points that bound the face.

2. An edge point is created for each interior edge of the polygonal surface. The point is the average of the midpoint of the edge and of the two face points on both sides of the edge.

3. A vertex point is generated for each interior vertex $\mathbf{P}$ of the original mesh. The point is the average of $\mathbf{Q}$, $2\mathbf{R}$, and $\mathbf{S}(n-3)/4$, where $\mathbf{Q}$ is the average of the face points on all the faces adjacent to $\mathbf{P}$, $\mathbf{R}$ is the average of the midpoints of all the edges incident on $\mathbf{P}$, and $\mathbf{S}$ is simply $\mathbf{P}$ itself.

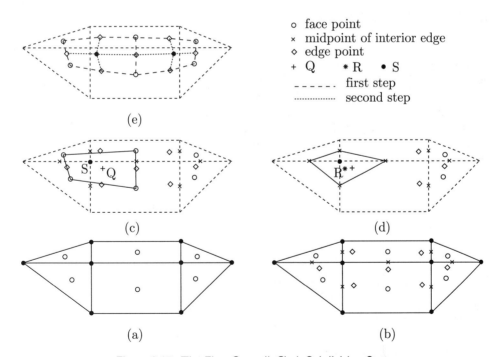

Figure 8.17: The First Catmull–Clark Subdivision Step.

Figure 8.17 shows an example. We start with a mesh of eight vertices defining six polygons, two rectangles and four triangles (notice that the polygons may have any number of sides, not just three or four). This surface has six faces, seven interior edges, and two interior vertices. The six new face points are shown in Figure 8.17a as small circles. Each is the average of the points bounding its face. Figure 8.17b shows the midpoints of the edges as small ×'s and the seven new edge points as diamonds. In Figure 8.17c, we select one of the two interior vertices as **S**, temporarily connect the four face points surrounding it (just to identify them), and calculate **Q** (shown as a small "+") as their average. In Figure 8.17d, we show how **R** (the asterisk) is computed as the average of four midpoints of edges (temporarily connected).

After the new points have been generated, they are connected according to the following rules:

1. Each face point is connected to all the edge points of the interior edges bounding its face. These are shown as long dashes in Figure 8.17e.

2. Each new vertex point is connected to all the edge points that were used in calculating it. These lines are shown as short dashes in Figure 8.17e.

Notice that even though the original polygonal mesh may have polygons with any number of sides, the new, subdivided mesh will consist of quadrilaterals (four-sided polygons) only.

8.10 Loop Surfaces

The Loop subdivision scheme subdivides a triangular mesh surface by performing several iterations, yet the term Loop refers to its developer, Charles Teorell Loop. In his MS thesis [Loop 87] Loop developed an algorithm that subdivides each triangle into four smaller triangles (Figure 8.18a). This is now referred to as a binary Loop subdivision. In [Loop 02] he extended this algorithm to subdivide each triangle into nine smaller ones (Figure 8.18b). The extended algorithm is termed Loop ternary subdivision. Notice that the polygons that make up the surface must be triangles, they cannot be arbitrary flat polygons.

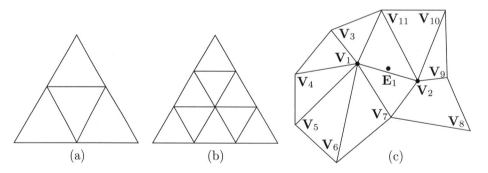

Figure 8.18: Binary and Ternary Loop Triangle Subdivisions.

The binary Loop algorithm starts with a set of points that are the vertices of triangles. Each iteration computes a new set of edge and vertex points that become the vertices of the new, smaller triangles. Specifically, A new edge point is computed for each edge and a new vertex point is computed for each vertex of the triangular mesh. If the original mesh has E edges and V vertices, the new mesh will have $E + V$ vertex points (and a number of edges that depends on the complexity of the original mesh). The new points become the vertices of the new, finer mesh, and more iterations may be applied to refine the mesh as much as needed.

To understand the rule for generating an edge point, consider the edge between vertices $\mathbf{V}_1$ and $\mathbf{V}_2$ of Figure 8.18c. This edge, like any other edge, connects two vertices. Like most edges, it is shared by two triangles. The new edge point $\mathbf{E}_1$ for this edge is constructed as the weighted sum $\frac{3}{8}(\mathbf{V}_1 + \mathbf{V}_2) + \frac{1}{8}(\mathbf{V}_{11} + \mathbf{V}_7)$. The two vertices connected by the edge are given the large weights 3/8, whereas the two edges of the triangles sharing the edge are assigned the small weights 1/8. The weights are barycentric. If the edge is on the boundary of the surface and is part of only one triangle, the edge point is computed as the average of the two vertices connected by the edge. Thus, if the edge connected by $\mathbf{V}_3$ and $\mathbf{V}_{11}$ is on the boundary (i.e., there is no triangle "above" it), then the new edge point for this edge is the average $(\mathbf{V}_3 + \mathbf{V}_{11})/2$ and is located on the edge.

Notice that even though $\mathbf{E}_1$ is called an edge point, it does not have to be located on an edge, and it becomes a vertex, not an edge, in the new, finer triangular mesh constructed after the iteration.

Similarly, a new vertex point is also constructed as a weighted sum. The new vertex for $\mathbf{V}_1$, for example, is computed as the sum $\frac{5}{8}\mathbf{V}_1 + \frac{3}{8}\mathbf{Q}_1$ where $\mathbf{Q}_1$ is the average of the vertices of all the triangles sharing $\mathbf{V}_1$, i.e.,

$$\mathbf{Q}_1 = (\mathbf{V}_3 + \mathbf{V}_4 + \mathbf{V}_5 + \mathbf{V}_6 + \mathbf{V}_7 + \mathbf{V}_2 + \mathbf{V}_{11})/7.$$

If a vertex is located on the boundary of the surface, the weights are slightly different. For example, if the three vertices $\mathbf{V}_3$, $\mathbf{V}_{11}$, and $\mathbf{V}_{10}$ are on the boundary of the surface, then the new vertex point for $\mathbf{V}_{11}$ is computed as the sum $\frac{6}{8}\mathbf{V}_{11} + \frac{1}{8}(\mathbf{V}_3 + \mathbf{V}_{10})$. Similarly, if $\mathbf{V}_8$ is a boundary point, then the new vertex for $\mathbf{V}_8$ is $\mathbf{V}_8$ itself.

A downside of this simple algorithm is that the fine mesh obtained after a few iterations may have several "extraordinary" points where the surface is not smooth. More precisely, the continuity of the tangent plane is lost at the extraordinary points. An improvement to the original algorithm computes each new vertex point as the weighted sum $\alpha_n \mathbf{V}_i + (1 - \alpha_n)\mathbf{Q}_i$, where $\mathbf{V}_i$ is a vertex shared by n triangles, $\mathbf{Q}_i$ is the average of the n vertices around $\mathbf{V}_i$, and

$$\alpha_n = \left(\frac{3}{8} + \frac{1}{4}\cos\frac{2\pi}{n}\right)^2 + \frac{3}{8}.$$

Figure 8.19 illustrates the principle of ternary triangle subdivision. Dividing a triangle into nine smaller triangles requires (Figure 8.19a) the construction of one face point, six edge points (two on each edge) and three vertex points. Part (b) of the figure shows how one edge point (labeled "b") is computed as a weighted sum of seven vertices from six triangles. The weights shown should be normalized by dividing them by their sum, 81. The other edge points are computed similarly. Part (c) of the figure shows how the face point "c" is computed as a weighted sum of six vertices in four different triangles. The weights should again be divided by their sum, 27. Computing a vertex point, such as "a" in part (a) of the figure, depends on the number n of the triangles sharing the vertex. Each vertex on an edge sharing "a" is assigned a weight of $(1-\alpha)/n$ and "a" itself is assigned weight α, where the value $\alpha = 5/9$ was found to work in most cases. Notice that the weights are barycentric.

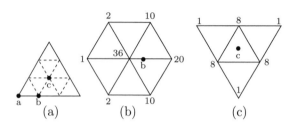

Figure 8.19: Edge and Face Points in Ternary Loop Subdivision.

As in the binary algorithm, the weights have to be modified for cases where a vertex or an edge is located on the boundary of the surface.

Subdivision is a powerful paradigm for the generation of surfaces of arbitrary topology. Given an initial triangular mesh the goal is to produce a smooth and visually pleasing surface whose shape is controlled by the initial mesh. Of particular interest are interpolating schemes since they match the original data exactly, and play an important role in fast multiresolution and wavelet techniques.

D. Zorin, P. Schröder and W. Sweldens

9
Sweep Surfaces

The surfaces described in this chapter are obtained by transforming a curve. They are not generated as interpolations or approximations of points or vectors and are consequently different from the surfaces described in previous chapters. A reader who wishes a full understanding of this chapter should be familiar with the important three-dimensional transformations (rotation, translation, scaling, reflection, and shearing) and how they are described mathematically by a 4×4 transformation matrix. This material is available in most texts on computer graphics, but the next paragraph is a short summary, for those who only need a refresher.

A three-dimensional point $\mathbf{P} = (x, y, z)$ is transformed to a point $\mathbf{P}^* = (x^*, y^*, z^*)$ by appending a fourth coordinate of 1 to it and then multiplying it by the 4×4 transformation matrix

$$\mathbf{T} = \begin{pmatrix} a & b & c & p \\ d & e & f & q \\ h & i & j & r \\ l & m & n & s \end{pmatrix}. \tag{9.1}$$

The product $(x, y, z, 1)\mathbf{T}$ is a 4-tuple (X, Y, Z, H), where $H = xp + yq + zr + s$. The three coordinates (x^*, y^*, z^*) of $\mathbf{P}^*$ are obtained by dividing (X, Y, Z) by H. Hence, $(x^*, y^*, z^*) = (X/H, Y/H, Z/H)$. The top left 3×3 submatrix of $\mathbf{T}$ is responsible for scaling and reflection (parameters a, e, and j), shearing (b, c, f, and d, h, i), and rotation (all nine). The three quantities l, m, and n are responsible for translation, and s is a global scale factor. The three parameters p, q, and r are used for perspective projection.

9.1 Sweep Surfaces

A sweep surface is obtained when a space curve $\mathbf{C}(u)$, termed the *profile*, is transformed by a transformation rule $\mathbf{T}(w)$. The transformation must include translation and/or rotation and may also include scaling and shearing. We say that the surface is *swept* by the profile curve when it (the curve) is transformed. The expression of the surface is simply the product $\mathbf{P}(u, w) = \mathbf{C}(u) \cdot \mathbf{T}(w)$. The transformation $\mathbf{T}$ is a 4×4 matrix, so vector $\mathbf{C}$ should be written in homogeneous coordinates, as the 4-tuple $\mathbf{C}(u) = (x(u), y(u), z(u), 1)$.

The simplest example is the translation of a straight line. The straight segment from the origin to $(1, 0, 0)$ is given by $\mathbf{C}(u) = (u, 0, 0, 1)$ where $0 \leq u \leq 1$. This segment is translated along the y axis by the transformation matrix

$$\mathbf{T}(w) = \begin{pmatrix} 1 & 0 & 0 & 0 \\ 0 & 1 & 0 & 0 \\ 0 & 0 & 1 & 0 \\ 0 & w & 0 & 1 \end{pmatrix},$$

where $0 \leq w \leq 1$. The surface $\mathbf{P}(u, w) = \mathbf{C}(u) \cdot \mathbf{T}(w) = (u, w, 0, 1)$ swept by this segment is (after dividing by the fourth element) $\mathbf{P}(u, w) = (u, w, 0)$. This surface is simply the square, on the xy plane, whose opposite corners are the origin and point $(1, 1, 0)$.

A more interesting example is the same segment $\mathbf{C}(u) = (u, 0, 0, 1)$, where $0 \leq u \leq 1$, translated a distance α along the z axis while being rotated $360°$ about that axis. The transformation matrix is

$$\mathbf{T}(w) = \begin{pmatrix} \cos(2\pi w) & \sin(2\pi w) & 0 & 0 \\ -\sin(2\pi w) & \cos(2\pi w) & 0 & 0 \\ 0 & 0 & 1 & 0 \\ 0 & 0 & \alpha w & 1 \end{pmatrix}, \quad \text{for} \quad 0 \leq w \leq 1.$$

The expression of the surface is $\mathbf{P}(u, w) = (u \cos(2\pi w), u \sin(2\pi w), \alpha w)$ and it is displayed in Figure 9.1a. For $w = 0.5$, it reduces to the segment $(0, u, 0.5\alpha)$ (in the y direction), and for $w = 1$, it becomes the segment $(u, 0, \alpha)$ [a segment in the original x direction, but at a height α on the z axis].

A more general example is a rectangular surface patch constructed as a sweep surface by translating an arbitrary profile along another curve, the *trajectory*. Given the two cubic Bézier curves

$$\begin{aligned} \mathbf{C}(t) &= (1-t)^3(0, 1, 1) + 3t(1-t)^2(1, 1, 0) + 3t^2(1-t)(4, 2, 0) + t^3(6, 1, 1) \\ &= (-3t^3 + 6t^2 + 3t, -3t^3 + 3t^2 + 1, 3t^2 - 3t + 1) \end{aligned}$$

and

$$\begin{aligned} \mathbf{Q}(t) &= (1-t)^3(0, 0, 0) + 3t(1-t)^2(1, 2, 1) + 3t^2(1-t)(3, 2, 2) + t^3(2, 0, 1) \\ &= (-4t^3 + 3t^2 + 3t, -6t^2 + 6t, -2t^3 + 3t) \end{aligned}$$

we can create a sweep surface $\mathbf{P}(u, w)$ by translating $\mathbf{C}(u)$ along $\mathbf{Q}(w)$. The expression of the surface is the product

$$\mathbf{P}(u, w) = (-3u^3 + 6u^2 + 3u, -3u^3 + 3u^2 + 1, 3u^2 - 3u + 1, 1)$$

$$\times \begin{pmatrix} 1 & 0 & 0 & 0 \\ 0 & 1 & 0 & 0 \\ 0 & 0 & 1 & 0 \\ -4w^3+3w^2+3w & -6w^2+6w & -2w^3+3w & 1 \end{pmatrix}$$

$$= (3u+6u^2-3u^3+3w+3w^2-4w^3, 1+3u^2-3u^3+6w-6w^2, 1-3u+3u^2+3w-2w^3, 1).$$

Figure 9.1b shows the resulting surface patch.

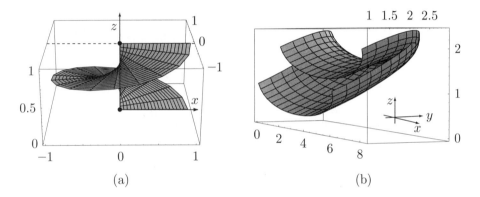

```
(* 2 sweep surface examples *)
alf=1;
ParametricPlot3D[
{u Cos[2Pi w],u Sin[2Pi w], alf w},{u,0,1},{w,0,1}, DefaultFont->{"cmr10", 10},
 Compiled->False, ViewPoint->{3.369, -2.693, 0.479}, PlotPoints->20]
m={-3u^3+6u^2+3u,-3u^3+3u^2+1,3u^2-3u+1,1}.
 {{1,0,0,0},{0,1,0,0},{0,0,1,0},{-4w^3+3w^2+3w,-6w^2+6w,-2w^3+3w,1}};
ParametricPlot3D[Drop[m,-1],{u,0,1},{w,0,1}, DefaultFont->{"cmr10", 10},
 Compiled->False, ViewPoint->{4.068, -1.506, 0.133}, PlotPoints->20]
```

Figure 9.1: Two Sweep Surfaces.

⬦ **Exercise 9.1:** Calculate the sweep surface obtained when line $\mathbf{C}(u) = (3u, 0, 0, 1)$ is translated along the z axis and at the same time translated in the y direction along a sine curve.

⬦ **Exercise 9.2:** Calculate the half-sphere produced when the quarter circle

$$\mathbf{C}(u) = \left(\frac{1-u^2}{1+u^2}, \frac{2u}{1+u^2}, 0 \right), \quad \text{where} \quad 0 \le u \le 1,$$

is rotated $360°$ about the y axis.

⬦ **Exercise 9.3:** Calculate the expression of a cone as a sweep surface. Assume that the cone is created by constructing the line from the origin to point $(R, 0, H)$, and rotating it 360° about the z axis.

> ... treat Nature by the sphere, the cylinder and the cone ...
>
> —Paul Cézanne

Example: A Möbius strip can be constructed as a sweep surface by rotating a short straight segment in a big circle (i.e., through an angle of 2π radians) while also rotating it about itself at half speed (i.e., through π radians). We start with the segment $\mathbf{segm}(t) = (t, 0, 0)$. When t is varied from, say, -3 to 3, this becomes a short segment along the x axis from $(-3, 0, 0)$ to $(3, 0, 0)$. Note that it is centered on the origin. The segment is rotated in steps about the z-axis by varying a variable ϕ from 0 to 2π. At each step of this rotation, the segment starts at its original position, it is rotated about the y-axis through an angle of $\phi/2$, it is then translated 20 units in the positive x direction, and is finally rotated by ϕ about the z-axis. Figure 9.2 shows the resulting surface swept by this segment and the code that does the computations.

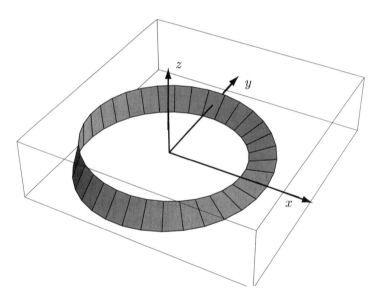

```
(* Mobius strip as a sweep surface *)
<<:Graphics:ParametricPlot3D.m
Clear[r,roty,rotz,segm];
segm[t_]:={t,0,0}; (* a short line segment *)
roty[phi_]:={{Cos[phi],0,-Sin[phi]},{0,1,0},{Sin[phi],0,Cos[phi]}};
rotz[phi_]:={{Cos[phi],-Sin[phi],0},{Sin[phi],Cos[phi],0},{0,0,1}};
ParametricPlot3D[Evaluate[rotz[phi].(roty[phi/2].segm[t]+{20,0,0})],
 {phi,0,2Pi}, {t,-3,3}, Boxed->True, PlotPoints->{35,2}, Axes->False]
Show[{%,Graphics3D[{AbsoluteThickness[1], (* show the 3 axes *)
 Line[{{0,0,30},{0,0,0},{30,0,0},{0,0,0},{0,30,0}}]}]}]
```

Figure 9.2: A Möbius Strip.

The basic sweep surface $\mathbf{C}(u)\mathbf{T}(w)$ can be extended to the product $\mathbf{C}(u,w)\mathbf{T}(w)$ of a surface and a transformation. This product is still a sweep surface, since $\mathbf{C}(u,w)$ reduces to a curve for any value of w. We can think of $\mathbf{C}(u,w)$ as a curve that's a function of the parameter u but whose shape depends on w. As w is varied, $\mathbf{C}(u,w)$ yields different curves and each is transformed differently.

Example: The lofted surface of Figure Ans.3 is multiplied by a transformation matrix that scales in the x dimension. The result is shown in Figure 9.3. (This is one way to obtain a triangular surface patch, but it looks bad as a wireframe because one family of curves converges to a point.)

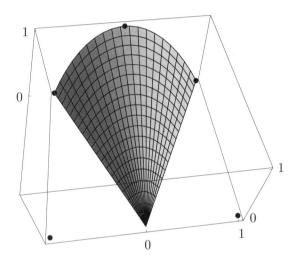

```
(* Sweep surface example. Lofted surface with scaling transform *)
<<:Graphics:ParametricPlot3D.m
pnts={{-1,-1,0},{1,-1,0},{-1,1,0},{0,1,1},{1,1,0}};
{2u-1,2w-1,4u w(1-u)}.{{w,0,0},{0,1,0},{0,0,1}};
g1=ParametricPlot3D[%, {u,0,1},{w,0,1}, Compiled->False,
DefaultFont->{"cmr10", 10},
  AspectRatio->Automatic, Ticks->{{0,1},{0,1},{0,1}}]
g2=Graphics3D[{AbsolutePointSize[4], Table[Point[pnts[[i]]],{i,1,5}]}]
Show[g1,g2, ViewPoint->{-0.139, -1.179, 1.475}]
```

Figure 9.3: A Lofted Swept Surface.

Example: A sweep surface that's a product of the surface $\mathbf{C}(u,w) = (u,1,u+2)w + (-u,1,u-2)(1-w)$ and a rotation about the z axis. Note that $\mathbf{C}(u,w)$ varies from the curve $\mathbf{C}(u,0) = (-u,1,u-2)$ to the straight line $\mathbf{C}(u,1) = (u,1,u+2)$ while being rotated. This is shown in Figure 9.4.

An even more general (and interesting) sweep surface is generated when a profile curve $\mathbf{C}(u)$ is swept along a trajectory curve $\mathbf{Q}(w) = (Q_x(w), Q_y(w), Q_z(w))$ and is also rotated about a certain axis by a rotation matrix $\mathbf{R}(w)$. Such a surface is called a *swung*

```
(* A Sweep Surface.
 Curve Cu[u,w] times matrix Trn[w] *)
<<:Graphics:ParametricPlot3D.m;
Clear[Cu,Trn];
Cu[u_,w_]:={u,1,u+2}w+{-u,1,u-2}(1-w);
Trn[w_]:={
 {Cos[2Pi w],Sin[2Pi w],0},
 {-Sin[2Pi w],Cos[2Pi w],0},
 {0,0,1}};
ParametricPlot3D[
 {Cu[u,w].Trn[w][[1]],Cu[u,w].Trn[w][[2]],
 Cu[u,w].Trn[w][[3]]},
{u,0,1,.2},{w,0,1,.2}, Ticks->None,
 PlotRange->All, AspectRatio->Automatic,
 RenderAll->False, Prolog->AbsoluteThickness[.4],
 ViewPoint->{-0.510, -1.365, 1.210}]
```

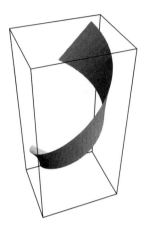

Figure 9.4: Sweeping while Rotating.

surface and its expression is

$$
\mathbf{P}(u, w) = \mathbf{C}(u) \begin{pmatrix} 1 & 0 & 0 & 0 \\ 0 & 1 & 0 & 0 \\ 0 & 0 & 1 & 0 \\ Q_x(w) & Q_y(w) & Q_z(w) & 1 \end{pmatrix} \mathbf{R}(w),
$$

where parameter w is related to the rotation angle θ in a simple way, such as $\theta = 2\pi w$ (for a 360° rotation when w varies from 0 to 1) or $\theta = \pi w$ (for a 180° rotation).

In order to construct a useful, meaningful surface, the profile, trajectory, and axis of rotation have to be selected with care. A simple example is a profile curve in the yz plane, a trajectory curve in the xy plane, and a rotation about the z-axis. Figure 9.5 is an example.

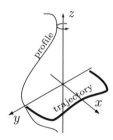

Figure 9.5: A Swung Surface.

9.2 Surfaces of Revolution

A surface of revolution is a special case of a swept surface. It is obtained when a space curve (termed the *profile* of the surface) is rotated about an axis $\mathbf{r} = (r_x, r_y, r_z)$ in space. The rotation angle can be 360° or less. A general rotation in three dimensions is fully specified by the axis of rotation (a vector) and the rotation angle (a number). If the rotation angle is θ and the rotation axis (as a unit vector) is $\mathbf{r}$, then the rotation matrix $\mathbf{T}(\theta)$ about $\mathbf{u}$ is given by

$$\begin{pmatrix} r_x^2 + \cos\theta(1 - r_x^2) & r_x r_y(1 - \cos\theta) - r_z\sin\theta & r_x r_z(1 - \cos\theta) + r_y\sin\theta \\ r_x r_y(1 - \cos\theta) + r_z\sin\theta & r_y^2 + \cos\theta(1 - r_y^2) & r_y r_z(1 - \cos\theta) - r_x\sin\theta \\ r_x r_z(1 - \cos\theta) - r_y\sin\theta & r_y r_z(1 - \cos\theta) + r_x\sin\theta & r_z^2 + \cos\theta(1 - r_z^2) \end{pmatrix}.$$

If the space curve is expressed by $\mathbf{P}(u)$, where $0 \leq u \leq 1$, then the surface of revolution has the form $\mathbf{P}(u, \theta) = \mathbf{P}(u)\mathbf{T}(\theta)$, where $0 \leq u \leq 1$ and $0 \leq \theta \leq 2\pi$. Varying u moves us along the curve and varying θ moves us in a circle (or a circular arc) about the rotation axis.

Example: Given the parametric curve $\mathbf{P}(u) = (f(u), 0, g(u))$ in the xz plane, we can revolve it around the z axis using the rotation matrix

$$\mathbf{T}_z(w) = \begin{pmatrix} \cos w & \sin w & 0 \\ -\sin w & \cos w & 0 \\ 0 & 0 & 1 \end{pmatrix} \tag{9.2}$$

to get the surface

$$\mathbf{P}(u)\mathbf{T}_z(w) = (f(u)\cos w, f(u)\sin w, g(u)), \quad \text{where } 0 \leq u \leq 1 \text{ and } 0 \leq w \leq 2\pi.$$

Example: Given the five points $\mathbf{P}_1 = (0, 1, 0)$, $\mathbf{P}_2 = (1, 1, 0)$, $\mathbf{P}_3 = (2, 2, 0)$, $\mathbf{P}_4 = (1.5, 3, 0)$, and $\mathbf{P}_5 = (1.5, 5, 0)$, we construct $\mathbf{P}(u)$ as their Bézier curve

$$\mathbf{P}(u) = \left(4t - 6t^3 + 2t^4 + (3/2)t^4, -4(t-1)^3 t + 12(t-1)^2 t^2 - 12(t-1)t^3 + 5t^4 + (t-1)^4, 0\right).$$

Since all the z coordinates are zero, the curve lies in the xy plane. We arbitrarily decide to rotate it about the y axis, so the rotation matrix is

$$\mathbf{T}_y(w) = \begin{pmatrix} \cos w & 0 & \sin w \\ 0 & 1 & 0 \\ -\sin w & 0 & \cos w \end{pmatrix}. \tag{9.3}$$

The surface expression is

$$\begin{aligned} \mathbf{P}(u)\mathbf{T}_y(w) = &\left((4t - 6t^3 + 7t^4/2)\cos w, \right. \\ &(t-1)^4 - 4(-1+t)^3 t + 12(t-1)^2 t^2 - 12(t-1)t^3 + 5t^4, \\ &\left. (4t - 6t^3 + 7t^4/2)\sin w\right). \end{aligned}$$

Such a surface is easy to display. To display it as a wire frame, just perform a double loop in which u is varied from 0 to 1, and w is varied from 0 to 2π, in any desired steps (Section 1.8.2). To display it as a solid surface, a similar double loop should cover every pixel (i.e., should iterate in very small steps) and should calculate the normal to the surface at the pixel and, from it, the intensity of light reflected from the pixel.

Following are other examples of surfaces of revolution (see also Exercise 1.34):

Example: A sphere of radius R is generated by rotating a half-circle 360° about the axis that passes through the half-circle's endpoints. Figure 9.6a shows the half-circle $\mathbf{P}(u) = (R\cos u, R\sin u, 0)$ in the xy plane. A sphere $\mathbf{P}(u, w)$ is obtained when this half-circle is rotated about the y axis:

$$\mathbf{P}(u, w) = \mathbf{P}(u)\mathbf{T}_y(w) = (R\cos u \cos w, R\sin u, R\cos u \sin w), \qquad (9.4)$$

where $-\pi/2 \le u \le \pi/2$ and $0 \le w \le 2\pi$. It is obvious, from Figure 9.6b, that curves of constant w are meridians of longitude. As u varies from $-\pi/2$ to $\pi/2$, we travel on a semicircle (the profile of the surface) on the sphere. Similarly, varying w for a constant u takes us along a latitude. The north pole is obtained for $u = \pi/2$ (and any w). The equator is the curve obtained when varying w for $u = 0$.

◇ **Exercise 9.4:** Derive the expression for the same sphere centered at (x_0, y_0, z_0).

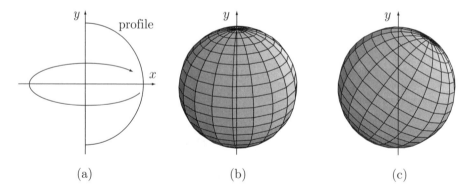

(a)   (b)   (c)

Figure 9.6: A Sphere as a Surface of Revolution.

◇ **Exercise 9.5:** Tilt the sphere of Equation (9.4) θ degrees about the z axis (Figure 9.6c).

◇ **Exercise 9.6:** Derive the expression of the sphere that's obtained when the half-circle in the xz plane is rotated 360° about the z axis.

Example: An ellipsoid with radii a and b is obtained by rotating, for example, the ellipse $\mathbf{P}(u) = (a\cos u, b\sin u, 0)$ about the y axis. After translating by (x_0, y_0, z_0), the result is

$$(x_0 + a\cos u \cos w, y_0 + b\sin u, z_0 + a\cos u \sin w),$$

where $-\pi/2 \le u \le \pi/2$ and $0 \le w \le 2\pi$.

$\diamond$ **Exercise 9.7:** Derive the equation of a torus as a surface of revolution. Assume that the torus is centered at the origin, and its two radii are R and r (Figure 9.7). The surface is created by drawing the circle of radius r centered at $(R, 0, 0)$, and rotating it $360°$ about the z axis.

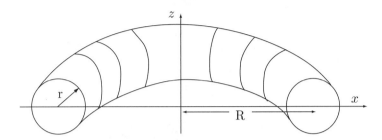

Figure 9.7: Torus as a Surface of Revolution.

Example: Figure 9.8a,b shows a chalice as a surface of revolution and its profile.

9.3 An Alternative Approach

Generating surfaces of revolution with a rotation matrix is simple but slow, since it requires the use of trigonometric functions. An alternative method is described here.

Two given curves $\mathbf{P}(u) = (P_x(u), P_y(u), P_z(u))$ and $\mathbf{C}(w) = (C_x(w), C_y(w), C_z(w))$ can be combined as follows:

$$\mathbf{S}(u, w) = \big(P_x(u)C_x(w), P_y(u)C_y(w), P_z(u)C_z(w)\big), \tag{9.5}$$

and it's easy to show that $\mathbf{S}(u, w)$ is a surface. When u is fixed at a value u_0, expression (9.5) becomes

$$\mathbf{S}(u_0, w) = \big(P_x(u_0)C_x(w), P_y(u_0)C_y(w), P_z(u_0)C_z(w)\big)$$
$$= \big(\alpha C_x(w), \beta C_y(w), \gamma C_z(w)\big),$$

which is a curve in the w direction. For each u_0 we therefore have a curve in the w direction. Similarly, for each value w_0 we have a curve going in the u direction. The only condition is that none of the components of the curves be identical to zero. If, for example, $C_x(w) \equiv 0$, then the x component of $\mathbf{S}(u_0, w)$ is always zero, so it degenerates from a surface to a curve in the yz plane.

Equation (9.5) can be used to construct a surface of revolution if $\mathbf{C}(w)$ is a circle or an arc. To explain our approach, let's first restrict the discussion to curves that are cubic polynomial segments. Such a curve has the form $\mathbf{P}(u) = (u^3, u^2, u, 1)\mathbf{MP}$, where

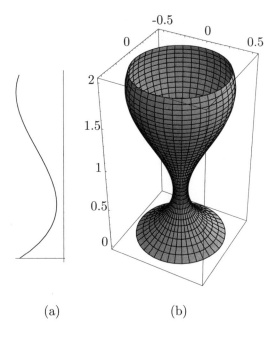

(a)   (b)

```
(* A Chalice *)
<<:Graphics:SurfaceOfRevolution.m
(* the profile *)
ParametricPlot[{.5u^3-.3u^2-.5u-.2,u+1},{u,-1,1},
 AspectRatio->Automatic]
(* the surface *)
SurfaceOfRevolution[{.5u^3-.3u^2-.5u-.2,u+1},{u,-1,1}, PlotPoints->40]
```

Figure 9.8: A Chalice as a Surface of Revolution.

$\mathbf{M}$ is a 4×4 basis matrix and $\mathbf{P}$ is a geometry vector, a 4-tuple of points and/or vectors. We can write such a curve in the form

$$\mathbf{P}(u) = \big(F_0(u), F_1(u), F_2(u), F_3(u)\big) \begin{pmatrix} \mathbf{P}_0 \\ \mathbf{P}_1 \\ \mathbf{P}_2 \\ \mathbf{P}_3 \end{pmatrix}$$

$$= F_0(u)\mathbf{P}_0 + F_1(u)\mathbf{P}_1 + F_2(u)\mathbf{P}_2 + F_3(u)\mathbf{P}_3$$

$$= \sum_{i=0}^{3} F_i(u)\mathbf{P}_i.$$

[See, for example, Equations (4.5), (6.7), and (7.11).] Similarly, curve $\mathbf{C}(w)$ can be expressed as

$$\mathbf{C}(w) = (w^3, w^2, w, 1)\mathbf{N}\mathbf{C}$$

$$= \big(G_0(w), G_1(w), G_2(w), G_3(w)\big) \begin{pmatrix} \mathbf{C}_0 \\ \mathbf{C}_1 \\ \mathbf{C}_2 \\ \mathbf{C}_3 \end{pmatrix}$$

$$= G_0(w)\mathbf{C}_0 + G_1(w)\mathbf{C}_1 + G_2(w)\mathbf{C}_2 + G_3(w)\mathbf{C}_3$$

$$= \sum_{i=0}^{3} G_i(w)\mathbf{C}_i.$$

Now, consider the x component of the surface resulting from the product of two such curves:

$$\mathbf{S}_x(u, w) = \left[\sum_{i=0}^{3} F_i(u)P_{xi}\right]\left[\sum_{j=0}^{3} G_j(w)C_{xj}\right]$$

$$= \sum_{i,j=0}^{3} F_i(u)P_{xi}C_{xj}G_j(w)$$

$$= \sum_{i,j=0}^{3} F_i(u)Q_{xij}G_j(w)$$

$$= \big(F_0(u), F_1(u), F_2(u), F_3(u)\big)\mathbf{Q}_x \begin{pmatrix} G_0(w) \\ G_1(w) \\ G_2(w) \\ G_3(w) \end{pmatrix},$$

where $Q_{xij} = P_{xi}C_{xj}$ and similarly for the y and z components. The elements $\mathbf{Q}_{ij}$ of matrix $\mathbf{Q}$ are therefore triplets of the form

$$\mathbf{Q}_{ij} = (Q_{xij}, Q_{yij}, Q_{zij}) = \big(P_{xi}C_{xj}, P_{yi}C_{yj}, P_{zi}C_{zj}\big) \tag{9.6}$$

and the entire surface can be expressed as a typical bicubic patch

$$\mathbf{S}(u, w) = \big(F_0(u), F_1(u), F_2(u), F_3(u)\big)\mathbf{Q} \begin{pmatrix} G_0(w) \\ G_1(w) \\ G_2(w) \\ G_3(w) \end{pmatrix}$$

$$= (u^3, u^2, u, 1)\mathbf{M}\mathbf{Q}\mathbf{N}^T \begin{pmatrix} w^3 \\ w^2 \\ w \\ 1 \end{pmatrix}. \tag{9.7}$$

Equation (9.7) can be generalized to cases where the constructing curves $\mathbf{C}(w)$ and $\mathbf{P}(u)$ are not cubic polynomials.

Once the designer has an idea of the shape of the surface, it may not be too hard to select two curves that will produce this shape. The problem is to place the surface at the right location in space. The location of the surface depends both on the types and the locations of the curves used. Imagine, for example, that two cubic Bézier curves are used to construct such a surface. One curve starts and ends at control points $\mathbf{P}_0$ and $\mathbf{P}_3$, and the other goes from $\mathbf{C}_0$ to $\mathbf{C}_3$. The resulting surface will be a bicubic Bézier patch anchored at the four corner points:

$$
\begin{aligned}
\mathbf{Q}_{00} &= (P_{x0}C_{x0}, P_{y0}C_{y0}, P_{z0}C_{z0}), \quad \mathbf{Q}_{01} = (P_{x0}C_{x1}, P_{y0}C_{y1}, P_{z0}C_{z1}), \\
\mathbf{Q}_{10} &= (P_{x1}C_{x0}, P_{y1}C_{y0}, P_{z1}C_{z0}), \quad \mathbf{Q}_{11} = (P_{x1}C_{x1}, P_{y1}C_{y1}, P_{z1}C_{z1}).
\end{aligned}
$$

There is no reason why these points will happen to be in the right locations and it may take some effort to vary the coordinates of all the control points to move the curves to other locations without changing their shape, in order to move points $\mathbf{Q}_{ij}$ to the right locations. The use of this surface method may therefore be limited, but it is useful for surfaces of revolution. Imagine the problem of designing a machine part with circular symmetry. If the part is to be manufactured under computer control, the location of the part in three-dimensional space may be irrelevant because the machine making it is only interested in its shape.

In order to apply Equation (9.7) to create a surface of revolution we need one curve $\mathbf{P}(u)$ to serve as a "profile" and another curve $\mathbf{C}(w)$ that's a circle, an ellipse, or an arc. As an example, consider the approximate circles obtained by cubic uniform B-splines of Section B.2. We place four points $\mathbf{C}_i$ in the way explained in that section to make curve $\mathbf{C}(w)$ an approximate circle or circular arc. If curve $\mathbf{P}(u)$ is also expressed as a cubic B-spline, then Equation (9.7) becomes the bicubic B-spline patch:

$$
\mathbf{S}(u, w) = \left[\frac{1}{6}\right]^2 [u^3, u^2, u, 1]
\begin{bmatrix} -1 & 3 & -3 & 1 \\ 3 & -6 & 3 & 0 \\ -3 & 0 & 3 & 0 \\ 1 & 4 & 1 & 0 \end{bmatrix}
\mathbf{Q}
\begin{bmatrix} -1 & 3 & -3 & 1 \\ 3 & -6 & 3 & 0 \\ -3 & 0 & 3 & 0 \\ 1 & 4 & 1 & 0 \end{bmatrix}^T
\begin{bmatrix} w^3 \\ w^2 \\ w \\ 1 \end{bmatrix}
\quad (9.8)
$$

[compare with Equation (7.50)]. The surface is created in two steps. In step 1, the surface control points $\mathbf{Q}_{ij}$ are calculated. If $\mathbf{P}(u)$ is based on the $n + 1$ points $\mathbf{P}_0$ through $\mathbf{P}_n$ and $\mathbf{C}(w)$ is based on the $m+1$ control points $\mathbf{C}_0$ through $\mathbf{C}_m$, then matrix $\mathbf{Q}$ is of order $(n+1) \times (m+1)$. In step 2, Equation (9.8) is applied $(n-1) \times (m-1)$ times to calculate all the surface patches. If the surface should make a complete revolution, then curve $\mathbf{C}(w)$ should be closed. The number of control points in this case is the same, but the number of patches is $(n-1) \times (m+1)$. If curve $\mathbf{P}(u)$ is also closed (as in a torus), then $(n+1) \times (m+1)$ surface patches are needed.

If $\mathbf{C}(w)$ should be a full circle, at least four control points $\mathbf{C}_i$ are needed and the (closed) curve consists of four segments. If curve $\mathbf{P}(u)$ (the "profile" of the surface) is open and is defined by $n+1$ points, it consists of $n-1$ segments. In such a case, the total number of surface control points $\mathbf{Q}_{ij}$ is $4 \times (n+1)$ and the entire surface of revolution consists of $4 \times (n-1)$ patches.

Example: We select the third quarter-circle segment $\mathbf{P}_4(t)$ of Equation (Ans.32) and denote it by $\mathbf{C}(w)$:

$$
\mathbf{C}(w) = \frac{1}{4}(2t^3 - 6t^2 + 4, -2t^3 + 6t, 1).
$$

It is defined by the four control points $\mathbf{C}_0 = (0, -3/2, 1)$, $\mathbf{C}_1 = (3/2, 0, 1)$, $\mathbf{C}_2 = (0, 3/2, 1)$, and $\mathbf{C}_3 = (-3/2, 0, 1)$ and it goes from $(1, 0, 1)$ to $(0, 1, 1)$. Notice that we have located $\mathbf{C}(w)$ on the $z = 1$ plane, so none of its components are identical to zero. For the curve profile $\mathbf{P}(u)$ we select the cubic B-spline segment defined by the four control points $\mathbf{P}_0 = (0, 0, 0)$, $\mathbf{P}_1 = (-1, 1, 0)$, $\mathbf{P}_2 = (-1, 1, 3)$, and $\mathbf{P}_3 = (0, 0, 3)$. These points are located on the $x = -y$ plane and go from $z = 0$ to $z = 3$, so none of the three components of $\mathbf{P}(u)$ is zero. Matrix $\mathbf{Q}$ is shown in Table 9.9. Figure 9.10 shows the surface itself and the code that generated it.

	$(0, 0, 0)$	$(-1, 1, 0)$	$(-1, 1, 3)$	$(0, 0, 3)$
$(0, -3/2, 1)$	$(0, 0, 0)$	$(0, -3/2, 0)$	$(0, -3/2, 3)$	$(0, 0, 3)$
$(3/2, 0, 1)$	$(0, 0, 0)$	$(-3/2, 0, 0)$	$(-3/2, 0, 3)$	$(0, 0, 3)$
$(0, 3/2, 1)$	$(0, 0, 0)$	$(0, 3/2, 0)$	$(0, 3/2, 3)$	$(0, 0, 3)$
$(3/2, 0, 1)$	$(0, 0, 0)$	$(3/2, 0, 0)$	$(3/2, 0, 3)$	$(0, 0, 3)$

Table 9.9: Matrix $\mathbf{Q}$ for Surface of Revolution Example.

The location of this surface in space may sometimes be a problem and should therefore be discussed. Since our quarter circle goes from $(1, 0, 1)$ to $(0, 1, 1)$, we intuitively expect the profile $\mathbf{P}(u)$ to be rotated from direction $(1, 0)$ (the positive x axis) to direction $(0, 1)$ (the positive y axis). A direct check, however, shows that the four corners of this patch are $\mathbf{S}(0, 0) = (-0.833, 0, 0.5)$, $\mathbf{S}(0, 1) = (-0.833, 0, 2.5)$, $\mathbf{S}(1, 0) = (0, 0.833, 0.5)$, and $\mathbf{S}(1, 1) = (0, 0.833, 2.5)$. Thus, the profile has been rotated from direction $(-0.833, 0)$ to direction $(0, 0.833)$ because of its particular original location (as defined, the profile is located on the $x = -y$ plane).

Because of the high symmetry of surfaces of revolution, especially those that go through a complete revolution, their precise position in space may not be important, so our method may be useful for this type of surface.

The method developed here can be used with any type of parametric curves, not just B-splines and not just PCs. Equation (9.9) shows how a standard quadratic Lagrange polynomial [Equation (3.14)] can be combined with a degree-4 Bézier curve to form a surface patch based on 3×5 points

$$\mathbf{Q}_{ij} = (Q_{xij}, Q_{yij}, Q_{zij}) = \left(P_{xi}C_{xj}, P_{yi}C_{yj}, P_{zi}C_{zj}\right).$$

The surface expression is

$$\mathbf{S}(u, w) = (u^2, u, 1) \begin{pmatrix} 2 & -4 & 2 \\ -3 & 4 & -1 \\ 1 & 0 & 0 \end{pmatrix} \mathbf{Q} \begin{pmatrix} 1 & -4 & 6 & -4 & 1 \\ -4 & 12 & -12 & 4 & 0 \\ 6 & -12 & 6 & 0 & 0 \\ -4 & 4 & 0 & 0 & 0 \\ 1 & 0 & 0 & 0 & 0 \end{pmatrix}^T \begin{pmatrix} w^4 \\ w^3 \\ w^2 \\ w \\ 1 \end{pmatrix},$$

(9.9)

where

$$\mathbf{Q} = \begin{pmatrix} \mathbf{Q}_{00} & \mathbf{Q}_{01} & \mathbf{Q}_{02} & \mathbf{Q}_{03} & \mathbf{Q}_{04} \\ \mathbf{Q}_{10} & \mathbf{Q}_{11} & \mathbf{Q}_{12} & \mathbf{Q}_{13} & \mathbf{Q}_{14} \\ \mathbf{Q}_{20} & \mathbf{Q}_{21} & \mathbf{Q}_{22} & \mathbf{Q}_{23} & \mathbf{Q}_{24} \end{pmatrix}.$$

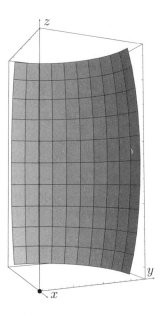

```
<<:Graphics:ParametricPlot3D.m;    (* Surface of revolution *)
Clear[basis,Cubi]; (* as a combination of 2 cubic B-splines *)
(* matrix 'basis' has dimensions 4x4x3 *)
basis={{{0,0,0},{0,-3/2,0},{0,-3/2,3},{0,0,3}}
 ,{{0,0,0},{-3/2,0,0},{-3/2,0,3},{0,0,3}}
 ,{{0,0,0},{0,3/2,0},{0,3/2,3},{0,0,3}},{{0,0,0}
 ,{3/2,0,0},{3/2,0,3},{0,0,3}}};
Cubi={{-1,3,-3,1},{3,-6,3,0},{-3,0,3,0},{1,4,1,0}};
prt[i_]:=basis[[Range[1,4],Range[1,4],i]];
(* 'prt' extracts component i from the 3rd dimen of 'basis' *)
coord[i_]:={u^3,u^2,u,1}.Cubi.prt[i].Transpose[Cubi].{w^3,w^2,w,1};
ParametricPlot3D[{coord[1],coord[2],coord[3]}/36,
{u,0,1,.1},{w,0,1,.1},
Prolog->AbsoluteThickness[.5],ViewPoint->{1.736, -0.751, -0.089}]
```

Figure 9.10: A Quarter-Circle Surface of Revolution made of B-Splines.

9.4 Skinned Surfaces

In many practical applications, the surface designer starts with only a rough idea of the shape of the required surface. The designer wants to compute and display the result of this idea, and then improve it interactively. A common example is a pipe that winds its way inside an engine avoiding hot parts. The pipe has to have a complex shape in order to go around the various parts of the engine and may even have to change its cross section as it travels through narrow passages. One approach to such a design problem is a *skinned surface*. The designer starts by specifying several curves $\mathbf{C}_i(u)$ that become profiles (or cross sections) of the surface, and the resulting surface $\mathbf{P}(u, w)$ is an interpolation of these cross sections.

It is intuitively clear that the precise shape of the surface depends on the method used to interpolate the cross sections. Thus, general-purpose software for skinned surfaces should give the user a choice of several interpolation and approximation methods. Three examples are discussed here.

Section 3.6.2 shows how to select four points on each of four given curves and employ bicubic interpolation (Section 3.6) to compute a bicubic surface that passes through the four curves. Such a surface is also a skinned surface that interpolates the four curves.

Given a set of $n + 1$ Bézier curves, each defined by a set of $m + 1$ control points, we can use the Bézier approximation method of Section 6.16 to compute a rectangular surface patch that's an interpolation of the curves. This surface passes only through the four corner points, but it passes through all $n + 1$ given curves. Figure 9.11 shows how a set of nine similar (but not identical) Bézier curves can be used as cross sections to construct the surface of a boat as a skinned surface. Each curve must be defined by the same number of control points (five in our example) and the fact that they are similar suggests that we can start by constructing one curve, and then duplicate it as many times as necessary and scale, move, and shear the copies as needed. In this type of work it makes sense to start with the most complex curve and use it as the basis of all the other cross sections. (Each of the curves in this example is actually two mirror image Bézier curves joined at one point.)

Figure 9.11: Nine Cross Sections of a Boat.

Similarly, given a set of $n + 1$ B-spline or NURBS curves, each defined by a set of $m + 1$ control points, we can use the approximation methods of Chapter 7 to compute a skinned surface with the curves as its cross sections. The computations in this case are more intensive, but the advantage is that such a surface may have sharp corners and edges.

Let everyone sweep in front of his own door, and the whole world will be clean.

Johann Wolfgang von Goethe

A
Conic Sections

The ellipse, hyperbola, and parabola (and also the circle, which is a special case of the ellipse) are called the *conic section* curves (or the *conic sections* or just *conics*), since they can be obtained by cutting a cone with a plane (i.e., they are the intersections of a cone and a plane).

The conics are easy to calculate and to display, so they are commonly used in applications where they can approximate the shape of other, more complex, geometric figures. Many natural motions occur along an ellipse, parabola, or hyperbola, making these curves especially useful. Planets move in ellipses; many comets move along a hyperbola (as do many colliding charged particles); objects thrown in a gravitational field follow a parabolic path.

There are several ways to define and represent these curves and this section uses a simple *geometric* definition that leads naturally to the parametric and the implicit representations of the conics.

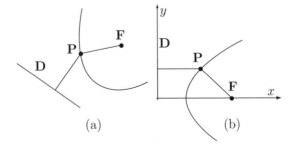

Figure A.1: Definition of Conic Sections.

Definition: A conic is the locus of all the points **P** that satisfy the following: The distance of **P** from a fixed point **F** (the *focus* of the conic, Figure A.1a) is proportional to its distance from a fixed line **D** (the *directrix*). Using set notation, we can write

$$\text{Conic} = \{\mathbf{P} | \mathbf{PF} = e\mathbf{PD}\},$$

where e is the *eccentricity* of the conic. It is easy to classify conics by means of their eccentricity:

$$e = \begin{cases} = 1, & \text{parabola,} \\ < 1, & \text{ellipse (the circle is the special case } e = 0), \\ > 1, & \text{hyperbola.} \end{cases}$$

In the special case where the directrix is the y axis ($x = 0$, Figure A.1b) and the focus is point $(k, 0)$, the definition results in

$$\frac{\sqrt{(x - k)^2 + y^2}}{|x|} = e, \quad \text{or } (1 - e^2)x^2 - 2kx + y^2 + k^2 = 0. \tag{A.1}$$

In this case, the conic is represented by a degree-2 equation. It can be shown that this is true for the general case, where the directrix and the focus can be located anywhere. It can also be shown that the inverse is also true, i.e., any degree-2 algebraic equation of the form

$$ax^2 + by^2 + 2hxy + 2fx + 2gy + c = 0 \tag{A.2}$$

represents a conic. Equation (A.2) can be used to classify the conics. If D is the determinant

$$D = \begin{vmatrix} a & h & f \\ h & b & g \\ f & g & c \end{vmatrix},$$

then Table A.2 provides a complete classification of the conics, including degenerate cases where the conic reduces to two lines (real or imaginary) or to a point.

$ab - h^2$	Conditions			Conic
$= 0$	$D \neq 0$			parabola
	$D = 0$	$b \neq 0$	$g^2 - bc > 0$	2 parallel lines
			$g^2 - bc = 0$	2 parallel coincident lines
			$g^2 - bc < 0$	2 parallel imaginary lines
		$b = h = 0$	$f^2 - ac > 0$	2 parallel real lines
			$f^2 - ac = 0$	2 parallel coincident lines
			$f^2 - ac < 0$	2 parallel imaginary lines
> 0	$D = 0$			point (degenerate ellipse)
	$D \neq 0$		$-bD > 0$	real ellipse
			$-bD < 0$	imaginary ellipse
< 0	$D = 0$			2 intersecting lines
	$D \neq 0$			hyperbola

Table A.2: Classification of Conics.

◇ **Exercise A.1:** Assume that the second-degree equation

$$Ax^2 + Bxy + Cy^2 + Dx + Ey + F = 0 \tag{A.3}$$

is given. Show how to use the six parameters to determine which conic is described by this equation.

Equation (A.1) can be used to generate the familiar implicit representations of the conics. We first treat the case $e \neq 1$ by transforming $x' = x - k/(1 - e^2)$. When this is substituted into Equation (A.1) (and the prime is eliminated), the result is

$$\frac{x^2}{a^2} + \frac{y^2}{b^2} = 1, \tag{A.4}$$

where $\quad a = \dfrac{ke}{1 - e^2} \quad$ and $\quad b^2 = a^2(1 - e^2).$

Case 1: Ellipse. The case $e < 1$ implies that both a and b are positive and $a > b$. In this case, Equation (A.4) represents the canonical ellipse. This ellipse is centered on the origin with the x and y axes being the major and minor axes of the ellipse, respectively. The major radius is a and the minor one is b. For $a = b$, this ellipse reduces to a circle. Hence, we can think of a circle as the limit of the ellipse when $e \to 0$ and $k \to \infty$.

Case 2: Hyperbola. The case $e > 1$ implies a negative a and a negative b^2 (hence an imaginary b). If we use the absolute value of the imaginary b, Equation (A.4) becomes

$$\frac{x^2}{a^2} - \frac{y^2}{b^2} = 1, \qquad a, b > 0. \tag{A.5}$$

This is a canonical hyperbola, where the x axis is the *traverse axis* and the y axis is called the *semiconjugate* or *imaginary axis*. The hyperbola consists of two distinct parts with the imaginary axis separating them. The two points $(-a, 0)$ and $(a, 0)$ are called the *vertices* of the hyperbola.

Case 3: Parabola ($e = 1$). The simple transformation $x' = x - k/2$ yields, when substituted into Equation (A.1), the canonical parabola

$$y^2 = 4ax, \qquad \text{where} \quad a = k/2 > 0, \tag{A.6}$$

with focus at $(a, 0)$ (thus, a is the focal distance) and directrix $x = -a$. The origin is the *vertex* of the canonical parabola.

All the conic sections can also be expressed (although not in their canonical forms) by

$$f(\theta) = \frac{K}{1 \pm e \cos(\theta)}.$$

For $e = 0$ this is a circle. For $0 < e < 1$ this is an ellipse. For $e = 1$ this is a parabola and for $e > 1$ it is a hyperbola.

The parametric representations of the conics are simple. We start with the ellipse. In order to show that the expression

$$\left(a\frac{1-t^2}{1+t^2}, b\frac{2t}{1+t^2}\right), \qquad -\infty < t < \infty, \tag{A.7}$$

traces out an ellipse we show that it satisfies Equation (A.4):

$$\frac{a^2\left(\frac{1-t^2}{1+t^2}\right)^2}{a^2} + \frac{b^2\left(\frac{2t}{1+t^2}\right)^2}{b^2} = \frac{1-2t^2+t^4+4t^2}{1+2t^2+t^4} = 1.$$

The first quadrant is obtained for $0 \le t \le 1$. To get the second quadrant, however, t has to vary from 1 to ∞. Quadrants 4 and 3 are obtained for $-\infty \le t \le 0$.

The canonical hyperbola is represented parametrically by

$$\left(a\frac{1+t^2}{1-t^2}, b\frac{2t}{1-t^2}\right), \qquad -\infty < t < \infty. \tag{A.8}$$

The right branch is traced out when $-1 \le t \le 1$, and the left branch is obtained when $-\infty \le t \le -1$ and $1 \le t \le \infty$. Thus, the two values $t = \pm 1$ represent hyperbola points at infinity.

The simple expression

$$(at^2, 2at), \qquad -\infty < t < \infty, \tag{A.9}$$

traces out the canonical parabola.

Equations (A.7) and (A.8) are called *rational parametrics* since they contain the parameter t in the denominator. Rational parametric curves are generally complex but can represent more shapes and are therefore more general than the nonrational ones. One disadvantage of the rational parametrics is variable velocity. Varying t in equal increments generally results in traveling along the curve in unequal steps.

In practice, it is sometimes necessary to have conics placed anywhere in three-dimensional space, not just on the xy plane. This is done by taking a general two-dimensional conic $\mathbf{P}(t)$ [one of Equations (A.7), (A.8), or (A.9)], adding a third co-ordinate $z = 0$ and transforming it with the general 4×4 transformation matrix $\mathbf{T}$ [Equation (9.1)]. Normally, such a curve is translated and rotated. It may also be scaled and sheared. The result is a three-dimensional curve of the form

$$\mathbf{P}^*(t) = \left(\frac{a_0 + a_1 t + a_2 t^2}{w_0 + w_1 t + w_2 t^2}, \frac{b_0 + b_1 t + b_2 t^2}{w_0 + w_1 t + w_2 t^2}, \frac{c_0 + c_1 t + c_2 t^2}{w_0 + w_1 t + w_2 t^2}\right)$$

$$= \left(\frac{\sum_{i=0}^{2} a_i t^i}{\sum_{i=0}^{2} w_i t^i}, \frac{\sum_{i=0}^{2} b_i t^i}{\sum_{i=0}^{2} w_i t^i}, \frac{\sum_{i=0}^{2} c_i t^i}{\sum_{i=0}^{2} w_i t^i}\right).$$

Denoting $x_i = a_i/w_i$, $y_i = b_i/w_i$, $z_i = c_i/w_i$, and $\mathbf{a}_i = (x_i, y_i, z_i)$, we can write this as

$$\mathbf{P}^*(t) = \frac{w_0\mathbf{a}_0 + w_1\mathbf{a}_1 t + w_2\mathbf{a}_2 t^2}{w_0 + w_1 t + w_2 t^2} = \frac{\sum_{i=0}^{2} w_i\mathbf{a}_i t^i}{\sum_{i=0}^{2} w_i t^i}. \tag{A.10}$$

This is the general rational form of the conic sections. It can also be shown that any rational parametric expression of the form (A.10) represents a conic.

> She could see at once by his degenerate conic and dissipative terms that he was bent on no good, "Arcsinh," she gasped. "Ho, Ho," he said. "What a symmetric little asymptote you have. I can see your angles have a lit of secs."
>
> Richard Woodman, *Impure Mathematics* (1981)

B
Approximate Circles

Parametric curves are general and can take many shapes. In principle, such a curve can be based on any functions, but in practice polynomials are virtually always used. It is well known, however, that one common, important curve, namely the circle, cannot be precisely represented by a polynomial. This short appendix discusses ways to compute approximate, albeit high precision, circles with Bézier and B-spline methods.

B.1 Circles and Bézier Curves

The equation of a circle is $x^2 + y^2 = r^2$ or $y = \pm\sqrt{r^2 - x^2}$. This is not a polynomial and, in fact, Exercise B.1 proves that a polynomial cannot represent a circle. Applying Bézier methods to circles can be done either by using rational Bézier curves (Section 6.15) or by deriving an approximation to the circle. This section discusses the latter approach.

We start with a three-point example. We select the three points $\mathbf{P}_0 = (1, 0)$, $\mathbf{P}_1 = (k, k)$, and $\mathbf{P}_2 = (0, 1)$ and attempt to find the value of k such that the quadratic Bézier curve defined by the points will best approximate a quarter circle of radius 1 (Figure B.1). The curve is given, of course, by

$$
\begin{aligned}
\mathbf{P}(t) &= (1-t)^2(1,0) + 2t(1-t)(k,k) + t^2(0,1) \\
&= \big(1 + 2t(k-1) + t^2(1-2k), 2kt + t^2(1-2k)\big) \\
&= \big(P_x(t), P_y(t)\big),
\end{aligned}
\tag{B.1}
$$

and it equals the circle at its start and end points. We need a constraint that will produce an equation whose solution will yield the value of k. A reasonable constraint is to require that the curve be identical to the circle at its midpoint. This can be expressed as $\mathbf{P}(0.5) = (1/\sqrt{2}, 1/\sqrt{2})$ and it produces the equation

$$\mathbf{P}(0.5) = \frac{1}{4}(1,0) + \frac{1}{2}(k,k) + \frac{1}{4}(0,1) = \left(\frac{1}{\sqrt{2}}, \frac{1}{\sqrt{2}}\right),$$

whose solution is

$$k = \frac{2\sqrt{2}-1}{2} \approx 0.914.$$

We also note that the tangent vector of Equation (B.1) is

$$\mathbf{P}^t(t) = \big(2(k-1) + 2t(1-2k), 2k + 2t(1-2k)\big). \tag{B.2}$$

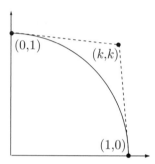

Figure B.1: A Quadratic Bézier Curve Approximating a Quarter Circle.

How much does this curve deviate from a true circle of radius 1? To answer this, we first notice that the distance of a point $\mathbf{P}(t)$ from the origin is

$$D(t) = \sqrt{P_x^2(t) + P_y^2(t)} = \sqrt{\big(1 + 2t(k-1) + t^2(1-2k)\big)^2 + \big(2kt + t^2(1-2k)\big)^2}.$$

To find the maximum distance, we differentiate $D(t)$:

$$\frac{d\,D(t)}{dt} = \frac{2P_x(t) \cdot P_x^t(t) + 2P_y(t) \cdot P_y^t(t)}{2\sqrt{P_x^2(t) + P_y^2(t)}}$$

and set the result equal to 0. This yields $P_x(t) \cdot P_x^t(t) + P_y(t) \cdot P_y^t(t) = \mathbf{P}(t) \cdot \mathbf{P}^t(t) = 0$. Applying Equations (B.1) and (B.2), we get the equation

$$2(k-1) + 2\big(1 + 2(k-1)^2\big)t - 6(1-2k)^2 t^2 + 4(1-2k)^2 t^3 = 0,$$

which has two roots in the interval $[0,1]$, namely $t_1 \approx 0.33179$ and $t_2 \approx 0.66821$, close to the expected values of $1/3$ and $2/3$. Simple computation shows the maximum distance of $\mathbf{P}(t)$ from the origin to be $D(t_1) = D(t_2) = 0.995685$. The maximum deviation of this from a circle of radius one is thus 0.432%, negligible for most purposes.

$\diamond$ **Exercise B.1:** Prove that the Bézier curve cannot be a circle.

$\diamond$ **Exercise B.2:** Consider the quarter circle from $\mathbf{P}_0 = (1, 0)$ to $\mathbf{P}_3 = (0, 1)$. Select two points $\mathbf{P}_1$ and $\mathbf{P}_2$ such that the Bézier curve defined by the four points would be the closest possible to a circle.

$\diamond$ **Exercise B.3:** Do the same for the oval (elliptic) arc from $(1, 0)$ to $(0, 1)$.

$\diamond$ **Exercise B.4:** Calculate the cubic Bézier curve that approximates the circular arc of Figure B.2 spanning an angle of 2θ. The calculation should be based on the requirement that the curve and the arc have the same endpoints and the same extreme tangent vectors.

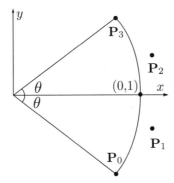

Figure B.2: A Cubic Bézier Curve Approximating an Arc.

Example: We approximate a sine wave by smoothly joining eight cubic Bézier segments (Figure B.3). The first segment requires four control points and each of the remaining seven segments requires three additional points. The total number of points is therefore 25. They are numbered $\mathbf{P}_0$ through $\mathbf{P}_{24}$, but because of the high symmetry of the sine wave, only the first seven points, $\mathbf{P}_0$ through $\mathbf{P}_6$, need be computed. The rest can be obtained from these by simple translations and reflections. We require that the following three points be on the sine curve, making it easy to find their coordinates:

$$\mathbf{P}_0 = (0, 0), \ \mathbf{P}_3 = \left(\frac{\pi}{4}, \sin\left(\frac{\pi}{4}\right)\right) \approx (0.785, 0.7071), \ \mathbf{P}_6 = \left(\frac{\pi}{2}, \sin\left(\frac{\pi}{2}\right)\right) \approx (1.57, 1).$$

The expression for segment i (where $i = 0, 3, 6, 9, 12, 15, 18,$ and 21) is

$$\mathbf{P}_i(t) = (1 - t)^3 \mathbf{P}_i + 3t(1 - t)^2 \mathbf{P}_{i+1} + 3t^2(1 - t)\mathbf{P}_{i+2} + t^3 \mathbf{P}_{i+3},$$

and its tangent vector is

$$\mathbf{P}_i^t(t) = -3(1 - t)^2 \mathbf{P}_i + (3 - 9t)(1 - t)\mathbf{P}_{i+1} + 3t(2 - 3t)\mathbf{P}_{i+2} + 3t^2 \mathbf{P}_{i+3}.$$

To calculate point $\mathbf{P}_1$, we require that the initial tangent $\mathbf{P}_0^t(0)$ of the first curve segment matches the initial slope of the sine wave, which is $45°$. We can therefore

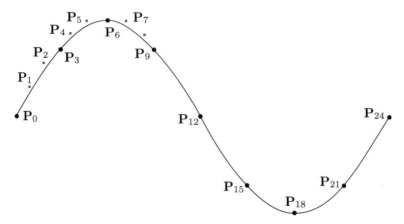

Figure B.3: A Sine Curve Approximated by Eight Cubic Bézier Segments.

write $\mathbf{P}_0^t(0) = (a, a)$ for any positive a and we select $a = 0.7071$ since this produces a normalized tangent vector. The result is

$$(0.7071, 0.7071) = \mathbf{P}_0^t(0) = -3\mathbf{P}_0 + 3\mathbf{P}_1 \text{ or } \mathbf{P}_1 = (0.7071, 0.7071)/3 = (0.2357, 0.2357).$$

To calculate points $\mathbf{P}_2$ and $\mathbf{P}_4$, we again require that the final tangent vector $\mathbf{P}_0^t(1)$ of the first segment match the slope of the sine wave at $x_3 = \pi/4$. That slope is 0.7071, so we select $(1, 0.7071)$ as the tangent vector, then normalize it to $(0.816, 0.577)$. We end up with

$$(0.816, 0.577) = \mathbf{P}_0^t(1) = -3\mathbf{P}_2 + 3\mathbf{P}_3 \text{ or } \mathbf{P}_2 = \mathbf{P}_3 - (0.816, 0.577)/3 = (0.513, 0.5151).$$

By symmetry we also get $\mathbf{P}_4 = \mathbf{P}_3 + (0.816, 0.577)/3 = (1.057, 0.899)$.

Only point $\mathbf{P}_5$ remains to be calculated. Again, we require that the final tangent vector $\mathbf{P}_3^t(1)$ of the second segment (segment 3) match the slope of the sine wave at $\mathbf{P}_6$, which is 0. Thus, the normalized tangent vector is $(1, 0)$, which produces the equation

$$(1, 0) = \mathbf{P}_3^t(1) = 3\mathbf{P}_6 - 3\mathbf{P}_5, \text{ or } \mathbf{P}_5 = \mathbf{P}_6 - (1, 0)/3 = (1.237, 1).$$

Points $\mathbf{P}_7$ through $\mathbf{P}_{24}$ can be obtained from the first seven points by translation and reflection. Alternatively, the first four cubic segments can be calculated and each pixel can be used to calculate one more pixel by translation and reflection.

B.2 The Cubic B-Spline as a Circle

The uniform B-spline, like the Bézier curve, cannot represent a precise circle. However, the cubic B-spline can provide an excellent approximation to a circle or a circular arc using just a few control points. The following discussion shows how to place those points in order to obtain a unit circle centered on the origin. Figure B.4b shows m equidistant control points $\mathbf{P}_i$ placed on a circle of radius R, where R has to be determined. The coordinates of those points are

$$
\mathbf{P}_i = (R\cos\theta_i, R\sin\theta_i) = \left(R\cos\frac{2\pi i}{m}, R\sin\frac{2\pi i}{m}\right) \quad \text{for} \quad i = 0, 1, \ldots, m-1.
$$

We divide the control points, as usual, into overlapping groups of four points each and calculate a cubic B-spline segment $\mathbf{P}_i(t)$ for each group. We require that the two terminal points $\mathbf{P}_i(0)$ and $\mathbf{P}_i(1)$ be at a distance of one unit from the origin.

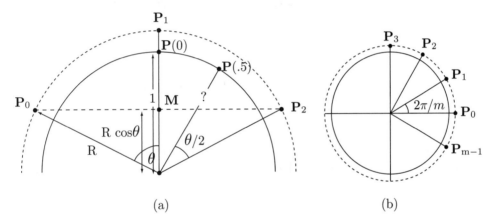

(a)   (b)

Figure B.4: A Cubic B-Spline and a Circle.

Exercise 7.2 shows that the start point $\mathbf{P}_1(0)$ of the first segment of this curve satisfies (see Figure B.4a)

$$
\mathbf{P}(0) = \frac{1}{3}\frac{\mathbf{P}_0 + \mathbf{P}_2}{2} + \frac{2}{3}\mathbf{P}_1 = \frac{1}{3}\mathbf{M} + \frac{2}{3}\mathbf{P}_1.
$$

The distance of $\mathbf{P}(0)$ from the origin is therefore

$$
\frac{1}{3}R\cos\theta + \frac{2}{3}R,
$$

and the same is true for the end point $\mathbf{P}(1)$. On the other hand, we require that this distance equals one unit, so the result is

$$
\frac{1}{3}R(\cos\theta + 2) = 1 \quad \text{or} \quad R = \frac{3}{2 + \cos\theta}. \tag{B.3}
$$

Our control points should therefore have coordinates

$$\mathbf{P}_i = \left(\frac{3 \cos \frac{2\pi i}{m}}{2 + \cos \frac{2\pi}{m}}, \frac{3 \sin \frac{2\pi i}{m}}{2 + \cos \frac{2\pi}{m}} \right) \quad \text{for} \quad i = 0, 1, \ldots, m - 1.$$

To estimate the number of control points necessary for a good approximation, we first estimate the error of this representation. Since the curve is identical to a circle at the control points, we assume that the worst approximation is obtained midway between control points, i.e., at points $\mathbf{P}_i(0.5)$. Figure B.4a shows one such point whose distance from the origin is labeled "?." The midpoint of a cubic segment, however, is easily calculated from Equation (7.11) to be

$$\mathbf{P}_1(0.5) = \frac{1}{6} \left(\frac{1}{8}\mathbf{P}_0 + \frac{23}{8}\mathbf{P}_1 + \frac{23}{8}\mathbf{P}_2 + \frac{1}{8}\mathbf{P}_3 \right)$$

$$= \left(\frac{(1 + \cos\theta)(11 + \cos\theta)}{8(2 + \cos\theta)}, \frac{\sin\theta(11 + \cos\theta)}{8(2 + \cos\theta)} \right),$$

where $\theta = 2\pi i/m$. The deviation from a true circle is therefore

$$1 - \sqrt{\mathbf{P}_1^2(0.5)} = \frac{(1 - \cos\frac{\pi}{m})^2(2 - \cos\frac{\pi}{m})}{2(2 + \cos\frac{2\pi}{m})}.$$

Even for $m = 4$, the deviation is only 2.77%. For $m = 5$, it is 0.94%, and for $m = 6$, it is 0.41%. The B-spline can therefore provide an excellent, fast approximation to a circle.

Example: We calculate the four segments for the case $m = 4$. The value of R is

$$R = \frac{3}{2 + \cos\frac{2\pi}{4}} = 3/2,$$

so the control points are

$$\mathbf{P}_0 = (R\cos 0, R\sin 0) = (3/2, 0),$$
$$\mathbf{P}_1 = (R\cos\frac{\pi}{2}, R\sin\frac{\pi}{2}) = (0, 3/2),$$
$$\mathbf{P}_2 = (R\cos\pi, R\sin\pi) = (-3/2, 0),$$
$$\mathbf{P}_3 = (R\cos\frac{3\pi}{2}, R\sin\frac{3\pi}{2}) = (0, -3/2).$$

Equation (7.11) is used to obtain the first segment:

$$\mathbf{P}_1(t) = \frac{1}{6}(t^3, t^2, t, 1) \begin{pmatrix} -1 & 3 & -3 & 1 \\ 3 & -6 & 3 & 0 \\ -3 & 0 & 3 & 0 \\ 1 & 4 & 1 & 0 \end{pmatrix} \begin{pmatrix} (3/2, 0) \\ (0, 3/2) \\ (-3/2, 0) \\ (0, -3/2) \end{pmatrix}$$

$$= \frac{1}{4}(2t^3 - 6t, 2t^3 - 6t^2 + 4).$$

This segment goes from $(0, 1)$ to $(-1, 0)$ and its midpoint is at $(-22/32, 22/32) = (-0.6875, 0.6875)$. The true circle is at $(-0.7071, 0.7071)$, so the difference is ≈ 0.02. Normally, a cubic B-spline curve based on four control points has two segments but our curve is closed, so it consists of four segments.

$\diamond$ **Exercise B.5:** Calculate the remaining three segments.

Example: Approximating a circular arc. We restrict our discussion to arcs on the unit circle centered on the origin. To specify such an arc, the user should input the coordinates of the two endpoints $\mathbf{S}$ and $\mathbf{E}$ (both at a distance of one unit from the origin) and the software should use them to calculate the coordinates of the four control points $\mathbf{C}_0$, $\mathbf{C}_1$, $\mathbf{C}_2$, and $\mathbf{C}_3$ that produce the best approximation for the arc $\mathbf{C}(t)$. Figure B.5a shows how $\mathbf{S}$ and $\mathbf{E}$ become the endpoints $\mathbf{C}(0)$ and $\mathbf{C}(1)$ of the arc. It also shows that $\cos\theta = \mathbf{E} \bullet \mathbf{S}$. Equation (B.3) gives the distance R of the four control points from the origin and shows how to compute the two interior points

$$\mathbf{C}_1 = R\mathbf{S} = \frac{3}{2 + \mathbf{E} \bullet \mathbf{S}}\mathbf{S}, \quad \mathbf{C}_2 = R\mathbf{E} = \frac{3}{2 + \mathbf{E} \bullet \mathbf{S}}\mathbf{E}.$$

Control point $\mathbf{C}_0$ is found by rotating $\mathbf{C}_1$ clockwise θ degrees and control point $\mathbf{C}_3$ is found by rotating $\mathbf{C}_2$ θ degrees counterclockwise. The rotation matrices are obtained from Equation (3.4) in [Salomon 99], bearing in mind that $\cos\theta = \mathbf{E} \bullet \mathbf{S}$ and $\sin\theta = \sqrt{1 - (\mathbf{E} \bullet \mathbf{S})^2}$:

$$\mathbf{C}_0 = \mathbf{C}_1 \begin{pmatrix} \mathbf{E} \bullet \mathbf{S} & \sqrt{1 - (\mathbf{E} \bullet \mathbf{S})^2} \\ -\sqrt{1 - (\mathbf{E} \bullet \mathbf{S})^2} & \mathbf{E} \bullet \mathbf{S} \end{pmatrix},$$

$$\mathbf{C}_3 = \mathbf{C}_2 \begin{pmatrix} \mathbf{E} \bullet \mathbf{S} & -\sqrt{1 - (\mathbf{E} \bullet \mathbf{S})^2} \\ \sqrt{1 - (\mathbf{E} \bullet \mathbf{S})^2} & \mathbf{E} \bullet \mathbf{S} \end{pmatrix}.$$

Once the four control points are known, the cubic B-spline segment can be constructed.

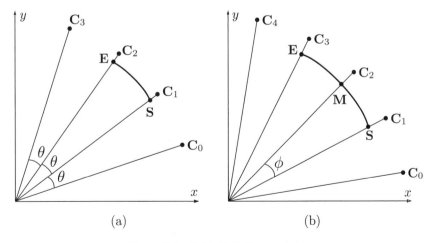

(a) (b)

Figure B.5: Cubic B-Splines and Arcs.

Approximating long arcs may require more than one spline segment and this can also be handled by our method. The user should again input the coordinates of the two endpoints S and E (both at a distance of one unit from the origin) and the software should use them to determine the coordinates of five control points C_0 through C_4 (Figure B.5b). The first step is to compute the midpoint M of S and E. Once M is known, the three interior control points C_1, C_2, and C_3 can easily be calculated. The two exterior points C_0 and C_4 are found by rotating C_1 and C_3, respectively. Once the five control points are known, two cubic spline segments can be calculated and, together, they constitute the arc.

◇ **Exercise B.6:** How is M calculated?

The discussion on page 307 shows how rational B-splines can be used to generate a circle precisely.

> The circle of the English language has a well-defined centre, but no discernible circumference.
>
> James Murray, *Oxford English Dictionary*

C
Graphics Gallery

This short appendix illustrates the quality of *Mathematica*-generated color images. The images shown here are taken from the book, except Figure C.3, which shows the implicit surface $-x^2 + 2y^2 + 2z^2 = 1$ and Figure C.1, that's a small detail from Seurat's painting *Circus Sideshow* (1887). Most of the images in the book were generated by simple *Mathematica* programs, but the triangular Bézier surfaces required more sophisticated code and were implemented by Garry Helzer.

Web sites [Gallery 05] and [Graphica 05], as well as the *Mathematica* book [Wolfram 03] should also be mentioned for their impressive graphics galleries that truly deserve a visit.

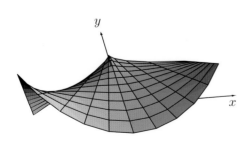

Figure 2.13: A Lofted Surface.

Figure C.1: Pointillism.

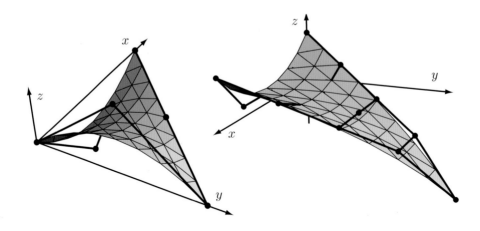

Figure C.2: Two Triangular Bézier Surfaces For $n = 2$ and $n = 4$.

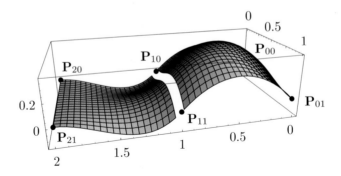

Figure 4.15: Two Ferguson Surface Patches.

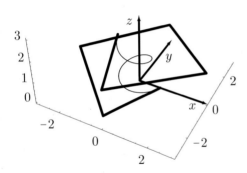

Figure 6.1: A Space Bézier Curve.

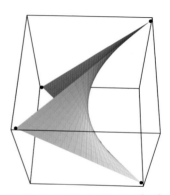

Figure 2.7: A Bilinear Surface.

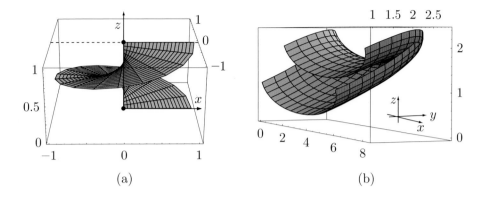

(a)   (b)

Figure 9.1: Two Sweep Surfaces.

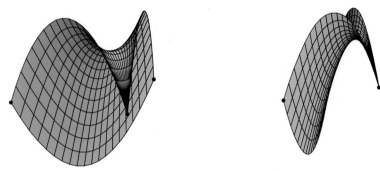

Figure 3.14: A Triangular Coons Surface.

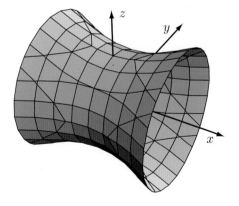

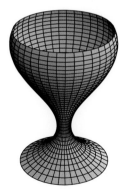

Figure C.3: An Implicit Surface.   Figure 9.8: A Surface of Revolution.

For a crowd is not company; and faces are but a gallery of
pictures; and talk but a tinkling cymbal, where there is no love.

—Francis Bacon

D
Mathematica Notes

One of the aims of this book is to give the reader confidence in writing *Mathematica* code for curves and surfaces. This chapter lists several of the *Mathematica* examples in the book and explains selected lines in each of them. The examples are all about curves and surfaces, which is why certain commands and techniques appear in several examples. Each command, technique, and approach is explained here once. *Mathematica* is an immense software system, with many commands, arguments, and options, which is why this short chapter often refers the reader to [Wolfram 03] (or the latest version of this excellent reference) for more details, more examples, and complete lists of options, data types, and directives.

The examples in this book have been written for ease of readability and are not the fastest or most sophisticated. They have all been run on version 3 of *Mathematica*. The first listing is the code for Figure 1.7 (effect of nonbarycentric weights).

```
1  (* non-barycentric weights example *)
2  Clear[p0,p1,g1,g2,g3,g4];
3  p0={0,0}; p1={5,6};
4  g1=ParametricPlot[(1-t)^3 p0+t^3 p1,{t,0,1}, PlotRange->All, Compiled->False,
5  DisplayFunction->Identity];
6  g3=Graphics[{AbsolutePointSize[4], {Point[p0],Point[p1]} }];
7  p0={0,-1}; p1={5,5};
8  g2=ParametricPlot[(1-t)^3 p0+t^3 p1,{t,0,1},PlotRange->All, Compiled->False,
9  PlotStyle->AbsoluteDashing[{2,2}], DisplayFunction->Identity];
10 g4=Graphics[{AbsolutePointSize[4], {Point[p0],Point[p1]} }];
11 Show[g2,g1,g3,g4, DisplayFunction->$DisplayFunction, DefaultFont->{"cmr10", 10}];
```

Line 1 is a comment. Anyone with any experience in computer coding, in any programming language, knows the importance of comments. The `Clear` command of line 2 is useful in cases where several programs are executed in different cells in one *Mathematica* session and should not affect each other. If a variable or a function is used by a program, and then used by another program without being redefined, it will have its original meaning. This is a useful feature where a large program can be divided into two parts ("cells" in *Mathematica* jargon) where the first part defines functions and the second part has the executable commands. However, if several cells are executed and there is no relation between them, a `Clear` command can save unnecessary errors and precious time spent on debugging.

Line 3 defines two variables of type "list." They are later used as points. Later examples show how to construct lists of control points or data points, either two-dimensional or three-dimensional. Line 4 is the first example of the `ParametricPlot` command (note the uppercase letters). This command plots a two-dimensional parametric curve (there is also a `ParametricPlot3D` version). It expects two or more arguments. The first argument is an expression (that normally depends on a parameter t) that evaluates to a pair of numbers for any value of t. Each pair is plotted as a point. If several curves should be plotted, this argument can be a list of expressions. The second argument is the range of values of t, written as {`t, tmin, tmax`}. The remaining arguments are options of `ParametricPlot`. This command has the same options as the low-level `Plot` command, and they are all listed in [Wolfram 03]. The options in this example are:

- `PlotRange->All`. Plot the entire curve. This option can be used to limit the plot to a certain rectangle.

- `Compiled->False`. Do not compile the parametric function.

- `DisplayFunction->Identity`. Do not display the graphics. Option `DisplayFunction` tells *Mathematica* how to display graphics. The value `Identity` implies no display.

The curve is not plotted immediately. Instead, it is assigned to variable g1, to be displayed later, with other graphics.

Line 6 prepares both p0 and p1 for display as points. Each is converted to an object of type `Point`, with an absolute size of four printer's points (there are 72 printer's points in an inch). There is also a `PointSize` option, where the size of a point is computed relative to the size of the entire display. The list of two points is assigned, as an object of type `Graphics`, to variable g3. Notice that the `Graphics` command accepts one argument that's a two-part list. The first part specifies the point size and the second part is the list of points. The following is a common mistake

```
Graphics[AbsolutePointSize[4], {Point[p0],Point[p1]} ]
```

which triggers the error message "Unknown Graphics option `AbsolutePointSize`." *Mathematica* doesn't recognize `AbsolutePointSize`, because it currently expects a single argument of type Graphics.

Line 7 assigns different coordinates to the two points, and lines 8 and 10 compute another curve and another list of two points and assign them to variables g2 and g4. Option `PlotStyle` receives the value `AbsoluteDashing`, which specifies the sizes of the dashes and spaces between them. In addition to dashing, plot styles may include graphics directives such as hue and thickness.

Finally, the `Show` command on line 11 displays the two curves and four points (variables g1 through g4). This command accepts any number of graphics arguments (two-dimensional or three-dimensional) followed by options, and displays the graphics. The options on line 11 are:

- `DisplayFunction->$DisplayFunction`.   This tells *Mathematica* to convert the graphics to Postscript and send it to the standard output.

■   DefaultFont->{"cmr10", 10}. Any text displayed will be in font cmr10 at a size
of 10 printer's point.

◇ **Exercise D.1:** Experiment to find out what happens if the semicolon following Show is
omitted.

The next listing is for Figure 2.7 (a bilinear Surface).

```
1  (* a bilinear surface patch *)
2  Clear[bilinear,pnts,u,w];
3  <<:Graphics:ParametricPlot3D.m;
4  pnts=ReadList["Points",{Number,Number,Number}, RecordLists->True];
5  bilinear[u_,w_]:=pnts[[1,1]](1-u)(1-w)+pnts[[1,2]]u(1-w) \
6  +pnts[[2,1]]w(1-u)+pnts[[2,2]]u w;
7  Simplify[bilinear[u,w]]
8  g1=Graphics3D[{AbsolutePointSize[5], Table[Point[pnts[[i,j]]],{i,1,2},{j,1,2}]}];
9  g2=ParametricPlot3D[bilinear[u,w],{u,0,1,.05},{w,0,1,.05}, Compiled->False,
10   DisplayFunction->Identity];
11 Show[g1,g2, ViewPoint->{0.063, -1.734, 2.905}];
```

Line 3 is the Get command, abbreviated "<<." It is followed by a file name. The
file is read and all the functions defined in it are evaluated, which makes it possible
to use them. The file specifies on this line is :Graphics:ParametricPlot3D.m, where
each colon indicates a folder or subdirectory. Line 4 reads data from file "Points" as
triplets of numbers into variable pnts. If option RecordLists is set to True, the list
in pnts will contain a sublist for each triplet read from the data file. Line 5 defines
the parametric function of the bilinear surface. Notice how the backslash "\" is the
Mathematica continuation symbol. When *Mathematica* gets to the end of an input line,
it sometimes cannot tell whether this is the end of a command. A continuation symbol
should be used to remove any ambiguity.

The Simplify command on line 7 displays the surface function in a simple form.
It does not contribute anything to the display. Line 8 is an example of Graphics3D, a
command that expects any number of graphics directives (from among Cuboid, Point,
Line, Polygon, and Text) followed by options. Line 9 is an example of the important
command ParametricPlot3D (note the uppercase letters). This command accepts a
parametric function (or a list of parametric functions) that evaluates to a triplet. This
is followed by one or two iterators of the form {u, umin, umax, du} where du is the
step size. If there is just one iterator, the result is a space curve. With two iterators,
this command generates a parametric surface.

◇ **Exercise D.2:** What is the effect of the iterators {u,0,1,.2},{w,0,1,.2}?

The Show command on line 11 employs the useful option ViewPoint which specifies
the point in space from which the three-dimensional object being displayed will be
viewed. ViewPoint->{x,y,z} specifies the position of the viewer relative to the center
of the bounding box (a three-dimensional box centered on the object).

Next, the code for Figure 6.27 (a rational Bézier surface patch) is listed. This
illustrates (1) the use of the If statement, (2) sums, (3) several commands to manipulate
lists, and (4) how the control polygon and coordinate axes can be included in a surface
display.

```
 1  (* A Rational Bezier Surface *)
 2  Clear[pwr,bern,spnts,n,m,wt,bzSurf,cpnts,patch,vlines,hlines,axes];
 3  <<:Graphics:ParametricPlot3D.m
 4  spnts={{{0,0,0},{1,0,1},{0,0,2}},
 5  {{1,1,0},{4,1,1},{1,1,2}}, {{0,2,0},{1,2,1},{0,2,2}}};
 6  m=Length[spnts[[1]]]-1; n=Length[Transpose[spnts][[1]]]-1;
 7  wt=Table[1, {i,1,n+1},{j,1,m+1}];
 8  wt[[2,2]]=5;
 9  pwr[x_,y_]:=If[x==0 && y==0, 1, x^y];
10  bern[n_,i_,u_]:=Binomial[n,i]pwr[u,i]pwr[1-u,n-i]
11  bzSurf[u_,w_]:=
12  Sum[wt[[i+1,j+1]]spnts[[i+1,j+1]]bern[n,i,u]bern[m,j,w], {i,0,n}, {j,0,m}]/
13  Sum[wt[[i+1,j+1]]bern[n,i,u]bern[m,j,w], {i,0,n}, {j,0,m}];
14  patch=ParametricPlot3D[bzSurf[u,w],{u,0,1}, {w,0,1},
15  Compiled->False, DisplayFunction->Identity];
16  cpnts=Graphics3D[{AbsolutePointSize[4], (* control points *)
17  Table[Point[spnts[[i,j]]], {i,1,n+1},{j,1,m+1}]}];
18  vlines=Graphics3D[{AbsoluteThickness[1], (* control polygon *)
19  Table[Line[{spnts[[i,j]],spnts[[i+1,j]]}], {i,1,n}, {j,1,m+1}]}];
20  hlines=Graphics3D[{AbsoluteThickness[1],
21  Table[Line[{spnts[[i,j]],spnts[[i,j+1]]}], {i,1,n+1}, {j,1,m}]}];
22  maxx=Max[Table[Part[spnts[[i,j]], 1], {i,1,n+1}, {j,1,m+1}]];
23  maxy=Max[Table[Part[spnts[[i,j]], 2], {i,1,n+1}, {j,1,m+1}]];
24  maxz=Max[Table[Part[spnts[[i,j]], 3], {i,1,n+1}, {j,1,m+1}]];
25  axes=Graphics3D[{AbsoluteThickness[1.5], (* the coordinate axes *)
26  Line[{{0,0,maxz},{0,0,0},{maxx,0,0},{0,0,0},{0,maxy,0}}]}];
27  Show[cpnts,hlines,vlines,axes,patch, PlotRange->All, DefaultFont->{"cmr10",10},
28  DisplayFunction->$DisplayFunction, ViewPoint->{2.783, -3.090, 1.243}];
```

Line 6 illustrates how array dimensions can be determined automatically and used later. Line 7 creates a table of weights that are all 1's and line 8 sets the center weight to 5. Line 9 defines function `pwr` that computes x^y, but returns a 1 in the normally-undefined case 0^0. Line 10 is an (inefficient) computation of the Bernstein polynomials and line 11–13 compute the rational Bézier surface as the ratio of two sums. Lines 18 and 20 prepare the segments of the control polygon. Several pairs of adjacent control points in array `spnts` are selected to form *Mathematica* objects of type `Line`. Lines 22–24 determine the maximum x, y, and z coordinates of the control points. These quantities are later used to plot the three coordinate axes. The construct

```
        Table[Part[spnts[[i,j]]], 1], {i,1,n+1}, {j,1,m+1}];
```

in line 22 creates a list with part 1 (i.e., the x coordinate) of every control point. The largest element of this list is selected, to become the length of the x axis. Line 26 shows how the `Line` command can have more than one pair of points. Finally, line 27 displays the surface with the control points, control polygon, and three coordinate axes.

The next example is a partial listing of the code for Figure 6.21 (a lofted Bézier surface patch). It illustrates one way of dealing with matrices whose elements are lists.

```
 1  pnts={{{0,1,0},{1,1,1},{2,1,0}},{{0,0,0},{1,0,0},{2,0,0}}};
 2  b1[w_]:={1-w,w}; b2[u_]:={(1-u)^2,2u(1-u),u^2};
 3  comb[i_]:=(b1[w].pnts)[[i]] b2[u][[i]];
 4  g1=ParametricPlot3D[comb[1]+comb[2]+comb[3], {u,0,1},{w,0,1}, Compiled->False,
 5  DefaultFont->{"cmr10", 10}, DisplayFunction->Identity,
 6  AspectRatio->Automatic, Ticks->{{0,1,2},{0,1},{0,.5}}];
```

The surface is computed as the product of the row vector `b1[w_]`, the matrix `pnts`, and the column `b2[u_]`. We first try the dot product `b1[w].pnts.b2[u]`, but this works only if the elements of matrix `pnts` are numbers. The following simple test

```
m={{m11,m12,m13},{m21,m22,m23}}; a={a1,a2}; b={b1,b2,b3};
a.m.b
```

produces the correct result

b1(a1 m11+a2 m21)+b2(a1 m12+a2 m22)+b3(a1 m13+a2 m23).

In our case, however, the elements of `pnts` are triplets, so the dot product `b1[w].pnts` produces a row of three triplets that we may denote by $((a, b, c), (d, e, f), (g, h, i))$. The dot product of this row by a column of the form (k, l, m) produces the triplet $(ka + lb + mc, kd + le + mf, kg + lh + mi)$ instead of the triplet $k(a, b, c) + l(d, e, f) + m(g, h, i)$. One way to obtain the correct result is to define a function `comb[i_]` that multiplies part i of `b1[w].pnts` by part i of `b2[u]`. The correct expression for the surface is then the sum `comb[1]+comb[2]+comb[3]`.

◇ **Exercise D.3:** When do we need the sum `comb[1]+comb[2]+comb[3]+comb[4]`?

Finally, the last listing is associated with Figure 6.24 (code for degree elevation of a rectangular Bézier surface). This code illustrates the extension of a smaller array `p` to an extended array `r`, some of whose elements are left undefined (they are set to the undefined symbol `a` and are never used). Array `r` is then used to compute the control points of a degree-elevated Bézier surface, and the point is that the undefined elements of `r` are not needed in this computation, but are appended to `r` (and also prepended to it) to simplify the computations.

```
1  (* Degree elevation of a rect Bezier surface from 2x3 to 4x5 *)
2  Clear[a,p,q,r];
3  m=1; n=2;
4  p={{p00,p01,p02},{p10,p11,p12}}; (* array of points *)
5  r=Array[a, {m+3,n+3}]; (* extended array, still undefined *)
6  Part[r,1]=Table[a, {i,-1,m+2}];
7  Part[r,2]=Append[Prepend[Part[p,1],a],a];
8  Part[r,3]=Append[Prepend[Part[p,2],a],a];
9  Part[r,n+2]=Table[a, {i,-1,m+2}];
10 MatrixForm[r] (* display extended array *)
11 q[i_,j_]:=({i/(m+1),1-i/(m+1)}. (* dot product *)
12  {{r[[i+1,j+1]],r[[i+1,j+2]]},{r[[i+2,j+1]],r[[i+2,j+2]]}}).
13  {j/(n+1),1-j/(n+1)}
14 q[2,3] (* test *)
```

Line 5 constructs array `r` two rows and two columns bigger than array `p`. Lines 6 and 9 fill up the first and last rows of `r` with the symbol `a`, while lines 7 and 8 move array `p` to the central area of `r` and then fill up the leftmost and rightmost columns of `r` with symbol `a`. Array `r` becomes the 4×5 matrix

$$\begin{bmatrix} a & a & a & a & a \\ a & p00 & p01 & p02 & a \\ a & p10 & p11 & p12 & a \\ a & a & a & a & a \end{bmatrix}.$$

Lines 11–13 compute the control points for the degree-elevated Bézier surface as described in Section 6.18. Each undefined symbol `a` corresponds to $i = 0$, $i = m+1$, $j = 0$, or $j = n + 1$, and is consequently multiplied by zero.

◇ **Exercise D.4:** Why is it important to clear the value of the undefined symbol a on line 2?

I have a number of notes about Mathematica, our products, and how I use them in this website and elsewhere. Please note, however, that beyond this there is no official connection whatsoever between Wolfram Research and this website. Everything on it is my personal opinion, not endorsed, controlled, or vetted any way by Wolfram Research. I am solely and entirely responsible for any and all errors, libels, liabilities, dangerous instructions, or just plain stupidities you may happen to find here.

Theodore Gray, *http://www.theodoregray.com*

Answers to Exercises

1: This is a row vector whose four elements are points and are therefore vectors (pairs in two dimensions and triplets in three dimensions).

1.1: Yes, because $(2, 2.5) = 0.5(1, 1) + 0.5(3, 4)$.

1.2: We can write $\mathbf{P}_1 = \mathbf{P}_0 + \alpha(\mathbf{P}_3 - \mathbf{P}_0)$ and similarly $\mathbf{P}_2 = \mathbf{P}_0 + \beta(\mathbf{P}_3 - \mathbf{P}_0)$. It is obvious that n collinear points can be represented by two points and $n-2$ real numbers.

1.3: Three two-dimensional points are independent if they are not collinear. The three corners of a triangle cannot, of course, be on the same line and are therefore independent. As a result, the three components of Equation (1.3), which are based on the coordinates of the corner points, are independent.

1.4: It is always true that $\mathbf{P}_0 = 1 \cdot \mathbf{P}_0 + 0 \cdot \mathbf{P}_1 + 0 \cdot \mathbf{P}_2$, so the barycentric coordinates of $\mathbf{P}_0$ are $(1, 0, 0)$. Points outside the triangle have barycentric coordinates, some of which are negative and others are greater than 1 (Figure Ans.1).

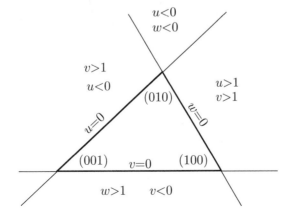

Figure Ans.1: Barycentric Coordinates Outside a Triangle.

1.5: This is easy. The centroid is given by $(1/3)\mathbf{P}_0 + (1/3)\mathbf{P}_1 + (1/3)\mathbf{P}_2$.

1.6: We can look at this sum in two ways:

1. As the sum $(\mathbf{P}+\mathbf{v})+(-\mathbf{Q}+\mathbf{w})$. We know that $\mathbf{P}+\mathbf{v}$ is a point and so is $-\mathbf{Q}+\mathbf{w}$. This is, therefore, the sum of points and it equals the vector from point $-\mathbf{Q}+\mathbf{w}$ to point $\mathbf{P}+\mathbf{v}$.

2. As the sum $(\mathbf{P}-\mathbf{Q})+\mathbf{v}+\mathbf{w}$. This is the sum of three vectors, so it is a vector.

1.7: (This is a long answer.) This is easily shown by showing that both dot products $(\mathbf{P}\times\mathbf{Q})\bullet\mathbf{P}$ and $(\mathbf{P}\times\mathbf{Q})\bullet\mathbf{Q}$ equal zero:

$$(\mathbf{P}\times\mathbf{Q})\bullet\mathbf{P} = P_1(P_2Q_3 - P_3Q_2) + P_2(-P_1Q_3 + P_3Q_1) + P_3(P_1Q_2 - P_2Q_1) = 0,$$

and similarly for $(\mathbf{P}\times\mathbf{Q})\bullet\mathbf{Q}$.

Perhaps the best proof is to construct the cross-product from first principles. Given the two vectors $\mathbf{P}$ and $\mathbf{Q}$, we are looking for a vector $\mathbf{R}$ perpendicular to both $\mathbf{P}$ and $\mathbf{Q}$. This requirement does not fully define $\mathbf{R}$, since both $\mathbf{R}$ and $-\mathbf{R}$ satisfy it, and since it says nothing about the magnitude of $\mathbf{R}$. We therefore extend our definition of the cross-product by requiring that the triplet $(\mathbf{P}, \mathbf{Q}, \mathbf{R})$ be a right-handed triad of vectors and also that the magnitude of $\mathbf{R}$ be the product $|\mathbf{P}|\,|\mathbf{Q}|\sin\theta$, where θ is the angle between $\mathbf{P}$ and $\mathbf{Q}$.

The derivation exploits the orthogonality of the three coordinate axes $\mathbf{i} = (1,0,0)$, $\mathbf{j} = (0,1,0)$, and $\mathbf{k} = (0,0,1)$ and also uses our definition. The definition implies that $\mathbf{i}\times\mathbf{i} = \mathbf{0}$, because the angle between $\mathbf{i}$ and itself is zero, and the same for $\mathbf{j}$ and $\mathbf{k}$. It also implies that the cross-product of any two of the three basis vectors is a unit vector, because the basis vectors are unit vectors and because $\sin 90° = 1$. Once we arrange the triplet $(\mathbf{i},\mathbf{j},\mathbf{k})$ as a right-handed triad, we can deduce the following: $\mathbf{i}\times\mathbf{j} = \mathbf{k}$, $\mathbf{j}\times\mathbf{k} = \mathbf{i}$, $\mathbf{k}\times\mathbf{i} = \mathbf{j}$, $\mathbf{j}\times\mathbf{i} = -\mathbf{k}$, $\mathbf{k}\times\mathbf{j} = -\mathbf{i}$, and $\mathbf{i}\times\mathbf{k} = -\mathbf{j}$.

Armed with this information we can easily derive the cross-product $\mathbf{R}$

$$\begin{aligned}
\mathbf{R} = \mathbf{P}\times\mathbf{Q} &= (P_1\mathbf{i} + P_2\mathbf{j} + P_3\mathbf{k})\times(Q_1\mathbf{i} + Q_2\mathbf{j} + Q_3\mathbf{k})\\
&= (P_1\mathbf{i} + P_2\mathbf{j} + P_3\mathbf{k})\times Q_1\mathbf{i} + (P_1\mathbf{i} + P_2\mathbf{j} + P_3\mathbf{k})\times Q_2\mathbf{j} + (P_1\mathbf{i} + P_2\mathbf{j} + P_3\mathbf{k})\times Q_3\mathbf{k}\\
&= (P_2Q_3 - P_3Q_2)\mathbf{i} + (-P_1Q_3 + P_3Q_1)\mathbf{j} + (P_1Q_2 - P_2Q_1)\mathbf{k}\\
&= (P_2Q_3 - P_3Q_2, -P_1Q_3 + P_3Q_1, P_1Q_2 - P_2Q_1).
\end{aligned}$$

The magnitude of $\mathbf{R}$ can be calculated explicitly

$$\begin{aligned}
|\mathbf{R}|^2 &= (P_2Q_3 - P_3Q_2)^2 + (-P_1Q_3 + P_3Q_1)^2 + (P_1Q_2 - P_2Q_1)\\
&= (P_1^2 + P_2^2 + P_3^2)(Q_1^2 + Q_2^2 + Q_3^2) - (P_1Q_1 + P_2Q_2 + P_3Q_3)^2\\
&= |\mathbf{P}|^2|\mathbf{Q}|^2 - (\mathbf{P}\cdot\mathbf{Q})^2 = |\mathbf{P}|^2|\mathbf{Q}|^2 - (|\mathbf{P}||\mathbf{Q}|\cos\theta)^2\\
&= |\mathbf{P}|^2|\mathbf{Q}|^2(1 - \cos^2\theta) = |\mathbf{P}|^2|\mathbf{Q}|^2\sin^2\theta.
\end{aligned}$$

To illustrate the magnitude, we can draw the parallelogram defined by $\mathbf{P}$ and $\mathbf{Q}$ (with an angle θ between them) and show that vector $\mathbf{Q}\sin\theta$ is perpendicular to $\mathbf{P}$.

For those who insist on learning the whole story, here is a short discussion of the cross-product of vectors in four dimensions.

In three dimensions, for any two linearly independent (i.e., nonparallel) vectors there is a vector (unique up to magnitude and direction) that's perpendicular to both. The cross-product in three dimensions is therefore a binary operation. In four dimensions the situation is more complex. For any three linearly independent vectors there is a unique vector that's orthogonal (a term more general than *perpendicular*) to all three. The cross-product in four dimensions is therefore a ternary operation that may be denoted by $\times(\mathbf{U}, \mathbf{V}, \mathbf{W})$. Given just two four-dimensional vectors $\mathbf{P}$ and $\mathbf{Q}$, there is an entire subspace of vectors that are orthogonal to both $\mathbf{P}$ and $\mathbf{Q}$, so the cross-product of two vectors in four dimensions is not well defined.

Based on our experience with the cross-product in three dimensions, it is reasonable to expect the definition of the four-dimensional cross product to satisfy the following requirements.

1. If the operands are linearly independent, the cross-product must be orthogonal to each of the operands.

2. Scaling must be conserved. The expressions $\times(\alpha\mathbf{U}, \mathbf{V}, \mathbf{W})$, $\times(\mathbf{U}, \alpha\mathbf{V}, \mathbf{W})$, and $\times(\mathbf{U}, \mathbf{V}, \alpha\mathbf{W})$ should equal $\alpha\times(\mathbf{U}, \mathbf{V}, \mathbf{W})$ for any real α.

3. Changing the order of two of the operands should reverse the direction of the result.

4. If the three operands are not linearly independent, the four-dimensional cross-product must be the zero vector.

It is easy to show that these four requirements are satisfied if the determinant notation of the three-dimensional cross-product is extended to four dimensions. Hence, we can write

$$
\times(\mathbf{U}, \mathbf{V}, \mathbf{W}) = \begin{vmatrix} \mathbf{i} & \mathbf{j} & \mathbf{k} & \mathbf{l} \\ U_0 & U_1 & U_2 & U_3 \\ V_0 & V_1 & V_2 & V_3 \\ W_0 & W_1 & W_2 & W_3 \end{vmatrix}
$$

$$
= \mathbf{i} \begin{vmatrix} U_1 & U_2 & U_3 \\ V_1 & V_2 & V_3 \\ W_1 & W_2 & W_3 \end{vmatrix} - \mathbf{j} \begin{vmatrix} U_0 & U_2 & U_3 \\ V_0 & V_2 & V_3 \\ W_0 & W_2 & W_3 \end{vmatrix} + \mathbf{k} \begin{vmatrix} U_0 & U_1 & U_3 \\ V_0 & V_1 & V_3 \\ W_0 & W_1 & W_3 \end{vmatrix} - \mathbf{l} \begin{vmatrix} U_0 & U_1 & U_2 \\ V_0 & V_1 & V_2 \\ W_0 & W_1 & W_2 \end{vmatrix}.
$$

(End of answer.)

1.8: In the special case where $\mathbf{i} = (1, 0, 0)$ and $\mathbf{j} = (0, 1, 0)$, it is easy to verify that the product $\mathbf{i} \times \mathbf{j}$ equals $(0, 0, 1) = \mathbf{k}$. Thus, the triplet $(\mathbf{i}, \mathbf{j}, \mathbf{i} \times \mathbf{j} = \mathbf{k})$ has the handedness of the coordinate system (it is either right-handed or left-handed, depending on the coordinate system). In a right-handed coordinate system, the right-hand rule makes it easy to predict the direction of $\mathbf{P} \times \mathbf{Q}$. The rule is as follows: If your thumb points in the direction of $\mathbf{P}$ and your second finger points in the direction of $\mathbf{Q}$, then your middle finger will point in the direction of $\mathbf{P} \times \mathbf{Q}$. In a left-handed coordinate system, a similar left-hand rule applies.

1.9: They either point in the same direction, or in opposite directions.

1.10:   We are looking for a vector $\mathbf{P}(t)$ that's linear in t and that satisfies $\mathbf{P}(0) = \mathbf{P}_1$ and $\mathbf{P}(1) = \mathbf{P}_2$. It is easy to guess that

$$\mathbf{P}(t) = (1 - t)\mathbf{P}_1 + t\mathbf{P}_2 = t(\mathbf{P}_2 - \mathbf{P}_1) + \mathbf{P}_1$$

satisfies both conditions. Note that this result is duplicated in Equation (2.1). It is an important relation and is used often in graphics.

1.11:   This is straightforward

$$\mathbf{c} = \frac{2 \cdot 1 + 1 \cdot 0 + 3 \cdot (-1)}{1^2 + 0^2 + (-1)^2}(1, 0, -1) = (-1/2, 0, 1/2),$$
$$\mathbf{d} = \mathbf{a} - \mathbf{c} = (2.5, 1, 2.5).$$

1.12:   This expression is an attempt to find a parametric cubic polynomial that's close to a circle in the first quadrant. The general form of such a polynomial is $\mathbf{P}(t) = \mathbf{a}t^3 + \mathbf{b}t^2 + \mathbf{c}t + \mathbf{d}$ where the coefficients $\mathbf{a}, \mathbf{b}, \mathbf{c}$, and $\mathbf{d}$ are pairs of numbers and t is a parameter varying in the interval $[0, 1]$. To determine the four coefficient pairs, we need four equations, so we require that the polynomial and the circle be identical at four points. For $t = 0$, we require that $\mathbf{P}(0) = (1, 0)$ and, for $t = 1$, that $\mathbf{P}(1) = (0, 1)$. In addition, we select the two equally-spaced values $t = 1/3$ and $2/3$ and require that $\mathbf{P}(1/3) = (\cos 30°, \sin 30°)$ and $\mathbf{P}(1/3) = (\cos 60°, \sin 60°)$. This results in the four equations

$$\mathbf{P}(0) = \mathbf{a}t^3 + \mathbf{b}t^2 + \mathbf{c}t + \mathbf{d}|_{t=0} = (1, 0),$$
$$\mathbf{P}(1/3) = \mathbf{a}t^3 + \mathbf{b}t^2 + \mathbf{c}t + \mathbf{d}|_{t=1/3} = (\cos 30°, \sin 30°),$$
$$\mathbf{P}(2/3) = \mathbf{a}t^3 + \mathbf{b}t^2 + \mathbf{c}t + \mathbf{d}|_{t=2/3} = (\cos 60°, \sin 60°),$$
$$\mathbf{P}(1) = \mathbf{a}t^3 + \mathbf{b}t^2 + \mathbf{c}t + \mathbf{d}|_{t=1} = (0, 1),$$

whose solutions are $\mathbf{a} = (0.441, -0.441)$, $\mathbf{b} = (-1.485, -0.162)$, $\mathbf{c} = (0.044, 1.603)$, and $\mathbf{d} = (1, 0)$. The cubic polynomial is therefore

$$\mathbf{P}(t) = (0.441, -0.441)t^3 + (-1.485, -0.162)t^2 + (0.044, 1.603)t + (1, 0).$$

This polynomial is just an approximation (see Appendix B for more circle approximations). At other values of t, it passes close to the circle but not on it.

1.13:   It is easy to see from Figure 1.6b that $d = R\cos(\pi/n)$, so $R - d = R(1 - \cos(\pi/n))$. This expression approaches zero for large n.

1.14:   This velocity is variable since it goes down from $\mathbf{P}^t(0) = (0, 2)$ (a speed of $\sqrt{0^2 + 2^2} = 2$) to $\mathbf{P}^t(1) = (-1, 0)$ (a speed of $\sqrt{(-1)^2 + 0^2} = 1$). Notice that the term "speed" refers to a scalar, whereas "velocity" is a vector, having both direction and magnitude.

1.15: We know that the two-dimensional parametric curve $(\cos t, \sin t)$ is a circle of radius 1, centered on the origin. As a result, the three-dimensional curve $(\cos t, \sin t, t)$ is a helix spiraling around the z axis upward from the origin.

1.16: The two simple curves $x(t)$ and $y(t)$ defined below are identical. When drawn in the xt or yt plane, each is a horizontal line followed by a $45°$ line:

$$x(t) = y(t) = \begin{cases} 0.5, & 0 \leq t \leq 0.5, \\ t, & 0.5 \leq t \leq 1. \end{cases}$$

The curve itself is now defined parametrically:

$$\mathbf{P}(t) = (x(t), y(t)) = \begin{cases} (0.5, 0.5), & 0 \leq t \leq 0.5, \\ (t, t), & 0.5 \leq t \leq 1. \end{cases}$$

In the range $0 \leq t \leq 0.5$ the curve stays at point $(0.5, 0.5)$, it is degenerate. Then, when $0.5 \leq t \leq 1$, the curve moves smoothly from $(0.5, 0.5)$ to $(1, 1)$.

1.17: The following functions are degree-3 polynomials in t and are not straight lines:

$$x(t) = 2t^3 - 3t^2 + 2, \quad y(t) = -4t^3 + 6t^2 + 1, \quad z(t) = -2t^3 + 3t^2 + 3.$$

When combined to form a parametric space curve, the result is

$$\mathbf{P}(t) = \big(x(t), y(t), z(t)\big) = (-1, 2, 1)(-2t^3 + 3t^2) + (2, 1, 3).$$

A simple change of parameter $T = -2t^3 + 3t^2$ yields $\mathbf{P}(T) = (-1, 2, 1)T + (2, 1, 3)$, a straight line from point $(2, 1, 3)$ to point $(-1, 2, 1) + (2, 1, 3) = (1, 3, 4)$. Notice that $t = 0 \rightarrow T = 0$ and $t = 1 \rightarrow T = 1$. The expression $(-2t^3 + 3t^2)$ also happens to be function $F_2(t)$ of Equation (4.6) and is plotted in Figure 4.3.

1.18: The curve can be written $\mathbf{P}(t) = \mathbf{P} + (\mathbf{Q} - \mathbf{P})[2\alpha t + (1 - 2\alpha)t^2]$. We define $T = 2\alpha t + (1 - 2\alpha)t^2$ and substitute T for t as the parameter. Note that $t = 0 \rightarrow T = 0$ and $t = 1 \rightarrow T = 1$. The curve can now be written $\mathbf{P}(T) = \mathbf{P} + (\mathbf{Q} - \mathbf{P})T$ (where $0 \leq T \leq 1$), which is linear in T and is therefore a straight line. This is a (sometimes baffling) property of parametric curves. A substitution of the parameter does not change the shape of the curve and can be used to shed light on its behavior. Intuitively, the reason our curve is a straight line is that the same vector $(\mathbf{Q} - \mathbf{P})$ is used in the coefficients of both t and t^2.

1.19: Such a polynomial is fully defined by three coefficients $\mathbf{A}$, $\mathbf{B}$, and $\mathbf{C}$ that can be considered three-dimensional points and any three points are on the same plane.

1.20: We can gain an insight into the shape of the n-degree polynomial $P(x) = \sum_{i=0}^{n} A_i x^i$ by writing the equation $P(x) = 0$. This is an nth-degree equation in the unknown x and consequently has n solutions (some may be identical or complex). Each solution is an x value for which the polynomial becomes zero. As x is varied, the polynomial crosses the x axis n times, so it oscillates between positive and negative values.

1.21: Because $\mathbf{P}_1(t)$ is expressed in Equation (1.12) with the same matrix $\mathbf{M}$ and the same four points as $\mathbf{P}(t)$.

1.22: The attributes "vertical" and "horizontal" are extrinsic. "Cusp" and "smooth," however, are intrinsic. The length of a curve and area of a polygon or a closed curve are extrinsic since they can be changed by scaling the coordinate system. If a certain point on a surface has a tangent plane in one coordinate system (i.e., the surface is smooth in the vicinity of the point), it will have such a plane (although perhaps a different one) in any other coordinate system. This property of a surface is therefore intrinsic.

1.23: The principal normal vector at point i points, by definition, in the direction the curve turns at the point. Since a straight line does not make any turns, its principal normal vector is undefined. We can also see this from Equation (1.18). The second derivative of a straight line is the zero vector, so vector $\mathbf{K}(t)$ is also zero, resulting in a principal normal vector of the form $0/0$.

1.24: The first two derivatives are $\mathbf{P}^t(t) = (-3,0)t^2 + (2,-2)t + (1,1)$ and $\mathbf{P}^{tt}(t) = (-6,0)t + (2,-2)$. The principal normal vector (still unnormalized) is therefore

$$\mathbf{N}(t) = \mathbf{P}^{tt}(t) - \frac{\mathbf{P}^{tt}(t) \bullet \mathbf{P}^t(t)}{|\mathbf{P}^t(t)^2|}\mathbf{P}^t(t) = \mathbf{P}^{tt}(t) - \left[\frac{18t^3 - 18t^2 + 2t}{9t^4 - 12t^3 + 2t^2 + 2}\right]\mathbf{P}^t(t).$$

Simple tests result in $\mathbf{N}(0) = (2,-2)$, $\mathbf{N}(.5) = (0,-2)$, and $\mathbf{N}(1) = (-4,0)$. Thus, vector $\mathbf{N}(t)$ starts in direction $(1,-1)$, changes to $(0,-1)$ (down) when $t = 0.5$ (this makes sense since $\mathbf{P}^t(0.5)$ is horizontal), and ends in direction $(-1,0)$ (i.e., in the negative x direction). It is always perpendicular to the direction of the curve.

1.25: The curve and its first two derivatives are given by

$$\begin{aligned}
\mathbf{P}(t) &= (1-t)^3(0,0,0) + 3t(1-t)^2(1,0,0) + 3t^2(1-t)(2,1,0) + t^3(3,0,1) \\
&= (3t, 3t^2(1-t), t^3), \\
\mathbf{P}^t(t) &= (3, 3t(2-3t), 3t^2), \\
\mathbf{P}^{tt}(t) &= (0, 6-18t, 6t).
\end{aligned}$$

The unnormalized principal normal vector is given by

$$\mathbf{N}(t) = \mathbf{P}^{tt}(t) - \left[\frac{18t(2-9t+10t^2)}{9+9t^2(2-3t)^2}\right]\mathbf{P}^t(t),$$

from which we get $\mathbf{N}(0) = (0,6,0)$, $\mathbf{N}(0.5) = (0,-3,3)$, and $\mathbf{N}(1) = (-9,-3,-3)$.

The osculating plane is the solution of $\det[((x,y,z) - \mathbf{P}(t))\,\mathbf{P}^t(t)\,\mathbf{P}^{tt}(t)] = 0$. The explicit determinant is

$$\begin{vmatrix} x-3t & y-3t^2(1-t) & z-t^3 \\ 3 & 3t(2-3t) & 3t^2 \\ 0 & 6-18t & 6t \end{vmatrix}.$$

Thus, the osculating plane is given by $t^2 x - ty + (1 - 3t)z - t^3 = 0$. At $t = 0, 0.5$ and 1 this plane has the equations $z = 0$, $0.25x - 0.5y - 0.5z - 0.125 = 0$, and $x - y - 2z - 1 = 0$, respectively.

1.26: Equation (1.24) becomes

$$\frac{d^2 x}{ds^2} = -R \frac{dy}{ds}, \quad \frac{d^2 y}{ds^2} = R \frac{dx}{ds}.$$

It is easy to guess that the solutions are $x(s) = R \cos(R \cdot s) + A$ and $y(s) = R \sin(R \cdot s) + B$. The curve is a circle of radius R with center at (A, B).

1.27: A surface is two-dimensional because it has no depth. Imagine a flat surface, such as the xy plane. Each point on this surface has just two coordinates (the third one, z, is zero) and can be located by means of these two numbers. Now, crumple this flat surface. Each surface point now has three coordinates (the z coordinate is no longer zero), but the same two numbers are still the distances of the point from the two edges of the surface and are therefore still sufficient to locate the point on the crumpled surface.

A surface is a two-dimensional structure embedded in three-dimensional space. Each point on the surface has three coordinates, but only two numbers are needed to specify the position of the point on the surface. In contrast, a solid object requires three parameters to be expressed. The surface function $\mathbf{P}(u, w)$ evaluates to a triplet (the three coordinates of a point on the surface) for every pair (u, w) of parameters.

1.28: It is easy to see that the corner points are $\mathbf{P}_{00} = (0, 0, 1)$, $\mathbf{P}_{10} = (1, 0, 0)$, $\mathbf{P}_{01} = (0.5, 1, 0)$, and $\mathbf{P}_{11} = (1, 1, 0)$. The boundary curves are also not hard to calculate. They are

$$\mathbf{P}(0, w) = (0.5w, w, 1 - w), \quad \mathbf{P}(u, 1) = (0.5(1 - u) + u, 1, 0),$$
$$\mathbf{P}(1, w) = (1, w, 0), \quad \mathbf{P}(u, 0) = (u, 0, 1 - u).$$

The two diagonals are

$$\mathbf{P}(u, 1 - u) = (0.5(1 - u)^2 + u, 1 - u, (1 - u)u),$$
$$\mathbf{P}(u, u) = (0.5(1 - u)u + u, u, (1 - u)^2).$$

1.29: The four boundary curves are

$$\mathbf{P}(u, 0) = ((c - a)u + a, b, 0), \quad \mathbf{P}(u, 1) = ((c - a)u + a, d, 0),$$
$$\mathbf{P}(0, w) = (a, (d - b)w + b, 0), \quad \mathbf{P}(1, w) = (c, (d - b)w + b, 0).$$

Obviously, they are straight lines. The four corner points can be obtained from the boundary curves

$$\mathbf{P}_{00} = (a, b, 0), \quad \mathbf{P}_{01} = (a, d, 0), \quad \mathbf{P}_{10} = (c, b, 0); \quad \mathbf{P}_{11} = (c, d, 0).$$

The surface patch is the flat rectangle on the xy plane delimited by these points.

1.30:   We can always write the explicit surface $z = f(x, y)$ as the implicit surface $f(x, y) - z = 0$. The normal is, therefore,

$$\left(\frac{\partial f(x_0, y_0)}{\partial x}, \frac{\partial f(x_0, y_0)}{\partial y}, -1 \right).$$

1.31:   The normal will point in the negative x direction.

1.32:   The face is given by $\mathbf{P}(u, w) = (a(2u - 1)(1 - w), a(w - 1), Hw)$. The two partial derivatives are

$$\frac{\partial \mathbf{P}}{\partial u} = (2a(1 - w), 0, 0), \quad \frac{\partial \mathbf{P}}{\partial w} = (a(1 - 2u), a, H).$$

The normal is the cross-product (Equation (1.5), page 7) $2a(1 - w)[0, -H, a]$.

  To understand this result, recall that the face in question is the triangle defined by the three points $(-a, -a, 0)$, $\mathbf{P}_2 = (a, -a, 0)$, and $(0, 0, H)$. This explains why the x component of the normal is zero. Note that its magnitude depends on w, but its direction does not. The direction of the normal can be expressed by saying "for each H units traveled in the negative y direction, we should travel a units in the positive z direction."

1.33:   The cone is defined by $\mathbf{P}(u, w) = (Ru \cos w, Ru \sin w, Hu)$. The two partial derivatives are

$$\frac{\partial \mathbf{P}}{\partial u} = (R \cos w, R \sin w, H), \quad \frac{\partial \mathbf{P}}{\partial w} = (- Ru \sin w, Ru \cos w, 0).$$

The normal is the cross-product [Equation (1.5)] $Ru(-H \cos w, -H \sin w, R)$. Note that its direction does not depend on u. When w varies, the normal rotates about the z axis, and it always has a positive component in the z direction.

1.34:   The line from $(-a, 0, R)$ to $(a, 0, R)$ is given by $(a(2u - 1), 0, R)$. The surface is given by the product of this line and the rotation matrix about the x axis:

$$\mathbf{P}(u, w) = (a(2u - 1), 0, R) \begin{pmatrix} 1 & 0 & 0 \\ 0 & \cos w & -\sin w \\ 0 & \sin w & \cos w \end{pmatrix}$$

$$= (a(2u - 1), R \sin w, R \cos w), \quad\quad\quad \text{(Ans.1)}$$

where $0 \le u \le 1$ and $0 \le w \le 2\pi$. The two partial derivatives are

$$\frac{\partial \mathbf{P}}{\partial u} = (2a, 0, 0), \quad \frac{\partial \mathbf{P}}{\partial w} = (0, R \cos w, -R \sin w).$$

The normal is the cross-product $2aR(0, \sin w, \cos w)$. Note that it is perpendicular to the x axis. When w varies, the normal rotates about the x axis.

2.1: Three approaches are discussed.

Approach 1: The general two-dimensional line $y = ax + b$ goes through point $(0, b)$ and its direction is the vector $(1, a)$ (for each step in the x direction, take a steps in the y direction). We can therefore express it as

$$\mathbf{P}(t) = \mathbf{P}_0 + t\mathbf{v} = (0, b) + t(1, a).$$

Applying Equation (2.2), we get point $\mathbf{Q}$:

$$\mathbf{Q} = \mathbf{P}_0 + \frac{(\mathbf{P} - \mathbf{P}_0) \bullet \mathbf{v}}{\mathbf{v} \bullet \mathbf{v}}\mathbf{v}$$

$$= (0, b) + \frac{(P_x, P_y - b) \bullet (1, a)}{(1, a) \bullet (1, a)}(1, a)$$

$$= \left(\frac{aP_y + P_x - ab}{a^2 + 1}, \frac{a^2 P_y + aP_x + b}{a^2 + 1}\right) \qquad \text{(Ans.2)}$$

Hence, the distance between $\mathbf{P}$ and $\mathbf{Q}$ is

$$D = \sqrt{(P_x - Q_x)^2 + (P_y - Q_y)^2}$$

$$= \sqrt{\left(P_x - \frac{aP_y + P_x - ab}{a^2 + 1}\right)^2 + \left(P_y - \frac{a^2 P_y + aP_x + b}{a^2 + 1}\right)^2}$$

$$= \sqrt{\left(\frac{a^2 P_x - aP_y + ab}{a^2 + 1}\right)^2 + \left(\frac{P_y - aP_x - b}{a^2 + 1}\right)^2}$$

$$= \sqrt{(P_y - aP_x - b)^2} = \frac{|P_y - aP_x - b|}{\sqrt{1 + a^2}};$$

same as Equation (2.3).

Approach 2: We denote the line $y = ax + b$ by L_1. We find $\mathbf{Q}$, the point on L_1 closest to $\mathbf{P}$, by calculating the equation of a line L_2 that's (1) perpendicular to L_1 and (2) goes through $\mathbf{P}$. Denote L_2 by $y = Ax + B$. Since L_2 is perpendicular to L_1, its slope is $-1/a$. The requirement that it goes through $\mathbf{P}$ gives us an equation for B:

$$P_y = -\frac{1}{a}P_x + B, \quad \text{whose solution is} \quad B = P_y + \frac{P_x}{a}.$$

Therefore, line L_2 is

$$y = -\frac{1}{a}x + \left(P_y + \frac{P_x}{a}\right).$$

The intersection of the two lines yields point $\mathbf{Q}$:

$$ax + b = -\frac{1}{a}x + \left(P_y + \frac{P_x}{a}\right) \quad \text{yields} \quad x = \frac{aP_y + P_x - ab}{a^2 + 1},$$

and

$$y = a \left(\frac{aP_y + P_x - ab}{a^2 + 1} \right) + b = \frac{a^2 P_y + aP_x + b}{a^2 + 1}.$$

Thus, point $\mathbf{Q}$ is

$$\mathbf{Q} = \left(\frac{aP_y + P_x - ab}{a^2 + 1}, \frac{a^2 P_y + aP_x + b}{a^2 + 1} \right),$$

which is the same as that given by Equation (Ans.2).

Approach 3: Any point $\mathbf{Q}$ on the line has coordinates $(x, ax + b)$. The distance D between $\mathbf{P}$ and a general point $\mathbf{Q}$ on the line is therefore

$$D(x) = \sqrt{(P_x - x)^2 + (P_y - ax - b)^2}.$$

This distance is a function of x and we are looking for that x value for which $D(x)$ has a minimum. Instead of differentiating $D(x)$ (tedious because of the square root), we differentiate $D^2(x)$, noting that both functions $D(x)$ and $D^2(x)$ have a minimum at the same value of x. The result is

$$\frac{d}{dx} D^2(x) = -2(P_x - x) - 2a(P_y - ax - b)$$
$$= -2P_x - 2aP_y + 2ab + 2x + 2a^2 x.$$

When this is equated to zero, we find that $D(x)$ has a minimum for

$$x = \frac{aP_y + P_x - ab}{a^2 + 1}.$$

Thus, point $\mathbf{Q}$ is

$$\mathbf{Q} = (x, ax + b) = \left(\frac{aP_y + P_x - ab}{a^2 + 1}, \frac{a^2 P_y + aP_x + b}{a^2 + 1} \right),$$

same as Equation (Ans.2). The distance is therefore given by Equation (2.3)

2.2: Direct calculation shows that in the former case, both α and β have the indefinite value $0/0$. In the latter case, they are both of the form $x/0$, where x is nonzero.

2.3: We select $(1, 0, 0)$ as our pivot point and calculate the three vectors $\mathbf{v}_1 = (0, 1, 0) - (1, 0, 0) = (-1, 1, 0)$, $\mathbf{v}_2 = (1, a, 1) - (1, 0, 0) = (0, a, 1)$, and $\mathbf{v}_3 = (0, -a, 0) - (1, 0, 0) = (-1, -a, 0)$. Next, we calculate the only scalar triple product

$$\mathbf{v}_1 \bullet (\mathbf{v}_2 \times \mathbf{v}_3) = \begin{vmatrix} -1 & 1 & 0 \\ 0 & a & 1 \\ -1 & -a & 0 \end{vmatrix} = a + 1.$$

It equals zero for $a = -1$.

2.4: The plane should pass through the three points $(0,0,0)$, $(0,0,1)$, and $(1,1,0)$. Equation (2.4) gives

$$A = \begin{vmatrix} 0 & 0 & 1 \\ 0 & 1 & 1 \\ 1 & 0 & 1 \end{vmatrix} = -1, \qquad B = - \begin{vmatrix} 0 & 0 & 1 \\ 0 & 1 & 1 \\ 1 & 0 & 1 \end{vmatrix} = 1,$$

$$C = \begin{vmatrix} 0 & 0 & 1 \\ 0 & 0 & 1 \\ 1 & 1 & 1 \end{vmatrix} = 0, \qquad D = - \begin{vmatrix} 0 & 0 & 0 \\ 0 & 0 & 1 \\ 1 & 1 & 0 \end{vmatrix} = 0.$$

The expression of the plane is, therefore, $-x + y = 0$

2.5: They are the points where the plane $x/a + y/b + z/c = 1$ intercepts the three coordinate axes.

2.6: $s = \mathbf{N} \bullet \mathbf{P}_1 = (1,1,1) \bullet (1,1,1) = 3$, so the plane is given by $x + y + z - 3 = 0$. It intercepts the three coordinate axes at points $(3,0,0)$, $(0,3,0)$, and $(0,0,3)$ (Figure Ans.2).

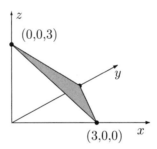

Figure Ans.2: A Plane.

2.7: The expression is

$$\mathbf{P}(u,w) = \mathbf{P}_1 + u(\mathbf{P}_2 - \mathbf{P}_1) + w(\mathbf{P}_3 - \mathbf{P}_1) = (3,0,0) + u(-3,3,0) + w(-3,0,3).$$

2.8: If the cross-product $\mathbf{a} \times \mathbf{b}$ points in the direction of $\mathbf{N}$, the angle between them is zero. Its cosine therefore equals 1, causing the dot product $\mathbf{N} \bullet (\mathbf{a} \times \mathbf{b})$ to be positive (since it is the product of the magnitudes of the vectors and the cosine of the angle between them).

2.9: If the line is parallel to the plane, then its direction vector $\mathbf{d}$ is parallel to the plane (i.e., perpendicular to the normal), resulting in $\mathbf{N} \bullet \mathbf{d} = 0$ (infinite t). If the line is also in the plane, then $\mathbf{P}_1$ is in the plane, resulting in $s = \mathbf{N} \bullet \mathbf{P}_1$ or $s - \mathbf{N} \bullet \mathbf{P}_1 = 0$ (in this case, t is of the form $0/0$, indefinite).

2.10: We first subtract $\mathbf{P}_2 - \mathbf{P}_1 = (-2, 1, -0.8)$ and $\mathbf{P}_3 - \mathbf{P}_1 = (-2, 9, -0.8)$. The triangle is therefore given by

$$(10, -5, 4) + u(-2, 1, -0.8) + w(-2, 9, -0.8)$$
$$= \big(10 - 2(u + w), -5 + u + 9w, 4 - 0.8(u + w)\big),$$

where $u \geq 0$, $w \geq 0$, and $u + w \leq 1$.

2.11: We first subtract $\mathbf{P}_2 - \mathbf{P}_1 = (-2, 1, -0.8)$ and $\mathbf{P}_3 - \mathbf{P}_1 = (2, -1, 0.8)$. The differences are related because the points are collinear; the triangle is therefore given by $(10, -5, 4) + (-2, 1, -0.8)(u - w) = \mathbf{P}_1 + (\mathbf{P}_2 - \mathbf{P}_1)(u - w)$. It depends only on the difference $u - w$. When u and w are varied independently, the difference between them changes from -1 to 1. The triangle therefore degenerates into the straight line $\mathbf{P}(t) = \mathbf{P}_1 + (\mathbf{P}_2 - \mathbf{P}_1)t$, where $-1 \leq t \leq 1$. This line goes from $\mathbf{P}(-1) = \mathbf{P}_1 - (\mathbf{P}_2 - \mathbf{P}_1) = 2\mathbf{P}_1 - \mathbf{P}_2 = \mathbf{P}_3$ to $\mathbf{P}(1) = \mathbf{P}_1 + (\mathbf{P}_2 - \mathbf{P}_1) = \mathbf{P}_2$.

2.12: 1. Equation (2.8) yields the expression of the surface

$$\mathbf{P}(u, w) = ((0,0,0)(1-u)(1-w) + (1,0,0)u(1-w) + (0,1,0)(1-u)w + (1,1,1)uw$$
$$= (u, w, uw).$$

2. The explicit representation is $z = xy$. This is easy to guess because the x coordinate equals u, the y coordinate is w, and the z coordinate equals uw.

3. The two conditions $z = k$ and $z = xy$ produce $k = xy$ or $y = k/x$. This curve is a hyperbola.

4. The plane through the three points $(0,0,0)$, $(0,0,1)$, and $(1,1,0)$ contains the z axis and is especially easy to calculate. Its equation is $x - y = 0$. Intersected with $z = xy$, it yields the curve $z = x^2$, a parabola.

This is the reason why the bilinear surface is sometimes called a *hyperbolic paraboloid*.

2.13: The two tangent vectors are

$$\frac{\partial \mathbf{P}(u, w)}{\partial u} = (1 - w, 0, w - 1), \quad \frac{\partial \mathbf{P}(u, w)}{\partial w} = (-u, 1, u - 1).$$

The normal vector is

$$\mathbf{N}(u, w) = \frac{\partial \mathbf{P}(u, w)}{\partial u} \times \frac{\partial \mathbf{P}(u, w)}{\partial w} = (1 - w, 1 - w, 1 - w) = (1 - w)(1, 1, 1).$$

This vector does not depend on u, it always points in the $(1, 1, 1)$ direction, and its magnitude varies from $(1, 1, 1)$ for $w = 0$ to the indefinite $(0, 0, 0)$ for $w = 1$ at the multiple point $\mathbf{P}_{01} = \mathbf{P}_{11}$. Thus, the surface does not posses a normal vector at $w = 1$ since the surface itself reduces to a point at this value. The reason that the normal does not depend on u is that this surface patch is flat. It is simply the triangle connecting the three points $(0, 0, 1)$, $(1, 0, 0)$, and $(0, 1, 0)$.

2.14: The rotation matrix for a $60°$ rotation about the y axis is

$$\begin{pmatrix} \cos 60° & 0 & -\sin 60° \\ 0 & 1 & 0 \\ \sin 60° & 0 & \cos 60° \end{pmatrix} = \begin{pmatrix} 0.5 & 0 & -0.866 \\ 0 & 1 & 0 \\ 0.866 & 0 & 0.5 \end{pmatrix}.$$

Applying this rotation to the original pair $\mathbf{P}_{00}\mathbf{P}_{10}$ yields the following six points (Figure 2.10), where the translation in the y direction has already been included

$$\mathbf{P}_{01} = (-0.5, 0, 0.866), \quad \mathbf{P}_{11} = (0.5, 0, -0.866),$$
$$\mathbf{P}_{02} = (0.5, 1, 0.866), \quad \mathbf{P}_{12} = (-0.5, 1, -0.866),$$
$$\mathbf{P}_{03} = (1, 2, 0), \quad \mathbf{P}_{13} = (-1, 2, 0).$$

The three bilinear patches are now easy to calculate:

$$\begin{aligned}
\mathbf{P}_1(u, w) &= (-1, -1, 0)(1 - u)(1 - w) + (-0.5, 0, .866)(1 - u)w \\
&\quad + (1, -1, 0)u(1 - w) + (0.5, 0, -0.866)uw \\
&= (-1 + 2u + 0.5uw + 0.5w - 1.5uw, -1 + w, -0.866uw + 0.866w - 0.866uw), \\
\mathbf{P}_2(u, w) &= (1, -1, 0)(1 - u)(1 - w) + (0.5, 1, .866)(1 - u)w \\
&\quad + (0.5, 0, -0.866)u(1 - w) + (-0.5, 1, -0.866)uw \\
&= (1 - 0.5u - 0.5uw - 0.5w, -1 + u + uw + 2w - 2uw, -0.866u - 0.866uw + 0.866w), \\
\mathbf{P}_3(u, w) &= (0.5, 1, 0.866)(1 - u)(1 - w) + (1, 2, 0)(1 - u)w \\
&\quad + (-0.5, 1, -0.866)u(1 - w) + (-1, 2, 0)uw \\
&= (0.5 - 1u - uw + 0.5w, 1 + 2uw + w - 2uw, 0.866 - 1.732u - 0.866w + 1.732uw).
\end{aligned}$$

2.15: (a) The straight line is

$$\mathbf{P}(u, 0) = \mathbf{P}_1(1 - u) + \mathbf{P}_2 u = \mathbf{P}_1 + (\mathbf{P}_2 - \mathbf{P}_1)u = (-1, -1, 0) + (2, 0, 0)u.$$

(b) For the quadratic, we set up the equations

$$\begin{aligned}
(-1, 1, 0) &= \mathbf{P}_3 = \mathbf{P}(0, 1) = \mathbf{C}, \\
(0, 1, 1) &= \mathbf{P}_4 = \mathbf{P}(0.5, 1) = 0.25\mathbf{A} + 0.5\mathbf{B} + \mathbf{C}, \\
(1, 1, 0) &= \mathbf{P}_5 = \mathbf{P}(1, 1) = \mathbf{A} + \mathbf{B} + \mathbf{C},
\end{aligned}$$

which are solved to yield $\mathbf{A} = (0, 0, -4)$, $\mathbf{B} = (2, 0, 4)$, and $\mathbf{C} = (-1, 1, 0)$. The top curve is therefore

$$\mathbf{P}(u, 1) = (0, 0, -4)u^2 + (2, 0, 4)u + (-1, 1, 0),$$

and the surface is $\mathbf{P}(u, w) = \mathbf{P}(u, 0)(1 - w) + \mathbf{P}(u, 1)w = (2u - 1, 2w - 1, 4uw(1 - u))$. The center point is $\mathbf{P}(0.5, 0.5) = (0, 0, 0.5)$. Figure Ans.3 shows this surface.

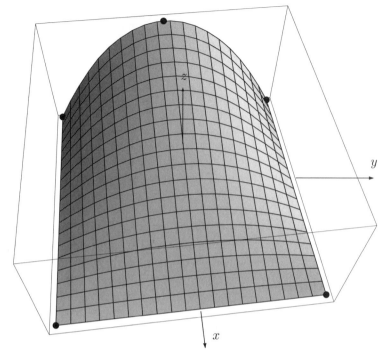

```
(* A lofted surface example. Bottom boundary curve is straight *)
pnts={{-1,-1,0},{1,-1,0},{-1,1,0},{0,1,1},{1,1,0}};
g1=Graphics3D[{AbsolutePointSize[5],
        Table[Point[pnts[[i]]],{i,1,5}]}]
g2=ParametricPlot3D[{2u-1,2w-1,4u w(1-u)}, {u,0,1},{w,0,1},
DefaultFont->{"cmr10", 10}, DisplayFunction->Identity,
 AspectRatio->Automatic, Ticks->{{0,1},{0,1},{0,1}}]
Show[g1,g2, ViewPoint->{-0.139, -1.179, 1.475}]
```

Figure Ans.3: A Lofted Surface.

2.16: The base can be considered one boundary curve. Its equation is $\mathbf{P}_1(u) = (R\cos u, R\sin u, H)$, where $0 \le u \le 2\pi$. The other boundary curve is the vertex $\mathbf{P}_2(u) = (0,0,0)$ (it is a degenerate curve). The entire surface is obtained, as usual, by

$$\mathbf{P}(u,w) = \mathbf{P}_1(u)w + \mathbf{P}_2(u)(1-w) = (Rw\cos u, Rw\sin u, Hw), \qquad \text{(Ans.3)}$$

where $0 \le w \le 1$ and $0 \le u \le 2\pi$.

2.17: The four corner points of the base are $(-a,-a,0)$, $(-a,a,0)$, $(a,-a,0)$, and $(a,a,0)$. We select the two points $\mathbf{P}_1 = (-a,-a,0)$ and $\mathbf{P}_2 = (a,-a,0)$. The straight segment connecting them is

$$\mathbf{P}(u) = (1-u)\mathbf{P}_1 + u\mathbf{P}_2 = (-a + 2ua, -a, 0).$$

The face defined by these points is therefore expressed by

$$\mathbf{P}(u, w) = \mathbf{P}(u)(1 - w) + (0, 0, H)w$$
$$= \big(a(2u - 1)(1 - w), a(w - 1), Hw\big).$$

(Ans.4)

The other three faces are calculated similarly.

2.18: The tangent vector in the u direction is

$$\frac{\partial \mathbf{P}(u, w)}{\partial u} = \Big(24u^2(1 - w) - 24u(1 - w) - 4w + 6,$$

$$12u^2(1 - w) - 18u(1 - w) + 8uw - 10w + 6, 0\Big).$$

At $u = 0.5$, this vector reduces to

$$\frac{\partial \mathbf{P}(0.5, w)}{\partial u} = \big(6(w - 1) - 4w + 6, 6(w - 1) - 6w + 6, 0\big) = (2w, 0, 0),$$

which implies that

$$\frac{\partial \mathbf{P}(0.5, 0)}{\partial u} = (0, 0, 0).$$

This shows that the surface does not have a tangent at the cusp, point $(0, 5/4, 0)$.

3.1: This is straightforward

$$\mathbf{P}(2/3) = (0, -9)(2/3)^3 + (-4.5, 13.5)(2/3)^2 + (4.5, -3.5)(2/3)$$
$$= (0, -8/3) + (-2, 6) + (3, -7/3)$$
$$= (1, 1) = \mathbf{P}_3.$$

3.2: We use the relations $\sin 30° = \cos 60° = 0.5$ and the approximation $\cos 30° = \sin 60° \approx 0.866$. The four points are $\mathbf{P}_1 = (1, 0)$, $\mathbf{P}_2 = (\cos 30°, \sin 30°) = (0.866, 0.5)$, $\mathbf{P}_3 = (0.5, 0.866)$, and $\mathbf{P}_4 = (0, 1)$. The relation $\mathbf{A} = \mathbf{NP}$ becomes

$$\begin{pmatrix} \mathbf{a} \\ \mathbf{b} \\ \mathbf{c} \\ \mathbf{d} \end{pmatrix} = \mathbf{A} = \mathbf{NP} = \begin{pmatrix} -4.5 & 13.5 & -13.5 & 4.5 \\ 9.0 & -22.5 & 18 & -4.5 \\ -5.5 & 9.0 & -4.5 & 1.0 \\ 1.0 & 0 & 0 & 0 \end{pmatrix} \begin{pmatrix} (1, 0) \\ (0.866, 0.5) \\ (0.5, 0.866) \\ (0, 1) \end{pmatrix}.$$

The solutions are

$$\mathbf{a} = -4.5(1, 0) + 13.5(0.866, 0.5) - 13.5(0.5, 0.866) + 4.5(0, 1) = (0.441, -0.441),$$
$$\mathbf{b} = 19(1, 0) - 22.5(0.866, 0.5) + 18(0.5, 0.866) - 4.5(0, 1) = (-1.485, -0.162),$$
$$\mathbf{c} = -5.5(1, 0) + 9(0.866, 0.5) - 4.5(0.5, 0.866) + 1(0, 1) = (0.044, 1.603),$$
$$\mathbf{d} = 1(1, 0) - 0(0.866, 0.5) + 0(0.5, 0.866) - 0(0, 1) = (1, 0).$$

Thus, the PC is $\mathbf{P}(t) = (0.441, -0.441)t^3 + (-1.485, -0.162)t^2 + (0.044, 1.603)t + (1, 0)$. The midpoint is $\mathbf{P}(0.5) = (0.7058, 0.7058)$, only 0.2% away from the midpoint of the arc, which is at $(\cos 45°, \sin 45°) \approx (0.7071, 0.7071)$.

(See also Exercise 1.12.)

3.3: From the definitions of the relative coordinates, we get $\mathbf{P}_2 = \Delta_1 + \mathbf{P}_1$, $\mathbf{P}_3 = \Delta_2 + \mathbf{P}_2 = \Delta_1 + \Delta_2 + \mathbf{P}_1$, and $\mathbf{P}_4 = \Delta_3 + \mathbf{P}_3 = \Delta_1 + \Delta_2 + \Delta_3 + \mathbf{P}_1$. When this is substituted in Equations (3.4) and (3.6), they become

$$\mathbf{P}(t) = \mathbf{G}(t)\,\mathbf{P} = \mathbf{T}(t)\,\mathbf{N}\,\mathbf{P}$$

$$= (t^3, t^2, t, 1) \begin{pmatrix} -4.5 & 13.5 & -13.5 & 4.5 \\ 9.0 & -22.5 & 18 & -4.5 \\ -5.5 & 9.0 & -4.5 & 1.0 \\ 1.0 & 0 & 0 & 0 \end{pmatrix} \begin{pmatrix} \mathbf{P}_1 \\ \mathbf{P}_2 \\ \mathbf{P}_3 \\ \mathbf{P}_4 \end{pmatrix}$$

$$= (t^3, t^2, t, 1) \begin{pmatrix} -4.5 & 13.5 & -13.5 & 4.5 \\ 9.0 & -22.5 & 18 & -4.5 \\ -5.5 & 9.0 & -4.5 & 1.0 \\ 1.0 & 0 & 0 & 0 \end{pmatrix} \begin{pmatrix} \mathbf{P}_1 \\ \mathbf{P}_1 + \Delta_1 \\ \mathbf{P}_1 + \Delta_1 + \Delta_2 \\ \mathbf{P}_1 + \Delta_1 + \Delta_2 + \Delta_3 \end{pmatrix}.$$

Selecting, for example, $\Delta_1 = (2, 0)$, $\Delta_2 = (0, 2)$, and $\Delta_3 = (1, 1)$ produces

$$\mathbf{P}(t) = (t^3, t^2, t, 1) \begin{pmatrix} -4.5 & 13.5 & -13.5 & 4.5 \\ 9.0 & -22.5 & 18 & -4.5 \\ -5.5 & 9.0 & -4.5 & 1.0 \\ 1.0 & 0 & 0 & 0 \end{pmatrix} \begin{pmatrix} \mathbf{P}_1 \\ \mathbf{P}_1 + (2, 0) \\ \mathbf{P}_1 + (2, 2) \\ \mathbf{P}_1 + (3, 3) \end{pmatrix}$$

$$= \mathbf{P}_1 + (12t - 22.5t^2 + 13.5t^3, -6t + 22.5t^2 - 13.5t^3).$$

It is now clear that the three relative coordinates fully determine the shape of the curve but do not fix its position in space. The value of $\mathbf{P}_1$ is needed for that.

3.4: The new equations are easy enough to set up. With the help of *Mathematica*, they are also easy to solve. The code

```
Solve[{d==p1,
a al^3+b al^2+c al+d==p2,
a be^3+b be^2+c be+d==p3,
a+b+c+d==p4},{a,b,c,d}];
ExpandAll[Simplify[%]]
```

(where **al** and **be** stand for α and β, respectively) produces the (messy) solutions

$$a = -\frac{\mathbf{P}_1}{\alpha\beta} + \frac{\mathbf{P}_2}{-\alpha^2 + \alpha^3 + \alpha\beta - \alpha^2\beta}$$

$$+ \frac{\mathbf{P}_3}{\alpha\beta - \beta^2 - \alpha\beta^2 + \beta^3} + \frac{\mathbf{P}_4}{1 - \alpha - \beta + \alpha\beta},$$

$$b = \mathbf{P}_1 \left(-\alpha + \alpha^3 + \beta - \alpha^3\beta - \beta^3 + \alpha\beta^3\right)/\gamma + \mathbf{P}_2 \left(-\beta + \beta^3\right)/\gamma$$

$$+ \mathbf{P}_3 \left(\alpha - \alpha^3\right)/\gamma + \mathbf{P}_4 \left(\alpha^3\beta - \alpha\beta^3\right)/\gamma,$$

$$c = -\mathbf{P}_1 \left(1 + \frac{1}{\alpha} + \frac{1}{\beta}\right) + \frac{\beta\mathbf{P}_2}{-\alpha^2 + \alpha^3 + \alpha\beta - \alpha^2\beta}$$

$$+ \frac{\alpha\mathbf{P}_3}{\alpha\beta - \beta^2 - \alpha\beta^2 + \beta^3} + \frac{\alpha\beta\mathbf{P}_4}{1 - \alpha - \beta + \alpha\beta},$$

$$d = \mathbf{P}_1,$$

where

$$\gamma = (-1 + \alpha)\alpha(-1 + \beta)\beta(-\alpha + \beta).$$

From here, the basis matrix immediately follows:

$$\begin{pmatrix} -\frac{1}{\alpha\beta} & \frac{1}{-\alpha^2+\alpha^3\alpha\beta-\alpha^2\beta} & \frac{1}{\alpha\beta-\beta^2-\alpha\beta^2+\beta^3} & \frac{1}{1-\alpha-\beta+\alpha\beta} \\ \frac{-\alpha+\alpha^3+\beta-\alpha^3\beta-\beta^3+\alpha\beta^3}{\gamma} & \frac{-\beta+\beta^3}{\gamma} & \frac{\alpha-\alpha^3}{\gamma} & \frac{\alpha^3\beta-\alpha\beta^3}{\gamma} \\ -\left(1 + \frac{1}{\alpha} + \frac{1}{\beta}\right) & \frac{\beta}{-\alpha^2+\alpha^3+\alpha\beta-\alpha^2\beta} & \frac{\alpha}{\alpha\beta-\beta^2-\alpha\beta^2+\beta^3} & \frac{\alpha\beta}{1-\alpha-\beta+\alpha\beta} \\ 1 & 0 & 0 & 0 \end{pmatrix}.$$

A direct check, again using *Mathematica*, for $\alpha = 1/3$ and $\beta = 2/3$ produces the basis matrix of Equation (3.6).

3.5: This is the case $n = 1$. The general form of the LP is, therefore, $y = \sum_{i=0}^{1} y_i L_i^1$. The weight functions are easy to calculate:

$$L_0^1 = \frac{x - x_1}{x_0 - x_1}, \qquad L_1^1 = \frac{x - x_0}{x_1 - x_0},$$

and the curve is therefore

$$y = y_0 L_0^1 + y_1 L_1^1 = y_0 \frac{x - x_1}{x_0 - x_1} + y_1 \frac{x - x_0}{x_1 - x_0}$$

$$= x \frac{y_0 - y_1}{x_0 - x_1} + \frac{y_1 x_0 - y_0 x_1}{x_0 - x_1} = ax + b.$$

This is a straight line.

3.6: Since there are just two points, the only knots are $t_0 = 0$ and $t_1 = 1$. The weight functions are

$$L_0^1 = \frac{t - t_1}{t_0 - t_1} = 1 - t, \qquad L_1^1 = \frac{t - t_0}{t_1 - t_0} = t,$$

and the curve is

$$\mathbf{P}(t) = \mathbf{P}_0 L_0^1 + \mathbf{P}_1 L_1^1 = (1 - t)\mathbf{P}_0 + t\mathbf{P}_1.$$

This is a straight line expressed parametrically.

3.7: Since the three points are approximately equally spaced, it makes sense to use knot values $t_0 = 0$, $t_1 = 1/2$, and $t_2 = 1$. The first step is to calculate the three basis functions $L_i^2(t)$:

$$L_0^2 = \frac{\Pi_{j\neq 0}^2(t - t_j)}{\Pi_{j\neq 0}^2(t_i - t_j)} = \frac{(t - t_1)(t - t_2)}{(t_0 - t_1)(t_0 - t_2)} = 2(t - 1/2)(t - 1),$$

$$L_1^2 = \frac{\Pi_{j\neq 1}^2(t - t_j)}{\Pi_{j\neq 1}^2(t_i - t_j)} = \frac{(t - t_0)(t - t_2)}{(t_1 - t_0)(t_1 - t_2)} = -4t(t - 1),$$

$$L_2^2 = \frac{\Pi_{j\neq 2}^2(t - t_j)}{\Pi_{j\neq 2}^2(t_i - t_j)} = \frac{(t - t_0)(t - t_1)}{(t_2 - t_0)(t_2 - t_1)} = 2t(t - 1/2).$$

The LP is now easy to calculate:

$$\mathbf{P}(t) = (0,0)2(t - 1/2)(t - 1) - (0,1)4t(t - 1) + (1,1)2t(t - 1/2)$$
$$= (2t^2 - t, -2t^2 + 3t). \tag{Ans.5}$$

This is a quadratic (degree-2) parametric polynomial and a simple test verifies that it passes through the three given points.

3.8: We set the knots to $t_0 = 0$, $t_1 = 1/3$, $t_2 = 2/3$, and $t_3 = 1$. The first step is to calculate the four basis functions $L_i^3(t)$:

$$L_0^3 = \frac{\Pi_{j\neq 0}^3(t - t_j)}{\Pi_{j\neq 0}^3(t_i - t_j)} = \frac{(t - t_1)(t - t_2)(t - t_3)}{(t_0 - t_1)(t_0 - t_2)(t_0 - t_3)} = -4.5t^3 + 9t^3 - 5.5t + 1,$$

$$L_1^3 = \frac{\Pi_{j\neq 1}^3(t - t_j)}{\Pi_{j\neq 1}^3(t_i - t_j)} = \frac{(t - t_0)(t - t_2)(t - t_3)}{(t_1 - t_0)(t_1 - t_2)(t_1 - t_3)} = 13.5t^3 - 22.5t^3 + 9t,$$

$$L_2^3 = \frac{\Pi_{j\neq 2}^3(t - t_j)}{\Pi_{j\neq 2}^3(t_i - t_j)} = \frac{(t - t_0)(t - t_1)(t - t_3)}{(t_2 - t_0)(t_2 - t_1)(t_2 - t_3)} = -13.5t^3 + 18t^3 - 4.5t,$$

$$L_3^3 = \frac{\Pi_{j\neq 3}^3(t - t_j)}{\Pi_{j\neq 3}^3(t_i - t_j)} = \frac{(t - t_0)(t - t_1)(t - t_2)}{(t_3 - t_0)(t_3 - t_1)(t_3 - t_2)} = 4.5t^3 - 4.5t^3 + t.$$

The LP is now easy to calculate:

$$\mathbf{P}(t) = (-4.5t^3 + 9t^2 - 5.5t + 1)\mathbf{P}_1 + (13.5t^3 - 22.5t^2 + 9t)\mathbf{P}_2$$
$$+ (-13.5t^3 + 18t^2 - 4.5t)\mathbf{P}_3 + (4.5t^3 - 4.5t^2 + t)\mathbf{P}_4.$$

This is identical to Equation (3.4).

3.9: See Section 3.5.1.

3.10: The first step is to calculate the basis functions

$$N_0(t) = 1, \quad N_1(t) = t - t_0 = t, \quad N_2(t) = (t - t_0)(t - t_1) = t(t - 1/2).$$

The next step is to compute the three coefficients

$$\mathbf{A}_0 = \mathbf{P}_0 = (0,0),$$
$$\mathbf{A}_1 = \frac{\mathbf{P}_1 - \mathbf{P}_0}{t_1 - t_0} = \frac{(0,1) - (0,0)}{1/2} = (0,2),$$
$$\mathbf{A}_2 = \frac{\dfrac{(1,1) - (0,1)}{1 - 1/2} - \dfrac{(0,1) - (0,0)}{1/2 - 0}}{1 - 0} = (2,-2).$$

The polynomial can now be calculated:

$$\mathbf{P}(t) = 1 \times (0,0) + t(0,2) + t(t-1/2)(2,-2) = (2t^2 - t, -2t^2 + 3t).$$

It is, of course, identical to the LP calculated in Exercise 3.7.

3.11: The curve is given by $\mathbf{P}(t) = (2t^2 - t, -2t^2 + 3t)$, so its derivative is $\mathbf{P}^t(t) = (4t - 1, -4t + 3)$. The three tangent vectors are $\mathbf{P}^t(t_0 = 0) = (-1,3)$, $\mathbf{P}^t(t_1 = 1/2) = (1,1)$, and $\mathbf{P}^t(t_2 = 1) = (3,-1)$. The direction of tangent vector $(-1,3)$ is described by saying "for every three steps in the y direction, the curve moves one step in the negative x direction."

The slopes are calculated by dividing the y coordinate of a tangent vector by its x coordinate. The slopes at the three points are therefore $-3/1$, $1/1$, and $-1/3$. They correspond to angles of $288.44°$, $45°$, and $-18.43°$, respectively.

3.12: The geometry matrix can be transposed without affecting the shape of the surface. The way the geometry matrix is written in Equation (3.23) implies that point $\mathbf{P}_{00}$ corresponds to $\mathbf{P}(1,1)$. If we transpose the matrix so that point $\mathbf{P}(0,0)$ becomes its top-left corner, the surface will be the same, with the only difference that point $\mathbf{P}_{00}$ will correspond to $\mathbf{P}(0,0)$.

3.13: Figure Ans.4a shows a diamond-shaped grid of 16 equally-spaced points. The eight points with negative weights are shown in black. Figure Ans.4b shows a cut (labeled xx in Figure Ans.4a) through four points in this surface. The cut is a curve that passes through pour data points. It is easy to see that when the two exterior (black) points are raised, the center of the curve (and, as a result, the center of the surface) is lowered. It is now clear that points with negative weights push the center of the surface in a direction opposite that of the points. The figure serves to make bicubic interpolation more intuitive.

Figure Ans.4c is a more detailed example that also shows why the four corner points should have positive weights. It shows a simple symmetric surface patch that interpolates the 16 points

$$
\begin{array}{llll}
\mathbf{P}_{00} = (0,0,0), & \mathbf{P}_{10} = (1,0,1), & \mathbf{P}_{20} = (2,0,1), & \mathbf{P}_{30} = (3,0,0), \\
\mathbf{P}_{01} = (0,1,1), & \mathbf{P}_{11} = (1,1,2), & \mathbf{P}_{21} = (2,1,2), & \mathbf{P}_{31} = (3,1,1), \\
\mathbf{P}_{02} = (0,2,1), & \mathbf{P}_{12} = (1,2,2), & \mathbf{P}_{22} = (2,2,2), & \mathbf{P}_{32} = (3,2,1), \\
\mathbf{P}_{03} = (0,3,0), & \mathbf{P}_{13} = (1,3,1), & \mathbf{P}_{23} = (2,3,1), & \mathbf{P}_{33} = (3,3,0).
\end{array}
$$

We first raise the eight boundary points from $z = 1$ to $z = 1.5$. Figure Ans.4d shows how the center point $\mathbf{P}(.5, .5)$ gets lowered from $(1.5, 1.5, 2.25)$ to $(1.5, 1.5, 2.10938)$. We next return those points to their original positions and instead raise the four corner points from $z = 0$ to $z = 1$. Figure Ans.4e shows how this raises the center point from $(1.5, 1.5, 2.25)$ to $(1.5, 1.5, 2.26563)$.

3.14: In such a case, the tangent vector of the surface along the degenerate boundary curve $\mathbf{P}(u, 1)$ is the weighted sum of the eight quantities

$$
\mathbf{T}_0 = \left.\frac{d\mathbf{P}(0, w)}{dw}\right|_{w=1}, \quad \mathbf{T}_1 = \left.\frac{d\mathbf{P}(1, w)}{dw}\right|_{w=1}, \quad \mathbf{P}(u, 0), \mathbf{P}(u, 1), \mathbf{P}_{00}, \mathbf{P}_{01}, \mathbf{P}_{10}, \mathbf{P}_{11},
$$

instead of being the simple linear combination $B_0(u)\mathbf{T}_0 + B_1(u)\mathbf{T}_1$ of Equation (3.41). As it swings from $\mathbf{T}_0$ to $\mathbf{T}_1$, this vector will not have to stay in the plane defined by $\mathbf{T}_0$ and $\mathbf{T}_1$ and may wiggle wildly in and out of this plane, causing the surface to be wrinkled in the vicinity of the common point.

3.15: We start with the boundary curves. They are straight lines, and so are obtained from Equation (2.1)

$$
\begin{aligned}
\mathbf{P}(0, w) = (1 - w)\mathbf{P}_{00} + w\mathbf{P}_{01}, \qquad & \mathbf{P}(1, w) = (1 - w)\mathbf{P}_{10} + w\mathbf{P}_{11}, \\
\mathbf{P}(u, 0) = (1 - u)\mathbf{P}_{00} + u\mathbf{P}_{10}, \qquad & \mathbf{P}(u, 1) = (1 - u)\mathbf{P}_{01} + u\mathbf{P}_{11}.
\end{aligned}
$$

The surface expression is now obtained from Equation (3.28). It is

$$
\begin{aligned}
\mathbf{P}(u, w) = {} & (1 - u)(1 - w)\mathbf{P}_{00} + (1 - u)w\mathbf{P}_{01} + u(1 - w)\mathbf{P}_{10} + uw\mathbf{P}_{11} \\
& + (1 - w)(1 - u)\mathbf{P}_{00} + (1 - w)u\mathbf{P}_{10} + w(1 - u)\mathbf{P}_{01} + wu\mathbf{P}_{11} \\
& - (1 - u)(1 - w)\mathbf{P}_{00} - u(1 - w)\mathbf{P}_{10} - (1 - u)w\mathbf{P}_{01} - uw\mathbf{P}_{11} \\
= {} & (0.5(1 - u)w + u, w, (1 - u)(1 - w)).
\end{aligned}
$$

Note that it is identical to the bilinear surface of Equation (2.11).

4.1: When the user specifies four points, the curve should pass through the original points. After a point is moved, the curve will no longer pass through the original point. When only the two endpoints are specified, the user is normally willing to consider different curves that pass through them, with different start and end directions.

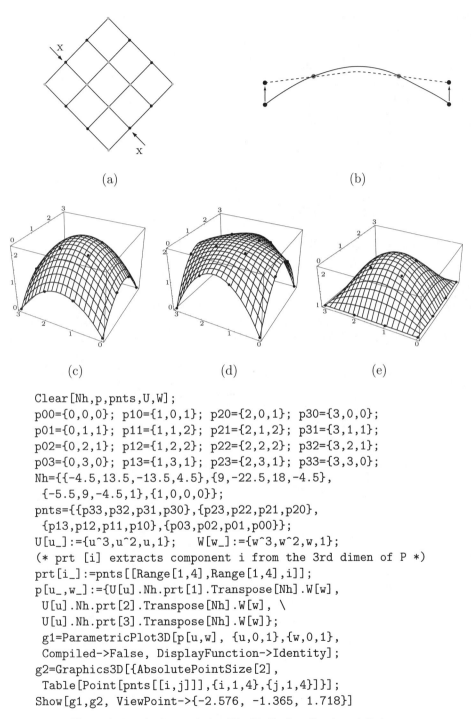

(a) (b)

(c) (d) (e)

```
Clear[Nh,p,pnts,U,W];
p00={0,0,0}; p10={1,0,1}; p20={2,0,1}; p30={3,0,0};
p01={0,1,1}; p11={1,1,2}; p21={2,1,2}; p31={3,1,1};
p02={0,2,1}; p12={1,2,2}; p22={2,2,2}; p32={3,2,1};
p03={0,3,0}; p13={1,3,1}; p23={2,3,1}; p33={3,3,0};
Nh={{-4.5,13.5,-13.5,4.5},{9,-22.5,18,-4.5},
 {-5.5,9,-4.5,1},{1,0,0,0}};
pnts={{p33,p32,p31,p30},{p23,p22,p21,p20},
 {p13,p12,p11,p10},{p03,p02,p01,p00}};
U[u_]:={u^3,u^2,u,1};    W[w_]:={w^3,w^2,w,1};
(* prt [i] extracts component i from the 3rd dimen of P *)
prt[i_]:=pnts[[Range[1,4],Range[1,4],i]];
p[u_,w_]:={U[u].Nh.prt[1].Transpose[Nh].W[w],
 U[u].Nh.prt[2].Transpose[Nh].W[w], \
 U[u].Nh.prt[3].Transpose[Nh].W[w]};
 g1=ParametricPlot3D[p[u,w], {u,0,1},{w,0,1},
 Compiled->False, DisplayFunction->Identity];
g2=Graphics3D[{AbsolutePointSize[2],
 Table[Point[pnts[[i,j]]],{i,1,4},{j,1,4}]}];
Show[g1,g2, ViewPoint->{-2.576, -1.365, 1.718}]
```

Figure Ans.4: An Interpolating Bicubic Surface Patch and Code.

4.2: Take one of these vectors, say, $(2, 1, 0.6)$ and divide it by its magnitude. The result is

$$\frac{(2, 1, 0.6)}{\sqrt{2^2 + 1^2 + 0.6^2}} \approx \frac{(2, 1, 0.6)}{2.93} = (0.7272, 0.3636, 0.2045).$$

The new vector points in the same direction but its magnitude is 1. Its components therefore satisfy

$$\sqrt{0.7272^2 + 0.3636^2 + 0.2045^2} = 1, \text{ or } 0.7272^2 + 0.3636^2 + 0.2045^2 = 1, \qquad \text{(Ans.6)}$$

so they are dependent. Any of them can be calculated from the other two with Equation (Ans.6).

4.3: Substituting $t = 0.5$ in Equation (4.4) yields

$$\mathbf{P}(0.5) = (2\mathbf{P}_1 - 2\mathbf{P}_2 + \mathbf{P}_1^t + \mathbf{P}_2^t)/8 + (-3\mathbf{P}_1 + 3\mathbf{P}_2 - 2\mathbf{P}_1^t - \mathbf{P}_2^t)/4 + \mathbf{P}_1^t/2 + \mathbf{P}_1$$
$$= \frac{1}{2}(\mathbf{P}_1 + \mathbf{P}_2) + \frac{1}{8}(\mathbf{P}_1^t - \mathbf{P}_2^t). \qquad \text{(Ans.7)}$$

The first part of this expression is the midpoint of the segment $\mathbf{P}_1 \to \mathbf{P}_2$ and the second part is the difference of the two tangents, divided by 8. Figure Ans.5 illustrates how adding $(\mathbf{P}_1^t - \mathbf{P}_2^t)/8$ to the midpoint of $\mathbf{P}_1 \to \mathbf{P}_2$ brings us to the midpoint of the curve.

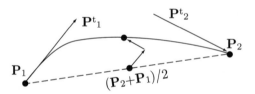

Figure Ans.5: The Midpoint $\mathbf{P}(0.5)$ of a Hermite Segment.

4.4: The Hermite segment is a cubic polynomial in t, so its third derivative is constant. It is easy to see, from Equation (4.6), that the third derivatives of the Hermite blending functions $F_i(t)$ are

$$F_1^{ttt}(t) = 12, \quad F_2^{ttt}(t) = -12, \quad F_3^{ttt}(t) = 6, \quad F_4^{ttt}(t) = 6.$$

The third derivative of the segment is therefore

$$\mathbf{P}^{ttt}(t) = (12\mathbf{P}_1 - 12\mathbf{P}_2 + 6\mathbf{P}_1^t + 6\mathbf{P}_2^t)$$
$$= (t^3, t^2, t, 1) \begin{bmatrix} 0 & 0 & 0 & 0 \\ 0 & 0 & 0 & 0 \\ 0 & 0 & 0 & 0 \\ 12 & -12 & 6 & 6 \end{bmatrix} \begin{bmatrix} \mathbf{P}_1 \\ \mathbf{P}_2 \\ \mathbf{P}_1^t \\ \mathbf{P}_2^t \end{bmatrix}$$
$$= \mathbf{T}(t)\mathbf{H}_{ttt}\mathbf{B} = \mathbf{H}_{ttt}\mathbf{B}.$$

$\mathbf{P}^{ttt}(t)$ is independent of t, because the top three rows of $\mathbf{H}_{ttt}$ are zero. This derivative is the constant vector $12(\mathbf{P}_1 - \mathbf{P}_2) + 6(\mathbf{P}_1^t + \mathbf{P}_2^t)$.

Here are the Hermite matrix and its derivatives side by side. Use your experience to explain how each is derived from its predecessor.

$$\mathbf{H} = \begin{pmatrix} 2 & -2 & 1 & 1 \\ -3 & 3 & -2 & -1 \\ 0 & 0 & 1 & 0 \\ 1 & 0 & 0 & 0 \end{pmatrix}, \quad \mathbf{H}_t = \begin{pmatrix} 0 & 0 & 0 & 0 \\ 6 & -6 & 3 & 3 \\ -6 & 6 & -4 & -2 \\ 0 & 0 & 1 & 0 \end{pmatrix},$$

$$\mathbf{H}_{tt} = \begin{pmatrix} 0 & 0 & 0 & 0 \\ 0 & 0 & 0 & 0 \\ 12 & -12 & 6 & 6 \\ -6 & 6 & -4 & -2 \end{pmatrix}, \quad \mathbf{H}_{ttt} = \begin{pmatrix} 0 & 0 & 0 & 0 \\ 0 & 0 & 0 & 0 \\ 0 & 0 & 0 & 0 \\ 12 & -12 & 6 & 6 \end{pmatrix}.$$

4.5: It's trivial to show that $\mathbf{P}(0) = (-1,0)0^3 + (1,-1)0^2 + (1,1)0 = (0,0)$ and $\mathbf{P}(1) = (-1,0)1^3 + (1,-1)1^2 + (1,1)1 = (1,0)$. The tangent vector of $\mathbf{P}(t)$ is

$$\frac{d\mathbf{P}(t)}{dt} = 3(-1,0)t^2 + 2(1,-1)t + (1,1),$$

so the two extreme tangent vectors are

$$\frac{d\mathbf{P}(0)}{dt} = 3(-1,0)0^2 + 2(1,-1)0 + (1,1) = (1,1),$$
$$\frac{d\mathbf{P}(1)}{dt} = 3(-1,0)1^2 + 2(1,-1) + (1,1) = (0,-1),$$

as should be.

4.6: Similar to the previous example, we get

$$\mathbf{P}(t) = (t^3, t^2, t, 1)\mathbf{H}\left((0,0),(1,0),(2,2),(0,-1)\right)^T$$
$$= (0,1)t^3 - (1,3)t^2 + (2,2)t.$$

It's a different polynomial and it has a different shape; yet a simple check shows that it passes through the same endpoints and has the same start and end directions.

4.7: Equation (4.7) becomes

$$\mathbf{P}(t) = (t^3, t^2, t, 1)\begin{bmatrix} 2 & -2 & 1 & 1 \\ -3 & 3 & -2 & -1 \\ 0 & 0 & 1 & 0 \\ 1 & 0 & 0 & 0 \end{bmatrix}\begin{bmatrix} \mathbf{P}_1 \\ \mathbf{P}_2 \\ (0,0) \\ (0,0) \end{bmatrix} = (3t^2 - 2t^3)(\mathbf{P}_2 - \mathbf{P}_1) + \mathbf{P}_1. \quad \text{(Ans.8)}$$

To find the type of the curve, we substitute $j = 3t^2 - 2t^3$ (note that $t = 0 \Rightarrow j = 0$ and $t = 1 \Rightarrow j = 1$). This results in the familiar expression $\mathbf{P}(t) = j(\mathbf{P}_2 - \mathbf{P}_1) + \mathbf{P}_1 = (1 - j)\mathbf{P}_1 + j\mathbf{P}_2$. The curve is therefore the straight segment from $\mathbf{P}_1$ to $\mathbf{P}_2$. The (important) conclusion is: If the initial and final directions of the Hermite segment are not specified, the curve will "choose" the shortest path from $\mathbf{P}_1$ to $\mathbf{P}_2$.

4.8:   For case 1, we use the notation $\mathbf{P}^t(0) = \mathbf{P}^t_1$, $\mathbf{P}^t(1/2) = \mathbf{P}^t_2$, and $\mathbf{P}^t(1) = \mathbf{P}^t_3$. From $\mathbf{P}(t) = \mathbf{a}t^3 + \mathbf{b}t^2 + \mathbf{c}t + \mathbf{d}$, we get $\mathbf{P}^t(t) = 3\mathbf{a}t^2 + 2\mathbf{b}t + \mathbf{c}$, resulting in the three equations

$$3\mathbf{a}\cdot 0^2 + 2\mathbf{b}\cdot 0 + \mathbf{c} = \mathbf{P}^t_1,$$
$$3\mathbf{a}\cdot(1/2)^2 + 2\mathbf{b}\cdot(1/2) + \mathbf{c} = \mathbf{P}^t_2,$$
$$3\mathbf{a}\cdot 1^2 + 2\mathbf{b}\cdot 1 + \mathbf{c} = \mathbf{P}^t_3,$$

where the unknowns are $\mathbf{a}$, $\mathbf{b}$, $\mathbf{c}$, and $\mathbf{d}$ (notice that $\mathbf{d}$ does not participate in our equations). It is clear that $\mathbf{c} = \mathbf{P}^t_1$. The other two unknowns are solved by the simple *Mathematica* code `Solve[{3a/4+2b/2+p1==p2, 3a+2b+p1==p3}, {a,b}]`, which yields $\mathbf{a} = \frac{2}{3}(\mathbf{P}^t_1 - 2\mathbf{P}^t_2 + \mathbf{P}^t_3)$ and $\mathbf{b} = \frac{1}{2}(-3\mathbf{P}^t_1 + 4\mathbf{P}^t_2 - \mathbf{P}^t_3)$. Thus, the curve is given by

$$\mathbf{P}(t) = \mathbf{a}t^3 + \mathbf{b}t^2 + \mathbf{c}t + \mathbf{d}$$
$$= \frac{2}{3}(\mathbf{P}^t_1 - 2\mathbf{P}^t_2 + \mathbf{P}^t_3)t^3 + \frac{1}{2}(-3\mathbf{P}^t_1 + 4\mathbf{P}^t_2 - \mathbf{P}^t_3)t^2 + \mathbf{P}^t_1 t + \mathbf{d},$$

which shows that the three given tangents fully determine the shape of the curve but not its position in space. The latter requires the value of $\mathbf{d}$.

For case 2, we denote $\mathbf{P}(1/3) = \mathbf{P}_1$, $\mathbf{P}(2/3) = \mathbf{P}_2$, $\mathbf{P}^t(0) = \mathbf{P}^t_1$, and $\mathbf{P}^t(1) = \mathbf{P}^t_2$. This results in the four equations

$$\mathbf{a}(1/3)^3 + \mathbf{b}(1/3)^2 + \mathbf{c}(1/3) + \mathbf{d} = \mathbf{P}_1,$$
$$\mathbf{a}(2/3)^3 + \mathbf{b}(2/3)^2 + \mathbf{c}(2/3) + \mathbf{d} = \mathbf{P}_2,$$
$$3\mathbf{a}\cdot 0^2 + 2\mathbf{b}\cdot 0 + \mathbf{c} = \mathbf{P}^t_1,$$
$$3\mathbf{a}\cdot 1^2 + 2\mathbf{b}\cdot 1 + \mathbf{c} = \mathbf{P}^t_2,$$

where the unknowns are again $\mathbf{a}$, $\mathbf{b}$, $\mathbf{c}$, and $\mathbf{d}$. It is again clear that $\mathbf{c} = \mathbf{P}^t_1$ and the other three unknowns are easily solved by the code

```
Solve[{a (1/3)^3+b (1/3)^2+p1t (1/3)+d==p1,
a (2/3)^3+b (2/3)^2+p1t (1/3)+d==p2, 3a+2b+p1t==p2t}, {a,b,d}],
```

which yields the solutions

$$\mathbf{a} = -\frac{9}{13}(-6\mathbf{P}_1 + \mathbf{P}^t_1 + 6\mathbf{P}_2 - \mathbf{P}^t_2),$$
$$\mathbf{b} = \frac{1}{13}(-81\mathbf{P}_1 + 7\mathbf{P}^t_1 + 81\mathbf{P}_2 - 7\mathbf{P}^t_2),$$
$$\mathbf{d} = \frac{1}{117}(180\mathbf{P}_1 - 43\mathbf{P}^t_1 - 63\mathbf{P}_2 + 4\mathbf{P}^t_2).$$

Thus, the PC segment is

$$\mathbf{P}(t) = \mathbf{a}t^3 + \mathbf{b}t^2 + \mathbf{c}t + \mathbf{d}$$
$$= -\frac{9}{13}(-6\mathbf{P}_1 + \mathbf{P}^t_1 + 6\mathbf{P}_2 - \mathbf{P}^t_2)t^3 + \frac{1}{13}(-81\mathbf{P}_1 + 7\mathbf{P}^t_1 + 81\mathbf{P}_2 - 7\mathbf{P}^t_2)t^2$$
$$+ \mathbf{P}^t_1 \cdot t + \frac{1}{117}(180\mathbf{P}_1 - 43\mathbf{P}^t_1 - 63\mathbf{P}_2 + 4\mathbf{P}^t_2).$$

Case 3 is similar to case 2 and is not shown here.

4.9: We are looking for a parametric curve $\mathbf{P}(t)$ that's a quadratic polynomial satisfying

$$\mathbf{P}(t) = \mathbf{a}t^2 + \mathbf{b}t + \mathbf{c}, \tag{Ans.9}$$
$$\mathbf{P}^t(t) = 2\mathbf{a}t + \mathbf{b}, \tag{Ans.10}$$
$$\mathbf{P}(0) = \mathbf{c} = \mathbf{P}_0, \tag{Ans.11}$$
$$\mathbf{P}(1) = \mathbf{a} + \mathbf{b} + \mathbf{c} = \mathbf{P}_2, \tag{Ans.12}$$
$$\mathbf{P}^t(0) = \mathbf{b} = 4\alpha(\mathbf{P}_1 - \mathbf{P}_0), \tag{Ans.13}$$
$$\mathbf{P}^t(1) = 2\mathbf{a} + \mathbf{b} = 4\alpha(\mathbf{P}_2 - \mathbf{P}_1). \tag{Ans.14}$$

Subtracting Equation (Ans.12) from Equation (Ans.14) yields

$$\mathbf{a} = 4\alpha(\mathbf{P}_2 - \mathbf{P}_1) + \mathbf{P}_2 - \mathbf{P}_0. \tag{Ans.15}$$

Substituting Eqs. (Ans.11), (Ans.13), and (Ans.15) in Equation (Ans.12) yields

$$\mathbf{a} + \mathbf{b} + \mathbf{c} = 4\alpha(\mathbf{P}_2 - \mathbf{P}_1) + \mathbf{P}_2 - \mathbf{P}_0 + 4\alpha(\mathbf{P}_1 - \mathbf{P}_0) + \mathbf{P}_0 = \mathbf{P}_2,$$

or $4\alpha(\mathbf{P}_2 - \mathbf{P}_0) = 2(\mathbf{P}_2 - \mathbf{P}_0)$, implying $\alpha = 0.5$. Once α is known, the curve is obtained from Equation (Ans.9) as

$$\mathbf{P}(t) = (1 - t)^2 \mathbf{P}_0 + 2t(1 - t)\mathbf{P}_1 + t^2 \mathbf{P}_2.$$

4.10: For $\theta = 90°$, we have $\sin\theta = 1$, $\cos\theta = 0$ and $a = 4$. Equation (4.15) becomes

$$\begin{aligned}
\mathbf{P}(t) &= (2t^3 - 3t^2 + 1)(0, -1) + (-2t^3 + 3t^2)(0, 1) \\
&\quad + (t^3 - 2t^2 + t)4(1, 0) + (t^3 - t^2)4(-1, 0) \\
&= (-4t^2 + 4t, -4t^3 + 6t^2 - 1).
\end{aligned}$$

It is easy to see that $\mathbf{P}(0) = (0, -1)$, $\mathbf{P}(1) = (0, 1)$, and $\mathbf{P}(0.5) = (1, 0)$. At $t = 0.25$, the curve passes through point $\mathbf{P}(0.25) = (12/16, -11/16)$, whose distance from the origin is

$$\sqrt{\left(\frac{12}{16}\right)^2 + \left(\frac{11}{16}\right)^2} \approx 1.0174.$$

The deviation from a true circle at this point is therefore about 1.74%, an excellent approximation for such a large arc.

4.11: Yes. Equation (4.19) was derived for any real values of a and b, not just positive. However, when a and b become negative, the tangent vectors reverse directions, and the curve changes its shape completely. Figure 4.7 shows the (dashed) curve for $\alpha = -0.4$. Another example of negative a and b is $\alpha = -1/4$, which yields $a = b = -1$, changing Equation (4.19) to $\mathbf{Q}(t) = -(6, 3)t^3 + (9, 5)t^2 - (1, 1)t$. It is easy to verify that $\mathbf{Q}(0) = (0, 0)$, $\mathbf{Q}(1) = (2, 1)$, and $\mathbf{Q}(0.5) = (1, 3/8)$.

4.12: The midpoint of our curve is always $(1, 5/8 + \alpha)$. The condition $\mathbf{Q}(0.5) = (1, 0)$ implies $5/8 + \alpha = 0$ or $\alpha = -5/8$. Since $a = b = 1 + 8\alpha$, we get $a = b = -4$, resulting in $\mathbf{Q}(t) = -(12, 6)t^3 + (18, 11)t^2 - (4, 4)t$.

4.13: Equation (4.8) yields the first derivative of the Hermite segment

$$\mathbf{P}^t(t) = (t^3, t^2, t, 1) \begin{bmatrix} 0 & 0 & 0 & 0 \\ 6 & -6 & 3 & 3 \\ -6 & 6 & -4 & -2 \\ 0 & 0 & 1 & 0 \end{bmatrix} \begin{bmatrix} (0,0) \\ (1,0) \\ \alpha(\cos\theta, \sin\theta) \\ \alpha(\cos\theta, -\sin\theta) \end{bmatrix}$$

$$= 3[(-2, 0) + \alpha(2\cos\theta, 0)]t^2 + 2[(3, 0) - \alpha(3\cos\theta, \sin\theta)]t + (\cos\theta, \sin\theta).$$

Because of the symmetry of the endpoints and vectors, a cusp can only occur in the middle of this curve. A cusp is the case where the tangent vector of the curve becomes indefinite, so we are looking for the value of α that's a solution of $\mathbf{P}^t(0.5) = (0, 0)$.

$$\mathbf{P}^t(0.5) = \frac{3}{4}(-2, 0) + \frac{3\alpha}{4}(2\cos\theta, 0) + (3, 0) - \alpha(3\cos\theta, \sin\theta) + \alpha(\cos\theta, \sin\theta)$$

$$= (3/2, 0) + (-\cos\theta/2, 0).$$

It is easy to figure out that $\mathbf{P}^t(0.5) = (0, 0)$ yields $\alpha = 3/\cos\theta$.

4.14: The two endpoints of $\mathbf{Q}(T)$ are $\mathbf{P}(0.25) = (0.3, 0.19)$ and $\mathbf{P}(0.75) = (0.89, 0.19)$. The two extreme tangents are $0.5\mathbf{P}^t(0.25) = (0.66, 0.25)$ and $.5\mathbf{P}^t(0.75) = (0.41, -0.25)$ [notice the 0.5 factor that equals $(t_j - t_i)$]. The new PC and its derivative are therefore

$$\mathbf{Q}(T) = (T^3, T^2, T, 1) \begin{bmatrix} 2 & -2 & 1 & 1 \\ -3 & 3 & -2 & -1 \\ 0 & 0 & 1 & 0 \\ 1 & 0 & 0 & 0 \end{bmatrix} \begin{bmatrix} (0.3, 0.19) \\ (0.89, 0.19) \\ (0.66, 0.25) \\ (0.41, -0.25) \end{bmatrix}$$

$$= (-0.125, 0)t^3 + (0.0625, -0.25)t^2 + (0.65625, 0.25)t + (0.296875, 0.1875).$$

$$\mathbf{Q}^T(T) = (-0.375, 0)t^2 + (0.125, -0.5)t + (0.65625, 0.25).$$

Direct checks verify that $\mathbf{Q}(0) = \mathbf{P}(0.25)$, $\mathbf{Q}(1) = \mathbf{P}(0.75)$, $\mathbf{Q}^T(0) = \mathbf{P}^t(0.25)$, and $\mathbf{Q}^T(1) = \mathbf{P}^t(0.75)$.

4.15: The tangent vector of Equation (4.27) is $\mathbf{P}^t(t) = (-6t^2 + 6t)(\mathbf{P}_2 - \mathbf{P}_1)$. Its absolute value (the speed of the curve) is therefore proportional to the function $-6t^2 + 6t$. When t varies from 0 to 1, this function goes up from 0 to a maximum of 1.5 at $t = 0.5$, then down to 0.

4.16: In this case, Equation (4.31) becomes

$$\mathbf{a} \cdot 0^2 + \mathbf{b} \cdot 0 + \mathbf{c} = \mathbf{P}_1,$$

$$\mathbf{a}\Delta^2 + \mathbf{b}\Delta + \mathbf{c} = \mathbf{P}_2, \qquad \text{(Ans.16)}$$

$$2\mathbf{a} \cdot 0 + \mathbf{b} = \mathbf{P}_1^t.$$

The solutions are

$$\mathbf{c} = \mathbf{P}_1, \quad \mathbf{b} = \mathbf{P}_1^t, \quad \text{and } \mathbf{a} = \frac{\mathbf{P}_2}{\Delta^2} - \frac{\mathbf{P}_1}{\Delta^2} - \frac{\mathbf{P}_1^t}{\Delta}$$

and the polynomial is therefore

$$\mathbf{P}(t) = (t^2, t, 1) \begin{pmatrix} \frac{-1}{\Delta^2} & \frac{1}{\Delta^2} & \frac{-1}{\Delta} \\ 0 & 0 & 1 \\ 1 & 0 & 0 \end{pmatrix} \begin{pmatrix} \mathbf{P}_1 \\ \mathbf{P}_2 \\ \mathbf{P}_1^t \end{pmatrix}. \tag{Ans.17}$$

It is easy to see that Equation (Ans.17) reduces to Equation (4.32) for $\Delta = 1$.

4.17: We are looking for a curve of the form $\mathbf{P}(t) = \mathbf{a}t^2 + \mathbf{b}t + \mathbf{c}$. Its tangent vector is the derivative $\mathbf{P}^t(t) = 2\mathbf{a}t + \mathbf{b}$. We denote the two known quantities by $\mathbf{P}^t(0) = \mathbf{P}_1^t$ and $\mathbf{P}^t(1) = \mathbf{P}_2^t$. The two equations $2\mathbf{a}\cdot 0 + \mathbf{b} = \mathbf{P}_1^t$ and $2\mathbf{a}\cdot 1 + \mathbf{b} = \mathbf{P}_2^t$ are easily solved to yield $\mathbf{b} = \mathbf{P}_1^t$ and $\mathbf{a} = (\mathbf{P}_2^t - \mathbf{P}_1^t)/2$. Thus, the curve is expressed by $\mathbf{P}(t) = \frac{1}{2}(\mathbf{P}_2^t - \mathbf{P}_1^t)t^2 + \mathbf{P}_1^t t + \mathbf{c}$ and its derivative is $\mathbf{P}^t(t) = (\mathbf{P}_2^t - \mathbf{P}_1^t)t + \mathbf{P}_1^t$ (the straight line from $\mathbf{P}_1^t$ to $\mathbf{P}_2^t$). Notice that the two extreme tangents fully define the shape of this curve but do not fix its position in space. To place such a curve in space, we have to know the value of $\mathbf{c}$. The two endpoints of this curve are $\mathbf{P}(0) = \mathbf{c}$ and $\mathbf{P}(1) = \mathbf{c} + \frac{1}{2}(\mathbf{P}_1^t + \mathbf{P}_2^t)$. The reader is encouraged to draw a diagram that shows the geometric meaning of adding the vector sum $\frac{1}{2}(\mathbf{P}_1^t + \mathbf{P}_2^t)$ to point $\mathbf{c}$.

4.18: The curve and its first two derivatives can be expressed as

$$\mathbf{P}(t) = \mathbf{a}t^2 + \mathbf{b}t^2 + \mathbf{c}t + \mathbf{d},$$
$$\mathbf{P}^t(t) = 3\mathbf{a}t^2 + 2\mathbf{b}t + \mathbf{c},$$
$$\mathbf{P}^{tt}(t) = 6\mathbf{a}t + 2\mathbf{b}.$$

In the standard case where $0 \le t \le 1$, the three conditions are expressed as $\mathbf{P}(0) = \mathbf{P}_1$, $\mathbf{P}(1) = \mathbf{P}_2$, $\mathbf{P}^t(1) = \mathbf{P}_2^t$, and $\mathbf{P}^{tt}(0) = 0$. The explicit equations are

$$\begin{aligned} \mathbf{a}\cdot 0^3 + \mathbf{b}\cdot 0^2 + \mathbf{c}\cdot 0 + \mathbf{d} &= \mathbf{P}_1, \\ \mathbf{a}\cdot 1^3 + \mathbf{b}\cdot 1^2 + \mathbf{c}\cdot 1 + \mathbf{d} &= \mathbf{P}_2, \\ 3\mathbf{a}\cdot 1^2 + 2\mathbf{b}\cdot 1 + \mathbf{c} &= \mathbf{P}_2^t, \\ 6\mathbf{a}\cdot 0 + 2\mathbf{b} &= 0. \end{aligned} \tag{Ans.18}$$

They are easy to solve and yield $\mathbf{a} = \frac{1}{2}\mathbf{P}_2^t - \frac{1}{2}(\mathbf{P}_2 - \mathbf{P}_1)$, $\mathbf{b} = 0$, $\mathbf{c} = \frac{3}{2}(\mathbf{P}_2 - \mathbf{P}_1) - \frac{1}{2}\mathbf{P}_2^t$, and $\mathbf{d} = \mathbf{P}_1$. The polynomial is therefore

$$\begin{aligned} \mathbf{P}_{std}(t) &= \left(\frac{1}{2}\mathbf{P}_2^t - \frac{1}{2}(\mathbf{P}_2 - \mathbf{P}_1) \right) t^3 + \left(\frac{3}{2}(\mathbf{P}_2 - \mathbf{P}_1) - \frac{1}{2}\mathbf{P}_2^t \right) t + \mathbf{P}_1 \\ &= \left(\frac{1}{2}t^3 - \frac{3}{2}t + 1 \right) \mathbf{P}_1 + \left(-\frac{1}{2}t^3 + \frac{3}{2}t \right) \mathbf{P}_2 + \left(\frac{1}{2}t^3 - \frac{1}{2}t \right) \mathbf{P}_2^t \end{aligned}$$

$$= (t^3, t^2, t, 1) \begin{pmatrix} 1/2 & -1/2 & 1/2 \\ 0 & 0 & 0 \\ -3/2 & 3/2 & -1/2 \\ 1 & 0 & 0 \end{pmatrix} \begin{pmatrix} \mathbf{P}_1 \\ \mathbf{P}_2 \\ \mathbf{P}_2^t \end{pmatrix}. \tag{Ans.19}$$

in the nonstandard case where $0 \le t \le \Delta$, Equation (Ans.19) is extended to

$$\mathbf{P}_{nstd}(t) = (t^3, t^2, t, 1) \begin{pmatrix} \dfrac{1}{2\Delta^3} & -\dfrac{1}{2\Delta^3} & \dfrac{1}{2\Delta^2} \\ 0 & 0 & 0 \\ -\dfrac{3}{2\Delta} & \dfrac{3}{2\Delta} & -\dfrac{1}{2} \\ 1 & 0 & 0 \end{pmatrix} \begin{pmatrix} \mathbf{P}_1 \\ \mathbf{P}_2 \\ \mathbf{P}_2^t \end{pmatrix}. \tag{Ans.20}$$

5.1: By using the *same* symbol, $\mathbf{P}_{k+1}^t$, for the end tangent of $\mathbf{P}_k(t)$ and the start tangent of $\mathbf{P}_{k+1}(t)$.

5.2: The three segments are

$$\begin{aligned} \mathbf{P}_1(t) &= (-\tfrac{1}{3}, -\tfrac{1}{5})t^3 + (\tfrac{1}{3}, \tfrac{6}{5})t^2 + (1, -1)t, \\ \mathbf{P}_2(t) &= (0, -\tfrac{2}{5})t^3 + (-\tfrac{2}{3}, \tfrac{3}{5})t^2 + (\tfrac{2}{3}, \tfrac{4}{5})t + (1, 0), \\ \mathbf{P}_3(t) &= (\tfrac{1}{3}, -\tfrac{1}{5})t^3 - (\tfrac{2}{3}, \tfrac{3}{5})t^2 + (-\tfrac{2}{3}, \tfrac{4}{5})t + (1, 1). \end{aligned}$$

The first intermediate point should be $\mathbf{P}_1(1)$ and also $\mathbf{P}_2(0)$. A simple calculation yields

$$\begin{aligned} \mathbf{P}_1(1) &= (-\tfrac{1}{3}, -\tfrac{1}{5})1^3 + (\tfrac{1}{3}, \tfrac{6}{5})1^2 + (1, -1) = (1, 0), \\ \mathbf{P}_2(0) &= (0, -\tfrac{2}{5})0^3 + (-\tfrac{2}{3}, \tfrac{3}{5})0^2 + (\tfrac{2}{3}, \tfrac{4}{5})0 + (1, 0) = (1, 0). \end{aligned}$$

The second intermediate point should be $\mathbf{P}_2(1)$ and also $\mathbf{P}_3(0)$. A similar calculation gives

$$\begin{aligned} \mathbf{P}_2(1) &= (0, -\tfrac{2}{5})1^3 + (-\tfrac{2}{3}, \tfrac{3}{5})1^2 + (\tfrac{2}{3}, \tfrac{4}{5})1 + (1, 0) = (1, 1), \\ \mathbf{P}_3(0) &= (\tfrac{1}{3}, -\tfrac{1}{5})0^3 - (\tfrac{2}{3}, \tfrac{3}{5})0^2 + (-\tfrac{2}{3}, \tfrac{4}{5})0 + (1, 1) = (1, 1). \end{aligned}$$

Both tangent vectors can be obtained from the second segment. Its derivative is

$$\mathbf{P}_2^t(t) = \frac{d\,\mathbf{P}_2(t)}{d\,t} = 3(0, -\tfrac{2}{5})t^2 + 2(-\tfrac{2}{3}, \tfrac{3}{5})t + (\tfrac{2}{3}, \tfrac{4}{5}).$$

So the two vectors are

$$\begin{aligned} \mathbf{P}_2^t(0) &= 3(0, -\tfrac{2}{5})0^2 + 2(-\tfrac{2}{3}, \tfrac{3}{5})0 + (\tfrac{2}{3}, \tfrac{4}{5}) = (\tfrac{2}{3}, \tfrac{4}{5}), \\ \mathbf{P}_2^t(1) &= 3(0, -\tfrac{2}{5})1^2 + 2(-\tfrac{2}{3}, \tfrac{3}{5})1 + (\tfrac{2}{3}, \tfrac{4}{5})(-\tfrac{2}{3}, \tfrac{10}{5}). \end{aligned}$$

Thus, the first tangent points in the direction $(5, 6)$ and the second one, in the direction $(-1, 3)$.

5.3: Equation (5.7) becomes

$$\begin{pmatrix} 1 & 4 & 1 & 0 \\ 0 & 1 & 4 & 1 \end{pmatrix} \begin{pmatrix} (0,0) \\ \mathbf{P}_2^t \\ \mathbf{P}_3^t \\ (-1,-1) \end{pmatrix} = \begin{pmatrix} 3[(1,1) - (0,0)] \\ 3[(0,1) - (1,0)] \end{pmatrix} = \begin{pmatrix} (3,3) \\ (-3,3) \end{pmatrix},$$

or explicitly

$$(0,0) + 4\mathbf{P}_2^t + \mathbf{P}_3^t = (3,3), \text{ and } \mathbf{P}_2^t + 4\mathbf{P}_3^t + (-1,-1) = (-3,3).$$

The solutions are $\mathbf{P}_2^t = (\frac{8}{15}, \frac{8}{15})$ and $\mathbf{P}_3^t = (\frac{37}{15}, \frac{37}{15})$. The first segment, from Equation (4.7), is

$$\mathbf{P}_1(t) = (t^3, t^2, t, 1) \begin{pmatrix} 2 & -2 & 1 & 1 \\ -3 & 3 & -2 & -1 \\ 0 & 0 & 1 & 0 \\ 1 & 0 & 0 & 0 \end{pmatrix} \begin{pmatrix} (0,0) \\ (1,0) \\ (0,0) \\ (\frac{8}{15}, \frac{8}{15}) \end{pmatrix}$$

$$= (-\frac{22}{15}, \frac{8}{15})t^3 + (\frac{37}{15}, -\frac{8}{15})t^2.$$

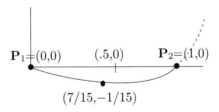

$$\mathbf{P}_1 = (0,0) \qquad (.5,0) \qquad \mathbf{P}_2 = (1,0)$$

$$(7/15, -1/15)$$

Figure Ans.6: An Indefinite Start Direction.

An initial direction of $(0,0)$ means that the curve will be the shortest possible (Figure Ans.6). It also means that the curve will start slowly and will speed up as it goes along. It is easy to see that

$$\mathbf{P}_1(0.5) = (-\frac{22}{15}, \frac{8}{15})\frac{1}{8} + (\frac{37}{15}, -\frac{8}{15})\frac{1}{4} = (\frac{7}{15}, -\frac{1}{15}).$$

At $t = 0.5$, the curve hasn't reached the midpoint between $\mathbf{P}_1$ and $\mathbf{P}_2$.

5.4: For the third segment, Equation (4.7) becomes

$$\mathbf{P}_3(t) = (t^3, t^2, t, 1) \begin{pmatrix} 2 & -2 & 1 & 1 \\ -3 & 3 & -2 & -1 \\ 0 & 0 & 1 & 0 \\ 1 & 0 & 0 & 0 \end{pmatrix} \begin{pmatrix} (1,1) \\ (0,1) \\ (-\frac{3}{5}, \frac{2}{3}) \\ (-\frac{6}{5}, -\frac{1}{3}) \end{pmatrix}$$

$$= (\frac{1}{5}, \frac{1}{3})t^3 - (\frac{3}{5}, 1)t^2 + (-\frac{3}{5}, \frac{2}{3})t + (1,1).$$

5.5: For the third segment, Equation (4.7) becomes

$$\mathbf{P}_3(t) = (t^3, t^2, t, 1)\begin{pmatrix} 2 & -2 & 1 & 1 \\ -3 & 3 & -2 & -1 \\ 0 & 0 & 1 & 0 \\ 1 & 0 & 0 & 0 \end{pmatrix}\begin{pmatrix} (0,1) \\ (-1,0) \\ (-\frac{3}{2},0) \\ (0,-\frac{3}{2}) \end{pmatrix}$$

$$= (\tfrac{1}{2}, \tfrac{1}{2})t^3 + (0, -\tfrac{3}{2})t^2 + (-\tfrac{3}{2}, 0)t + (0, 1).$$

For the fourth segment, Equation (4.7) becomes

$$\mathbf{P}_4(t) = (t^3, t^2, t, 1)\begin{pmatrix} 2 & -2 & 1 & 1 \\ -3 & 3 & -2 & -1 \\ 0 & 0 & 1 & 0 \\ 1 & 0 & 0 & 0 \end{pmatrix}\begin{pmatrix} (-1,0) \\ (0,-1) \\ (0,-\frac{3}{2}) \\ (\frac{3}{2},0) \end{pmatrix}$$

$$= (-\tfrac{1}{2}, \tfrac{1}{2})t^3 + (\tfrac{3}{2}, 0)t^2 + (0, -\tfrac{3}{2})t + (-1, 0).$$

5.6: Equation (5.15) gives

$$\mathbf{P}_1^t = -\mathbf{P}_3^t = \frac{3}{4}\left(\mathbf{P}_2 - \mathbf{P}_1 - \mathbf{P}_3 + \mathbf{P}_2\right) - \frac{1}{4}\left(\mathbf{P}_2^t - \mathbf{P}_2^t\right) = \frac{3}{4}(2\mathbf{P}_2 - \mathbf{P}_1 - \mathbf{P}_3) = (0, 3/2).$$

We next substitute the anticyclic end condition in Equation (5.14), which becomes

$$(1, 4, 1)\begin{bmatrix} (0, 3/2) \\ \mathbf{P}_2^t \\ (0, -3/2) \end{bmatrix} = 3(\mathbf{P}_3 - \mathbf{P}_1) = (6, 0). \tag{Ans.21}$$

The solution is $\mathbf{P}_2^t = (3/2, 0)$.

The first spline segment can now be calculated from Equation (4.7):

$$\mathbf{P}_1(t) = (t^3, t^2, t, 1)\begin{pmatrix} 2 & -2 & 1 & 1 \\ -3 & 3 & -2 & -1 \\ 0 & 0 & 1 & 0 \\ 1 & 0 & 0 & 0 \end{pmatrix}\begin{pmatrix} (-1,0) \\ (0,1) \\ (0,3/2) \\ (3/2,0) \end{pmatrix}$$

$$= (-\tfrac{1}{2}, -\tfrac{1}{2})t^3 + (\tfrac{3}{2}, 0)t^2 + (0, \tfrac{3}{2})t + (-1, 0).$$

Its derivative is

$$\mathbf{P}_1^t(t) = (-3/2, -3/2)t^2 + (3, 0)t + (0, 3/2),$$

so $\mathbf{P}_1^t(0) = (0, 3/2)$ and $\mathbf{P}_1^t(1) = (3/2, 0)$.

The second spline segment is similarly calculated:

$$\mathbf{P}_2(t) = (t^3, t^2, t, 1)\begin{pmatrix} 2 & -2 & 1 & 1 \\ -3 & 3 & -2 & -1 \\ 0 & 0 & 1 & 0 \\ 1 & 0 & 0 & 0 \end{pmatrix}\begin{pmatrix} (0,1) \\ (1,0) \\ (3/2,0) \\ (0,-3/2) \end{pmatrix}$$

$$= (-\tfrac{1}{2}, \tfrac{1}{2})t^3 + (0, -\tfrac{3}{2})t^2 + (\tfrac{3}{2}, 0)t + (0, 1).$$

Its derivative is
$$\mathbf{P}_2^t(t) = (-3/2, 3/2)t^2 + (0, -3)t + (3/2, 0),$$
so $\mathbf{P}_2^t(0) = (3/2, 0)$ and $\mathbf{P}_2^t(1) = (0, -3/2)$.

To compare this anticyclic cubic spline to the clamped cubic spline for the same points, we have to select the same start and end tangents, namely $\mathbf{P}_1^t = (0, 3/2)$ and $\mathbf{P}_3^t = (0, -3/2)$. When these tangents are substituted in Equation (5.7), it becomes identical to Equation (Ans.21), showing that for this particular choice of points, the clamped and anticyclic cubic splines are identical.

5.7: When $T > 1$, s becomes negative, causing the two tangent vectors to reverse directions. This changes the shape of the curve completely. However, large negative values of s still produce a loose curve.

5.8: Differentiating Equation (5.32) and substituting $s = 0$ results in $\mathbf{P}^t(t) = 6t(1 - t)(\mathbf{P}_3 - \mathbf{P}_2)$ [see Equation (6.26)]. This expression is zero for both $t = 0$ and $t = 1$, but is nonzero for any other values of t. It has a maximum for $t = 0.5$.

6.1: Figure Ans.7 lists the points and the code for this computation. Notice how the sharp corner at the top-center of the heart is obtained by the particular placement of points 3 through 6 and how parameter `ppr` determines the width of the heart.

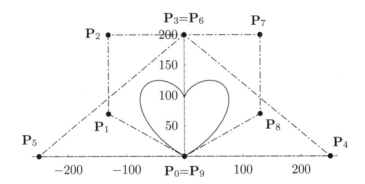

```
(* Heart-shaped Bezier curve *)
n=9; ppr=130;
pnts={{0,0},{-ppr,70},{-ppr,200},{0,200},{250,0},{-250,0},{0,200},
  {ppr,200},{ppr,70},{0,0}};
pwr[x_,y_]:=If[x==0 && y==0, 1, x^y];
bern[n_,i_,t_]:=Binomial[n,i]pwr[t,i]pwr[1-t,n-i]
bzCurve[t_]:=Sum[pnts[[i+1]]bern[n,i,t], {i,0,n}]
g1=ListPlot[pnts, Prolog->AbsolutePointSize[4], PlotRange->All,
  AspectRatio->Automatic, DisplayFunction->Identity]
g2=ParametricPlot[bzCurve[t], {t,0,1}, Compiled->False,
  PlotRange->All, AspectRatio->Automatic, DisplayFunction->Identity]
g3=Graphics[{AbsoluteDashing[{1,2,5,2}], Line[pnts]}]
Show[g1,g2,g3, DisplayFunction->$DisplayFunction,
  DefaultFont->{"cmr10", 10}]
```

Figure Ans.7: A Heart-Shaped Bézier Curve.

6.2: We simply calculate the quadratic Bézier curve for the three points. As a quadratic parametric polynomial it is a parabola (see second paragraph of Section 1.5). Since this is a Bézier curve, its extreme tangents point in the desired directions:

$$\mathbf{P}(t) = \mathbf{P}_1(1-t)^2 + 2\mathbf{P}_2(1-t)t + \mathbf{P}_3 t^2 = (\mathbf{P}_1 - 2\mathbf{P}_2 + \mathbf{P}_3)t^2 + 2(\mathbf{P}_2 - \mathbf{P}_1)t + \mathbf{P}_1.$$

(See also Section 4.2.4.)

6.3: A simple procedure is to compute

$$\mathbf{P}_0 = \mathbf{P}(0) = (1,0), \quad \mathbf{P}_1 = \mathbf{P}(1/3) = (13/9, 1/27),$$
$$\mathbf{P}_2 = \mathbf{P}(2/3) = (19/9, 8/27), \quad \mathbf{P}_3 = \mathbf{P}(1) = (3,1).$$

6.4: The substitution is $u = 2t-1$, from which we get $t = (1+u)/2$ and $1-t = (1-u)/2$. The curve of Equation (6.8) can now be written

$$\mathbf{P}(t) = \frac{1}{8}(1-u)^3 \mathbf{P}_0 + \frac{1}{4}(1+u)(1-u)^2 \mathbf{P}_1 + 2\left(\frac{1+u}{2}\right)^2 \left(\frac{1-u}{2}\right)\mathbf{P}_2 + \frac{1}{8}(1+u)^3 \mathbf{P}_3$$

$$= \frac{1}{8}(u^3, u^2, u, 1)\begin{pmatrix} -1 & 2 & -2 & 1 \\ 3 & -2 & -2 & 3 \\ -3 & -2 & 2 & 3 \\ 1 & 2 & 2 & 1 \end{pmatrix}\begin{pmatrix} \mathbf{P}_0 \\ \mathbf{P}_1 \\ \mathbf{P}_2 \\ \mathbf{P}_3 \end{pmatrix}.$$

The only difference is the basis matrix.

6.5: Direct calculation of $B_{4,i}(t)$ for $0 \le i \le 4$ yields the five functions

$$B_{4,0} = (1-t)^4, \ B_{4,1} = 4t(1-t)^3, \ B_{4,2} = 6t^2(1-t)^2, \ B_{4,3} = 4t^3(1-t), \text{ and } B_{4,4} = t^4.$$

6.6: The weights are $B_{1,0}(t) = \binom{1}{0}t^0(1-t)^{1-0} = (1-t)$ and $B_{1,1}(t) = \binom{1}{1}t^1(1-t)^{1-1} = t$, and the curve is $\mathbf{P}(t) = \mathbf{P}_0(1-t) + \mathbf{P}_1 t$, the straight segment from $\mathbf{P}_0$ to $\mathbf{P}_1$.

6.7: Three collinear points are dependent, which means that any of the three can be expressed as a linear combination (a weighted sum) of the other two, with barycentric weights. We therefore assume that $\mathbf{P}_1 = (1-\alpha)\mathbf{P}_0 + \alpha\mathbf{P}_2$ for some real α. The general Bézier curve for three points,

$$\mathbf{P}(t) = \mathbf{P}_0(1-t)^2 + \mathbf{P}_1 2t(1-t) + \mathbf{P}_2 t^2,$$

now becomes

$$\mathbf{P}(t) = \mathbf{P}_0(1-t)^2 + [(1-\alpha)\mathbf{P}_0 + \alpha\mathbf{P}_2]2t(1-t) + \mathbf{P}_2 t^2,$$

which is easily simplified to

$$
\begin{aligned}
\mathbf{P}(t) &= \mathbf{P}_0 + 2\alpha(\mathbf{P}_2 - \mathbf{P}_0)t + (1 - 2\alpha)(\mathbf{P}_2 - \mathbf{P}_0)t^2 \\
&= \mathbf{P}_0 + (\mathbf{P}_2 - \mathbf{P}_0)[2\alpha t + (1 - 2\alpha)t^2] \\
&= \mathbf{P}_0 + (\mathbf{P}_2 - \mathbf{P}_0)T.
\end{aligned}
\tag{Ans.22}
$$

This is linear in T and therefore represents a straight line.

This case does not contradict the fact that the Bézier curve does not pass through the intermediate points. We have considered three *collinear* points, which really are only two points. The Bézier curve for two points is a straight line. Note that even with four collinear points, only two are really independent.

We continue this discussion by examining two cases. The first is the special case of uniformly-spaced collinear points and the second is the case of three collinear points $\mathbf{P}_0$, $\mathbf{P}_1$, and $\mathbf{P}_2$ where $\mathbf{P}_1$ is not between $\mathbf{P}_0$ and $\mathbf{P}_2$ but is one of the endpoints.

Case 1. Consider the case of $n+1$ points that are equally spaced along the straight segment from $\mathbf{P}_0$ to $\mathbf{P}_n$. We show that the Bézier curve for these points is the straight segment from $\mathbf{P}_0$ to $\mathbf{P}_n$. We start with two auxiliary relations;

1. Point $\mathbf{P}_k$ (for $k = 0,\ 1,\ldots,\ n$) can be expressed in this case as the blend $(1 - k/n)\mathbf{P}_0 + (k/n)\mathbf{P}_n$.

2. It can be proved by induction that $\sum_{i=0}^{n} iB_{n,i}(t) = nt$.

Based on these relations, the Bézier curve for uniformly-spaced collinear points is

$$
\begin{aligned}
\mathbf{P}(t) &= \sum_{i=0}^{n} B_{n,i}(t)\mathbf{P}_i = \sum_{i=0}^{n} B_{n,i}(t)\big[(1 - i/n)\mathbf{P}_0 + (i/n)\mathbf{P}_n\big] \\
&= \mathbf{P}_0 \sum B_{n,i}(t) - \frac{\mathbf{P}_0}{n} \sum iB_{n,i}(t) + \frac{\mathbf{P}_n}{n} \sum iB_{n,i}(t) \\
&= \mathbf{P}_0 - t\mathbf{P}_0 + t\mathbf{P}_n = (1 - t)\mathbf{P}_0 + t\mathbf{P}_n.
\end{aligned}
$$

Case 2. $\mathbf{P}_1$ is not located between $\mathbf{P}_0$ and $\mathbf{P}_2$ but is one of the endpoints. The two cases $\alpha = 0$ and $\alpha = 1$ imply that point $\mathbf{P}_1$ is identical to $\mathbf{P}_0$ or $\mathbf{P}_2$, respectively. The case $\alpha = 0.5$ means that $\mathbf{P}_1$ is midway between $\mathbf{P}_0$ and $\mathbf{P}_2$. The cases $\alpha < 0$ and $\alpha > 1$ are special. The former means that $\mathbf{P}_1$ "precedes" $\mathbf{P}_0$. The latter means that $\mathbf{P}_1$ "follows" $\mathbf{P}_2$. In these cases, the curve is no longer a straight line but goes from $\mathbf{P}_0$ toward $\mathbf{P}_1$, reverses direction without reaching $\mathbf{P}_1$, and continues to $\mathbf{P}_2$. The point where it reverses direction becomes a cusp (a sharp corner), where the curve has an indefinite tangent vector (Figure Ans.8).

Figure Ans.8: Bézier Straight Segments.

Analysis. We first show that in these cases the curve does not go through point $\mathbf{P}_1$. Equation (Ans.22) can be written

$$\mathbf{P}(t) = \mathbf{P}_0 \left(1 - 2\alpha t - t^2 + 2\alpha t^2\right) + \mathbf{P}_2 \left(2\alpha t + t^2 - 2\alpha t^2\right).$$

Let's see for what value of t the curve passes through point $\mathbf{P}_1 = (1 - \alpha)\mathbf{P}_0 + \alpha\mathbf{P}_2$. The conditions are

$$1 - 2\alpha t - t^2 + 2\alpha t^2 = 1 - \alpha \quad \text{and} \quad 2\alpha t + t^2 - 2\alpha t^2 = \alpha.$$

These conditions yield the following quadratic equations for t:

$$\alpha - 2\alpha t + (2\alpha - 1)t^2 = 0 \quad \text{and} \quad -\alpha + 2\alpha t - (2\alpha - 1)t^2 = 0.$$

These equations are identical and their solutions are

$$t = \frac{\alpha \pm \sqrt{\alpha(\alpha - 1)}}{\alpha} \quad \text{and} \quad t = \frac{-\alpha \pm \sqrt{\alpha(1 - \alpha)}}{-\alpha}.$$

The first solution has no real values for negative α and the second one has no real values for $\alpha > 1$. For these values of α, the curve does not pass through control point $\mathbf{P}_1$.

We now calculate the value of t for which the curve has a cusp (a sharp corner). The tangent vector of the curve is

$$\mathbf{P}^t(t) = \mathbf{P}_0 \left(-2\alpha - 2t + 4\alpha t\right) + \mathbf{P}_2 \left(2\alpha + 2t - 4\alpha t\right) = (2\alpha + 2t - 4\alpha t)(\mathbf{P}_2 - \mathbf{P}_0).$$

The condition for an indefinite tangent vector is therefore $2\alpha + 2t - 4\alpha t = 0$, which happens for $t = \alpha/(2\alpha - 1)$.

The following three special cases are particularly interesting:

1. $\alpha \ll 0$. This is the case where $\mathbf{P}_1$ is far away from both $\mathbf{P}_0$ and $\mathbf{P}_2$. The limit of $\alpha/(2\alpha - 1)$ in this case is $1/2$, which means that the curve changes direction at its midpoint.

2. $\alpha = -1$. In this case point $\mathbf{P}_0$ is exactly between $\mathbf{P}_1$ and $\mathbf{P}_2$. The value of $\alpha/(2\alpha - 1)$ in this case is $1/3$ (Figure Ans.8a illustrates why this makes sense).

3. $\alpha \gg 1$. Here, $\mathbf{P}_1$ is again far from both $\mathbf{P}_0$ and $\mathbf{P}_2$, but in the other direction (Figure Ans.8b). The limit of $\alpha/(2\alpha - 1)$ in this case is, again, $1/2$.

(End of long answer.)

6.8: The condition $\alpha \mathbf{Q}^t(0) = \mathbf{P}^t(1)$ is equivalent to $\alpha m \mathbf{Q}_1 - \alpha m \mathbf{Q}_0 = n\mathbf{P}_n - n\mathbf{P}_{n-1}$. Equation (6.15) shows that this can be written

$$\mathbf{P}_n = \frac{\alpha m}{\alpha m + n}\mathbf{Q}_1 + \frac{n}{\alpha m + n}\mathbf{P}_{n-1}.$$

The three points should be collinear, but the weights in this case are different. In the special case where $n = m$, this condition reduces to

$$\mathbf{P}_n = \frac{\alpha}{\alpha + 1}\mathbf{Q}_1 + \frac{1}{2}\mathbf{P}_{n-1}.$$

6.9: The process is the same, regardless of the placement of the control points and the shape of the final curve. Figure Ans.9 illustrates the scaffolding (thin segments), the intermediate points (squares), and the final point (circle).

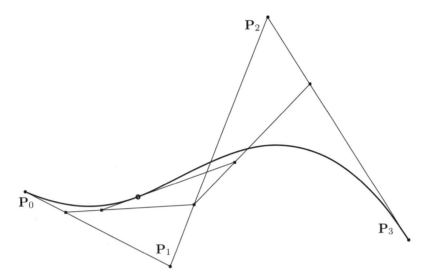

Figure Ans.9: Scaffolding With An Inflection Point.

6.10: The curve is easy to compute from Equation (6.5)

$$\mathbf{P}(t) = (1-t)^3(0,1,1) + 3t(1-t)^2(1,1,0) + 3t^2(1-t)(4,2,0) + t^3(6,1,1)$$
$$= (-3t^3 + 6t^2 + 3t, -3t^3 + 3t^2 + 1, 3t^2 - 3t + 1).$$

The three associated blossoms are obtained immediately from Equation (6.16). They are

$$f_x(u,v,w) = -3uvw + 2(uv + uw + vw) + (u + v + w)$$
$$f_y(u,v,w) = -3uvw + (uv + uw + vw) + 1,$$
$$f_z(u,v,w) = (uv + uw + vw) - (u + v + w) + 1.$$

The four control points are given as the four special values of the blossoms as follows:

$$\big(f_x(000), f_y(000), f_z(000)\big) = (0,1,1), \quad \big(f_x(001), f_y(001), f_z(001)\big) = (1,1,0),$$
$$\big(f_x(011), f_y(011), f_z(011)\big) = (4,2,0), \quad \big(f_x(111), f_y(111), f_z(111)\big) = (6,1,1).$$

It is also trivial to verify that

$$(f_x(ttt), f_y(ttt), f_z(ttt)) = (-3t^3 + 6t^2 + 3t, -3t^3 + 3t^2 + 1, 3t^2 - 3t + 1).$$

6.11: Figure 6.8 shows the new points. For an arbitrary α their values are

$$\mathbf{P}_{01} = \alpha\mathbf{P}_0 + (1-\alpha)\mathbf{P}_1, \quad \mathbf{P}_{12} = \alpha\mathbf{P}_1 + (1-\alpha)\mathbf{P}_2, \quad \mathbf{P}_{23} = \alpha\mathbf{P}_2 + (1-\alpha)\mathbf{P}_3,$$
$$\mathbf{P}_{012} = \alpha^2\mathbf{P}_0 + 2\alpha(1-\alpha)\mathbf{P}_1 + (1-\alpha^2)\mathbf{P}_2, \quad \mathbf{P}_{123} = \alpha^2\mathbf{P}_1 + 2\alpha(1-\alpha)\mathbf{P}_2 + (1-\alpha)^2\mathbf{P}_3,$$
$$\mathbf{P}_{0123} = \alpha^3\mathbf{P}_0 + 3\alpha^2(1-\alpha)\mathbf{P}_1 + 3\alpha(1-\alpha)^2\mathbf{P}_2 + (1-\alpha)^3\mathbf{P}_3.$$

Using matrix notation, this can be expressed as,

$$\begin{pmatrix} \mathbf{P}_0 \\ \mathbf{P}_{01} \\ \mathbf{P}_{012} \\ \mathbf{P}_{0123} \end{pmatrix} = \begin{pmatrix} 1 & 0 & 0 & 0 \\ \alpha & 1-\alpha & 0 & 0 \\ \alpha^2 & 2\alpha(1-\alpha) & (1-\alpha)^2 & 0 \\ \alpha^3 & 3\alpha^2(1-\alpha) & 3\alpha(1-\alpha)^2 & (1-\alpha)^3 \end{pmatrix} \begin{pmatrix} \mathbf{P}_0 \\ \mathbf{P}_1 \\ \mathbf{P}_2 \\ \mathbf{P}_3 \end{pmatrix} = \mathbf{M}_L(\alpha)\mathbf{G},$$

$$\begin{pmatrix} \mathbf{P}_{0123} \\ \mathbf{P}_{123} \\ \mathbf{P}_{23} \\ \mathbf{P}_3 \end{pmatrix} = \begin{pmatrix} \alpha^3 & 3\alpha^2(1-\alpha) & 3\alpha(1-\alpha)^2 & (1-\alpha)^3 \\ 0 & \alpha^2 & 2\alpha(1-\alpha) & (1-\alpha)^2 \\ 0 & 0 & \alpha & 1-\alpha \\ 0 & 0 & 0 & 1 \end{pmatrix} \begin{pmatrix} \mathbf{P}_0 \\ \mathbf{P}_1 \\ \mathbf{P}_2 \\ \mathbf{P}_3 \end{pmatrix} = \mathbf{M}_R(\alpha)\mathbf{G},$$

where $\mathbf{G}$ is the column consisting of the four original control points of the segment. Notice that the elements of each row of matrices $\mathbf{M}_L(\alpha)$ and $\mathbf{M}_R(\alpha)$ are barycentric. For the special case $\alpha = 0.5$, these expressions reduce to

$$\begin{pmatrix} \mathbf{P}_0 \\ \mathbf{P}_{01} \\ \mathbf{P}_{012} \\ \mathbf{P}_{0123} \end{pmatrix} = \frac{1}{8}\begin{pmatrix} 8 & 0 & 0 & 0 \\ 4 & 4 & 0 & 0 \\ 2 & 4 & 2 & 0 \\ 1 & 3 & 3 & 1 \end{pmatrix} \begin{pmatrix} \mathbf{P}_0 \\ \mathbf{P}_1 \\ \mathbf{P}_2 \\ \mathbf{P}_3 \end{pmatrix}, \quad \begin{pmatrix} \mathbf{P}_{0123} \\ \mathbf{P}_{123} \\ \mathbf{P}_{23} \\ \mathbf{P}_3 \end{pmatrix} = \frac{1}{8}\begin{pmatrix} 1 & 3 & 3 & 1 \\ 0 & 2 & 4 & 2 \\ 0 & 0 & 4 & 4 \\ 0 & 0 & 0 & 8 \end{pmatrix} \begin{pmatrix} \mathbf{P}_0 \\ \mathbf{P}_1 \\ \mathbf{P}_2 \\ \mathbf{P}_3 \end{pmatrix}.$$

6.12: Four control points implies $n = 3$. The two original interior points $\mathbf{P}_1$ and $\mathbf{P}_2$ are deleted. Three of the six new points are obtained from Equation (6.17)a

$$\mathbf{P}_{01} = \sum_{j=0}^{1} B_{1j}(\tfrac{1}{3})\mathbf{P}_j = B_{10}(\tfrac{1}{3})\mathbf{P}_0 + B_{11}(\tfrac{1}{3})\mathbf{P}_1 = (\tfrac{1}{3}, \tfrac{3}{3}, \tfrac{2}{3}),$$

$$\mathbf{P}_{012} = \sum_{j=0}^{2} B_{2j}(\tfrac{1}{3})\mathbf{P}_j = B_{20}(\tfrac{1}{3})\mathbf{P}_0 + B_{21}(\tfrac{1}{3})\mathbf{P}_1 + B_{22}(\tfrac{1}{3})\mathbf{P}_2 = (\tfrac{8}{9}, \tfrac{10}{9}, \tfrac{4}{9}),$$

$$\mathbf{P}_{0123} = \sum_{j=0}^{3} B_{3j}(\tfrac{1}{3})\mathbf{P}_j = B_{30}(\tfrac{1}{3})\mathbf{P}_0 + B_{31}(\tfrac{1}{3})\mathbf{P}_1 + B_{32}(\tfrac{1}{3})\mathbf{P}_2 + B_{33}(\tfrac{1}{3})\mathbf{P}_3 = (\tfrac{14}{9}, \tfrac{11}{9}, \tfrac{3}{9}),$$

and the other three are obtained from Equation (6.17)b

$$\mathbf{P}_{0123} = \sum_{j=0}^{3} B_{3j}(\tfrac{1}{3})\mathbf{P}_{3-3+j} = (\tfrac{14}{9}, \tfrac{11}{9}, \tfrac{3}{9}),$$

$$\mathbf{P}_{123} = \sum_{j=0}^{2} B_{2j}(\tfrac{1}{3})\mathbf{P}_{3-2+j} = (\tfrac{26}{9}, \tfrac{13}{9}, \tfrac{1}{9}),$$

$$\mathbf{P}_{23} = \sum_{j=0}^{1} B_{1j}(\tfrac{1}{3})\mathbf{P}_{3-1+j} = (\tfrac{14}{3}, \tfrac{5}{3}, \tfrac{1}{3}).$$

Figure Ans.10 lists the code for this computation.

```
(* New points for Bezier curve subdivision exercise *)
pnts={{0,1,1},{1,1,0},{4,2,0},{6,1,1}};
t=1/3;
pwr[x_,y_]:=If[x==0 && y==0, 1, x^y];
bern[n_,i_,t_]:=Binomial[n,i]pwr[t,i]pwr[1-t,n-i]
p01=Sum[pnts[[i+1]]bern[1,i,t], {i,0,1}]
p012=Sum[pnts[[i+1]]bern[2,i,t], {i,0,2}]
p0123=Sum[pnts[[i+1]]bern[3,i,t], {i,0,3}]
p0123=Sum[pnts[[3-3+i+1]]bern[3,i,t], {i,0,3}]
p123=Sum[pnts[[3-2+i+1]]bern[2,i,t], {i,0,2}]
p23=Sum[pnts[[3-1+i+1]]bern[1,i,t], {i,0,1}]
```

Figure Ans.10: Code to Compute Six New Points.

6.13: Applying Equation (6.18) to the original three points yields the four points

$$\mathbf{P}_0, \quad (\mathbf{P}_0 + 2\mathbf{P}_1)/3, \quad (2\mathbf{P}_1 + \mathbf{P}_2)/3, \quad \text{and} \quad \mathbf{P}_2.$$

Applying the same equation to these points results in the five points

$$\mathbf{P}_0, \quad (\mathbf{P}_0 + 3(\mathbf{P}_0 + 2\mathbf{P}_1)/3)/4 = (\mathbf{P}_0 + \mathbf{P}_1)/2,$$
$$\big(2(\mathbf{P}_0 + 2\mathbf{P}_1)/3 + 2(2\mathbf{P}_1 + \mathbf{P}_2)/3\big)/4 = (\mathbf{P}_0 + 4\mathbf{P}_1 + \mathbf{P}_2)/6,$$
$$\big(2(2\mathbf{P}_1 + \mathbf{P}_2)/3 + \mathbf{P}_2\big)/4 = (\mathbf{P}_1 + \mathbf{P}_2)/2, \quad \text{and} \quad \mathbf{P}_2.$$

6.14: Equation (6.18) gives the five new control points

$$\mathbf{Q}_0 = \mathbf{P}_0 = (0,0), \quad \mathbf{Q}_1 = \frac{\mathbf{P}_0 + 3\mathbf{P}_1}{4} = \left(\frac{3}{4}, \frac{3}{2}\right), \quad \mathbf{Q}_2 = \frac{2\mathbf{P}_1 + 2\mathbf{P}_2}{4} = (2,2),$$

$$\mathbf{Q}_3 = \frac{3\mathbf{P}_2 + \mathbf{P}_3}{4} = \left(\frac{11}{4}, \frac{3}{2}\right), \quad \text{and} \quad \mathbf{Q}_4 = \mathbf{P}_3 = (2,0).$$

The original curve is

$$\mathbf{P}_3(t) = (1-t)^3(0,0) + 3t(1-t)^2(1,2) + 3t^2(1-t)(3,2) + t^3(2,0),$$

and the new one is

$$\mathbf{P}_4(t) = (1-t)^4(0,0) + 4t(1-t)^3(3/4, 3/2) + 6t^2(1-t)^2(2,2)$$
$$+ 4t^3(1-t)(11/4, 3/2) + t^4(2,0).$$

These polynomials seem different, but a closer look reveals that both equal the polynomial $t^3(-4,0) + t^2(3,-6) + t(3,6)$. The two curves $\mathbf{P}_3(t)$ and $\mathbf{P}_4(t)$ therefore have the same shape, but $\mathbf{P}_4(t)$ is easier to reshape because it depends on five control points.

6.15: A direct check shows that the elements of every row of matrix $\mathbf{B}$ add up to 1, regardless of the values of a and b. This guarantees that each new control point $\mathbf{Q}_i$ will be a barycentric sum of the $\mathbf{P}_i$'s.

6.16: (1) For $a = 1$ and $b = a + x$, matrix $\mathbf{B}$ is

$$\mathbf{B} = \begin{pmatrix} 0 & 0 & 0 & 1 \\ 0 & 0 & -x & 1+x \\ 0 & x^2 & -2x(1+x) & (1+x)^2 \\ -x^3 & 3x^2(1+x) & -3x(1+x)^2 & (1+x)^3 \end{pmatrix}.$$

(2) The new control points are

$$\mathbf{Q}_0 = \mathbf{P}_3,$$
$$\mathbf{Q}_1 = -(\mathbf{P}_2 x) + \mathbf{P}_3(1+x),$$
$$\mathbf{Q}_2 = \mathbf{P}_1 x^2 + \mathbf{P}_3(1+x)^2 + \mathbf{P}_2(1+x)(3+x-3(1+x)),$$
$$\mathbf{Q}_3 = -(\mathbf{P}_0 x^3) + 3\mathbf{P}_1 x^2(1+x) - 3\mathbf{P}_2 x(1+x)^2 + \mathbf{P}_3(1+x)^3.$$

(3) For $x = 0.75$, they become

$$\mathbf{Q}_0 = \mathbf{P}_3,$$
$$\mathbf{Q}_1 = -0.75\mathbf{P}_2 + 1.75\mathbf{P}_3,$$
$$\mathbf{Q}_2 = 0.5625\mathbf{P}_1 - 2.625\mathbf{P}_2 + 3.0625\mathbf{P}_3,$$
$$\mathbf{Q}_3 = -0.421875\mathbf{P}_0 + 2.953125\mathbf{P}_1 - 6.890625\mathbf{P}_2 + 5.359375\mathbf{P}_3.$$

Notice how each $\mathbf{Q}_i$ is a barycentric combination of the $\mathbf{P}_i$'s.

6.17: For $a = 0$ and $b = 0.5$, matrix $\mathbf{B}$ becomes

$$\mathbf{B} = \begin{pmatrix} 1 & 0 & 0 & 0 \\ 0.5 & 0.5 & 0 & 0 \\ 0.25 & 0.5 & 0.25 & 0 \\ 0.125 & 0.375 & 0.375 & 0.125 \end{pmatrix},$$

and the new control points are

$$\mathbf{Q}_0 = \mathbf{P}_0,$$

$$\mathbf{Q}_1 = \frac{1}{2}\mathbf{P}_0 + \frac{1}{2}\mathbf{P}_1,$$

$$\mathbf{Q}_2 = \frac{1}{4}\mathbf{P}_0 + \frac{1}{2}\mathbf{P}_1 + \frac{1}{4}\mathbf{P}_2,$$

$$\mathbf{Q}_3 = \frac{1}{8}\mathbf{P}_0 + \frac{3}{8}\mathbf{P}_1 + \frac{3}{8}\mathbf{P}_2 + \frac{1}{8}\mathbf{P}_3.$$

6.18: This is easy to verify directly. We substitute the two points $\mathbf{P}_0$ and $\mathbf{P}_3$ and the two tangents $3(\mathbf{P}_1 - \mathbf{P}_0)$ and $3(\mathbf{P}_3 - \mathbf{P}_2)$ in Equation (4.5)

$$\mathbf{P}(t) = (2t^3 - 3t^2 + 1)\mathbf{P}_0 + (-2t^3 + 3t^2)\mathbf{P}_3 + (t^3 - 2t^2 + t)3(\mathbf{P}_1 - \mathbf{P}_0) + (t^3 - t^2)3(\mathbf{P}_3 - \mathbf{P}_2).$$

After rearranging, we get

$$\mathbf{P}(t) = (1-t)^3\mathbf{P}_0 + 3t(1-t)^2\mathbf{P}_1 + 3t^2(1-t)\mathbf{P}_2 + t^3\mathbf{P}_3,$$

which is the cubic Bézier curve defined by the four points.

6.19: This is obvious. Equation (6.26) describes a vector whose direction is from $\mathbf{P}_0$ to $\mathbf{P}_3$. Varying t changes the magnitude of this vector but not its direction.

6.20: Figure Ans.11 shows the curve, the points, and the code that produced them.

6.21: Figure Ans.12 lists the *Mathematica* code for Figure 6.19.

6.22: The weight functions in the u direction are

$$B_{20}(u) = (1-u)^2, \quad B_{21}(u) = 2u(1-u), \quad B_{22}(u) = u^2.$$

Those in the w direction are

$$B_{30}(w) = (1-w)^3, \quad B_{31}(w) = 3w(1-w)^2,$$
$$B_{32}(w) = 3w^2(1-w), \quad B_{33}(w) = w^3.$$

The surface patch is, therefore,

$$\mathbf{P}(u, w)$$
$$= \sum_{i=0}^{2}\sum_{j=0}^{3} B_{2i}(u)\mathbf{P}_{ij}B_{3j}(w)$$
$$= B_{20}(u)[\mathbf{P}_{00}B_{30}(w) + \mathbf{P}_{01}B_{31}(w) + \mathbf{P}_{02}B_{32}(w) + \mathbf{P}_{03}B_{33}(w)]$$

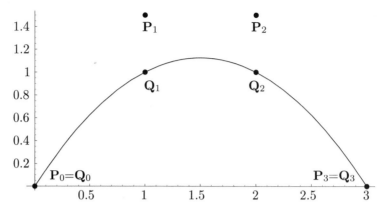

```
q0={0,0}; q1={1,1}; q2={2,1}; q3={3,0};
p0=q0; p1={1,3/2}; p2={2,3/2}; p3=q3;
c[t_]:=(1-t)^3 p0+3t(1-t)^2 p1+3t^2(1-t) p2+t^3 p3
g1=ListPlot[{p0,p1,p2,p3,q1,q2},
 Prolog->AbsolutePointSize[4], PlotRange->All,
 AspectRatio->Automatic, DisplayFunction->Identity]
g2=ParametricPlot[c[t], {t,0,1}, DisplayFunction->Identity]
Show[g1,g2, DisplayFunction->$DisplayFunction]
```

Figure Ans.11: An Interpolating Bézier Curve.

$$
\begin{aligned}
& + B_{21}(u)[\mathbf{P}_{00}B_{30}(w) + \mathbf{P}_{01}B_{31}(w) + \mathbf{P}_{02}B_{32}(w) + \mathbf{P}_{03}B_{33}(w)] \\
& + B_{22}(u)[\mathbf{P}_{00}B_{30}(w) + \mathbf{P}_{01}B_{31}(w) + \mathbf{P}_{02}B_{32}(w) + \mathbf{P}_{03}B_{33}(w)] \\
= \; & (1-u)^2[(1-w)^3(0,0,0) + 3w(1-w)^2(1,0,1) \\
& + 3w^2(1-w)(2,0,1) + w^3(3,0,0)] \\
& + 2u(1-u)[(1-w)^3(0,1,0) + 3w(1-w)^2(1,1,1) \\
& + 3w^2(1-w)(2,1,1) + w^3(3,1,0)] \\
& + u^2[(1-w)^3(0,2,0) + 3w(1-w)^2(1,2,1) + 3w^2(1-w)(2,2,1) + w^3(3,2,0)] \\
= \; & (3w, 2u, 3w(1-w)). \tag{Ans.23}
\end{aligned}
$$

6.23: The result is shown in Figure Ans.13.

6.24: Here is one such loop. It displays one family of 11 curves (see also Figure Ans.1).

```
for u:=0 step 0.1 to 1 do (* 11 curves *)
 for v:=0 step 0.01 to 1-u do (* 100 pixels per curve *)
 w:=1-u-v;
 Calculate & project point P(u,v,w)
 endfor;
endfor;
```

The second family consists of the curves parallel to the base of the triangle (the main loop is on v), and the third family consists of the curves parallel to the right side.

```
(* Effects of varying weights in Rational Cubic Bezier curve *)
Clear[RatCurve,g1,g2,w];
pnts={{0,0},{.2,1},{.8,1},{1,0}};
w={1,1,1,1}; (* Four weights for a cubic curve *)
pwr[x_,y_]:=If[x==0 && y==0, 1, x^y];
bern[n_,i_,t_]:=Binomial[n,i]pwr[t,i]pwr[1-t,n-i] (* t^i*(1-t)^(n-i) *)
RatCurve[t_]:=Sum[(w[[i+1]]pnts[[i+1]]bern[3,i,t])/(Sum[w[[j+1]]bern[3,j,t],
  {j,0,3}]), {i,0,3}];
g1=ListPlot[pnts, Prolog->AbsolutePointSize[4], PlotRange->All,
  AspectRatio->Automatic, DisplayFunction->Identity]
g2=ParametricPlot[RatCurve[t], {t,0,1}, Compiled->False,
  PlotRange->All, AspectRatio->Automatic, DisplayFunction->Identity]
w={1,2,1,1}; (* change weights *)
g3=ParametricPlot[RatCurve[t], {t,0,1}, Compiled->False,
  PlotRange->All, AspectRatio->Automatic, DisplayFunction->Identity]
w={1,3,1,1}; (* increase w1 *)
g4=ParametricPlot[RatCurve[t], {t,0,1}, Compiled->False,
  PlotRange->All, AspectRatio->Automatic, DisplayFunction->Identity]
w={1,4,1,1}; (* increase w1 *)
g5=ParametricPlot[RatCurve[t], {t,0,1}, Compiled->False,
  PlotRange->All, AspectRatio->Automatic, DisplayFunction->Identity]
Show[g1,g2,g3,g4,g5, DisplayFunction->$DisplayFunction,
  DefaultFont->{"cmr10",10}]

(* Effects of moving a control point in Rational Cubic Bezier curve *)
Clear[RatCurve,g1,g2,w];
pnts={{0,0},{.2,.8},{.8,.8},{1,0}};
w={1,1,1,1}; (* Four weights for a cubic curve *)
pwr[x_,y_]:=If[x==0 && y==0, 1, x^y];
bern[n_,i_,t_]:=Binomial[n,i]pwr[t,i]pwr[1-t,n-i] (* t^i*(1-t)^(n-i) *)
RatCurve[t_]:=Sum[(w[[i+1]]pnts[[i+1]]bern[3,i,t])/(Sum[w[[j+1]]bern[3,j,t],
  {j,0,3}]), {i,0,3}];
g1=ListPlot[pnts, Prolog->AbsolutePointSize[4], PlotRange->All,
  AspectRatio->Automatic, DisplayFunction->Identity]
g2=ParametricPlot[RatCurve[t], {t,0,1}, Compiled->False,
  PlotRange->All, AspectRatio->Automatic, DisplayFunction->Identity]
pnts={{0,0},{.2,.8},{.86,.86},{1,0}};
g3=ParametricPlot[RatCurve[t], {t,0,1}, Compiled->False,
  PlotRange->All, AspectRatio->Automatic, DisplayFunction->Identity]
pnts={{0,0},{.2,.8},{.93,.93},{1,0}};
g4=ParametricPlot[RatCurve[t], {t,0,1}, Compiled->False,
  PlotRange->All, AspectRatio->Automatic, DisplayFunction->Identity]
pnts={{0,0},{.2,.8},{1,1},{1,0}};
g5=ParametricPlot[RatCurve[t], {t,0,1}, Compiled->False,
  PlotRange->All, AspectRatio->Automatic, DisplayFunction->Identity]
Show[g1,g2,g3,g4,g5, DisplayFunction->$DisplayFunction,
  DefaultFont->{"cmr10",10}]
```

Figure Ans.12: Code for Figure 6.19.

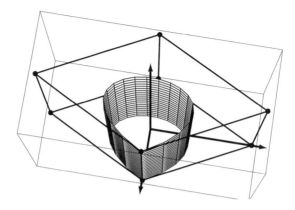

```
(* A Rational closed Bezier Surface *)
Clear[pwr,bern,spnts,n,m,wt,bzSurf,cpnts,patch,vlines,hlines,axes];
<<:Graphics:ParametricPlot3D.m
r=1; h=3; (* radius & height of cylinder *)
spnts={{{r,0,0},{0,2r,0},{-r,0,0},{0,-2r,0},{r,0,0}},
 {{r,0,h},{0,2r,h},{-r,0,h},{0,-2r,h},{r,0,h}}};
m=Length[spnts[[1]]]-1; n=Length[Transpose[spnts][[1]]]-1;
wt=Table[1, {i,1,n+1},{j,1,m+1}];
pwr[x_,y_]:=If[x==0 && y==0, 1, x^y];
bern[n_,i_,u_]:=Binomial[n,i]pwr[u,i]pwr[1-u,n-i]
bzSurf[u_,w_]:=
 Sum[wt[[i+1,j+1]]spnts[[i+1,j+1]]bern[n,i,u]bern[m,j,w], {i,0,n}, {j,0,m}]/
 Sum[wt[[i+1,j+1]]bern[n,i,u]bern[m,j,w], {i,0,n}, {j,0,m}];
patch=ParametricPlot3D[bzSurf[u,w], {u,0,1}, {w,0,1},
 Compiled->False, DisplayFunction->Identity];
cpnts=Graphics3D[{AbsolutePointSize[4], (* control points *)
 Table[Point[spnts[[i,j]]], {i,1,n+1},{j,1,m+1}]}];
vlines=Graphics3D[{AbsoluteThickness[1], (* control polygon *)
 Table[Line[{spnts[[i,j]],spnts[[i+1,j]]}], {i,1,n}, {j,1,m+1}]}];
hlines=Graphics3D[{AbsoluteThickness[1],
 Table[Line[{spnts[[i,j]],spnts[[i,j+1]]}], {i,1,n+1}, {j,1,m}]}];
maxx=Max[Flatten[Table[Part[spnts[[i,j]], 1], {i,1,n+1}, {j,1,m+1}]]];
maxy=Max[Flatten[Table[Part[spnts[[i,j]], 2], {i,1,n+1}, {j,1,m+1}]]];
maxz=Max[Flatten[Table[Part[spnts[[i,j]], 3], {i,1,n+1}, {j,1,m+1}]]];
axes=Graphics3D[{AbsoluteThickness[1.5], (* the coordinate axes *)
 Line[{{0,0,maxz},{0,0,0},{maxx,0,0},{0,0,0},{0,maxy,0}}]}];
Show[cpnts,hlines,vlines,axes,patch, PlotRange->All,DefaultFont->{"cmr10", 10},
 DisplayFunction->$DisplayFunction, ViewPoint->{0.998, 0.160, 4.575},Shading->False];
```

Figure Ans.13: A Closed Rational Bézier Surface Patch.

6.25:   The 15 original control points are listed in Figure 6.28. The first step of the algorithm produces the 10 intermediate points for $n = 3$ (Figure Ans.14)

$$\mathbf{P}^1_{003} = u\mathbf{P}^0_{103} + v\mathbf{P}^0_{013} + w\mathbf{P}^0_{004}, \quad \mathbf{P}^1_{102} = u\mathbf{P}^0_{202} + v\mathbf{P}^0_{112} + w\mathbf{P}^0_{103},$$

$$\mathbf{P}^1_{201} = u\mathbf{P}^0_{301} + v\mathbf{P}^0_{211} + w\mathbf{P}^0_{202}, \quad \mathbf{P}^1_{300} = u\mathbf{P}^0_{400} + v\mathbf{P}^0_{310} + w\mathbf{P}^0_{301},$$

$$\mathbf{P}^1_{012} = u\mathbf{P}^0_{112} + v\mathbf{P}^0_{022} + w\mathbf{P}^0_{013}, \quad \mathbf{P}^1_{111} = u\mathbf{P}^0_{211} + v\mathbf{P}^0_{121} + w\mathbf{P}^0_{112},$$

$$\mathbf{P}^1_{210} = u\mathbf{P}^0_{310} + v\mathbf{P}^0_{220} + w\mathbf{P}^0_{211}, \quad \mathbf{P}^1_{021} = u\mathbf{P}^0_{121} + v\mathbf{P}^0_{031} + w\mathbf{P}^0_{022},$$

$$\mathbf{P}^1_{120} = u\mathbf{P}^0_{220} + v\mathbf{P}^0_{130} + w\mathbf{P}^0_{121}, \quad \mathbf{P}^1_{030} = u\mathbf{P}^0_{130} + v\mathbf{P}^0_{040} + w\mathbf{P}^0_{031}.$$

the second step of the algorithm produces the six intermediate points for $n = 2$

$$\mathbf{P}^2_{002} = u\mathbf{P}^1_{102} + v\mathbf{P}^1_{012} + w\mathbf{P}^1_{003}, \quad \mathbf{P}^2_{101} = u\mathbf{P}^1_{201} + v\mathbf{P}^1_{111} + w\mathbf{P}^1_{102},$$
$$\mathbf{P}^2_{200} = u\mathbf{P}^1_{300} + v\mathbf{P}^1_{210} + w\mathbf{P}^1_{201}, \quad \mathbf{P}^2_{011} = u\mathbf{P}^1_{111} + v\mathbf{P}^1_{021} + w\mathbf{P}^1_{012},$$
$$\mathbf{P}^2_{110} = u\mathbf{P}^1_{210} + v\mathbf{P}^1_{120} + w\mathbf{P}^1_{111}, \quad \mathbf{P}^2_{020} = u\mathbf{P}^1_{120} + v\mathbf{P}^1_{030} + w\mathbf{P}^1_{021}.$$

The third step produces the three intermediate points for $n = 1$

$$\mathbf{P}^3_{001} = u\mathbf{P}^2_{101} + v\mathbf{P}^2_{011} + w\mathbf{P}^2_{002},$$
$$\mathbf{P}^3_{100} = u\mathbf{P}^2_{200} + v\mathbf{P}^2_{110} + w\mathbf{P}^2_{101},$$
$$\mathbf{P}^3_{010} = u\mathbf{P}^2_{110} + v\mathbf{P}^2_{020} + w\mathbf{P}^2_{011}.$$

And the fourth step produces the single point

$$\mathbf{P}^4_{000} = u\mathbf{P}^3_{100} + v\mathbf{P}^3_{010} + w\mathbf{P}^3_{001}.$$

This is the point that corresponds to the particular triplet (u, v, w) on the triangular patch defined by the 15 original control points.

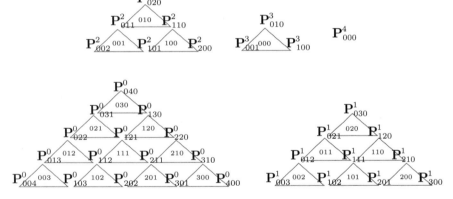

Figure Ans.14: Scaffolding in a Triangular Bézier Patch.

6.26: The code of Figure Ans.15 does the computations and yields the surface point $(2, 1, 1/2)$.

6.27: For $n = 4$ and $r = 3$, point $\mathbf{P}^3_{001}$ is computed directly from the control points as the sum

$$\mathbf{P}^3_{001} = \sum_{a+b+c=3} B^3_{abc}(u, v, w)\mathbf{P}_{0+a,0+b,1+c}$$
$$= w^3\mathbf{P}_{004} + 3uw^2\mathbf{P}_{103} + 3u^2w\mathbf{P}_{202} + u^3\mathbf{P}_{301} + 3vw^2\mathbf{P}_{013}$$
$$6uvw\mathbf{P}_{112} + 3u^2v\mathbf{P}_{211} + 3v^2w\mathbf{P}_{022} + 3uv^2\mathbf{P}_{121} + v^3\mathbf{P}_{031}.$$

```
P0300={3,3,0};
P0210={2,2,0}; P1200={4,2,1};
P0120={1,1,0}; P1110={3,1,1}; P2100={5,1,2};
P0030={0,0,0}; P1020={2,0,1}; P2010={4,0,2}; P3000={6,0,3};
n=3; u=1/6; v=2/6; w=3/6;
P0021=u P1020+v P0120+w P0030;
P1011=u P2010+v P1110+w P1020;
P2001=u P3000+v P2100+w P2010;
P0111=u P1110+v P0210+w P0120;
P1101=u P2100+v P1200+w P1110;
P0201=u P1200+v P0300+w P0210;
P0012=u P1011+v P0111+w P0021;
P1002=u P2001+v P1101+w P1011;
P0102=u P1101+v P0201+w P0111;
P0003=u P1002+v P0102+w P0012
B[i_,j_,k_]:=(n!/(i! j! k!))u^i v^j w^k;
P0030 B[0,0,3]+P1020 B[1,0,2]+P2010 B[2,0,1]+P3000 B[3,0,0]+
P0120 B[0,1,2]+P1110 B[1,1,1]+P2100 B[2,1,0]+
P0210 B[0,2,1]+P1200 B[1,2,0]+P0300 B[0,3,0]
```

Figure Ans.15: Triangular Bézier Patch Subdivision Exercise.

For $n = 4$ and $r = 1$, point $\mathbf{P}^1_{111}$ is computed directly from the control points as the sum

$$\mathbf{P}^1_{111} = \sum_{a+b+c=1} B^1_{abc}(u,v,w)\mathbf{P}_{1+a,1+b,1+c} = u\mathbf{P}_{211} + v\mathbf{P}_{121} + w\mathbf{P}_{112}.$$

6.28: Figure Ans.16 shows the initial triangle with 15 control points and the final result, with 31 points.

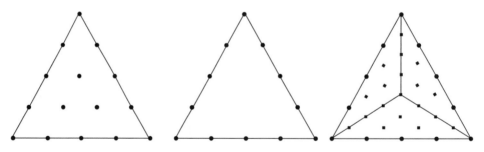

Figure Ans.16: Subdividing the Triangular Bézier Patch for $n = 4$.

7.1: The second quadratic spline segment is also obtained from Equation (7.6)

$$\mathbf{P}_2(t) = \frac{1}{2}(t^2, t, 1) \begin{pmatrix} 1 & -2 & 1 \\ -2 & 2 & 0 \\ 1 & 1 & 0 \end{pmatrix} \begin{pmatrix} \mathbf{P}_1 \\ \mathbf{P}_2 \\ \mathbf{P}_3 \end{pmatrix}$$

$$= \frac{1}{2}(t^2 - 2t + 1)(1,1) + \frac{1}{2}(-2t^2 + 2t + 1)(2,1) + \frac{t^2}{2}(2,0)$$

$$= (-t^2/2 + t + 3/2, -t^2/2 + 1).$$

It starts at joint $\mathbf{K}_2 = \mathbf{P}_2(0) = (\frac{3}{2}, 1)$ and ends at joint $\mathbf{K}_3 = \mathbf{P}_2(1) = (2, \frac{1}{2})$. The tangent vector is $\mathbf{P}_2^t(t) = (-t+1, -t)$, showing that this segment starts going in direction $\mathbf{P}_2^t(0) = (1, 0)$ and ends going in direction $\mathbf{P}_2^t(1) = (0, -1)$ (down).

7.2: We write

$$
\mathbf{P}_i(0) = \frac{1}{6}(\mathbf{P}_{i-1} + 4\mathbf{P}_i + \mathbf{P}_{i+1}) = \frac{1}{3}\frac{\mathbf{P}_{i-1} + \mathbf{P}_{i+1}}{2} + \frac{2}{3}\mathbf{P}_i = \frac{1}{3}\mathbf{M} + \frac{2}{3}\mathbf{P}_i,
$$

where $\mathbf{M}$ is the midpoint between $\mathbf{P}_{i-1}$ and $\mathbf{P}_{i+1}$. This shows that $\mathbf{P}_i(0)$ is located on the straight segment connecting $\mathbf{M}$ to $\mathbf{P}_i$, two-thirds of the way from $\mathbf{M}$ (Figure Ans.17). Similarly,

$$
\mathbf{P}_i(1) = \frac{1}{3}\frac{\mathbf{P}_i + \mathbf{P}_{i+2}}{2} + \frac{2}{3}\mathbf{P}_{i+1} = \frac{1}{3}\mathbf{M} + \frac{2}{3}\mathbf{P}_{i+1}.
$$

This is called the 2/3 rule.

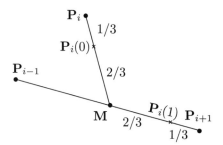

Figure Ans.17: The 2/3 Rule.

7.3: The second cubic segment is given by Equation (7.11)

$$
\begin{aligned}
\mathbf{P}_1(t) &= \frac{1}{6}(-t^3 + 3t^2 - 3t + 1)(0, 1) + \frac{1}{6}(3t^3 - 6t^2 + 4)(1, 1) \\
&\quad + \frac{1}{6}(-3t^3 + 3t^2 + 3t + 1)(2, 1) + \frac{t^3}{6}(2, 0) \\
&= (-t^3/6 + t + 1, -t^3/6 + 1).
\end{aligned}
$$

It goes from joint $\mathbf{K}_2 = \mathbf{P}_2(0) = (1, 1)$ to joint $\mathbf{K}_3 = \mathbf{P}_2(1) = (11/6, 5/6)$. The tangent vector is

$$
\begin{aligned}
\mathbf{P}_2^t(t) &= \frac{1}{6}(-3t^2 + 6t - 3)(0, 1) + \frac{1}{6}(9t^2 - 12t)(1, 1) \\
&\quad + \frac{1}{6}(-9t^2 + 6t + 3)(2, 1) + \frac{t^2}{2}(2, 0) \\
&= (-t^2/2 + 1, -t^2/2).
\end{aligned}
$$

The two extreme tangents are $\mathbf{P}_2^t(0) = (1, 0)$ and $\mathbf{P}_2^t(1) = (1/2, -1/2)$. Figure 7.4 shows this segment.

7.4: Each of the three quadratic segments is given by Equation (7.6). The first segment is

$$\mathbf{P}_1(t) = \frac{1}{2}(t^2 - 2t + 1)(0, 0) + \frac{1}{2}(-2t^2 + 2t + 1)(0, 1) + \frac{t^2}{2}(1, 1)$$
$$= (t^2/2, -t^2/2 + t + 1/2).$$

It goes from $\mathbf{K}_1 = \mathbf{P}_1(0) = (0, 1/2)$ to $\mathbf{K}_2 = \mathbf{P}_1(1) = (1/2, 1)$.

The second segment is

$$\mathbf{P}_2(t) = \frac{1}{2}(t^2 - 2t + 1)(0, 1) + \frac{1}{2}(-2t^2 + 2t + 1)(1, 1) + \frac{t^2}{2}(2, 1)$$
$$= (t + 1/2, 1).$$

It goes from $\mathbf{K}_2 = \mathbf{P}_2(0) = (1/2, 1)$ to $\mathbf{K}_3 = \mathbf{P}_2(1) = (3/2, 1)$. Notice that this segment is horizontal.

The third segment is

$$\mathbf{P}_3(t) = \frac{1}{2}(t^2 - 2t + 1)(1, 1) + \frac{1}{2}(-2t^2 + 2t + 1)(2, 1) + \frac{t^2}{2}(2, 0)$$
$$= (-t^2/2 + t + 3/2, -t^2/2 + 1).$$

It goes from $\mathbf{K}_3 = \mathbf{P}_3(0) = (3/2, 1)$ to $\mathbf{K}_4 = \mathbf{P}_3(1) = (2, 1/2)$.

Figure 7.4 shows these segments (the solid curves).

7.5: The two segments are easy to calculate from Equation (7.11). They are

$$\mathbf{P}_3(t) = \frac{1}{6}(2t^3 - 3t^2 - 3t + 5)\mathbf{P}_2 + \frac{1}{6}(-2t^3 + 3t^2 + 3t + 1)\mathbf{P}_4,$$
$$\mathbf{P}_4(t) = \frac{1}{6}(-t^3 + 3t^2 - 3t + 1)\mathbf{P}_3 + \frac{1}{6}(t^3 - 3t^2 + 3t + 5)\mathbf{P}_4.$$

Their extreme points are therefore

$$\mathbf{P}_3(0) = \frac{5}{6}\mathbf{P}_2 + \frac{1}{6}\mathbf{P}_4, \quad \mathbf{P}_3(1) = \frac{1}{6}\mathbf{P}_2 + \frac{5}{6}\mathbf{P}_4,$$
$$\mathbf{P}_4(0) = \frac{1}{6}\mathbf{P}_3 + \frac{5}{6}\mathbf{P}_4, \quad \mathbf{P}_4(1) = \mathbf{P}_4.$$

They are indicated by small crosses in Figure 7.6.

7.6: Given the four control points $\mathbf{P}_0 = \mathbf{P}_1 = \mathbf{P}_2 \neq \mathbf{P}_3$, we use Equation (7.11) to construct such a segment:

$$\mathbf{P}_1(t) = \frac{1}{6}(-t^3 + 6)\mathbf{P}_0 + \frac{t^3}{6}\mathbf{P}_3 = (1 - u)\mathbf{P}_0 + u\mathbf{P}_3, \quad \text{for } u = t^3/6,$$

which shows the segment to be straight and the start point to be $\mathbf{P}_1(0) = \mathbf{P}_0$.

7.7: There are five segments. The first is defined by points $\mathbf{P}_0$ through $\mathbf{P}_3$. The second is defined by $\mathbf{P}_1$ through $\mathbf{P}_4$, and so on until the last segment which is defined by $\mathbf{P}_4$ through $\mathbf{P}_7$.

$$\mathbf{P}_1(t) = \frac{1}{6}(t^3 + 6, t^3), \quad \text{a straight line,}$$

$$\mathbf{P}_2(t) = \frac{1}{6}(3t^2 + 3t + 7, -3t^3 + 3t^2 + 3t + 1),$$

$$\mathbf{P}_3(t) = \frac{1}{6}(-3t^3 + 3t^2 + 9t + 13, 4t^3 - 6t^2 + 4),$$

$$\mathbf{P}_4(t) = \frac{1}{6}(2t^3 - 6t^2 + 6t + 22, -3t^3 + 6t^2 + 2),$$

$$\mathbf{P}_5(t) = \frac{1}{6}(24, t^3 - 3t^2 + 3t + 5), \quad \text{a vertical straight line.}$$

They meet at the four joints $(7/6, 1/6)$, $(13/6, 4/6)$, $(22/6, 2/6)$, and $(24/6, 5/6)$. Notice that the fifth segment is vertical. Figure Ans.18 shows these curves (slightly separated to indicate the joint points).

The cubic Bézier curve defined by the same points (where only four distinct points are used) is

$$\mathbf{P}(t) = (1-t)^3(1,0) + 3t(1-t)^2(2,1) + 3t^2(1-t)(4,0) + t^3(4,1)$$
$$= (-3t^3 + 3t^2 + 3t + 1, 4t^3 - 6t^2 + 3t).$$

It goes from $(1,0)$ to $(4,1)$ but is different from the B-spline because, for example, it is never vertical. It is shown dashed in Figure Ans.18. The degree-7 Bézier curve defined by the same points is also shown (dot-dashed) in the same figure for comparison. It is clear that it is tight because of its strong attraction to the multiple points.

7.8: Equation (7.17) can be written $\mathbf{P}^t(t) = (t^2 - t)[(\mathbf{P}_0 - \mathbf{P}_3) + 3(\mathbf{P}_1 - \mathbf{P}_2)]$. This is the sum of two differences of points. The first difference is the vector from $\mathbf{P}_3$ to $\mathbf{P}_0$ and the second is the vector from $\mathbf{P}_2$ to $\mathbf{P}_1$ (multiplied by 3). The tangent vector of Equation (7.17) therefore points in the direction of the sum of these vectors, and this direction does not depend on t. The size of the tangent vector depends on t, but the size affects just the speed of the spline segment, not its shape.

7.9: By substituting, for example, $t + 1$ for t in the expression for $N_{13}(t)$.

7.10: The tangent vectors of the three segments are

$$\mathbf{P}_1^t(t) = 2(t - 1)\mathbf{P}_0 + (2 - 3t)\mathbf{P}_1 + t\mathbf{P}_2,$$
$$\mathbf{P}_2^t(t) = (t - 2)\mathbf{P}_1 + (3 - 2t)\mathbf{P}_2 + (t - 1)\mathbf{P}_3,$$
$$\mathbf{P}_3^t(t) = (t - 3)\mathbf{P}_2 + (7 - 3t)\mathbf{P}_3 + 2(t - 2)\mathbf{P}_4.$$

They satisfy $\mathbf{P}_1^t(1) = \mathbf{P}_2^t(1) = \mathbf{P}_2 - \mathbf{P}_1$, and $\mathbf{P}_2^t(2) = \mathbf{P}_3^t(2) = \mathbf{P}_3 - \mathbf{P}_2$.

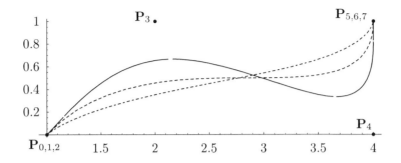

```
(* Exercise. 8 points, 5-segment uniform B-spline curve, compared to the Bezier
curve for the same 8 points *)
Clear[p1,p2,p3,p4,p5,bez,l1,g1,g2,g3,g4,g5,g6];
pnts={{1,0},{2,1},{4,0},{4,1}};
p1[t_]:={t^3+6,t^3}/6;
p2[t_]:={3t^2+3t+7,-3t^3+3t^2+3t+1}/6;
p3[t_]:={-3t^3+3t^2+9t+13,4t^3-6t^2+4}/6;
p4[t_]:={2t^3-6t^2+6t+22,-3t^3+6t^2+2}/6;
p5[t_]:={24,t^3-3t^2+3t+5}/6;
bez[t_]:={-3t^3+3t^2+3t+1,4t^3-6t^2+3t};
l1=ListPlot[pnts, Prolog->PointSize[.01], DisplayFunction->Identity];
g1=ParametricPlot[p1[t], {t,0,.97}, Compiled->False, DisplayFunction->Identity];
g2=ParametricPlot[p2[t], {t,0,.97}, Compiled->False, DisplayFunction->Identity];
g3=ParametricPlot[p3[t], {t,0,.97}, Compiled->False, DisplayFunction->Identity];
g4=ParametricPlot[p4[t], {t,0,.97}, Compiled->False, DisplayFunction->Identity];
g5=ParametricPlot[p5[t], {t,0,.97}, Compiled->False, DisplayFunction->Identity];
g6=ParametricPlot[bez[t], {t,0,1}, PlotStyle->AbsoluteDashing[{2,2}],
Compiled->False, DisplayFunction->Identity];
(* Now the degree-7 Bezier curve *)
pnts={{1,0},{1,0},{1,0},{2,1},{4,0},{4,1},{4,1},{4,1}};
pwr[x_,y_]:=If[x==0 && y==0, 1, x^y];
bern[n_,i_,t_]:=Binomial[n,i]pwr[t,i]pwr[1-t,n-i] (* t^i x (1-t)^(n-i) *)
bzCurve[t_]:=Sum[pnts[[i+1]]bern[7,i,t], {i,0,7}]
g7=ParametricPlot[bzCurve[t], {t,0,1}, Compiled->False,
PlotStyle->AbsoluteDashing[{1,2,2,2}], PlotRange->All,
AspectRatio->Automatic, DisplayFunction->Identity];
Show[l1,g1,g2,g3,g4,g5,g6,g7, PlotRange->All, DisplayFunction->$DisplayFunction,
AspectRatio->Automatic, DefaultFont->{"cmr10", 10}];
```

Figure Ans.18: Comparing a Uniform B-spline and a Bézier Curve for Eight Points.

7.11: For the curve of Figure 7.19c, the knot vector is

$$(-3, -2, -1, 0, 1, 1, 1, 2, 3, 4, 5, 6).$$

The range of the parameter t is from $t_3 = 0$ to $t_8 = 3$ and we obtain the blending functions by direct calculations (only the last group N_{i4} of blending functions is shown):

$$N_{04}(t) = \frac{t - t_0}{t_3 - t_0} N_{03} + \frac{t_4 - t}{t_4 - t_1} N_{13} = \frac{1}{6}(1 - t)^3 \qquad \text{for } t \in [0, 1),$$

$$N_{14}(t) = \frac{t - t_1}{t_4 - t_1} N_{13} + \frac{t_5 - t}{t_5 - t_2} N_{23} = \frac{1}{12}(11t^3 - 15t^2 - 3t + 7) \qquad \text{for } t \in [0, 1),$$

$$N_{24}(t) = \frac{t - t_2}{t_5 - t_2} N_{23} + \frac{t_6 - t}{t_6 - t_3} N_{33} = \frac{1}{4}(-7t^3 + 3t^2 + 3t + 1) \qquad \text{for } t \in [0, 1),$$

$$N_{34}(t) = \frac{t - t_3}{t_6 - t_3} N_{33} + \frac{t_7 - t}{t_7 - t_4} N_{43} = \begin{cases} t^3 & \text{for } t \in [0, 1), \\ (2 - t)^3 & \text{for } t \in [1, 2), \end{cases}$$

$$N_{44}(t) = \frac{t - t_4}{t_7 - t_4} N_{43} + \frac{t_8 - t}{t_8 - t_5} N_{53} = \frac{1}{4} \begin{cases} (7t^3 - 39t^2 + 69t - 37) & \text{for } t \in [1, 2), \\ (3 - t)^3 & \text{for } t \in [2, 3), \end{cases}$$

$$N_{54}(t) = \frac{t - t_5}{t_8 - t_5} N_{53} + \frac{t_9 - t}{t_9 - t_6} N_{63} = \frac{1}{12} \begin{cases} (-11t^3 + 51t^2 - 69t + 29) & \text{for } t \in [1, 2), \\ (7t^3 - 57t^2 + 147t - 115) & \text{for } t \in [2, 3), \end{cases}$$

$$N_{64}(t) = \frac{t - t_6}{t_9 - t_6} N_{63} + \frac{t_{10} - t}{t_{10} - t_7} N_{73} = \frac{1}{6} \begin{cases} (t - 1)^3 & \text{for } t \in [1, 2), \\ (-3t^3 + 21t^2 - 45t + 31) & \text{for } t \in [2, 3), \end{cases}$$

$$N_{74}(t) = \frac{t - t_7}{t_{10} - t_7} N_{73} + \frac{t_{11} - t}{t_{11} - t_8} N_{83} = \frac{1}{6}(t - 2)^3 \qquad \text{for } t \in [2, 3).$$

This group of blending functions can now be used to construct the five spline segments

$$\mathbf{P}_3(t) = N_{04}(t)\mathbf{P}_0 + N_{14}(t)\mathbf{P}_1 + N_{24}(t)\mathbf{P}_2 + N_{34}(t)\mathbf{P}_3$$
$$= \frac{1}{6}(1 - t)^3 \mathbf{P}_0 + \frac{1}{12}(11t^3 - 15t^2 - 3t + 7)\mathbf{P}_1$$
$$+ \frac{1}{4}(-7t^3 + 3t^2 + 3t + 1)\mathbf{P}_2 + t^3 \mathbf{P}_3, \qquad t \in [0, 1),$$

$$\mathbf{P}_4(t) = N_{14}(1)\mathbf{P}_1 + N_{24}(1)\mathbf{P}_2 + N_{34}(1)\mathbf{P}_3 + N_{44}(1)\mathbf{P}_4$$
$$= \mathbf{P}_3 \quad \text{(a point)}, \qquad t \in [1, 1),$$

$$\mathbf{P}_5(t) = N_{24}(1)\mathbf{P}_2 + N_{34}(1)\mathbf{P}_3 + N_{44}(1)\mathbf{P}_4 + N_{54}(1)\mathbf{P}_5$$
$$= \mathbf{P}_3 \quad \text{(a point)}, \qquad t \in [1, 1),$$

$$\mathbf{P}_6(t) = N_{34}(t)\mathbf{P}_3 + N_{44}(t)\mathbf{P}_4 + N_{54}(t)\mathbf{P}_5 + N_{64}(t)\mathbf{P}_6$$
$$= (2 - t)^3 \mathbf{P}_3 + \frac{1}{4}(7t^3 - 39t^2 + 69t - 37)\mathbf{P}_4$$
$$+ \frac{1}{12}(-11t^3 + 51t^2 - 69t + 29)\mathbf{P}_5 + \frac{1}{6}(t - 1)^3 \mathbf{P}_6, \qquad t \in [1, 2),$$

$$\mathbf{P}_7(t) = N_{44}(t)\mathbf{P}_4 + N_{54}(t)\mathbf{P}_5 + N_{64}(t)\mathbf{P}_6 + N_{74}(t)\mathbf{P}_7$$
$$= (3 - t)^3 \mathbf{P}_4 + \frac{1}{12}(7t^3 - 57t^2 + 147t - 115)\mathbf{P}_5$$
$$+ (-3t^3 + 21t^2 - 45t + 31)\mathbf{P}_6 + \frac{1}{6}(t - 2)^3 \mathbf{P}_7, \qquad t \in [2, 3).$$

A direct check verifies that each segment has barycentric weights. The entire curve starts at $\mathbf{P}_3(0) = \mathbf{P}_0/6 + 7\mathbf{P}_1/12 + \mathbf{P}_2/4$ and ends at $\mathbf{P}_7(3) = (\mathbf{P}_5 + 4\mathbf{P}_6 + \mathbf{P}_7)/6$. The two join points between the segments are

$$\mathbf{P}_3(1) = \mathbf{P}_6(1) = \mathbf{P}_3, \quad \mathbf{P}_6(2) = \mathbf{P}_7(2) = \mathbf{P}_4/4 + 7\mathbf{P}_5/12 + \mathbf{P}_6/6.$$

Both segments $\mathbf{P}_4(t)$ and $\mathbf{P}_5(t)$ reduce to the single control point $\mathbf{P}_3$.

For the curve of Figure 7.19d, the knot vector is

$$(-3, -2, -1, 0, 1, 1, 1, 1, 2, 3, 4, 5).$$

The range of the parameter t is from $t_3 = 0$ to $t_8 = 2$ and we get by direct calculations (again only the last group N_{i4} of blending functions is shown)

$$N_{04}(t) = \frac{t - t_0}{t_3 - t_0} N_{03} + \frac{t_4 - t}{t_4 - t_1} N_{13} = \frac{1}{6}(1 - t)^3 \qquad \text{for } t \in [0, 1),$$

$$N_{14}(t) = \frac{t - t_1}{t_4 - t_1} N_{13} + \frac{t_5 - t}{t_5 - t_2} N_{23} = \frac{1}{12}(11t^3 - 15t^2 - 3t + 7) \qquad \text{for } t \in [0, 1),$$

$$N_{24}(t) = \frac{t - t_2}{t_5 - t_2} N_{23} + \frac{t_6 - t}{t_6 - t_3} N_{33} = \frac{1}{4}(-7t^3 + 3t^2 + 3t + 1) \qquad \text{for } t \in [0, 1),$$

$$N_{34}(t) = \frac{t - t_3}{t_6 - t_3} N_{33} + \frac{t_7 - t}{t_7 - t_4} N_{43} = t^3 \qquad \text{for } t \in [0, 1),$$

$$N_{44}(t) = \frac{t - t_4}{t_7 - t_4} N_{43} + \frac{t_8 - t}{t_8 - t_5} N_{53} = (2 - t)^3 \qquad \text{for } t \in [1, 2),$$

$$N_{54}(t) = \frac{t - t_5}{t_8 - t_5} N_{53} + \frac{t_9 - t}{t_9 - t_6} N_{63} = \frac{1}{4}(7t^3 - 39t^2 + 69t - 37) \qquad \text{for } t \in [1, 2),$$

$$N_{64}(t) = \frac{t - t_6}{t_9 - t_6} N_{63} + \frac{t_{10} - t}{t_{10} - t_7} N_{73} = \frac{1}{12}(-11t^3 + 51t^2 - 69t + 29) \qquad \text{for } t \in [1, 2),$$

$$N_{74}(t) = \frac{t - t_7}{t_{10} - t_7} N_{73} + \frac{t_{11} - t}{t_{11} - t_8} N_{83} = \frac{1}{6}(t - 1)^3 \qquad \text{for } t \in [1, 2).$$

This group of blending functions can now be used to construct the five spline segments

$$\mathbf{P}_3(t) = N_{04}(t)\mathbf{P}_0 + N_{14}(t)\mathbf{P}_1 + N_{24}(t)\mathbf{P}_2 + N_{34}(t)\mathbf{P}_3$$
$$= \frac{1}{6}(1 - t)^3 \mathbf{P}_0 + \frac{1}{12}(11t^3 - 15t^2 - 3t + 7)\mathbf{P}_1$$
$$+ \frac{1}{4}(-7t^3 + 3t^2 + 3t + 1)\mathbf{P}_2 + t^3 \mathbf{P}_3, \qquad t \in [0, 1),$$

$$\mathbf{P}_4(t) = N_{14}(1)\mathbf{P}_1 + N_{24}(1)\mathbf{P}_2 + N_{34}(1)\mathbf{P}_3 + N_{44}(1)\mathbf{P}_4$$
$$= \mathbf{P}_3 + \mathbf{P}_4 \quad \text{(undefined)}, \qquad t \in [1, 1),$$

$$\mathbf{P}_5(t) = N_{24}(1)\mathbf{P}_2 + N_{34}(1)\mathbf{P}_3 + N_{44}(1)\mathbf{P}_4 + N_{54}(1)\mathbf{P}_5$$
$$= \mathbf{P}_3 + \mathbf{P}_4 \quad \text{(undefined)}, \qquad t \in [1, 1),$$

$$\mathbf{P}_6(t) = N_{34}(t)\mathbf{P}_3 + N_{44}(t)\mathbf{P}_4 + N_{54}(t)\mathbf{P}_5 + N_{64}(t)\mathbf{P}_6$$
$$= \mathbf{P}_3 + \mathbf{P}_4 \quad \text{(undefined)}, \qquad t \in [1, 1),$$

$$\mathbf{P}_7(t) = N_{44}(t)\mathbf{P}_4 + N_{54}(t)\mathbf{P}_5 + N_{64}(t)\mathbf{P}_6 + N_{74}(t)\mathbf{P}_7$$
$$= (2 - t)^3 \mathbf{P}_4 + \frac{1}{4}(7t^3 - 39t^2 + 69t - 37)\mathbf{P}_5$$
$$+ \frac{1}{12}(-11t^3 + 51t^2 - 69t + 29)\mathbf{P}_6 + \frac{1}{6}(t - 1)^3 \mathbf{P}_7, \qquad t \in [1, 2).$$

A direct check verifies that each segment has barycentric weights. The curve consists of the two separate segments $\mathbf{P}_3(t)$ and $\mathbf{P}_7(t)$. The former goes from $\mathbf{P}_0/6 + 7\mathbf{P}_1/12 + \mathbf{P}_2/4$ to $\mathbf{P}_3$ and the latter from $\mathbf{P}_4$ to $\mathbf{P}_5/4 + 7\mathbf{P}_6/12 + \mathbf{P}_7/6$. The three segments $\mathbf{P}_4(t)$, $\mathbf{P}_5(t)$, and $\mathbf{P}_6(t)$ get the undefined value $\mathbf{P}_3 + \mathbf{P}_4$ at $t = 1$.

7.12: We use the parameter substitution $T = t^2/((1-t)^2 + t^2)$ to write this curve in the form $(1-T)\mathbf{P}_0 + T\mathbf{P}_2$. It is now clear that this is the required line. It is also easy to see that $t = 0 \to T = 0$ and $t = 1 \to T = 1$.

7.13: It is obvious from the figure that $\mathbf{P}_0 = (0, -1)R$. To figure out the coordinates of $\mathbf{P}_2$ we notice the following:

1. The point is on the circle $x^2 + y^2 = R^2$, so it satisfies

$$x_2^2 + y_2^2 = R^2. \tag{Ans.24}$$

2. The point is on line L. The equation of this line can be written $y = ax + b$, where the slope a equals $\tan 60° = \sqrt{3} \approx 1.732$, so we have

$$y_2 = ax_2 + b, \tag{Ans.25}$$

where b still has to be determined.

3. $\mathbf{P}_2$ is located on the circle at a point where the tangent has a slope of $60°$. We differentiate the equation of the circle $x^2 + y^2 = R^2$ with respect to x to obtain $2x + 2y(dy/dx) = 0$ or $x = -y \cdot y'$. A slope of $60°$ means that $y' = \tan 60 = a$, so $\mathbf{P}_2$ also satisfies

$$x_2 = -y_2 a. \tag{Ans.26}$$

Equations (Ans.24) through (Ans.26) are easy to solve. The three solutions are $y_2 = R/\sqrt{a^2+1} = 0.5R$, $x_2 = -ay_2 = -0.866R$, and $b = y_2 - ax_2 = R(1+a^2)/\sqrt{a^2+1} = 2R$.

To figure out the coordinates of $\mathbf{P}_1$, we notice that it is located on line L and its y coordinate equals $-R$. It therefore satisfies $-R = ax_1 + b$, so $x_1 = -aR = -1.732R$.

7.14: The base angle of the triangle defined by the three points is $\theta = 45°$, so a circular arc is obtained when we set $w_1 = \cos\theta = 0.7071$. Substituting the points in Equation (7.45) and setting $w_1 = 0.7071$ yields the $90°$ circular arc that goes from $\mathbf{P}_0$ to $\mathbf{P}_2$:

$$\begin{aligned}
\mathbf{P}(t) &= \frac{(1-t)^2\mathbf{P}_0 + 2w_1 t(1-t)\mathbf{P}_1 + t^2\mathbf{P}_2}{(1-t)^2 + 2w_1 t(1-t) + t^2} \\
&= \frac{(1-t)^2(1,0) + 1.414t(1-t)(0,0) + t^2(0,1)}{(1-t)^2 + 1.414t(1-t) + t^2}R \\
&= \frac{((1-t)^2, t^2)R}{(1-t)^2 + 1.414t(1-t) + t^2}.
\end{aligned} \tag{Ans.27}$$

7.15: This is point $\mathbf{P}(0.5, 0.5) = (1, 1, (-1-2/2+2/4)(-1-2/2+2/4)/4) = (1, 1, 9/16)$.

7.16: In the case of equally-spaced control points, we have

$$(\mathbf{P}_{00} + \mathbf{P}_{20})/2 = \mathbf{P}_{10} \Rightarrow \frac{1}{6}(\mathbf{P}_{00} + 4\mathbf{P}_{10} + \mathbf{P}_{20}) = \mathbf{P}_{10},$$

$$(\mathbf{P}_{01} + \mathbf{P}_{21})/2 = \mathbf{P}_{11} \Rightarrow \frac{1}{6}(\mathbf{P}_{01} + 4\mathbf{P}_{11} + \mathbf{P}_{21}) = \mathbf{P}_{11},$$

$$(\mathbf{P}_{02} + \mathbf{P}_{22})/2 = \mathbf{P}_{12} \Rightarrow \frac{1}{6}(\mathbf{P}_{02} + 4\mathbf{P}_{12} + \mathbf{P}_{22}) = \mathbf{P}_{12},$$

so $\mathbf{K}_{00} = \frac{1}{6}\mathbf{P}_{10} + \frac{4}{6}\mathbf{P}_{11} + \frac{1}{6}\mathbf{P}_{12} = \mathbf{P}_{11}$.

8.1: No, since it does not pass through the first and last control points. (However, the text shows that this is a Bézier curve, but one defined by different control points.)

8.2: The first step of Chaikin's algorithm selects points

$$\frac{1}{4}\mathbf{P}_0 + \frac{3}{4}\mathbf{P}_1 = \frac{1}{2}\left(\frac{\mathbf{P}_0 + \mathbf{P}_1}{2} + \mathbf{P}_1\right) = \frac{\mathbf{A} + \mathbf{B}}{2} = \mathbf{M}_{ab},$$

$$\frac{3}{4}\mathbf{P}_1 + \frac{1}{4}\mathbf{P}_2 = \frac{1}{2}\left(\mathbf{P}_1 + \frac{\mathbf{P}_1 + \mathbf{P}_2}{2}\right) = \frac{\mathbf{B} + \mathbf{C}}{2} = \mathbf{M}_{bc}.$$

8.3: Given three control points $\mathbf{P}_0$, $\mathbf{P}_1$, and $\mathbf{P}_2$, the quadratic B-spline segment defined by them is identical to the quadratic Bézier segment that goes from the midpoint of $\mathbf{P}_0\mathbf{P}_1$ to the midpoint of $\mathbf{P}_1\mathbf{P}_2$. Both segments round out the corner created by the control points with a *parabolic fillet*.

8.4: Often, the coordinates of pixels are integers and we know that integer division truncates the result to the nearest integer. Comparing the distance between $\mathbf{P}_1$ and $\mathbf{P}_4$ to two pixels may result in a situation where they become identical. If this happens, then the assignment $\mathbf{P}_1 \leftarrow \mathbf{P}_4$ does not do anything, and the flow chart of Figure 8.7d shows that this results in a loop that empties the stack without doing anything useful.

8.5: This is straightforward and the triplets are $\mathbf{P}^4_{012} = \mathbf{AP}^3_{012}$, $\mathbf{P}^4_{234} = \mathbf{AP}^3_{123}$, $\mathbf{P}^4_{456} = \mathbf{AP}^3_{234}$, $\mathbf{P}^4_{678} = \mathbf{AP}^3_{345}$, $\mathbf{P}^4_{89\,10} = \mathbf{AP}^3_{456}$, $\mathbf{P}^4_{10\,11\,12} = \mathbf{AP}^3_{567}$, $\mathbf{P}^4_{12\,13\,14} = \mathbf{AP}^3_{678}$, $\mathbf{P}^4_{14\,15\,16} = \mathbf{AP}^3_{789}$, and $\mathbf{P}^4_{16\,17\,18} = \mathbf{AP}^3_{89\,10}$.

8.6: The problem is to compute $\lim_{k\to\infty}\mathbf{P}^3_{678} = \frac{1}{6}(\mathbf{P}^3_6 + 4\mathbf{P}^3_7 + \mathbf{P}^3_8)$. The *Mathematica* code of Figure Ans.19 does the calculations and produces the result $(\mathbf{P}^0_0 + 121\mathbf{P}^0_1 + 235\mathbf{P}^0_2 + 27\mathbf{P}^0_3)/384$. A comparison with Equation (7.11) shows that this is point $\mathbf{P}(3/4)$ of the B-spline curve segment.

8.7: The top left corner after one subdivision is

$$\mathbf{P}^1(0,0) = \frac{1}{36}(\mathbf{P}^1_{00} + \mathbf{P}^1_{02} + 4\mathbf{P}^1_{10} + 4\mathbf{P}^1_{12} + \mathbf{P}^1_{20} + 4\mathbf{P}^1_{01} + 16\mathbf{P}^1_{11} + 4\mathbf{P}^1_{21} + \mathbf{P}^1_{22})$$

```
a={{4,4,0},{1,6,1},{0,4,4}}/8; {p10,p11,p12}=a.{p00,p01,p02};
{p12,p13,p14}=a.{p01,p02,p03}; {p20,p21,p22}=a.{p10,p11,p12};
{p22,p23,p24}=a.{p11,p12,p13}; {p24,p25,p26}=a.{p12,p13,p14};
{p30,p31,p32}=a.{p20,p21,p22}; {p32,p33,p34}=a.{p21,p22,p23};
{p34,p35,p36}=a.{p22,p23,p24}; {p36,p37,p38}=a.{p23,p24,p25};
{p38,p39,p310}=a.{p24,p25,p26}; Simplify[(p36+4 p37+p38)/6]
```

Figure Ans.19: Code for Exercise 8.6

$$
\begin{aligned}
&= \frac{1}{36}\Big[\frac{1}{4}(\mathbf{P}_{00}^0 + \mathbf{P}_{10}^0 + \mathbf{P}_{01}^0 + \mathbf{P}_{11}^0) + \frac{1}{4}(\mathbf{P}_{01}^0 + \mathbf{P}_{11}^0 + \mathbf{P}_{02}^0 + \mathbf{P}_{12}^0) \\
&\quad + \frac{1}{4}(\mathbf{P}_{00}^0 + \mathbf{P}_{01}^0 + 6\mathbf{P}_{10}^0 + 6\mathbf{P}_{11}^0 + \mathbf{P}_{20}^0 + \mathbf{P}_{21}^0) \\
&\quad + \frac{1}{4}(\mathbf{P}_{00}^0 + 6\mathbf{P}_{10}^0 + \mathbf{P}_{20}^0 + 6\mathbf{P}_{01}^0 + 36\mathbf{P}_{11}^0 + 6\mathbf{P}_{21}^0 + \mathbf{P}_{02}^0 + 6\mathbf{P}_{12}^0 + \mathbf{P}_{22}^0) \\
&\quad + \frac{1}{4}(\mathbf{P}_{10}^0 + \mathbf{P}_{20}^0 + 6\mathbf{P}_{11}^0 + 6\mathbf{P}_{21}^0 + \mathbf{P}_{12}^0 + \mathbf{P}_{22}^0) + \frac{1}{4}(\mathbf{P}_{11}^0 + \mathbf{P}_{21}^0 + \mathbf{P}_{12}^0 + \mathbf{P}_{22}^0)\Big] \\
&= \frac{1}{36}(\mathbf{P}_{00}^0 + 4\mathbf{P}_{01}^0 + 16\mathbf{P}_{11}^0 + \mathbf{P}_{02}^0 + 4\mathbf{P}_{12}^0 + \mathbf{P}_{20}^0 + 4\mathbf{P}_{21}^0 + \mathbf{P}_{22}^0) \\
&= \mathbf{P}(0,0).
\end{aligned}
$$

9.1: We just need to multiply $\mathbf{C}(u)$ by the two translation matrices:

$$
(u,0,0,1)
\begin{pmatrix}
1 & 0 & 0 & 0 \\
0 & 1 & 0 & 0 \\
0 & 0 & 1 & 0 \\
0 & 0 & w & 1
\end{pmatrix}
\begin{pmatrix}
1 & 0 & 0 & 0 \\
0 & 1 & 0 & 0 \\
0 & 0 & 1 & 0 \\
0 & \sin w & 0 & 1
\end{pmatrix}.
$$

The result is the surface $\mathbf{P}(u,w) = (u, \sin w, w)$, displayed in Figure Ans.20.

9.2: This quarter circle starts at point $\mathbf{C}(0) = (1,0,0)$ on the x axis and ends at point $\mathbf{C}(1) = (0,1,0)$ on the y axis. A 360° rotation about the y axis is expressed by the $3{\times}3$ matrix

$$
\mathbf{T}(w) =
\begin{pmatrix}
\cos(2\pi w) & 0 & \sin(2\pi w) \\
0 & 1 & 0 \\
-\sin(2\pi w) & 0 & \cos(2\pi w)
\end{pmatrix}.
$$

This example involves a rotation but no translation, which is why there is no need for homogeneous coordinates. Multiplying $\mathbf{C}(u){\cdot}\mathbf{T}(w)$ yields the surface

$$
\frac{1}{1+u^2}\left((1-u^2)\cos(2\pi w), 2u, (1-u^2)\sin(2\pi w), 1\right).
$$

For $u = 0$, this reduces to $(\cos(2\pi w), 0, \sin(2\pi w))$, that's the unit circle in the xz plane. For $u = 1$, the same expression reduces to $1/2(0,2,0) = (0,1,0)$. This is the top of the half-sphere, a point that does not depend on w.

```
<<:Graphics:ParametricPlot3D.m;
ParametricPlot3D[{3u,Sin[w],w}, {u,0,1},{w,0,4Pi},
Ticks->False, AspectRatio->Automatic]
```

Figure Ans.20: A Sweep Surface.

9.3: The equation of the line is

$$(1-u)\mathbf{P}_1 + u\mathbf{P}_2 = (1-u)(0,0,0) + u(R,0,H) = (Ru,0,Hu),$$

where $0 \le u \le 1$. Multiplying by the rotation matrix about the z axis yields

$$(Ru,0,Hu)\begin{pmatrix} \cos w & -\sin w & 0 \\ \sin w & \cos w & 0 \\ 0 & 0 & 1 \end{pmatrix} = (Ru\cos w, -Ru\sin w, Hu),$$

where $0 \le w \le 2\pi$. [Compare with Equation (Ans.3).]

9.4: This is trivial. Just translate each coordinate:

$$\mathbf{P}(u,w) = \mathbf{P}(u)\mathbf{T}_y(w) = (x_0 + R\cos u \cos w, y_0 + R\sin u, z_0 + R\cos u \sin w),$$

where $-\pi/2 \le u \le \pi/2$ and $0 \le w \le 2\pi$.

9.5: This is done by multiplying the surface of Equation (9.4) by rotation matrix $\mathbf{T}_z(\theta)$. The result is

$$\mathbf{P}(u,w) = \mathbf{P}(u)\mathbf{T}_y(w)\mathbf{T}_z(\theta)$$
$$= (R\cos u \cos w \cos\theta - R\sin u \sin\theta, R\cos u \cos w \sin\theta + R\sin u \cos\theta, R\cos u \sin w),$$

where $-\pi/2 \le u \le \pi/2$ and $0 \le w \le 2\pi$. Notice how the z coordinates of this sphere don't depend on θ.

9.6: The half-circle is $\mathbf{P}(u) = (R\cos u, 0, R\sin u)$. Multiplying this by $\mathbf{T}_z(w)$ yields

$$\mathbf{P}(u, w) = (R\cos u \cos w, R\cos u \sin w, R\sin u).$$

9.7: We start with the circle $(R + r\cos u, 0, r\sin u)$, where $0 \le u \le 2\pi$. The torus is generated when this circle is rotated $360°$ about the z axis, by means of rotation matrix $\mathbf{T}_z(w)$:

$$\mathbf{P}(u, w) = (R + r\cos u, 0, r\sin u)\begin{pmatrix} \cos w & -\sin w & 0 \\ \sin w & \cos w & 0 \\ 0 & 0 & 1 \end{pmatrix}$$

$$= \big((R + r\cos u)\cos w, -(R + r\cos u)\sin w, r\sin u\big),$$

where $0 \le u, w \le 2\pi$, or

$$\mathbf{P}(u, w) = \big((R + r\cos(2\pi u))\cos(2\pi w), -(R + r\cos(2\pi u))\sin(2\pi w), r\sin(2\pi u)\big),$$

where $0 \le u, w \le 1$ (Figure Ans.21a). In general, R is greater than r, but there are two special cases. The case $R = r$ creates a horn torus and the case $0 \le R \le r$ becomes a spindle torus (Figure Ans.21b,c, respectively, where only half the torus is shown).

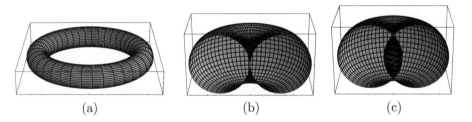

| (a) | (b) | (c) |

```
R=10; r=2;   (* The Torus as a surface of revolution *)
ParametricPlot3D[
  {(R+r Cos[2Pi u])Cos[2Pi w],-(R+r Cos[2Pi u])Sin[2Pi w],
  r Sin[2Pi u]},{u,0,1},{w,0,1},
  ViewPoint->{-0.028, -4.034, 1.599}]
```

Figure Ans.21: The Torus as a Surface of Revolution.

A.1: The particular conic generated by Equation (A.3) is determined by the sign of the discriminant $B^2 - 4AC$. The exact shape of the curve is determined by the values of all six parameters. The general rule is

$$B^2 - 4AC \begin{cases} < 0, & \text{ellipse (or circle),} \\ = 0, & \text{parabola,} \\ > 0, & \text{hyperbola.} \end{cases}$$

Example: Three examples are shown.

1. The canonical circle is obtained for $A = C = 1$, $F = -R^2$, and $B = D = E = 0$.
2. A straight line is the result of $A = B = C = 0$.
3. The canonical parabola $y^2 = 2ax$ is the result of $A = B = E = F = 0$, $C = 1$, and $D = -2a$.

B.1: The Bézier curve can be written $\mathbf{P}(t) = (f(t), g(t))$, where f and g are the (x, y) coordinates of points on the curve. Both f and g are polynomials in t, so they can be written as

$$f(t) = a_0 + a_1 t + \cdots + a_k t^k, \qquad g(t) = b_0 + b_1 t + \cdots + b_k t^k.$$

We now suppose that the curve $\mathbf{P}(t) = (f(t), g(t))$ produces the circle $(x-p)^2 + (y-q)^2 = R^2$ and will prove that this implies $a_i = b_i = 0$ for $i = 1, \ldots, k$.

The assumption implies that

$$\left(a_0 + a_1 t + \cdots + a_k t^k - p\right)^2 + \left(b_0 + b_1 t + \cdots + b_k t^k - q\right)^2 = R^2 \qquad \text{(Ans.28)}$$

for all values of t. For $t = 0$, we get $(a_0 - p)^2 + (b_0 - q)^2 = R^2$, so now we can write Equation (Ans.28) as

$$\left(a_1 t + \cdots + a_k t^k\right)^2 + \left(b_1 t + \cdots + b_k t^k\right)^2 = 0. \qquad \text{(Ans.29)}$$

Carrying out the multiplications produces an expression of the type

$$\left(\cdots + (a_k^2 + b_k^2) t^{2k}\right) = 0.$$

This implies that the sum $(a_k^2 + b_k^2)$ is zero, and since it is the sum of squares, each must be zero. We can now write Equation (Ans.29) as

$$\left(a_1 t + \cdots + a_{k-1} t^{k-1}\right)^2 + \left(b_1 t + \cdots + b_{k-1} t^{k-1}\right)^2 = 0,$$

and use similar arguments to prove that $a_{k-1} = b_{k-1} = 0$. In this way, we can show that all the coefficients of $f(t)$ and $g(t)$ are zero (except a_0 and b_0, which may create a circle consisting of one point if they satisfy $(a_0 - p)^2 + (b_0 - q)^2 = R^2$).

B.2: Because of the symmetry of a circle, the two interior points must have coordinates $\mathbf{P}_1 = (1, k)$ and $\mathbf{P}_2 = (k, 1)$. To set up an equation that will allow us to solve for k, we arbitrarily require that the midpoint of the curve $\mathbf{P}(0.5)$ coincide with the midpoint of the quarter circle $(1/\sqrt{2}, 1/\sqrt{2})$ (Figure Ans.22a). The equation becomes

$$\mathbf{P}(0.5) = \left(\frac{1}{\sqrt{2}}, \frac{1}{\sqrt{2}}\right),$$

or

$$\mathbf{P}(0.5) = \sum_{i=0}^{3} \mathbf{P}_i B_{3,i}(0.5)$$

$$= \frac{1}{8}\mathbf{P}_0 + \frac{3}{8}\mathbf{P}_1 + \frac{3}{8}\mathbf{P}_2 + \frac{1}{8}\mathbf{P}_3$$

$$= \frac{1}{8}(1,0) + \frac{3}{8}(1,k) + \frac{3}{8}(k,1) + \frac{1}{8}(0,1)$$

$$= \left(\frac{3k+4}{8}, \frac{3k+4}{8}\right),$$

and the solution is $k = 4(\sqrt{2} - 1)/3 \approx 0.5523$.

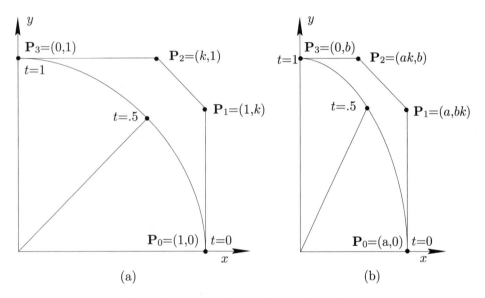

Figure Ans.22: An Almost-Circular Bézier Curve.

The final expression of the curve is

$$\mathbf{P}(t) = \big((1-t)^3 + 3t(1-t)^2 + 3t^2(1-t)k,$$
$$3t(1-t)^2k + 3t^2(1-t) + t^3\big) \qquad \text{(Ans.30)}$$
$$\approx \big(0.3431t^3 - 1.3431t^2 + 1, -0.3431t^3 - 0.3138t^2 + 1.6569t\big). \quad \text{(Ans.31)}$$

(See Sections A.2 and 5.12.3 of [Salomon 99] for applications of this expression.) For a circle of radius R, the expression above is simply multiplied by R.

The maximum deviation of this cubic curve from a true circle can be calculated similar to the quadratic case. The tangent vector of Equation (Ans.30) is

$$\mathbf{P}^t(t) = \big(6(k-1)t + 3(2-3k)t^2, 3k + 6(1-2k)t + 3(3k-2)t^2\big),$$

so the condition $\mathbf{P}(t) \cdot \mathbf{P}^t(t) = 0$ becomes

$$(9k^2 + 6k - 6)t + (-54k^2 + 18k + 6)t^2 + (126k^2 - 132k + 36)t^3$$
$$-15(3k - 2)^2 t^4 + 6(3k - 2)t^5 = 0.$$

Numerical solution gives the five roots $t_1 = 0$, $t_2 = 0.211$, $t_3 = 0.5$, $t_4 = 0.789$, and $t_5 = 1$. Thus, the maximum distance between the origin and a point $\mathbf{P}(t)$ on this curve is $D(t_2) = D(t_4) = 1.00027$. The maximum deviation of this cubic curve from a true circle is this just 0.027%, much better than the quadratic approximation, although the latter is preferable in practice since it is simpler.

B.3: The implicit expression for the arc is simply the equation of the ellipse

$$\frac{x^2}{a^2} + \frac{y^2}{b^2} = 1, \quad \text{where} \quad 0 \le x \le a.$$

The two endpoints are $\mathbf{P}_0 = (a, 0)$ and $\mathbf{P}_3 = (0, b)$. Based on the symmetry shown in Figure Ans.22b, we select the other two control points with coordinates $\mathbf{P}_1 = (a, bk)$ and $\mathbf{P}_2 = (ak, b)$. To set up an equation that will allow us to solve for k, we require that the midpoint of the curve, $\mathbf{P}(0.5)$, coincide with the midpoint of the quarter arc $(a/\sqrt{2}, b/\sqrt{2})$ (Figure Ans.22b). The equation becomes

$$\mathbf{P}(.5) = \left(\frac{a}{\sqrt{2}}, \frac{b}{\sqrt{2}} \right),$$

or

$$= \sum_{i=0}^{3} \mathbf{P}_i B_{3,i}(.5)$$
$$= \frac{1}{8}\mathbf{P}_0 + \frac{3}{8}\mathbf{P}_1 + \frac{3}{8}\mathbf{P}_2 + \frac{1}{8}\mathbf{P}_3$$
$$= \frac{1}{8}(a, 0) + \frac{3}{8}(a, bk) + \frac{3}{8}(ak, b) + \frac{1}{8}(0, b)$$
$$= \left(\frac{a(3k + 4)}{8}, \frac{b(3k + 4)}{8} \right),$$

and the solution is, again, $k = 4(\sqrt{2} - 1)/3 \approx 0.5523$.

The final expression of the curve is

$$\mathbf{P}(t) = (1 - t)^3(a, 0) + 3t(1 - t)^2(a, bk) + 3t^2(1 - t)(ak, b) + t^3(0, b).$$

B.4: The parametric equation of the circular arc is $\mathbf{a}(u) = (\cos u, \sin u)$ for $-\theta \le u \le \theta$. The expression of the curve is

$$\mathbf{P}(t) = (1 - t)^3 \mathbf{P}_0 + 3t(1 - t)^2 \mathbf{P}_1 + 3t^2(1 - t)\mathbf{P}_2 + t^3 \mathbf{P}_3,$$

where the four control points have to be calculated. To calculate $\mathbf{P}_0$ and $\mathbf{P}_3$, we require that the curve passes through the first and last points of the arc. This implies $\mathbf{P}_0 = (\cos\theta, -\sin\theta)$ and $\mathbf{P}_3 = (\cos\theta, \sin\theta)$. To calculate $\mathbf{P}_1$ and $\mathbf{P}_2$, we require the curve and the arc to have the same tangent vectors at their start and end points, i.e.,

$$\frac{d\mathbf{a}(-\theta)}{du} = \frac{d\mathbf{P}(0)}{dt} \quad \text{and} \quad \frac{d\mathbf{a}(\theta)}{du} = \frac{d\mathbf{P}(1)}{dt}.$$

The tangent vectors are

$$\frac{d\mathbf{a}(u)}{du} = (-\sin u, \cos u),$$

$$\frac{d\mathbf{P}(t)}{dt} = -3(1-t)^2\mathbf{P}_0 + (3-9t)(1-t)\mathbf{P}_1 + 3t(2-3t)\mathbf{P}_2 + 3t^2\mathbf{P}_3.$$

Equating them at the start point yields

$$(-\sin(-\theta), \cos(-\theta)) = -3\mathbf{P}_0 + 3\mathbf{P}_1 = -3(\cos\theta, -\sin\theta) + 3\mathbf{P}_1,$$

so

$$\mathbf{P}_1 = (\sin\theta + 3\cos\theta, -3\sin\theta - \cos\theta)/3,$$

and, by symmetry,

$$\mathbf{P}_2 = (\sin\theta + 3\cos\theta, 3\sin\theta + \cos\theta)/3.$$

Thus, the Bézier curve is

$$\mathbf{P}(t) = (1-t)^3(\cos\theta, -\sin\theta) + 3t(1-t)^2(\sin\theta + 3\cos\theta, -3\sin\theta - \cos\theta)/3$$
$$+ 3t^2(1-t)(\sin\theta + 3\cos\theta, 3\sin\theta + \cos\theta)/3 + t^3(\cos\theta, \sin\theta).$$

The midpoint of the arc is $\mathbf{a}(0.5) = (1, 0)$ and that of the curve is

$$\mathbf{P}(0.5) = \frac{1}{8}(\cos\theta, -\sin\theta) + \frac{3}{8}(\sin\theta + 3\cos\theta, -3\sin\theta - \cos\theta)/3$$

$$+ \frac{3}{8}(\sin\theta + 3\cos\theta, 3\sin\theta + \cos\theta)/3 + \frac{1}{8}(\cos\theta, \sin\theta)$$

$$= (\cos\theta + \frac{1}{4}\sin\theta, 0).$$

For small θ, the deviation of this curve from the true arc is small (for angles up to 45° the deviation is less than 12%). For larger angles, the curve deviates much from the arc (for $\theta = 90°$, the midpoint of the curve is $(0.25, 0)$, so it is very different from the arc).

B.5: They are all computed from Equation (7.11) by rotating the four control points to the left in each segment. The result is

$$\mathbf{P}_2(t) = \frac{1}{6}[(-t^3 + 3t^2 - 3t + 1)(0, 3/2) + (3t^3 - 6t^2 + 4)(-3/2, 0)$$

$$+ (-3t^3 + 3t^2 + 3t + 1)(0, -3/2) + t^3(3/2, 0)]$$

$$= \frac{1}{4}(-2t^3 + 6t^2 - 4, 2t^3 - 6t),$$

$$\mathbf{P}_3(t) = \frac{1}{6}[(-t^3 + 3t^2 - 3t + 1)(-3/2, 0) + (3t^3 - 6t^2 + 4)(0, -3/2)$$

$$+ (-3t^3 + 3t^2 + 3t + 1)(3/2, 0) + t^3(0, 3/2)]$$

$$= \frac{1}{4}(-2t^3 + 6t, -2t^3 + 6t^2 - 4),$$

$$\mathbf{P}_4(t) = \frac{1}{6}[(-t^3 + 3t^2 - 3t + 1)(0, -3/2) + (3t^3 - 6t^2 + 4)(3/2, 0)$$

$$+ (-3t^3 + 3t^2 + 3t + 1)(0, 3/2) + t^3(-3/2, 0)]$$

$$= \frac{1}{4}(2t^3 - 6t^2 + 4, -2t^3 + 6t). \tag{Ans.32}$$

B.6: Compute the midpoint $(\mathbf{S} + \mathbf{E})/2$ and normalize its coordinates.

D.1: Normally a semicolon following a command suppresses any output. In our case, the `DisplayFunction` option forces output, and the only effect of omitting the semicolon is that *Mathematica* generates in this case another output cell with the word `Graphics`.

D.2: The surface patch will be displayed as a set of 6×6 small flat rectangles.

D.3: When the matrix of points has four rows or four columns as, for example, in Figure 3.6 (a bicubic surface patch example).

D.4: Because a previous evaluation of another cell in a *Mathematica* notebook may have defined a. Specifically, if `a` has been defined as a list, matrix `r` would have a mixture of scalar and nonscalar elements, and the evaluation of lines 11–13 would result in an error message.

Solutions are not the answer.

Richard M. Nixon

Bibliography

> Some books leave us free and some books make us free.
> —Ralph Waldo Emerson

Adobe (2004) `http://www.adobe.com/products/illustrator/main.html`.

Beach, Robert C. (1991) *An Introduction to the Curves and Surfaces of Computer-Aided Design*, New York, Van Nostrand Reinhold.

Berrut, Jean-Paul, and Lloyd N. Trefethen (2004) "Barycentric Lagrange Interpolation," *SIAM Review*, **46**(3)501–517.

Bézier, Pierre (1986) *The Mathematical Basis of the UNISURF CAD System*, Newton, Mass., Butterworth-Heinemann.

Blender (2005) `http://www.blender.org/`.

CAD (2004) `http://www.sciencedirect.com/science/journal/00104485`.

CAGD (2004) `http://www.sciencedirect.com/science/journal/01678396`.

Catmull, E., and J. Clark (1978) "Recursively Generated B-Spline Surfaces on Arbitrary Topological Meshes," *Computer-Aided Design* **10**(6):350–355, Sept.

Catmull, E. and R. Rom (1974) "A Class of Interpolating Splines," in R. Barnhill and R. Riesenfeld, editors, *Computer Aided Geometric Design*, Academic Press, pages 317–326.

Chaikin, G. (1974) "An Algorithm for High-Speed Curve Generation," *Computer Graphics and Image Processing*, **3**:346–349.

Cohen, E., et al., (1980) "Discrete B-Splines and Subdivision Techniques in Computer Aided Geometric Design and Computer Graphics," *Computer Graphics and Image Processing*, **14**:87–111.

Cohen, E., et al. (1985) "Algorithms For Degree Raising of Splines," *ACM Transactions on Graphics*, **4**:171–181.

Coons, Steven A. (1964) "Surfaces for Computer-Aided Design of Space Figures," Cambridge, MA, MIT Project MAC, report MAC-M-253, January.

Coons, Steven A. (1967) "Surfaces for Computer-Aided Design of Space Forms," Cambridge, MA, MIT Project MAC TR-41, June.

Crow, Frank (1987) "The Origins of the Teapot," *IEEE Computer Graphics and Applications*, **7**(1):8–19, January.

Davis, Philip J. (1963) *Interpolation and Approximation*, Waltham, MA, Blaisdell Publishing, and New York, Dover Publications, 1975.

DeBoor, Carl, (1972) "On Calculating With B-Splines," *Journal of Approximation Theory*, **6**:50–62.

DeRose T., and C. Loop (1989) "The S-Patch: A New Multisided Patch Scheme," *ACM Transactions on Graphics*, **8**(3):204–234.

DesignMentor (2005) `http://www.cs.mtu.edu/~shene/NSF-2/DM2-BETA/index.html`.

Doo, Donald, and M. Sabin (1978) "Behavior of Recursive Division Surfaces Near Extraordinary Points," *Computer-Aided Design*, **10**(6):356–360, Sept.

Dupuy, M. (1948) "Le Calcul Numérique des Fonctions par l'Interpolation Barycentrique," *Comptes Rendus de l'Académie des Sciences*, Paris, 158–159.

Farin, Gerald (1999) *NURBS: From Projective Geometry to Practical Use*, 2nd edition, Wellesley, MA, AK Peters.

Farin, Gerald (2001) *Curves and Surfaces for CAGD (Computer Aided Graphics and Design)*, San Diego, Academic Press.

Farin, Gerald (2004) *A History of Curves and Surfaces in CAGD*, in G. Farin, J. Hoschek, and M. S. Kim, editors, *Handbook of CAGD*, pages 1–22. Elsevier, 2002. (Available, in PDF format from the author of this book.)

Ferguson, J. (1964) "Multivariate Curve Interpolation," *Journal of the ACM*, **11**(2):221–228.

Free Software Foundation (2004), 59 Temple Place, Suite 330, Boston MA 02111-1307 USA. `http://www.fsf.org/`.

Freeman, H. (ed.) (1980) *Tutorial and Selected Readings in Interactive Computer Graphics*, Silver Springs, MD, IEEE Computer Society Press.

Gallery (2005) `http://gallery.wolfram.com/`.

Gallier, Jean (2000) *Curves and Surfaces in Geometric Modeling*, San Francisco, Morgan Kaufmann.

GIMP (2005) `http://www.gimp.org/`.

Gouraud, Henri (1971) "Continuous-Shading of Curved Surfaces," *IEEE Transactions on Computers*, C-20(6):623–629, June. (Reprinted in [Freeman 80].)

Graphica (2005) `http://www.graphica.com/see-it/`.

Kimberling, C., (1994) "Central Points and Central Lines in the Plane of a Triangle," *Mathematical Magazine* **67**:163–187.

Knuth, Donald E., (1986) *The Metafont Book*, Reading, MA, Addison-Wesley.

Kochanek, D. H. U., and R. H. Bartels (1984) "Interpolating Splines with Local Tension, Continuity, and Bias Control," *Computer Graphics* **18**(3):33–41 (Proceedings SIGGRAPH '84).

Lagrange, J. L. (1877) "Leçons Élémentaires Sur Les Mathématiques, Donées à l'Ecole Normale en 1795," in *Ouvres*, VII, Paris, Gauthier-Villars, 183–287.

Lee, E. (1986) "Rational Bézier Representations for Conics," in *Geometric Modeling*, Farin, G., editor, Philadelphia, SIAM Publications, pp. 3–27.

Liu, D., and J. Hoschek (1989) "GC^1 Continuity Conditions Between Adjacent Rectangular and Triangular Bézier Surface Patches," *Computer-Aided Design*, **21**:194–200.

Loop, Charles T. (1987) "Smooth Subdivision Surfaces Based on Triangles," M.S. thesis, University of Utah, Mathematics. See also `http://research.microsoft.com/~cloop`

Loop, Charles T. (2002) "Smooth Ternary Subdivision of Triangle Meshes," *Curve and Surface Fitting: St. Malo 2002.*

MathSource (2005) `http://library.wolfram.com/infocenter/MathSource/4930/`.

Mathworks (2005) `http://www.mathworks.com/`.

Piegl L., and W. Tiller (1997) *The NURBS Book*, 2nd edition, Berlin, Springer-Verlag.

Prautzsch, H., (1984) "A Short Proof of the Oslo Algorithm," *Computer Aided Geometric Design*, **1**:95–96.

Press, W. H., B. P. Flannery, et al. (1988) *Numerical Recipes in C: The Art of Scientific Computing*, Cambridge University Press. (Also available online at `http://www.nr.com/`.)

Ramshaw, Lyle (1987) "Blossoming: A Connect-the-Dots Approach to Splines," Digital Equipment Corporation, Research Report 19, June 21.

Riesenfeld, R. (1975) "On Chaikin's Algorithm," *IEEE Computer Graphics and Applications*, **4**(3):304–310.

Rogers, David (2001) *An Introduction To NURBS: With Historical Perspective*, San Francisco, Morgan Kaufmann.

Salomon, David (1999) *Computer Graphics and Geometric Modeling*, New York, Springer-Verlag.

Salomon, David (2004) *Data Compression: The Complete Reference*, 3rd edition, New York, Springer-Verlag.

Schumaker, L. (1981) *Spline Functions: Basic Theory*, New York, John Wiley.

Späth, Helmuth (1983) *Spline Algorithmen zur Konstruktion glatter Kurven und Flächen*, 3rd edition, Munich, Vienna, Oldenbourg Wissenschaftsverlag.

Triangles (2004) `http://faculty.evansville.edu/ck6/index.html`

Turnbull, Herbert W. et al. (editors) (1959) *The Correspondence of Isaac Newton*, Seven volumes, Cambridge, Cambridge University Press.

Wings3D `http://www.wings3d.com/`.

Wolfram Research (2005) `http://www.wolfram.com`.

Wolfram, Stephen (2003) *The Mathematica Book*, Fifth Edition, Champaign, IL., Wolfram Media.

Yamaguchi, F. (1988) *Curves and Surfaces in Computer Aided Geometric Design*, Berlin, Springer-Verlag.

There are eight words or references that light up employers' eyes: languages, computer, experience, achievement, hard-working, overseas experience, flexible, and task-oriented.

Carol Kleiman

Index

The index has been prepared in several steps. While writing the book, the author had to flag all the index items. When the book was typeset, special TEX macros, written by the author, wrote the items on a raw (`.idx`) index file in the format required by the *MakeIndex* program. This program was then run to produce the final (`.ind`) index file, that in turn was processed by further TEX macros (influenced by the macros on page 417 of *The TEXbook*) to typeset the pages of the index. Any errors and omissions found should be brought to the attention of the author.

Because he did not have time to read every new book in his field, the great Polish anthropologist Bronislaw Malinowski used a simple and efficient method of deciding which ones were worth his attention: Upon receiving a new book, he immediately checked the index to see if his name was cited, and how often. The more "Malinowski" the more compelling the book. No "Malinowski," and he doubted the subject of the book was anthropology at all.

Neil Postman

Colophon

Most of the material in this book was written during 2004. Some of the presentations, examples, and exercises were taken from [Salomon 99] and improved or extended. The book was designed by the author and was typeset by him in plain TeX. The figures and diagrams were computed by Adobe Illustrator, also on the Macintosh. Diagrams that require calculations were done in *Mathematica*, but even those were "polished" by Adobe Illustrator. The following points illustrate the amount of work that went into the book:

- The book contains about 200,000 words, consisting of about 1,100,000 characters.

- The text is typeset mainly in font cmr10, but about 30 other fonts were used.

- The raw index file has about 1700 items.

- There are about 670 cross references in the book.

I perceived that, to describe these impressions, to write that essential book, the only true book, a great writer does not need to invent it, in the current sense of the term, since it already exists in each one of us, but merely to translate it. The duty and task of a writer are those of the translator.

—Marcel Proust, *Time Regained* (1921)